T0183161

Lecture Notes in Computer Science   **10569**

Commenced Publication in 1973
Founding and Former Series Editors:
Gerhard Goos, Juris Hartmanis, and Jan van Leeuwen

More information about this series at http://www.springer.com/series/7409

Athman Bouguettaya · Yunjun Gao
Andrey Klimenko · Lu Chen
Xiangliang Zhang · Fedor Dzerzhinskiy
Weijia Jia · Stanislav V. Klimenko
Qing Li (Eds.)

Web Information Systems Engineering – WISE 2017

18th International Conference
Puschino, Russia, October 7–11, 2017
Proceedings, Part I

 Springer

Editors

Athman Bouguettaya
University of Sydney
Darlington, NSW
Australia

Yunjun Gao
Zhejiang University
Hangzhou
China

Andrey Klimenko
Institute of Computing for Physics
  and Technology
Protvino
Russia

Lu Chen
Nanyang Technological University
Singapore
Singapore

Xiangliang Zhang
King Abdullah University of Science
  and Technology
Thuwal
Saudi Arabia

Fedor Dzerzhinskiy
Institute of Computing for Physics
  and Technology
Protvino
Russia

Weijia Jia
Shanghai Jiao Tong University
Minhang Qu
China

Stanislav V. Klimenko
Institute of Computing for Physics
  and Technology
Protvino
Russia

Qing Li
City University of Hong Kong
Kowloon
Hong Kong

ISSN 0302-9743   ISSN 1611-3349 (electronic)
Lecture Notes in Computer Science
ISBN 978-3-319-68782-7   ISBN 978-3-319-68783-4   (eBook)
DOI 10.1007/978-3-319-68783-4

Library of Congress Control Number: 2017955787

LNCS Sublibrary: SL3 – Information Systems and Applications, incl. Internet/Web, and HCI

Printed on acid-free paper

This Springer imprint is published by Springer Nature
The registered company is Springer International Publishing AG
The registered company address is: Gewerbestrasse 11, 6330 Cham, Switzerland

Preface

Welcome to the proceedings of the 18th International Conference on Web Information Systems Engineering (WISE 2017), held in Moscow, Russia, during October 7–11, 2017. The series of WISE conferences aims to provide an international forum for researchers, professionals, and industrial practitioners to share their knowledge in the rapidly growing area of Web technologies, methodologies, and applications. The first WISE event took place in Hong Kong, SAR China (2000). Then the trip continued to Kyoto, Japan (2001); Singapore (2002); Rome, Italy (2003); Brisbane, Australia (2004); New York, USA (2005); Wuhan, China (2006); Nancy, France (2007); Auckland, New Zealand (2008); Poznan, Poland (2009); Hong Kong, SAR China (2010); Sydney, Australia (2011); Paphos, Cyprus (2012); Nanjing, China (2013); Thessaloniki, Greece (2014); Miami, USA (2015); Shanghai, China (2016); and this year, WISE 2017 was held in Moscow, Russia, supported by the Institute of Computing for Physics and Technology and the Moscow Institute of Physics and Technology, Russia.

A total of 196 research papers were submitted to the conference for consideration, and each paper was reviewed by at least three reviewers. Finally, 49 submissions were selected as full papers (with an acceptance rate of 25% approximately) plus 24 as short papers. The research papers cover the areas of microblog data analysis, social network data analysis, data mining, pattern mining, event detection, cloud computing, query processing, spatial and temporal data, graph theory, crowdsourcing and crowdsensing, Web data model, language processing and Web protocols, Web-based applications, data storage and generator, security and privacy, sentiment analysis, and recommender systems.

In addition to regular and short papers, the WISE 2017 program also featured a special session on "Security and Privacy." The special session is a forum for presenting and discussing novel ideas and solutions related to the problems of security and privacy. Experts and companies were invited to present their reports in this forum. The objective of this forum is to provide forward-looking ideas and views for research and application of security and privacy, which will promote the development of techniques in security and privacy, and further facilitate the innovation and industrial development of big data. The forum was organized by Prof. Xiangliang Zhang, Prof. Fedor Dzerzhinskiy, Prof. Weijia Jia, and Prof. Hua Wang.

We also wish to take this opportunity to thank the honorary the general co-chairs, Prof. Stanislav V. Klimenko, Prof. Qing Li; the program co-chairs, Prof. Athman Bouguettaya, Prof. Yunjun Gao, and Prof. Andrey Klimenko; the local arrangements chair, Prof. Maria Berberova; the special area chairs, Prof. Xiangliang Zhang, Prof. Fedor Dzerzhinskiy, Prof. Weijia Jia, and Prof. Hua Wang; the workshop co-chairs, Prof. Reynold C.K. Cheng and Prof. An Liu; the tutorial and panel chair, Prof. Wei Wang; the publication chair, Dr. Lu Chen; the publicity co-chairs, Prof. Jiannan Wang, Prof. Bin Yao, and Prof. Daria Marinina; the website co-chairs, Mr. Rashid Zalyalov,

Mr. Ravshan Burkhanov, and Mr. Boris Strelnikov; the WISE Steering Committee representative, Prof. Yanchun Zhang. The editors and chairs are grateful to Ms. Sudha Subramani and Mr. Sarathkumar Rangarajan for their help with preparing the proceedings and updating the conference website.

We would like to sincerely thank our keynote and invited speakers:

- Professor Beng Chin Ooi, Fellow of the ACM, IEEE, and Singapore National Academy of Science (SNAS), NGS faculty member and Director of Smart Systems Institute, National University of Singapore, Singapore
- Professor Lei Chen, Department of Computer Science and Engineering, Hong Kong University, Hong Kong, SAR China
- Professor Jie Lu, Associate Dean (Research Excellence) in the Faculty of Engineering and Information Technology, University of Technology Sydney, Sydney, Australia

In addition, special thanks are due to the members of the international Program Committee and the external reviewers for a rigorous and robust reviewing process. We are also grateful to the Moscow Institute of Physics and Technology, Russia, the Institute of Computing for Physics and Technology, Russia, City University of Hong Kong, SAR China, University of Sydney, Australia, Zhejiang University, China, Victoria University, Australia, University of New South Wales, Australia, and the International WISE Society for supporting this conference. The WISE Organizing Committee is also grateful to the special session organizers for their great efforts to help promote Web information system research to a broader audience.

We expect that the ideas that emerged at WISE 2017 will result in the development of further innovations for the benefit of scientific, industrial, and social communities.

October 2017

Athman Bouguettaya
Yunjun Gao
Andrey Klimenko
Lu Chen
Xiangliang Zhang
Fedor Dzerzhinskiy
Weijia Jia
Stanislav V. Klimenko
Qing Li

Organization

General Co-chairs

Stanislav V. Klimenko   Moscow Institute of Physics and Technology, Russia
Qing Li   City University of Hong Kong, SAR China

Program Co-chairs

Athman Bouguettaya   University of Sydney, Australia
Yunjun Gao   Zhejiang University, China
Andrey Klimenko   Institute of Computing for Physics and Technology, Russia

Special Area Chairs

Xiangliang Zhang   KAUST, Saudi Arabia
Fedor Dzerzhinskiy   Institute of Computing for Physics and Technology, Russia
Weijia Jia   Shanghai JiaoTong University, China
Hua Wang   Victoria University, Australia

Tutorial and Panel Chair

Wei Wang   The University of New South Wales, Australia

Workshop Co-chairs

Reynold C.K. Cheng   The University of Hong Kong, SAR China
An Liu   Soochow University, China

Publication Chair

Lu Chen   Nangyang Technological University, Singapore

Publicity Co-chairs

Jiannan Wang   Simon Fraser University, Canada
Bin Yao   Shanghai Jiao Tong University, China
Daria Marinina   Moscow Institute of Physics and Technology, Russia
Mikhail Pochkaylov   Moscow Institute of Physics and Technology, Russia
Anton Semenistyy   Moscow Institute of Physics and Technology, Russia

Conference Website Co-chairs

Rashid Zalyalov   Institute of Computing for Physics and Technology, Russia
Ravshan Burkhanov   Moscow Institute of Physics and Technology, Russia
Boris Strelnikov   Moscow Institute of Physics and Technology, Russia

Local Arrangements Chair

Maria Berberova   Moscow Institute of Physics and Technology, Russia

WISE Steering Committee Representative

Yanchun Zhang   Victoria University, Australia

Program Committee

Karl Aberer   EPFL, Switzerland
Mohammed Eunus Ali   Bangladesh University of Engineering and Technology,
                             Bangladesh
Toshiyuki Amagasa   University of Tsukuba, Japan
Athman Bouguettaya   University of Sydney, Australia
Yi Cai   South China University of Technology, China
Xin Cao   UNSW, Australia
Bin Cao   Zhejiang University of Technology, China
Richard Chbeir   LIUPPA Laboratory, France
Lisi Chen   Hong Kong Baptist University, SAR China
Jinchuan Chen   Renmin University of China, China
Cindy Chen   University of Massachusetts Lowell, USA
Jacek Chmielewski   Poznań University of Economics and Business, Poland
Alex Delis   University of Athens, Greece
Ting Deng   Beihang University, China
Hai Dong   RMIT University, Australia
Schahram Dustdar   TU Wien, Austria
Fedor Dzerzhinskiy   Promsvyazbank, Russia
Islam Elgedawy   Middle East Technical University, Turkey
Hicham Elmongui   Alexandria University, Egypt
Yunjun Gao   Zhejiang University, China
Thanaa Ghanem   Metropolitan State University, USA
Azadeh Ghari Neiat   University of Sydney, Australia
Daniela Grigori   Laboratoire LAMSADE, Université Paris Dauphine,
                             France
Viswanath Gunturi   Indian Institute of Technology Ropar, India
Hakim Hacid   Bell Labs, USA
Armin Haller   Australian National University, Australia
Tanzima Hashem   Bangladesh University of Engineering and Technology,
                             Bangladesh

Md Rafiul Hassan	King Fahd University of Petroleum and Minerals, Saudi Arabia
Xiaofeng He	East China Normal University, China
Yuh-Jong Hu	National Chengchi University, Taiwan
Peizhao Hu	Rochester Institute of Technology, USA
Hao Huang	Wuhan University, China
Yoshiharu Ishikawa	Nagoya University, Japan
Adam Jatowt	Kyoto University, Japan
Weijia Jia	Shanghai Jiao Tong University, China
Dawei Jiang	Zhejiang University, China
Wei Jiang	Missouri University of Science and Technology, USA
Peiquan Jin	University of Science and Technology of China, China
Andrey Klimenko	Institute of Computing for Physics and Technology, Russia
Stanislav Klimenko	Institute of Computing for Physics and Technology, Russia
Jiuyong Li	University of South Australia, Australia
Hui Li	Xidian University, China
Qing Li	City University of Hong Kong, SAR China
Xiang Lian	Kent State University, USA
Dan Lin	Missouri University of Science and Technology, USA
Sebastian Link	The University of Auckland, New Zealand
Qing Liu	Zhejiang University, China
Wei Lu	Renmin University of China, China
Hui Ma	Victoria University of Wellington, New Zealand
Zakaria Maamar	Zayed University, United Arab Emirates
Murali Mani	University of Michigan-Flint, USA
Xiaoye Miao	Zhejiang University, China
Sajib Mistry	University of Sydney, Australia
Natwar Modani	Adobe Research, India
Wilfred Ng	Hong Kong University of Science and Technology, SAR China
Mitsunori Ogihara	University of Miami, USA
George Pallis	University of Cyprus, Cyprus
Tieyun Qian	Wuhan University, China
Shaojie Qiao	Southwest Jiaotong University, China
Lie Qu	University of Sydney, Australia
Jarogniew Rykowski	Poznań University of Economics, Poland
Shuo Shang	KAUST, Saudi Arabia
Yanyan Shen	Shanghai Jiao Tong University, China
Wei Shen	Nankai University, China
Yain-Whar Si	University of Macau, China
Dandan Song	Tsinghua University, China
Shaoxu Song	Tsinghua University, China
Weiwei Sun	Fudan university, China
Dimitri Theodoratos	New Jersey Institute of Technology, USA
Yicheng Tu	University of South Florida, USA
Leong Hou U.	University of Macau, China

Athena Vakali	Aristotle University of Thessaloniki, Greece
Hua Wang	Victoria University, Australia
Junhu Wang	Griffith University, Australia
Ingmar Weber	Qatar Computing Research Institute, Qatar
Adam Wojtowicz	Poznań University of Economics, Poland
Raymond Chi-Wing Wong	The Hong Kong University of Science and Technology, SAR China
Mingjun Xiao	University of Science and Technology of China, China
Takehiro Yamamoto	Kyoto University, Japan
Yanfang Ye	West Virginia University, USA
Hongzhi Yin	The University of Queensland, Australia
Tetsuya Yoshida	Nara Women's University, Japan
Ge Yu	Northeastern University, China
Rashid Zalyalov	Institute of Computing for Physics and Technology, Russia
Yanchun Zhang	Victoria University, Australia
Detian Zhang	Jiangnan University, China
Xiangliang Zhang	King Abdullah University of Science and Technology, Saudi Arabia
Ying Zhang	University of Technology Sydney, Australia
Qi Zhang	Fudan University, China
Chao Zhang	University of Illinois at Urbana-Champaign
Lei Zhao	Soochow University, China
Xiangmin Zhou	RMIT University, Australia
Xingquan Zhu	Florida Atlantic University, USA
Lizhen Wang	Yunnan University, China

Special Area Program Committee Co-chairs

Hua Wang	Victoria University, Australia
Xun Yi	RMIT University, Australia

Special Area Organizing Committee Co-chairs

Lili Sun	University of Southern Queensland, Australia
Surya Nepal	CSIRO Data61, Australia

Special Area Program Committee

Xu Yang	RMIT University, Australia
Hui Cui	RMIT University, Australia
Xuechao Yang	RMIT University, Australia
Yali Zeng	Fujian Normal University, China
Marios Anagnostopoulos	Singapore University of Technology and Design, Singapore
Georgios Kambourakis	University of the Aegean, Greece

Panagiotis Drakatos   University of the Aegean, Greece
Enamul Kabir   University of Southern Queensland, Australia
Uday Tupakula   The University of Newcastle, Australia
Vijay Varadharajan   The University of Newcastle, Australia

Contents – Part I

Microblog Data Analysis

A Refined Method for Detecting Interpretable and Real-Time Bursty
Topic in Microblog Stream.   3
    Tao Zhang, Bin Zhou, Jiuming Huang, Yan Jia, Bing Zhang, and Zhi Li

Connecting Targets to Tweets: Semantic Attention-Based Model
for Target-Specific Stance Detection   18
    Yiwei Zhou, Alexandra I. Cristea, and Lei Shi

A Network Based Stratification Approach for Summarizing
Relevant Comment Tweets of News Articles   33
    Roshni Chakraborty, Maitry Bhavsar, Sourav Dandapat,
    and Joydeep Chandra

Interpreting Reputation Through Frequent Named Entities in Twitter.   49
    Nacéra Bennacer, Francesca Bugiotti, Moditha Hewasinghage,
    Suela Isaj, and Gianluca Quercini

Social Network Data Analysis

Discovering and Tracking Active Online Social Groups.   59
    Md Musfique Anwar, Chengfei Liu, Jianxin Li, and Tarique Anwar

Dynamic Relationship Building: Exploitation Versus Exploration
on a Social Network   75
    Bo Yan, Yang Chen, and Jiamou Liu

Social Personalized Ranking Embedding for Next POI Recommendation   91
    Yan Long, Pengpeng Zhao, Victor S. Sheng, Guanfeng Liu, Jiajie Xu,
    Jian Wu, and Zhiming Cui

Assessment of Prediction Techniques: The Impact of Human Uncertainty . . .   106
    Kevin Jasberg and Sergej Sizov

Data Mining

Incremental Structural Clustering for Dynamic Networks   123
    Yazhong Chen, Rong-Hua Li, Qiangqiang Dai, Zhenjun Li,
    Shaojie Qiao, and Rui Mao

Extractive Summarization via Overlap-Based Optimized Picking.   135
  Gaokun Dai and Zhendong Niu

Spatial Information Recognition in Web Documents
Using a Semi-supervised Machine Learning Method   150
  Hendi Lie, Richi Nayak, and Gordon Wyeth

When Will a Repost Cascade Settle Down?   165
  Chi Chen, HongLiang Tian, Jie Tang, and ChunXiao Xing

Pattern Mining

Mining Co-location Patterns with Dominant Features.   183
  Yuan Fang, Lizhen Wang, Xiaoxuan Wang, and Lihua Zhou

Maximal Sub-prevalent Co-location Patterns and Efficient
Mining Algorithms   199
  Lizhen Wang, Xuguang Bao, Lihua Zhou, and Hongmei Chen

Overlapping Communities Meet Roles and Respective Behavioral
Patterns in Networks with Node Attributes.   215
  Gianni Costa and Riccardo Ortale

Efficient Approximate Entity Matching Using Jaro-Winkler Distance.   231
  Yaoshu Wang, Jianbin Qin, and Wei Wang

Cloud Computing

Long-Term Multi-objective Task Scheduling with Diff-Serv
in Hybrid Clouds   243
  Puheng Zhang, Chuang Lin, Wenzhuo Li, and Xiao Ma

Online Cost-Aware Service Requests Scheduling in Hybrid Clouds
for Cloud Bursting   259
  *Yanhua Cao, Li Lu, Jiadi Yu, Shiyou Qian, Yanmin Zhu, Minglu Li,
  Jian Cao, Zhong Wang, Juan Li, and Guangtao Xue*

Adaptive Deployment of Service-Based Processes into Cloud Federations . . .   275
  *Chahrazed Labba, Nour Assy, Narjès Bellamine Ben Saoud,
  and Walid Gaaloul*

Towards a Public Cloud Services Registry   290
  *Ahmed Mohammed Ghamry, Asma Musabah Alkalbani, Vu Tran,
  Yi-Chan Tsai, My Ly Hoang, and Farookh Khadeer Hussain*

Query Processing

Location-Based Top-*k* Term Querying over Sliding Window   299
*Ying Xu, Lisi Chen, Bin Yao, Shuo Shang, Shunzhi Zhu, Kai Zheng,
and Fang Li*

A Kernel-Based Approach to Developing Adaptable and Reusable
Sensor Retrieval Systems for the Web of Things.   315
Nguyen Khoi Tran, Quan Z. Sheng, M. Ali Babar, and Lina Yao

Reliable Retrieval of Top-k Tags   330
Yong Xu, Reynold Cheng, and Yudian Zheng

Estimating Support Scores of Autism Communities in Large-Scale
Web Information Systems   347
Nguyen Thin, Nguyen Hung, Svetha Venkatesh, and Dinh Phung

Spatial and Temporal Data

DTRP: A Flexible Deep Framework for Travel Route Planning   359
*Jie Xu, Chaozhuo Li, Senzhang Wang, Feiran Huang, Zhoujun Li,
Yueying He, and Zhonghua Zhao*

Taxi Route Recommendation Based on Urban Traffic Coulomb's Law   376
*Zheng Lyu, Yongxuan Lai, Kuan-Ching Li, Fan Yang, Minghong Liao,
and Xing Gao*

Efficient Order-Sensitive Activity Trajectory Search   391
*Kaiyang Guo, Rong-Hua Li, Shaojie Qiao, Zhenjun Li, Weipeng Zhang,
and Minhua Lu*

Time Series Classification by Modeling the Principal Shapes   406
Zhenguo Zhang, Yanlong Wen, Ying Zhang, and Xiaojie Yuan

Effective Caching of Shortest Travel-Time Paths for Web Mapping
Mashup Systems.   422
Detian Zhang, An Liu, Gangyong Jia, Fei Chen, Qing Li, and Jian Li

Graph Theory

Discovering Hierarchical Subgraphs of K-Core-Truss.   441
*Zhen-jun Li, Wei-Peng Zhang, Rong-Hua Li, Jun Guo, Xin Huang,
and Rui Mao*

Efficient Subgraph Matching on Non-volatile Memory.   457
Yishu Shen and Zhaonian Zou

Influenced Nodes Discovery in Temporal Contact Network 472
    Jinjing Huang, Tianqiao Lin, An Liu, Zhixu Li, Hongzhi Yin,
    and Lei Zhao

Tracking Clustering Coefficient on Dynamic Graph via Incremental
Random Walk 488
    Qun Liao, Lei Sun, Yunpeng Yuan, and Yulu Yang

Event Detection

Event Cube – A Conceptual Framework for Event Modeling and Analysis. . . . 499
    Qing Li, Yun Ma, and Zhenguo Yang

Cross-Domain and Cross-Modality Transfer Learning for Multi-domain
and Multi-modality Event Detection 516
    Zhenguo Yang, Min Cheng, Qing Li, Yukun Li, Zehang Lin,
    and Wenyin Liu

Determining Repairing Sequence of Inconsistencies
in Content-Related Data. 524
    Yuefeng Du, Derong Shen, Tiezheng Nie, Yue Kou, and Ge Yu

Author Index 541

Contents – Part II

Crowdsourcing and Crowdsensing

Real-Time Target Tracking Through Mobile Crowdsensing   3
  Jinyu Shi and Weijia Jia

Crowdsourced Entity Alignment: A Decision Theory Based Approach   19
  Yan Zhuang, Guoliang Li, and Jianhua Feng

A QoS-Aware Online Incentive Mechanism for Mobile Crowd Sensing   37
  Hui Cai, Yanmin Zhu, and Jiadi Yu

Iterative Reduction Worker Filtering for Crowdsourced Label Aggregation . . .   46
  Jiyi Li and Hisashi Kashima

Web Data Model

Semantic Web Datatype Inference: Towards Better RDF Matching   57
  Irvin Dongo, Yudith Cardinale, Firas Al-Khalil, and Richard Chbeir

Cross-Cultural Web Usability Model   75
  Rukshan Alexander, David Murray, and Nik Thompson

How Fair Is Your Network to New and Old Objects?: A Modeling
of Object Selection in Web Based User-Object Networks   90
  Anita Chandra, Himanshu Garg, and Abyayananda Maiti

Modeling Complementary Relationships of Cross-Category Products
for Personal Ranking   98
  Wenli Yu, Li Li, Fei Hu, Fan Li, and Jinjing Zhang

Language Processing and Web Protocols

Eliminating Incorrect Cross-Language Links in Wikipedia   109
  *Nacéra Bennacer, Francesca Bugiotti, Jorge Galicia, Mariana Patricio,
  and Gianluca Quercini*

Combining Local and Global Features in Supervised Word
Sense Disambiguation   117
  Xue Lei, Yi Cai, Qing Li, Haoran Xie, Ho-fung Leung, and Fu Lee Wang

A Concurrent Interdependent Service Level Agreement Negotiation
Protocol in Dynamic Service-Oriented Computing Environments.   132
   Lei Niu, Fenghui Ren, and Minjie Zhang

A New Static Web Caching Mechanism Based on Mutual Dependency
Between Result Cache and Posting List Cache   148
   Thanh Trinh, Dingming Wu, and Joshua Zhexue Huang

Web-Based Applications

A Large-Scale Visual Check-In System for TV Content-Aware Web
with Client-Side Video Analysis Offloading   159
   Shuichi Kurabayashi and Hiroki Hanaoka

A Robust and Fast Reputation System for Online Rating Systems   175
   Mohsen Rezvani and Mojtaba Rezvani

The Automatic Development of SEO-Friendly Single Page Applications
Based on HIJAX Approach   184
   Siamak Hatami

Towards Intelligent Web Crawling – A Theme Weight and Bayesian
Page Rank Based Approach   192
   Yan Tang, Lei Wei, Wangsong Wang, and Pengcheng Xuan

Data Storage and Generator

Efficient Multi-version Storage Engine for Main Memory Data Store   205
   Jinwei Guo, Bing Xiao, Peng Cai, Weining Qian, and Aoying Zhou

WeDGeM: A Domain-Specific Evaluation Dataset Generator for
Multilingual Entity Linking Systems   221
   Emrah Inan and Oguz Dikenelli

Extracting Web Content by Exploiting Multi-Category Characteristics   229
   Qian Wang, Qing Yang, Jingwei Zhang, Rui Zhou, and Yanchun Zhang

Security and Privacy

PrivacySafer: Privacy Adaptation for HTML5 Web Applications.   247
   Georgia M. Kapitsaki and Theodoros Charalambous

Anonymity-Based Privacy-Preserving Task Assignment
in Spatial Crowdsourcing   263
   Yue Sun, An Liu, Zhixu Li, Guanfeng Liu, Lei Zhao, and Kai Zheng

Understanding Evasion Techniques that Abuse Differences
Among JavaScript Implementations............................. 278
    Yuta Takata, Mitsuaki Akiyama, Takeshi Yagi, Takeo Hariu,
    and Shigeki Goto

Mining Representative Patterns Under Differential Privacy.............. 295
    Xiaofeng Ding, Long Chen, and Hai Jin

A Survey on Security as a Service 303
    Wenyuan Wang and Sira Yongchareon

Sentiment Analysis

Exploring the Impact of Co-Experiencing Stressor Events for Teens
Stress Forecasting 313
    Qi Li, Liang Zhao, Yuanyuan Xue, Li Jin, and Ling Feng

SGMR: Sentiment-Aligned Generative Model for Reviews.............. 329
    He Zou, Litian Yin, Dong Wang, and Yue Ding

An Ontology-Enhanced Hybrid Approach to Aspect-Based
Sentiment Analysis 338
    Daan de Heij, Artiom Troyanovsky, Cynthia Yang,
    Milena Zychlinsky Scharff, Kim Schouten, and Flavius Frasincar

DARE to Care: A Context-Aware Framework to Track Suicidal Ideation
on Social Media... 346
    Bilel Moulahi, Jérôme Azé, and Sandra Bringay

Recommender Systems

Local Top-*N* Recommendation via Refined Item-User Bi-Clustering 357
    Yuheng Wang, Xiang Zhao, Yifan Chen, Wenjie Zhang,
    and Weidong Xiao

HOMMIT: A Sequential Recommendation for Modeling
Interest-Transferring via High-Order Markov Model 372
    Yang Xu, Xiaoguang Hong, Zhaohui Peng, Yupeng Hu, and Guang Yang

Modeling Implicit Communities in Recommender Systems.............. 387
    Lin Xiao and Gu Zhaoquan

Coordinating Disagreement and Satisfaction in Group Formation
for Recommendation 403
    Lin Xiao and Gu Zhaoquan

Factorization Machines Leveraging Lightweight Linked Open
Data-Enabled Features for Top-N Recommendations   420
    Guangyuan Piao and John G. Breslin

A Fine-Grained Latent Aspects Model for Recommendation:
Combining Each Rating with Its Associated Review   435
    Xuehui Mao, Shizhong Yuan, Weimin Xu, and Daming Wei

Auxiliary Service Recommendation for Online Flight Booking   450
    Hongyu Lu, Jian Cao, Yudong Tan, and Quanwu Xiao

How Does Fairness Matter in Group Recommendation   458
    Lin Xiao and Gu Zhaoquan

Exploiting Users' Rating Behaviour to Enhance the Robustness
of Social Recommendation.   467
    Zizhu Zhang, Weiliang Zhao, Jian Yang, Surya Nepal, Cecile Paris,
    and Bing Li

Special Sessions on Security and Privacy

A Study on Securing Software Defined Networks   479
    Raihan Ur Rasool, Hua Wang, Wajid Rafique, Jianming Yong,
    and Jinli Cao

A Verifiable Ranked Choice Internet Voting System   490
    Xuechao Yang, Xun Yi, Caspar Ryan, Ron van Schyndel, Fengling Han,
    Surya Nepal, and Andy Song

Privacy Preserving Location Recommendations.   502
    Shahriar Badsha, Xun Yi, Ibrahim Khalil, Dongxi Liu, Surya Nepal,
    and Elisa Bertino

Botnet Command and Control Architectures Revisited: Tor Hidden
Services and Fluxing.   517
    Marios Anagnostopoulos, Georgios Kambourakis, Panagiotis Drakatos,
    Michail Karavolos, Sarantis Kotsilitis, and David K.Y. Yau

My Face is Mine: Fighting Unpermitted Tagging on Personal/Group
Photos in Social Media   528
    Lihong Tang, Wanlun Ma, Sheng Wen, Marthie Grobler, Yang Xiang,
    and Wanlei Zhou

Cryptographic Access Control in Electronic Health Record Systems:
A Security Implication.   540
    Pasupathy Vimalachandran, Hua Wang, Yanchun Zhang,
    Guangping Zhuo, and Hongbo Kuang

SDN-based Dynamic Policy Specification and Enforcement
for Provisioning SECaaS in Cloud   550
    Uday Tupakula, Vijay Varadharajan, and Kallol Karmakar

Topic Detection with Locally Weighted Semi-supervised
Collective Learning   562
    Ye Wang, Yong Quan, Bin Zhou, Yanchun Zhang, and Min Peng

Author Index   573

Microblog Data Analysis

A Refined Method for Detecting Interpretable and Real-Time Bursty Topic in Microblog Stream

Tao Zhang[1($\boxtimes$)], Bin Zhou[1], Jiuming Huang[1], Yan Jia[1], Bing Zhang[2], and Zhi Li[2]

[1] National University of Defense Technology, Changsha, Hunan, China
towermxt@gmail.com, binzhou@nudt.edu.cn,
jiuming.huang@qq.com, jiayanjy@vip.sina.com
[2] Hunan Eefung Software Co., Ltd., Changsha, Hunan, China
{zhangbing,lizhi}@eefung.com

Abstract. The real-time detection of bursty topics on microblog has acquired much research efforts in recent years, due to its wide use in a range of user-focused tasks such as information recommendation, trend analysis, and document search. Most existing methods can achieve good performance on real-time detection, but unfortunately, lack of much consideration on topic coherence and topic granularity for better semantic interpretability, which often results in odd topics hard to be interpreted. Therefore, it demands much more efforts on evaluation and improvement of the intrinsic quality of detected topics at their very early stages. In this paper, we propose a refined tensor decomposition model to effectively detect bursty topics, and at the same time, evaluate topic coherence and provide informative bursty topics with different burst levels. We evaluated our method over 7 million microblog stream. The experiment results demonstrate both efficiency in topic detection and effectiveness in topic interpretability. Specifically, our method on a single machine can consistently handle millions of microblogs per day and present ranked interpretable topics with different burst levels.

Keywords: Bursty topic real-time detection · Topic interpretability · Topic coherence · Word intrusion

1 Introduction

Microblog (such as Twitter, Snapchat, Sina weibo, etc.), as one of the most prevalent social media, allows users to share and exchange small digital contents (tweets, blog, photos, etc.) in a real-time manner. Usually, some new and interesting events spread vary fast on microblog, and also cause a myriad of discussion posts. For the purpose of relationship crisis management, product marketing, or even emergency management, many different microblog users (no matter organizational or personal) prefer to be informed or alerted as soon as bursty topics start to grow viral or dramatically. Tracking the microblog stream in a real-time manner can detect those headlines or breaking news as early as possible.

© Springer International Publishing AG 2017
A. Bouguettaya et al. (Eds.): WISE 2017, Part I, LNCS 10569, pp. 3–17, 2017.
DOI: 10.1007/978-3-319-68783-4_1

Bursty topic detection on real-time streams has acquired much research efforts in recent years, and is increasingly used in many user-focused tasks, such as information recommendation (Diao et al. [1], Kleinberg [2], Xie et al. [3], Xie et al. [4], Zhu and Shasha [5]), trend analysis (Huang et al. [6]), and document search (Magdy et al. [7]). Those detection tasks have been categorized as feature-pivot techniques in some survey works (Atefeh and Khreich [8]). Bursty topics on real-time microblog streams have bursty features of not only short-term surged keywords, but also sharply increasing tweet volume.

In bursty topic detection task, researchers have to face two main challenges, topic interpretability and memory scalability. Most of the effective prior works [2–4, 7, 9–12] take tweet volume, words frequency, or co-occurrence words frequency in the data stream as topic bursty features. When tracking the bursty features on real-time microblog streams, memory scalability is also a big challenge. Sketch-based methods, such as TopicSketch [3, 4, 13] and SigniTrend [9] surpass the rest with efficient performance in memory scalability.

Unfortunately, the word intrusion and topic overlap are always detrimental to the quality of detected bursty topic. Besides, topic words coherence is also sensitive to the fixed value N of the picked top-N topic words. Therefore, topic quality with fine coherence and granularity is another great challenge. In previous studies, a typical way [10, 14, 15] for this task is to detect bursty words and then cluster them. However, two drawbacks cause it been substituted, complicated heuristic tuning and post-processing, since noisy words and words ambiguity are unavoidable. Another attempt is to discover bursty topic via topic models, like TopicSketch [4] and LDA [16]. But when choose the top-N words in a detected topic, there always no consensus solutions for general topics.

In this paper, we propose a novel detection framework to detect bursty topics soon after they start burst, and devise an automatic evaluation on detected topics to provide coherent topic words with fine granularity. We summarize our major contributions as follows:

- We proposed a refined version of TopicSketch, a up-to-date and efficient detection method using tensor decomposition and dimension reduction [3, 17] for real-time bursty topic detection. Our main improvement is with the evalution of word intrusion and topic coherence, making use of clustering and fuzzy set theory jointly to facilitate the process of extracting informative and interpretable bursty topics and their bursty scores.
- We proposed a novel topic quality measure, sketch-based PMI method to estimate word intrusion and topic coherence based on pairwise pointwise mutual information (PMI) among topic words. We take the words sketch statistics for PMI reference corpus, in which words are dynamically sampled over consecutive sliding window on real-time data stream, and fresh word probability feeding into PMI, gives estimation of topic coherence much more reasonable and precise.
- We also conduct extensive experiments on real-world data from Eefung.com[1] to demonstrate the efficiency in real-time bursty topic detection, the soundness of the coherence of the detected topics, and the effectiveness in bursty topic interpretability.

---

[1] http://www.eefung.com/.

This paper is organized as follows: Sect. 2 briefly reviews the related work. Solution overview is specified in Sect. 3. Section 4 explains our topic refinement model based on tensor decomposition with topic evaluation. The experimental results are discussed in Sect. 5. The conclusion is summarized in Sect. 6.

2 Related Work

For early bursty topic detection, Kleinberg [2] propose an infinite-state automaton to model the arrival times of documents in a stream to identify bursts that have high intensity over limited durations of time. The states of the probabilistic automaton correspond to the frequencies of individual words, while the state transitions capture the burst, which correspond to a significant change in word frequency. Twevent [10] detectes bursty tweet segments as event segments and then clusters the event segments into events considering both their frequency distribution and content similarity. Wikipedia is exploited to identify the realistic events. Statistic based methods generate the bursty topic based on bursty features trend over real-time data stream. TopicSketch [3, 4, 13] monitors the acceleration of three quantities to provide early signals of popularity surge, and estimates the topic words probability distribution and topic acceleration. EMA/MACD [18], trend indicator wildly used in stock market, and sketch structure contribute to remarkable performance on memory scalability. SigniTrend [5] proposes a significance measure to detect emerging topics early, and can track even all keyword pairs using only a fixed amount of memory. At last, it aggregates the detected co-trends into larger topics. Huang et al. [6] extract high quality microblog by transforming some important social media features into wavelet domain and fuse further to get a weighted ensemble value, which filter much noisy documents, and then get bursty topic by LDA in new time window data stream.

Research efforts on topic quality evaluation become impressive a lot to approach or even surpass human levels of accuracy. Newman et al. [19] introduce the notion of topic "coherence", and propose an automatic method for estimating topic coherence based on pairwise PMI between the topic words. Aletras and Stevenson [20] calculate the distributional similarity between semantic vectors for the top-N topic words using a range of distributional similarity measures such as cosine similarity and the Dice coefficient. They show that their method correlates well with the observed coherence rated by human judges taking Wikipedia as the reference corpus. Lau et al. [21] explore two tasks of automatic evaluation of single topics and automatic evaluation of whole topic models, and provide recommendations on the best strategy for performing the two tasks. They can perform automatic evaluation of the human-interpretability of topics, as well as topic models. Besides, they have systematically compared different existing methods and found appreciable differences between them. For reasonable topic granularity, Lau and Baldwin [22], following Lau et al. [21], investigate the impact of the cardinality hyper-parameter, parameter N of top-N words, on topic coherence evaluation.

3 Solution Overview

3.1 Problem Formulation

Just like TopicSketch [3, 4], we follow two criteria in defining a bursty topic: (1) Bursty topic has to be a sudden surge of related tweets size in a short time, to avoid continuing hot topics blended into the detection. (2) The size of bursty topic related microblog would be large enough to filter away the trivial topics.

For topics generated by a topic model, extrinsic evaluation and intrinsic evaluation demonstrate efficiency and effectiveness of these detected topics. Extrinsic evaluation explains early detection and the importance of the discoveries. Intrinsic evaluation of the topics contribute to quantify interpretability via scoring word intrusion and topic coherence using the top-N topic words [19, 21, 23].

3.2 Solution Overview

Our solution can be divided into three parts. Firstly, like TopicSketch [3, 4], a sketch structure is used for fast word tokens indexing and token frequency updating with dimension reduction techniques. Secondly, a tensor decomposition based topic model with fuzzy theory is designed to provide better informative and interpretable bursty topics with burst scores. In recent years, tensor decomposition [17] has been adopted in TopicSketch [3, 13] to develop CLEar[2], an efficient real-time bursty topic detection system. But a tensor decomposition topic model has limited performance on topic quality due to word noise and spam [3, 4]. In our method, clustering and fuzzy set theory refine the tensor decomposition model by preserving topic interpretability. Clustered topics usually have different cardinality, which contribute to fine granularity topics with filtering away trivial topics naturally. Finally, automatic topic evaluation will contribute to better bursty topic recommendation. Performing PMI based on sketch in real time can automatically estimate topic coherence and word intrusion to quantify topic interpretability.

Figure 1 gives the overview of our bursty topic detection model. The real-time detection flow is as follows: (1) Data preprocessing, including word segmentation and word TFIDF estimation. (2) Updating Sketch with tokens from (1). (3) Upon bursting in microblog size, we will trigger refined topic model. (4) Refining results derived from tensor decomposition component to provide interpretable bursty topics. Then we will discuss step (2), (3), (4) in Sect. 4.1. (5) At last, we automatically evaluate detected topics by word intrusion and topic coherence based on PMI for better recommendation, which will be detailed in Sect. 4.2.

4 Refined Sketch-Based Topic Model and Evaluation

We first discuss how to extract bursty topics based on the refinement topic model, and then explain how to evaluate the detected bursty topics automatically.

---

[2] http://research.pinnacle.smu.edu.sg/clear/.

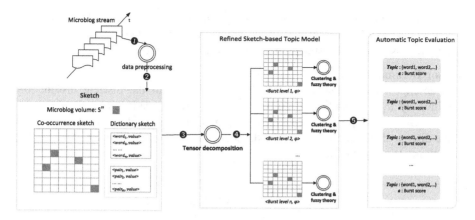

Fig. 1. The framework of solution overview.

4.1 Real-Time Detection

Sketch

In computing, sketch and its variant, count-min sketch [24], both are probabilistic data structures that serve as frequency tables of events in a stream of data[3]. Sketch in our method, also a variant designed for capturing the trend of word tokens frequency, has three components: the trend of microblog volume, co-occurrence sketch, dictionary sketch.

The trend of microblog volume is a valuable indicator for a burst stream containing bursty topics. We estimate the volume trend by EMA (Exponential Moving Average) and MACD (Moving Average Convergence/Divergence) [18], widely accepted stock market trend analysis techniques. Denote D_t means all microblogs at timestamp t, and $|D_t|$ is size of D_t. For a time interval Δt, microblog volume rate is $v = |\Delta D_{\Delta t}|/\Delta t$, and we form a discrete time series $V = \{v_t|t = 0, 1, \ldots\}$. The n-interval EMA with smoothing factor α is

$$EMA(n)[v]_t = \alpha v_t + (1 - \alpha)EMA(n - 1)[v]_{t-1} = \sum_{k \geq 0}^{n} \alpha(1 - \alpha)^k v_{t-k} \qquad (1)$$

The n is called the window size in EMA. Usually, the MACD is used to estimate the acceleration of v_t when defined by the difference of its n_1 and $n_2 - interval$ moving averages:

$$MACD(n_1, n_2) = EMA(n_1) - EMA(n_2) \qquad (2)$$

The co-occurrence sketch contains word pairs acceleration M_2 and word triples acceleration M_3. Their definitions are same with TopicSketch [3]. The acceleration are

---

[3] https://en.wikipedia.org/wiki/Count%E2%80%93min_sketch.

the trends of the frequency of word pairs and word triples, respectively. The dictionary sketch is statistics for probabilities of all words and pairs on the current data stream, and it is devised for PMI estimation at topic evaluation stage.

Tensor Decomposition Model

[17] describes that k distinct topics, drawn according to the discrete distribution specified by the probability vector $w = (w_1, w_2, \ldots, w_k)$, called burst level in our method. Given the topic k, the document's l words are drawn independently according to the discrete distribution specified by the probability vector ϕ_k. The sketches M_2 and M_3 [3] are demonstrated as:

$$M_2 = \sum_{k=1}^{K} w_k \phi_k \otimes \phi_k = \sum_{k=1}^{K} w_k \phi_k \phi_k^{\mathrm{T}} \tag{3}$$

$$M_3 = \sum_{k=1}^{K} w_k \phi_k \otimes \phi_k \otimes \phi_k \tag{4}$$

Algrithom 1 Tensor Decomposition

Input K : the number of topics.

M_2, M_3: the acceleration of word pair frequency and word triple frequency

Output topic ϕ_k and their corresponding burst level w_k

1 /*SVD on M_2 to form whitening matrix W */
2 (U, Λ) = eigs(M_2; K); /* the largest K magnitude eigenvalues and corresponding eigenvectors.*/
3 W = U$\Lambda^{-1/2}$
4 $T_3 = W^T M_3(\eta) W$
5 Compute generalised vectors v_k of M_3; /* SVD on T_3 */
6 /* recovery */
7 **for** k=1 to K **do**
8 $\phi_k = \dfrac{W(W^{\mathrm{T}}W)^{-1} v_k}{1_N^{\mathrm{T}} W(W^{\mathrm{T}}W)^{-1} v_k}$
9 $w_k = \dfrac{1}{(W^{\mathrm{T}}\phi_k)(W^{\mathrm{T}}\phi_k)^{\mathrm{T}}}$
10 **end for**
11 **return** $\{\phi_k, w_k\}$

Considering the memory consumed by M_3 and the computational complexity of tensor decomposition, M_3 has to be projected to a matrix $M_3(\eta)$ with a random vector η, where $\eta \in R^N$.

$$M_3(\eta) = \sum_{k=1}^{K} w_k \phi_k \phi_k^{\mathrm{T}} \langle \eta, \phi_k \rangle \tag{5}$$

Algorithm 1 from TopicSketch [3] explains how the tensor composition works to generate topics. First, The SVD method performs on M_2 to find a whitening matrix to whiten $M_3(\eta)$ and get the matrix T_3. Next SVD performs on T_3 to find the eigenvector v_k to recover topic words vector. The procedure contains two SVD work stages and Recovery, and the most time consumption is transforming $M_3(\eta)$ from a $N \times N$ matrix to a $K \times K$ matrix T_3, which take time in the order of $O(KN^2)$. And the method detailed proved in [3].

Refinement Model

Despite having been adopted in an efficient real-time bursty topic detection system [13], tensor decomposition [17] topic model does not perform well on topic quality due to word noise and spam. So, our goal in this part is to preserve bursty topic interpretability and filter away trivial topics by refining the tensor decomposition model.

In the result of word intrusion and topic overlap, the topic (ϕ_k) derived from tensor decomposition, cannot typically interpret a single real event. We implement clustering on co-occurred words to avoid these problems. Equation 6 is the notion of each cluster at each burst level w_k.

$$C_{km} = \{\{p_i, p_j\} \cap C_{km} | p_i, p_j \in \phi_k, PD(i,j) > \delta, PD(i,j) \in PairDictionary\} \tag{6}$$

Where p_i is word i probability in ϕ_k. The frequency of co-occurred word pairs $(word_i, word_j)$ is $PD(i, j)$, which is stored in the pair dictionary of the dictionary sketch. δ is the threshold of frequency for pairs to be picked. Each pair of words in ϕ_k, will be clustered into one cluster C_{km} once they co-occur.

For each word i, we adopt the fuzzy set theory to estimate the membership grade for each cluster, as described in Eq. 7. We then throw the word into the most co-occurred cluster according to the maximal membership grade.

$$\max_{m} Membership\_C_{km}(i) = \left\{ \sum_{j} PD(i,j) \,\Big|\, \{p_i, p_j\} \in C_{km} \right\} \tag{7}$$

Each word in cluster C_{km} retains the word probability p_i in ϕ_k, which helps to estimate burst score a_{km} for the cluster C_{km} on burst level w_k.

$$a_{km} = w_k \bullet \sum_{i} p_i, p_i \in \phi_k \cap C_{km} \tag{8}$$

The refinement model contains two steps, as described at Algorithm 2. The first step is for clustering at each burst level w_k. Top-N words according to ϕ_k are clustered into M clusters according to co-occurred word pairs in the pair dictionary sketch. And the variable size of clusters can help to provide flexible topic granularity. Besides, obviously the most of the word pairs that come from a bursty topic related microblog will be clustered into one cluster. Meanwhile the clusters preserve the topic interpretability quite well. In the second step, we can obtain a burst score for each cluster in step 1.

We estimate the burst score for each cluster according to Eq. 8. At last, we ranked all the clusters C_{km} in order of their burst score a_{km}. Consequently, the highest scoring clusters of the ranked C_{km} list are the bursty topics in the current stream. In this part, time consumption is O(n).

Algrithom 2 Refinement model

Input ϕ_k: word distribution of topic k
w_k: burst level of topic k

Output Clustered topic $\{topic_z\}$ and their corresponding burst scores $\{a_z\}$

1 **for** $k = 1$ to K **do**
2   ϕ_k={top n words}
3   **for** n=1 to N **do**
4   Estimate word i' membership grade for all clusters at burst level w_k.
5   Cluster the word i into C_{km} with $Membership\_C_{km}(i)$ is the maximum. If has no maximum, word i creates a new cluster for itself.
6   **end for**
7   **for** $c = 1$ to C **do**
8   /* burst score of the cluster */
9          $a_{km} = \sum_i w_k \times p_i , p_i \in \phi_k \cap \{C_{km}\}$
10   **end for**
11 **end for**

12 ($\{topic_z\}, \{a_z\}$)= rank C_{km} by a_{km}
13 **return** $\{\{topic_z\}, \{a_z\}\}$

4.2   Topic Evaluation – Sketch-Based PMI

In this part, we will consider word intrusion and topic coherence to evaluate topic interpretability. Lau et al. [25] proposed some features to learn the most representative or best topic word that summarises the semantics of the topic, which is the task of evaluating topic coherence. On the contrary, the task of word intrusion [23] targets detecting the least representative word.

Pointwise mutual information (PMI) is widely used in these two topic evaluation tasks. A large amount of reference corpus are needed to learn word probability and word co-occurrence pair probability. But information in microblog streams is real-time data which is rapidly updating. So considering the freshness of word and word pair probability, we discard the reference corpus on a full scope of data streams and conventional reference corpus, like Wikipedia. Especially, our dictionary sketch can typically provide real-time word probabilities and co-occurred word pair probabilities on current microblog streams for effective evaluation of detected bursty topics.

Word Intrusion
Word intrusion works as follows: for each detected bursty topic, we compute the word association features in [21] for each of the topic words, and learn the intruder words by

a ranking support vector regression model over the association features. Following Lau et al. [21], we use four association measures:

$$PMI(w_i) = \sum_{j}^{N-1} \log \frac{p(w_i, w_j)}{p(w_i)p(w_j)} \tag{9}$$

$$CP(w_i) = \sum_{j}^{N-1} \frac{p(w_i, w_j)}{p(w_j)} + \sum_{j}^{N-1} \frac{p(w_i, w_j)}{p(w_i)} \tag{7}$$

$$NPMI(w_i) = \sum_{j}^{N-1} \frac{\log \frac{p(w_i, w_j)}{p(w_i)p(w_j)}}{-\log p(w_i, w_j)} \tag{8}$$

Pointwise mutual information (PMI) between pairs can measure word association. Conditional probability (CP) contributes evaluating co-occurrence between the words with the rest. Normalised pointwise mutual information (NPMI) is an enhanced version of PMI. For NPMI, value 1 means two words only occur together; Value 0 means they are distributed as expected under independence; Value −1 means the two words occur separately without any encounter.

We score each word in a topic by these three methods, to evaluate the word coherence with its topic. If a word has a low score, it probably is an intruder word, and vice versa.

Topic Coherence

Topic coherence is the evaluation of co-occurrence over top-N topic words for the detected topic. We also follow Lau et al. [21] to experiment with the following methods for the topic coherence estimation of $topic_k$.

Pairwise PMI for each word pairs in top-N topic words:

$$PMI(topic_k) = \sum_{j=2}^{N} \sum_{i=1}^{j-1} \log \frac{p(w_i, w_j)}{p(w_i)p(w_j)} \tag{9}$$

Pairwise NPMI is a variant of pairwise PMI:

$$NPMI(topic_k) = \sum_{j=2}^{N} \sum_{i=1}^{j-1} \frac{\log \frac{p(w_i, w_j)}{p(w_i)p(w_j)}}{-\log p(w_i, w_j)} \tag{10}$$

Pairwise log conditional probability of top-N topic words:

$$LCP(topic_k) = \sum_{j=2}^{N} \sum_{i=1}^{j-1} \frac{p(w_j, w_i)}{p(w_i)} \tag{11}$$

We combine the three methods to score a single topic for topic coherence evaluation. If the topic has fine semantic interpretability, topic words will preserve satisfied association and interrelationship, and the topic coherence will score high.

5 Experiments and Evaluation

In this section, our experiments are designed to verify the effectiveness and efficiency of the proposed refined methods to discover bursty topics in Sina Weibo, the biggest microblog service in China.

5.1 Experiments Setting

We used the crawling data of Sina Weibo containing approximately 7 million blogs sampled in 62 days from Aug. 1 to Sep. 30, 2014. We used 10 min as experiment time interval in which blog size ranged from 10 to 7000. After emotions filtering and stopwords removing, we obtained 7,318,010 blogs and 54,980,668 tokens. We computed TF-IDF value for each token in a single microblog. Only top-N (e.g., 5) tokens would feed into our sketch. At last 26,342,873 high TF-IDF value tokens remained.

In our experiment, we set 4 gigabytes for sketch to track data on microblog streams. For the threshold of microblog volume trend, we set it empirically at 80 which is enough to indicate bursts in current streams.

5.2 Sketch-Based PMI Evaluation

Before using sketch-based PMI to evaluate the word intrusion and topic coherence, we designed an experiment to test the performance of methods introduced in Sect. 4.2. For TopicSketch and refinement model, five settings of latent topics (T = 10, 20, 30, 40, 50) were used to manually annotate the intruder words and score their topic coherence instinctively.

To compare our automatic methods with human annotation, Table 1 gives the Pearson Correlation coefficient for all methods in word intrusion and topic coherence. PMI based methods perform poorly for both of the two tasks, and their low stability leads itself easily to go to negative, due to poor numeric value of word probability in data streams. Fortunately, the variants, NPMI based methods, achieve much better results. Conditional probability based methods are the best performers and approach the human annotating level. Considering the performance of each sketch-based PMI methods, we just evaluated our detected topics by NPMI-based and CP-based methods.

Table 1. Pearson correlation of human and the sketch-based automated methods—WI-PMI, WI-CP, WI-NPMI, TC-PMI, TC-LCP and TC-NPMI

Pearson's with human annotation					
Word intrusion			Topic coherence		
WI-PMI	WI-CP	WI-NPMI	TC-PMI	TC-LCP	TC-NPMI
0.3746	**0.8140**	0.6582	0.2148	**0.8765**	0.7834

5.3   Results Analysis

Table 2 shows the detection results of TopicSketch and our refinement model for several annotated burst events. After comparison, we summarised our analysis of the results as follows. First, we can see that TopicSketch is not only hit one event in its single detected topic, and the bold words (like "#ice bucket challenge, #Mr.Bean, FBIcr") are the real event related key words that come from one detected topic, but semantically include more than one real event. In contrast, our refinement model topics have outstanding performance on providing coherent topic words for one event and avoiding word intrusion as much as possible. Second, our refinement model also can sometimes act as accuracy supplement for TopicSketch, because TopicSketch's detection may miss some bursty topics, such as first burst point 16:20 on Aug. 13 with annotated event "#xunlu brothers story". Even though our method can extract this ignored topic, the topic burst score is not satisfying. Because the refinement part of our method is the subsequent step of tensor decomposition that used in TopicSketch. Our outcomes are closely linked with what tensor decomposition have detected. Third, Table 2 shows the topic quality for each topic derived from the two methods. The comparison reveals that our refinement model preserves topic interpretability much better than TopicSketch. Finally, the burst start time and detected time are given in Table 2. From them, both the two detection methods can efficiently learn bursty topics shortly after they start burst, even the faster ones can be no more than an hour.

Figure 2 compares the topic quality based on word intrusion and topic coherence. Two groups are shown in Fig. 2. Group (a) is for WI (word intrusion), and Group (b) is for TC (topic coherence), and each group illustrates the figures of average CP, average NPMI, and average WI or TC (CP * 0.5 + NPMI * 0.5). For word intrusion, topic words in our method scored at a much higher level than TopicSketch, which means they have little probability to be intruder words. For topic coherence, our method also has remarkable performance, which also means topics detected by our method can preserve topic interpretability with reasonable interrelationship and higher topic words co-occurrence. Plus, CP based methods are more stable than those NPMI based, as NPMIs are sensitive to poor value of words or pairs probability. If topic words have low frequency rate on data streams, they will receive bad performance on NPMI scores.

From Table 3, obviously, our method can attain outstanding scores on model accuracy, due to the contribution of refinement methods in Sect. 4.1. While TopicSketch based on tensor decomposition encounters much word intrusion and topic overlap, which cause its weakness in accuracy. When we choose the top 5 burst topics for each method, some trivial topics and spam for advertisement intruded into the results. So we have a slight decrease for recall scores, and TopicSketch also faces this problem. To sum up, our method can efficiently and effectively detect informative bursty topic on real-time data streams.

Table 2. Comparison of TopicSketch and refinement model on bursty topic detection performance. Quality estimation is by TC&WI. Each topic has Top-8 words and topic number is 5 for TopicSketch.

Burst events	TopicSketch topics	Quality	Refinement model topics	Burst score	Quality	Burst start
[8/13 16:20] #xunlu brothers story	–	0.39	Luyi, Beier, fashion, health	0.244	12.23	8/13 15:42
			Cry, go to school, exciting, Helicopter	0.201	11.99	8/13 16:09
			Brother, one life, Luhan	**0.081**	**12.25**	**8/13 15:08**
[8/13 16:30] #Don't go to Shijiazhuang #Thank you for everything, Miroslav Klose	**Amsterdam, Jesse Lewis, ShiJiazhuang,** Brothers, **Miroslav, Klose, People at Shijiazhuang,** #linyuner	1.2	**Thank, everything, Miroslav, Klose**	**3.217**	**14.98**	**8/13 15:14**
			#Don't go to ShIjiazhuang, texi, breakfast, camera"	**3.098**	**17.45**	**8/13 15:27**
			Cry, go to school, exciting, helicopter"	0.739	12.11	8/13 16:09
[8/19 21:10] #ice bucket challenge #Mr.Bean in Shanghai Airport	**#ice bucket challenge, #Mr. Bean,** #xunlu, politeness, Libra, vedio, #Yuanhong 823 birthday, **FBIcr**	1.17	**#Mr.Bean, Shanghai, PuDong airprt, Rowan Atkinson**	**1.091**	**12.99**	**8/19 19:49**
			Libra, constellation, champion, third winner in contest	0.520	10.28	8/19 20:07
			ASL, #ice bucket challenge, ice bucket, spread	**0.500**	**13.09**	**8/19 15:18**

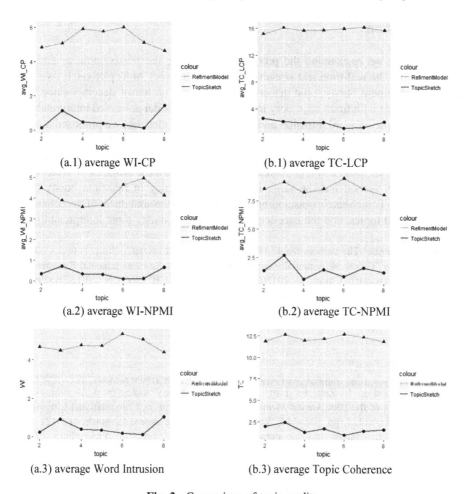

(a.1) average WI-CP

(b.1) average TC-LCP

(a.2) average WI-NPMI

(b.2) average TC-NPMI

(a.3) average Word Intrusion

(b.3) average Topic Coherence

Fig. 2. Comparison of topic quality

Table 3. Comparison on effectiveness of TopicSketch and refinement model

Latent topic	TopicSketch			Refinement model		
	Accuracy	Recall	F1	Accuracy	Recall	F1
10	0.51	0.63	0.55	0.91	0.87	0.88
20	0.49	0.69	0.57	0.91	0.88	0.89
30	0.51	0.69	0.58	0.93	0.88	0.90
40	0.48	0.64	0.54	0.92	0.86	0.88

6 Conclusion

In this paper, we re-examine the problem of detecting the bursty topic as early as possible from the real-time text streams, following the work of TopicSketch. We proposed a refinement bursty topic detection model based on tensor decomposition, and clustering fused with fuzzy set theory in the refinement model preserved interpretability of detected topic on word intrusion and topic coherence. Besides, we proposed a novel topic evaluation measure, sketch-based PMI method, to perform PMI-based (Pointwise mutual information) methods and CP-based (Conditional probability) methods for evaluate word intrusion and topic coherence by using real-time statistics in sketch. The results of our method has the same remarkable efficiency in real-time bursty topic detection with TopicSketch, outstanding performance on the soundness of the coherence of the detected topics, and the excellent effectiveness in bursty topic interpretability.

Acknowledgments. The authors would like to thank the joint research efforts between NUDT and Eefung.com. This work is partially supported by National Key Fundamental Research and Development Program of China (No. 2013CB329601, No. 2013CB329604, No. 2013CB329606), and National Natural Science Foundation of China (No. 61502517, No. 61372191, No. 61572492). This work is also funded by the major pre-research project of National University of Defense Technology (NUDT).

References

1. Diao, Q., Jiang, J., Zhu, F., Lim, E.-P.: Finding bursty topics from microblogs. In: Proceedings of the 50th Annual Meeting of the Association for Computational Linguistics: Long Papers, vol. 1, pp. 536–544. Association for Computational Linguistics (2012)
2. Kleinberg, J.: Bursty and hierarchical structure in streams. In: Proceedings of the Eighth ACM SIGKDD International Conference on Knowledge Discovery and Data Mining, pp. 91–101. ACM (2002)
3. Xie, W., Zhu, F., Jiang, J., Lim, E.-P., Wang, K.: TopicSketch: real-time bursty topic detection from Twitter. In: 2013 IEEE 13th International Conference on Data Mining (ICDM), pp. 837–846. IEEE (2013)
4. Xie, W., Zhu, F., Jiang, J., Lim, E.-P., Wang, K.: Topicsketch: real-time bursty topic detection from Twitter. IEEE Trans. Knowl. Data Eng. **28**(8), 2216–2229 (2016)
5. Zhu, Y., Shasha, D.: Efficient elastic burst detection in data streams. In: Proceedings of the Ninth ACM SIGKDD International Conference on Knowledge Discovery and Data Mining, pp. 336–345. ACM (2003)
6. Huang, J., Peng, M., Wang, H., et al.: A probabilistic method for emerging topic tracking in microblog stream. World Wide Web **20**(2), 325–350 (2017)
7. Magdy, A., et al.: GeoTrend: spatial trending queries on real-time microblogs. In: Proceedings of the 24th ACM SIGSPATIAL International Conference on Advances in Geographic Information Systems, p. 7. ACM (2016)
8. Atefeh, F., Khreich, W.: A survey of techniques for event detection in Twitter. Comput. Intell. **31**(1), 132–164 (2015)
9. Schubert, E., Weiler, M., Kriegel, H.-P.: Signitrend: scalable detection of emerging topics in textual streams by hashed significance thresholds. In: Proceedings of the 20th ACM SIGKDD International Conference on Knowledge Discovery and Data Mining, pp. 871–880. ACM (2014)

10. Li, C., Sun, A., Datta, A.: Twevent: segment-based event detection from tweets. In: Proceedings of the 21st ACM International Conference on Information and Knowledge Management, pp. 155–164. ACM (2012)
11. Schubert, E., Weiler, M., Kriegel, H.-P.: SPOTHOT: scalable detection of geo-spatial events in large textual streams. In: Proceedings 28th International Conference on Scientific and Statistical Database Management (SSDBM) (2016)
12. Kim, D., Kim, D., Hwang, E., Rho, S.: TwitterTrends: a spatio-temporal trend detection and related keywords recommendation scheme. Multimedia Syst. **21**(1), 73–86 (2015)
13. Xie, R., Zhu, F., Ma, H., Xie, W., Lin, C.: CLEar: a real-time online observatory for bursty and viral events. Proc. VLDB Endow. **7**(13), 1637–1640 (2014)
14. Mathioudakis, M., Koudas, N.: Twittermonitor: trend detection over the twitter stream. In: Proceedings of the 2010 ACM SIGMOD International Conference on Management of Data, pp. 1155–1158. ACM (2010)
15. Cataldi, M., Di Caro, L., Schifanella, C.: Emerging topic detection on Twitter based on temporal and social terms evaluation. In: Proceedings of the Tenth International Workshop on Multimedia Data Mining, p. 4. ACM (2010)
16. Blei, D.M., Ng, A.Y., Jordan, M.I.: Latent dirichlet allocation. J. Mach. Learn. Res. **3**(Jan), 993–1022 (2003)
17. Anandkumar, A., Ge, R., Hsu, D.J., Kakade, S.M., Telgarsky, M.: Tensor decompositions for learning latent variable models. J. Mach. Learn. Res. **15**(1), 2773–2832 (2014)
18. He, D., Parker, D.S.: Topic dynamics: an alternative model of bursts in streams of topics. In: Proceedings of the 16th ACM SIGKDD International Conference on Knowledge Discovery and Data Mining, pp. 443–452. ACM (2010)
19. Newman, D., Lau, J.H., Grieser, K., Baldwin, T.: Automatic evaluation of topic coherence. In: Human Language Technologies: The 2010 Annual Conference of the North American Chapter of the Association for Computational Linguistics, pp. 100–108. Association for Computational Linguistics (2010)
20. Aletras, N., Stevenson, M.: Evaluating topic coherence using distributional semantics. In: Proceedings of the 10th International Conference on Computational Semantics (IWCS 2013)–Long Papers, pp. 13–22 (2013)
21. Lau, J.H., Newman, D., Baldwin, T.: Machine reading tea leaves: automatically evaluating topic coherence and topic model quality. In: EACL, pp. 530–539 (2014)
22. Lau, J.H., Baldwin, T.: The Sensitivity of topic coherence evaluation to topic cardinality. In: Proceedings of NAACL-HLT, pp. 483–487 (2016)
23. Chang, J., Boyd-Graber, J.L., Gerrish, S., Wang, C., Blei, D.M.: Reading tea leaves: how humans interpret topic models. In: NIPS, vol. 31, pp. 1–9 (2009)
24. Cormode, G., Muthukrishnan, S.: An improved data stream summary: the count-min sketch and its applications. J. Algorithms **55**(1), 58–75 (2005)
25. Lau, J.H., Newman, D., Karimi, S., Baldwin, T.: Best topic word selection for topic labelling. In: Proceedings of the 23rd International Conference on Computational Linguistics: Posters, pp. 605–613. Association for Computational Linguistics (2010)

Connecting Targets to Tweets: Semantic Attention-Based Model for Target-Specific Stance Detection

Yiwei Zhou[1]([✉]), Alexandra I. Cristea[1], and Lei Shi[2]

[1] Department of Computer Science, University of Warwick, Coventry, UK
{Yiwei.Zhou,A.I.Cristea}@warwick.ac.uk
[2] University of Liverpool, Liverpool, UK
Lei.Shi@liverpool.ac.uk

Abstract. Understanding what people say and really mean in tweets is still a wide open research question. In particular, understanding the *stance* of a tweet, which is determined not only by its content, but also by the given *target*, is a very recent research aim of the community. It still remains a challenge to construct a tweet's vector representation with respect to the target, especially when the target is only *implicitly mentioned*, or *not mentioned at all* in the tweet. We believe that better performance can be obtained by incorporating the information of the target into the tweet's vector representation. In this paper, we thus propose to embed a *novel attention mechanism at the semantic level* in the bi-directional GRU-CNN structure, which is more fine-grained than the existing token-level attention mechanism. This novel attention mechanism allows the model to automatically attend to useful semantic features of informative tokens in deciding the target-specific stance, which further results in a conditional vector representation of the tweet, with respect to the given target. We evaluate our proposed model on a recent, widely applied benchmark Stance Detection dataset from Twitter for the SemEval-2016 Task 6.A. Experimental results demonstrate that the proposed model substantially outperforms several strong baselines, which include the state-of-the-art token-level attention mechanism on bi-directional GRU outputs and the SVM classifier.

Keywords: Target-specific Stance Detection · Text classification · Neural network · Attention mechanism

1 Introduction

Target-specific Stance Detection is a problem that can be formulated as follows: given a tweet X and a target Y, the aim is to classify the stance of X towards Y into three categories, *Favour*, *None* or *Against*. The target may be a

---

Y. Zhou—Work performed while at The Alan Turing Institute.

© Springer International Publishing AG 2017
A. Bouguettaya et al. (Eds.): WISE 2017, Part I, LNCS 10569, pp. 18–32, 2017.
DOI: 10.1007/978-3-319-68783-4_2

person, an organisation, a government policy, a movement, a product, etc. [8]. Target-specific Stance Detection is a different problem from Aspect-level Sentiment Analysis [11,15] in the following ways: the same stance can be expressed through positive, negative or neutral sentiment [9]; the target of interest of the Stance Detection does not necessarily have to occur in the tweet, as the target-specific stance can be expressed by mentioning the target implicitly, or by talking about other relevant targets. Besides typical tweets characteristics, such as being short and noisy, the main challenge in this task is that the decision made by the classifier has to be target-specific, *whilst having very little contextual information or supervision provided*. Example training data from the benchmark target-specific Stance Detection dataset for SemEval-2016 Task 6 [8] can be found in Table 1. Deep neural networks enable the continuous vector representations of underlying semantic and syntactic information in natural language texts, and save researchers the efforts of feature engineering [14,15]. Recently, they have achieved significant improvements in various natural language processing tasks, such as Machine Translation [2,3], Question Answering [14], Sentiment Analysis [6,11,15,18], etc. However, applying deep neural networks on target-specific Stance Detection has not been successful, as their performances have, up to now, been slightly worse than traditional machine learning algorithms with manual feature engineering, such as Support Vector Machines (SVM) [8].

Table 1. Examples of target-specific stance detection.

Target	Tweet	Stance
Donald Trump	#DonaldTrump my tell it like it is but his comments speaks to a prejudice and cold heart	*Against*
Hillary Clinton	I love the smell of Hillary in the morning. It smells like Republican Victory	*Against*
Hillary Clinton	Just think how many emails Hillary Clinton can delete with today's #leapsecond	*Against*
Climate Change	Coldest and wettest summer in memory	*Favour*

In this work, the above challenges are tackled, based on our intuition that the target information is vital for the Stance Detection, and that the vector representations for the tweets should be "aware" of the given targets. Since not all parts in the tweet are equally helpful for the Stance Detection task towards the specified target, we firstly apply the state-of-the-art token-level attention mechanism [2]. This allows neural networks to automatically pay more attention to the tokens that are more relevant to the target and more informative for detecting the target-specific stance. Importantly, a given token can be interpreted differently, according to different targets, and the semantic features in the token's vector representation can be of different levels of importance, conditional on the given target. We propose a novel attention mechanism, which extends the current attention mechanism, from the *token level*, to the *semantic level*, through a *gated structure*, whereby the tokens can be encoded adaptively, according to the target. We compare the models we propose based on the token-level attention mechanism and the novel semantic-level attention mechanism with several baselines, on the target-specific Stance Detection dataset for the SemEval-2016 Task 6.A [8], which is currently the most widely applied dataset on target-specific Stance Detection in tweets. The experimental results show that substantial improvements can be achieved on this task, compared with all previous neural network-based models, by inferencing conditional tweet vector representations with respect to the given targets; the neural network model with semantic-level attention also outperforms the SVM algorithm, which achieved the previous best performance in this task [8]. Additionally, it should be noted that our results are obtained with a *minimum of supervision*, with *no external domain corpus collected* to pre-train target-specific word embeddings, and *no extra sentiment information annotated*. Moreover, there are *no target-specific configurations or hand-engineered features involved*, thus *the proposed models can be easily generalised to other targets*, with no additional efforts.

2 Neural Network Models for Target-Specific Stance Detection in Tweets

In this section, we first describe two baseline models, the bi-directional Gated Recurrent Unit (biGRU) model, and the model that stacks a Convolutional Neural Network (CNN) structure on the outputs of the biGRU (biGRU-CNN) model. We then show how we extend these two baseline models, by incorporating the target information through *token-level* and *semantic-level attention mechanisms*, obtaining the AT-biGRU model and the AS-biGRU-CNN model, respectively. Finally, we demonstrate methods to generate the target embedding, and how to obtain the stance detection result based on the tweet vector representation, as well as other model training details.

2.1 biGRU Model

GRU [3] aims at solving the gradient vanishing or exploding problems, by introducing a gating mechanism. It adaptively captures dependencies in sequences, without introducing extra memory cells. GRU maps an input sequence of length N, $[x_1, x_2, \cdots, x_N]$ into a set of hidden states $[h_1, h_2, \cdots, h_N]$ as follows:

$$r_n = \sigma(W_r x_n + U_r h_{n-1} + b_r) \tag{1}$$

$$z_n = \sigma(W_z x_n + U_z h_{n-1} + b_z) \tag{2}$$

$$\tilde{h_n} = \tanh(W_h x_n + U_h(r_n \odot h_{n-1}) + b_h) \tag{3}$$

$$h_n = (1 - z_n) \odot h_{n-1} + z_n \odot \tilde{h_n}. \tag{4}$$

where $n \in \{1, \ldots, N\}$; r_n is the reset gate and z_n is the update gate; $\tilde{h_n} \in \mathbb{R}^{d_1}$ represents the "candidate" hidden state generated by the GRU; $h_n \in \mathbb{R}^{d_1}$

represents the real hidden state generated by the GRU; $x_n \in \mathbb{R}^{d_0}$ represents the word embedding vector of a token in the tweet; W_r, W_z, $W_h \in \mathbb{R}^{d_1 \times d_0}$ and U_r, U_z, $U_h \in \mathbb{R}^{d_1 \times d_1}$ represent the weight matrices; b_r, b_z, $b_h \in \mathbb{R}^{d_1}$ represent the bias terms; $\sigma(\cdot)$ represents the sigmoid function; $\odot$ represents the Hadamard product operation (element-wise multiplication).

To capture the information from both the past and the future sequence, the bi-directional GRU (biGRU), which processes the sequence in both the forward and backward directions, has proven to be successful in various applications [2,18]. In biGRU, the hidden states generated by processing the sequence in opposite directions are concatenated as the new output: $[\overrightarrow{h_1} \| \overleftarrow{h_1}, \overrightarrow{h_2} \| \overleftarrow{h_2}, \cdots, \overrightarrow{h_N} \| \overleftarrow{h_N}]$, where $\overrightarrow{h_n} \| \overleftarrow{h_n} \in \mathbb{R}^{2d_1}$, and the arrow represents the direction of the processing.

In the biGRU model, the final hidden states of the input sequence, when processing it in opposite directions, are concatenated, to form the vector representation of the tweet s:

$$s = \overrightarrow{h_N} \| \overleftarrow{h_1}. \tag{5}$$

2.2   biGRU-CNN Model

The biGRU model attempts to propagate all the semantic and syntactic information in a tweet into two fixed hidden state vectors, which could become a bottleneck, when there exist some long-distance dependencies in the tweet. In [14], Recurrent Neural Network (RNN) outputs were fed into a CNN structure, to generate a vector representation, based on all the hidden states of the RNN, rather than just the final hidden state. Specifically, a filter $w_f \in \mathbb{R}^{2kd_1}$ is applied to k concatenated consecutive hidden states $h_{i:i+k-1} \in \mathbb{R}^{2kd_1}$ to compute c_i, one value in the feature map corresponding to this filter:

$$c_i = f(w_f^T h_{i:i+k-1} + b_f), \tag{6}$$

where f is the rectified linear unit function and $b_f \in \mathbb{R}$ is a bias term. A max-pooling operation is further applied over the feature map $\mathbf{c} = (c_1, c_2, \cdots, c_{N-k+1})$, to capture the most important semantic feature $\hat{c}$ in each feature map:

$$\hat{c} = \max\{\mathbf{c}\}. \tag{7}$$

$\hat{c}$ is the feature generated by filter w_f. Filters with varying sliding window sizes k can be applied, to obtain multiple features. The features generated by different filters are concatenated, to form the vector representation of the tweet s.

2.3   AT-biGRU Model

Whilst they solve specific problems as above, neither the biGRU model nor the biGRU-CNN model takes into account the target information. However, when human annotators are asked to label the stance of a tweet towards a given target, they are likely to keep the information about the target in their mind, and pay more attention to the parts relevant to the target. The *token-level attention mechanism*, firstly proposed in [2] for Machine Translation, allowed the neural network to automatically search for tokens of a source sentence that were relevant to predicting a target word, and mask irrelevant tokens; it released the burden on RNN in compressing the entire source sentence into a static, fixed representation. The attention mechanism has been successfully applied in Question Answering [14], Caption Generation [17], Sentiment Analysis [18], etc.

In this paper, we propose to apply the attention mechanism to the biGRU model, to enable the model to automatically compute proper alignments in the tweet, which reflect the importance levels of different tokens in deciding the tweet's stance towards the given target, as shown in Fig. 1.

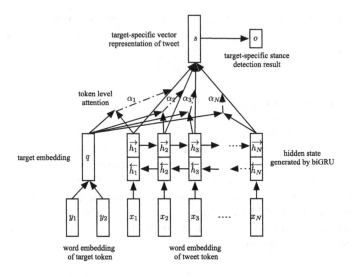

Fig. 1. The AT-biGRU model for target-specific stance detection.

In the AT-biGRU model, the vector representation s of the tweet is calculated as the weighted sum of the hidden states:

$$s = \sum_{n=1}^{N} \alpha_n h_n. \tag{8}$$

In the above equation, the weight α_n of each hidden state h_n is computed by:

$$\alpha_n = \frac{\exp(e_n)}{\sum_{n=1}^{N} \exp(e_n)}, \tag{9}$$

where $e_n \in \mathbb{R}$ is calculated through a multi-layer perceptron that takes h_n and the target embedding q as input, specifically:

$$e_n = att(h_n, q) = w_m^T(\tanh(W_{ah}h_n + W_{aq}q + b_a)) + b_m. \tag{10}$$

where $W_{ah} \in \mathbb{R}^{2d_1 \times 2d_1}$; $W_{aq} \in \mathbb{R}^{2d_1 \times d_2}$; b_a, $w_m \in \mathbb{R}^{2d_1}$; $b_m \in \mathbb{R}$ are token-level attention parameters to optimise. In Sect. 2.5, we explore various ways to generate the target embedding $q \in \mathbb{R}^{d_2}$, based on the embeddings of the tokens in the target Y, denoted by y_1, $y_2 \in \mathbb{R}^{d_0}$. The weight α_n can be interpreted as the degree to which the model attends to token x_n in the tweet, while deciding the stance of the tweet towards the given target.

2.4   AS-biGRU-CNN Model

The model we propose above is an improvement on prior research. However, it can be further refined, as follows. The AT-biGRU model applies the attention mechanism at the token level, which enables the model to pay more attention to the tokens that have contributed to the stance decision towards specified targets. However, in the AT-biGRU model, the vector representations of the tokens do not have direct interaction with the vector representation of the target, which is against the intuition that the target can influence the human annotators' interpretation of each token. For example, the token 'email' in Table 1 implies an *Against* stance towards the target "Hillary Clinton", but has no obvious influence on stances towards other targets; the token "cold" can either reveal the user's *Favour* stance towards the target "Climate Change is a Real Concern", or suggest the user's *Against* stance towards the target "Donald Trump".

Thus, we use a gated structure to extend the current token-level attention mechanism to a more fine grained semantic level, by introducing the direct interaction between the hidden states and the vector representation of the target. The gated structure can be embedded into the biGRU-CNN model, which results in the AS-biGRU-CNN model, as shown in Fig. 2.

In Fig. 2, we introduce the *target-specific hidden state* h'_n, to replace the original hidden state h_n generated by biGRU. The target-specific hidden state is calculated as follows:

$$h'_n = a_n \odot h_n. \tag{11}$$

The attention vector $a_n \in \mathbb{R}^{2d_1}$ decides which semantic features in each hidden state are meaningful specifically towards the target, which is calculated through a gated structure, as follows:

$$a_n = \sigma(W_m(\tanh(W_{ah}h_n + W_{aq}q + b_a)) + b_m). \tag{12}$$

where W_{ah}, $W_m \in \mathbb{R}^{2d_1 \times 2d_1}$; $W_{aq} \in \mathbb{R}^{2d_1 \times d_2}$; b_a, $b_m \in \mathbb{R}^{2d_1}$ are semantic-level attention parameters, to optimise in the gated structure. The methods to derive the target embedding $q \in \mathbb{R}^{d_2}$ based on the embeddings of the tokens in the target Y, denoted by y_1, $y_2 \in \mathbb{R}^{d_0}$, will be explained in Sect. 2.5. The elements in the attention vector a_n can be understood as the degrees to which the model attends to the semantic features of token x_n in the tweet, while deciding the stance of the tweet towards the given target.

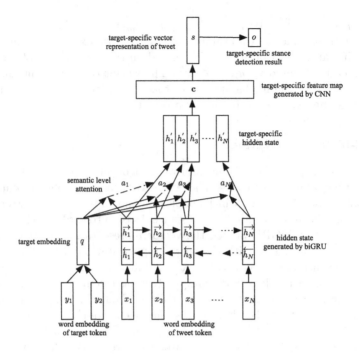

Fig. 2. The AS-biGRU-CNN model for target-specific stance detection.

2.5   Target Embedding

The models proposed in Sects. 2.3 and 2.4 employ the embedding of the given target $q \in \mathbb{R}^{d_2}$, derived from the embeddings of the tokens in the given target y_1, $y_2 \in \mathbb{R}^{d_0}$. Without loss of generality, here we use a target with two tokens, as an example. However, the methods can be directly applied on targets with any number of tokens. To generate target embeddings of the same dimensionality for the targets with different token numbers, we propose to use a separate biGRU model, described in Sect. 2.1, with the target token embeddings y_1 and y_2 as inputs. For this scenario, the dimensionality of q, denoted by d_2 in Sects. 2.3 and 2.4, equals to the dimensionality of the concatenated final hidden states of the biGRU model, denoted by $2d_1$. Results of the AT-biGRU model and the AS-biGRU-CNN model using the biGRU target embedding are reported in Sect. 3.4. In some aspect-level Sentiment Analysis works, researchers have been using the average of the aspect token embeddings to encode the aspect [11,15]. We also use the averaging method as a baseline target encoding approach to derive the target embedding q, by averaging the target token embeddings y_1 and y_2. For this scenario, d_2 equals to the dimensionality of the target token embeddings, denoted by d_0. Results of the AT-biGRU model and the AS-biGRU-CNN model using the averaging target embedding are reported in Sect. 3.5.

2.6   Model Training

The vector representation of the tweet s is fed as input to a softmax layer, after a linear transformation step that transforms it into a vector, whose length is equal to the number of possible stance categories. The outputs of the softmax layer (denoted by o in Figs. 1 and 2) are the probabilities of the tweet X belonging to the stance category z, given the target Y, denoted by $P(z|X,Y)$. The stance category with the maximum probability is selected as the *predicted category, z^**:

$$z^* = argmax_{z \in \mathbf{z}} P(z|X,Y). \tag{13}$$

All the models are smooth and differentiable, and they can be trained in an end-to-end manner, with standard back-propagation. We use the cross-entropy loss as the *objective function $L(\theta)$*, which is defined as follows:

$$L(\theta) = - \sum_{X \in \mathbf{X}} \sum_{z \in \mathbf{z}} P^{'}(z|X,Y) \cdot \log(P(z|X,Y)). \tag{14}$$

where $\mathbf{X}$ is the set of training data; $\mathbf{z}$ is the set of stance categories; $P^{'}(z|X,Y)$ denotes the target stance distribution z given X and Y; θ is the set of parameters.

3   Experimental Results

3.1   Dataset Description

As said, we evaluated the effectiveness of the proposed models on the benchmark Stance Detection dataset for the SemEval-2016 Task 6.A [8]. We used the exact same data as provided to the contestants for this task, with no extra labelled data [4] or domain corpus [1,9] employed. The benchmark Stance Detection training dataset contained 2,914 tweets relevant to five targets: "Atheism" (**A**), "Climate Change is a Real Concern" (**CC**), "Feminist Movement" (**FM**), "Hillary Clinton" (**HC**) and "Legalisation of Abortion" (**LA**). Each tweet was annotated as *Favour, Neither* or *Against* towards one of the five targets. The benchmark Stance Detection test dataset contained 1,249 tweets, as well as the interested targets. Detailed statistics about the dataset can be found in Table 2, where "**#**" represents the number of tweets, "**%F**", "**%A**" and "**%N**" represent the percentages of tweets with *Favour, Against* and *Neither* stances towards the targets, respectively.

3.2   Comparison Models

We compared the proposed models with the two best performing models in the SemEval-2016 Task 6.A: (1) MITRE [19], which trained separate Long Short-Term Memory (LSTM) networks with a voting scheme for different targets—the LSTM networks were pre-trained, by an auxiliary hashtag prediction task on 298,973 self-collected tweets; (2) pkudblab [16], which also trained

Table 2. Statistics of the benchmark target-specific stance detection dataset.

Target	Training				Test			
	#	%F	%A	%N	#	%F	%A	%N
A	513	17.9	59.3	22.8	220	14.5	72.7	12.7
CC	395	53.7	3.8	42.5	169	72.8	6.5	20.7
FM	664	31.6	49.4	19.0	285	20.4	64.2	15.4
HC	689	17.1	57.0	25.8	295	15.3	58.3	26.4
LA	653	18.5	54.4	27.1	280	16.4	67.5	16.1
All	2914	25.8	47.9	26.3	1249	24.3	57.3	18.4

separate CNN classifiers for different targets, with a voting scheme employed both in and out of each epoch, to improve the performance. We also compared against the SVM classifiers trained on the corresponding training datasets for the five targets, using word n-grams and character n-grams features, as reported in [8], representing the previous best performer for this task. Additionally, to illustrate the influence of the token-level and semantic-level attention mechanism, we included the performance comparison between the biGRU model (Sect. 2.1) and the AT-biGRU model (Sect. 2.3), the biGRU-CNN model (Sect. 2.2) and the AS-biGRU-CNN model (Sect. 2.4).

3.3   Experimental Settings and Model Configuration

In line with former works, we first trained separate classifiers for different targets. To obtain a fair comparison, we employed the *only* evaluation metric in the SemEval-2016 Task 6.A, which was the macro-average of the F1 scores for the *Favour* and *Against* stance categories. This evaluation metric will be referred to as "macro-average F1 score" in this paper for simplicity purpose. In the evaluation stage of SemEval-2016 Task 6.A, the target information of each tweet was ignored, in order to measure each team's overall performance, rather than performance on each separate target. This was because the training datasets for different targets had different percentages of tweets with Favour, Against and Neither stances, as well as different percentages of tweets expressing stances by mentioning the given target and by mentioning other targets. Thus, this evaluation metric can reflect each team's overall ability in dealing with different scenarios. It should be noted that even though separate classifiers were trained for different targets, we used the same configurations for target-specific classifiers, to make sure our proposed models can be easily applied to any other target, as well as effectively demonstrate the advantages of target-specific tweet vector representation, by eliminating the effects of target-specific model settings. Various methods were applied to avoid overfitting. We performed a standard 5-fold cross-validation. For each round of cross-validation, we experimentally set the maximum number of epochs to 50, and located the epoch that achieved the best performance on the validation dataset. The post-softmax probabilities of the 5 trained classifiers were averaged, to obtain the probabilities of a tweet in the test dataset belonging to the three stance categories.

We implemented the proposed models using Theano[1] and Keras[2].

For comparison fairness, all the neural network-based models in the experiments also used the same hyper-parameters (as illustrated below), which were selected using grid search on the baseline biGRU model. In the experiments, all the word embeddings were initialised by the Glove [10] 100-dimensional pre-trained embeddings on Wikipedia data, i.e., $d_0 = 100$. We applied dropout [13] with probability 0.2 on the embedding layer. The word embeddings were fine-tuned during the training process, to capture the stance information. From the preliminary experiments, we observed that the models that shared the embedding layer between the tweets and the targets performed significantly better than the models that did not. We chose the dimensionality of hidden states (d_1) of both the GRU encoding the tweet and the GRU encoding the target to be 64, and the GRU weights are initialised from a uniform distribution $U(-\epsilon, \epsilon)$. Following [5], we added a dropout level of 0.3 between each recurrent connection in the GRU that encoded the tweets. We further selected the hyper-parameters for the CNN structure on top of the fixed hyper-parameters of the biGRU model. Following [6], we used filters of $k \in \{3, 4, 5\}$, with widths equal to the dimensionality of the outputs of the biGRU, which was 128 in this case. There were 100 filters for each size. To increase the robustness of the models to overfitting, a dropout level of 0.5 was further applied before the softmax layer.

We used the Adam optimiser [7] for back-propagation with the two momentum parameters set to 0.9 and 0.999, respectively. The mini-batch size was set to 16. The code for the experiments is available at https://github.com/zhouyiwei/tsd.

3.4   Using the biGRU Target Embedding

The experimental results are shown in Table 3. Besides the evaluation metric of the SemEval-2016 Task6.A, we also provide the macro-average F1 scores of different targets, as references. From the comparison between the biGRU model and the biGRU-CNN model, it can be seen that the CNN structure on top of the biGRU model can help to generate more compact and abstract vector representations of the tweets for Stance Detection.

Both neural network-based models that incorporate target information when generating vector representations for the tweets, i.e., the AT-biGRU and AS-biGRU-CNN, outperform other neural network-based models that did not, i.e., MITRE, pkudblab, biGRU and biGRU-CNN. Specifically, the state-of-the-art token-level attention mechanism helps to increase the performance of the biGRU model by 0.32 in the overall macro-average F1 score. The injection of target information through the proposed semantic-level attention mechanism in the biGRU-CNN model, which results in the AS-biGRU-CNN model, leads to a

---

[1] http://deeplearning.net/software/theano/.

[2] https://keras.io/.

more significant improvement (1.71) on the basis of the biGRU-CNN model, which makes it the best performing model among all the neural network-based models. This demonstrates the effectiveness of attention mechanisms in constructing a composite vector representation between the target and contextual information provided in the tweet. The proposed AS-biGRU-CNN model with semantic-level attention, however, has stronger capability in modelling the complex interaction between the target and each token in the tweet, and generating an expressive conditional vector representation of the tweet, with respect to the target, compared with the AT-biGRU model with the token-level attention.

Moreover, the AS-biGRU-CNN model outperforms the traditional SVM algorithm, with word n-grams and character n-grams features reported in [8] by a substantial margin, in the absence of feature engineering and target-specific tuning, which justifies the motivation to automatically intensify the features that are essential to the target, and "dilute" the features that are not.

Table 3. Performance of target-specific stance detection based on the macro-average F1 score, using separate classifiers.

Model	Target					Overall
	A	CC	FM	HC	LA	
SVM	65.19	42.35	57.46	58.63	66.42	68.98
MITRE	61.47	41.63	62.09	57.67	57.28	67.82
pkudblab	63.34	52.69	51.33	64.41	61.09	67.33
biGRU	65.26	43.08	56.53	55.60	61.39	67.65
biGRU-CNN	63.42	42.91	58.69	55.11	60.55	67.71
AT-biGRU	62.32	43.89	54.15	57.94	64.05	67.97
AS-biGRU-CNN	66.76	43.40	58.83	57.12	65.45	**69.42**

3.5   Using the Averaging Target Embedding

In Table 3, we used biGRU to generate the vector representations for the targets. Additionally, we further experimented with the AT-biGRU and AS-biGRU-CNN models, using the averaging target embeddings. The overall macro-average F1 score of the AT-biGRU model increases from 67.97 to 68.30, while the macro-average F1 score of the AS-biGRU-CNN model decreases from 69.42 to 68.35. One possible explanation could be that a simple averaging approach is insufficient to capture the semantic meanings of the targets, thus for the biGRU-CNN model, which has stronger expressive power than the biGRU model in target-specific Stance Detection, it is helpful to use more flexible target embeddings to perform complex inference. However, for the AT-biGRU model, the target embeddings generated by biGRU surpass its capability to learn and generalise. *This is also the reason why stacking the CNN structure on top of the AT-biGRU model cannot help to improve the performance, as it does in the AS-biGRU-CNN model.*

3.6 Using Combined Classifiers

In the Stance Detection dataset for the SemEval-2016 Task 6.A, the training data for all the targets were of similar sizes, except for the target "Climate Change is a Real Concern". There were only 395 items in its training data and they were highly biased, with only 3.8% of them coming from the *Against* category. As a result of this, all the models in Table 3 cannot achieve a comparable performance on this target, when compared with other targets. When there was not enough training data for some targets, or the training data for some targets was highly biased, it was not possible to guarantee the performance of independent classifiers for these targets. For this case, we hypothesised that a combined classifier of all the targets can alleviate this problem, through jointly modelling the interaction between the stances and contexts of all the available targets. This way, when performing Stance Detection on the "Climate Change is a Real Concern" target, the classifier can employ—or even transfer—the knowledge about the intricate connection between the stances and contexts learnt from the training data of other targets. Motivated by this idea, we further trained combined classifiers based on the proposed models, using all the training data, rather than trained separate classifiers for different targets. The combined classifiers' performances are shown in Table 4.

Table 4. Performance of target-specific stance detection based on the macro-average F1 score, using combined classifiers.

Model	CC	Overall
SVM	47.76	62.06
biGRU	54.14	62.82
biGRU-CNN	54.57	62.70
AT-biGRU	55.69	63.36
AS-biGRU-CNN	**58.24**	**67.40**

In Table 4, we use the combined SVM classifier reported in [8] as a baseline. For combined classifiers, richer semantic and syntactic information was needed in the tweets' vector representations, as it was necessary to additionally encode the relatedness and diversity of different targets in stance expressions. This was a much harder task, as the combined classifier had to employ useful knowledge from other targets and avoid the impairment of useless information. For this reason, we continued to employ the biGRU model to generate the target embeddings, which had stronger expressive power than the averaging method. The difficulty level of this task is illustrated by the significant diminished overall macro-average F1 score of the SVM combined classifier in Table 4, compared with the overall macro-average F1 score of the SVM separate classifiers in Table 3. We experimentally increased the dimensionality of the pre-trained word embedding vectors from 100 to 300, and the dimensionality of the hidden states of GRU from 64 to 256, to satisfy the above requirements. All the other hyper-parameters were kept the same, as illustrated in Sect. 3.3.

From Table 4, it can be observed that for the target "Climate Change is a Real Concern", it is helpful for all models to employ the training data from other targets. Comparatively, combined classifiers using models based on neural networks achieve much better macro-average F1 scores on this target than the combined classifiers using the traditional SVM algorithm. This is because the neural network-based models employed continuous vector representations of tweets, which allows them to more easily incorporate information from other domains, compared with the traditional SVM algorithm, which employs sparse and discrete vector representations, based on feature engineering. The combined classifier using the proposed AS-biGRU-CNN model yields the best performance so far on the "Climate Change is a Real Concern" target, which further illustrates the model's strong ability to capture the generality in stance expressions of different targets. However, the overall performance of the combined classifiers all decreases. This is because the performances for targets with sufficient training data can be negatively influenced by the redundant information from other targets. Nevertheless, the AS-biGRU-CNN model still yields the best overall performance, using only combined classifiers, which shows the model's power in modelling the differences in stance expressions of different targets.

4   Related Work

Very few recent researches attempted to tackle the target-specific Stance Detection task on tweets, such as [1,4,9,16,19]. [1] focused on *predicting the stances towards targets with no training data provided*, which was the SemEval-2016 Task 6.B, a different task to the one studied here. For the problem we tackled in this work, there was a training dataset for each specified target, to effectively update the states and memories of the encoders. [4] studied the correlation between sentiment and stance, and the sentiment labels of the tweets were additionally needed to train the model. Thus, *the settings of both above researches were different from the settings of the SemEval-2016 Task 6.A.* [16,19] ignored the target information while performing classification, whereas our experiments have clearly proven that the target-specific vector representation of tweets can substantially boost the performance. [9] relied on feature engineering and large domain corpus, to perform feature selection, which was hard to generalise to other targets; and the collection of domain corpus additionally added difficulty, because of the limitations of the Twitter API. *The attention-based models proposed in this work, on the contrary, are fully automatic, with minimum supervision. We did not collect any extra domain corpus or use any linguistic tools, and no feature engineering was needed. Since no target-specific configurations are involved, the proposed models can be directly applied to other targets.*

Another track of relevant research is aspect-level Sentiment Analysis on texts [11,12,15]. In this task, the text to be analysed, or at least part of the text, focuses on the aspects of interest, by explicitly mentioning the aspects, which renders the problem of modelling the importance and relatedness of tokens with respect to the aspects, easier. *However, this is not the case for the target-specific*

Stance Detection task. Thus, a deeper integration between the target and the tweet, and a more complex inference mechanism, are needed, as proposed in our research.

5 Conclusion

To the best of our knowledge, we are the first ones to effectively apply the traditional token-level attention mechanism to the problem of target-specific Stance Detection in tweets, which achieves better performance than other neural network-based models. Moreover, we propose to use a gated structure on the basis of the biGRU-CNN model, to embed target information into the tweet's vector representation, aiming at introducing the direct semantic interaction between the target and each token in the tweet, to perform *target-specific Stance Detection*. The proposed model employs a *semantic-level attention mechanism*, which is more fine-grained than the token-level attention mechanism. The proposed semantic-level attention mechanism searches for certain semantic features of each token in the tweet, based on the information contribution these semantic features have, in deciding the stance of the tweet, towards the given target. For the resulting AS-biGRU-CNN model, not only the tweet's representation vector, but also the representation vectors of the tokens are target-specific. The experimental results demonstrates that the proposed model outperforms several state-of-the-art baselines, in terms of macro-average F1 score, on the benchmark target-specific Stance Detection dataset of tweets, for both the scenario when separate classifiers are allowed for different targets and the scenario when only one combined classifier is allowed. Thus, the AS-biGRU-CNN model has stronger expressive power, and higher generalising capability, to extract target-specific knowledge from annotated datasets, to perform target-specific Stance Detection on tweets. Importantly, unlike previous works on target-specific detection in tweets, the models employed in this work do not rely on any extra annotation, domain corpus or feature engineering, and can be easily generalised to other targets of interest.

Acknowledgments. This work was supported by The Alan Turing Institute under the EPSRC grant EP/N510129/1.

References

1. Augenstein, I., Rocktäschel, T., Vlachos, A., Bontcheva, K.: Stance detection with bidirectional conditional encoding. In: Proceedings of 2016 Conference on Empirical Methods in Natural Language Processing, pp. 876–885. ACL (2016)
2. Bahdanau, D., Cho, K., Bengio, Y.: Neural machine translation by jointly learning to align and translate. In: Proceedings of 3rd International Conference on Learning Representations (2015)
3. Cho, K., van Merrienboer, B., Gülçehre, Ç., Bahdanau, D., Bougares, F., Schwenk, H., Bengio, Y.: Learning phrase representations using RNN encoder-decoder for statistical machine translation. In: Proceedings of 2014 Conference on Empirical Methods in Natural Language Processing, pp. 1724–1734. ACL (2014)

4. Ebrahimi, J., Dou, D., Lowd, D.: A joint sentiment-target-stance model for stance classification in tweets. In: Proceedings of 26th International Conference on Computational Linguistics, pp. 2656–2665. ACL (2016)
5. Gal, Y., Ghahramani, Z.: A theoretically grounded application of dropout in recurrent neural networks. In: Proceedings of Advances in Neural Information Processing Systems, vol. 29, pp. 1019–1027 (2016)
6. Kim, Y.: Convolutional neural networks for sentence classification. In: Proceedings of 2014 Conference on Empirical Methods in Natural Language Processing, pp. 1746–1751. ACL (2014)
7. Kingma, D., Ba, J.: Adam: a method for stochastic optimization. In: Proceedings of 3rd International Conference on Learning Representations: Poster Session (2015)
8. Mohammad, S.M., Kiritchenko, S., Sobhani, P., Zhu, X., Cherry, C.: SemEval-2016 task 6: detecting stance in tweets. In: Proceedings of 10th International Workshop on Semantic Evaluation (2016)
9. Mohammad, S.M., Sobhani, P., Kiritchenko, S.: Stance and sentiment in tweets. ACM Trans. Internet Technol. **17**(3), 26 (2017)
10. Pennington, J., Socher, R., Manning, C.D.: Glove: global vectors for word representation. In: Proceedings of 2014 Conference on Empirical Methods in Natural Language Processing, vol. 14, pp. 1532–1543. ACL (2014)
11. Ruder, S., Ghaffari, P., Breslin, J.G.: A hierarchical model of reviews for aspect-based sentiment analysis. In: Proceedings of 2016 Conference on Empirical Methods in Natural Language Processing, pp. 999–1005. ACL (2016)
12. Schouten, K., Baas, F., Bus, O., Osinga, A., van de Ven, N., van Loenhout, S., Vrolijk, L., Frasincar, F.: Aspect-based sentiment analysis using lexico-semantic patterns. In: Cellary, W., Mokbel, M.F., Wang, J., Wang, H., Zhou, R., Zhang, Y. (eds.) WISE 2016. LNCS, vol. 10042, pp. 35–42. Springer, Cham (2016). doi:10.1007/978-3-319-48743-4_3
13. Srivastava, N., Hinton, G.E., Krizhevsky, A., Sutskever, I., Salakhutdinov, R.: Dropout: a simple way to prevent neural networks from overfitting. J. Mach. Learn. Res. **15**(1), 1929–1958 (2014)
14. Tan, M., Xiang, B., Zhou, B.: LSTM-based deep learning models for non-factoid answer selection. In: Proceedings of 4th International Conference on Learning Representations: Workshop Track (2016)
15. Tang, D., Qin, B., Feng, X., Liu, T.: Effective LSTMs for target-dependent sentiment classification. In: Proceedings of 26th International Conference on Computational Linguistics, pp. 3298–3307. ACL (2016)
16. Wei, W., Zhang, X., Liu, X., Chen, W., Wang, T.: pkudblab at SemEval-2016 task 6: a specific convolutional neural network system for effective stance detection. In: Proceedings of 10th International Workshop on Semantic Evaluation (2016)
17. Xu, K., Ba, J., Kiros, R., Cho, K., Courville, A.C., Salakhutdinov, R., Zemel, R.S., Bengio, Y.: Show, attend and tell: neural image caption generation with visual attention. In: Proceedings of 32nd International Conference on Machine Learning, pp. 2048–2057. ACM (2015)
18. Yang, Z., Yang, D., Dyer, C., He, X., Smola, A.J., Hovy, E.H.: Hierarchical attention networks for document classification. In: Proceedings of 2016 Conference of the North American Chapter of the Association for Computational Linguistics - Human Language Technologies, pp. 1480–1489. ACL (2016)
19. Zarrella, G., Marsh, A.: MITRE at SemEval-2016 task 6: transfer learning for stance detection. In: Proceedings of 10th International Workshop on Semantic Evaluation (2016)

A Network Based Stratification Approach for Summarizing Relevant Comment Tweets of News Articles

Roshni Chakraborty$^{(\boxtimes)}$, Maitry Bhavsar, Sourav Dandapat, and Joydeep Chandra

Indian Institute of Technology, Patna, Patna, India
{roshni.pcs15,bhavsar.mtcs15,sourav,joydeep}@iitp.ac.in

Abstract. Social media platforms like Twitter have become extremely popular for exchanging information and opinions. The opinions expressed through Twitter can be exploited by news media sources to obtain user reactions centered around different news articles. A comprehensive summary of the user reactions with respect to a news article can be crucial due to various reasons like: (i) obtaining insights about the diverse opinions of the readers with respect to the news and (ii) understanding the key aspects that draw the interest of the readers. However extracting the relevant opinions from tweets is a challenging task due to the enormous volume of contents generated and difference in vocabulary of social media contents from the published article. Existing supervised learning based techniques yield poor accuracy due to unavailability of sufficient training data and large heterogeneity in the features of various news articles, while the unsupervised techniques fail to handle the noise and diversity of the tweets.

In this paper, we propose a network community based unsupervised approach that effectively handles the problem of noise and diversity in tweet feeds to capture the relevant and the diverse opinions with respect to a news article. Using a combined metric that considers both relevance and diversity, we show that our proposed approach produces 16–25% improvement over existing schemes. Results based on human annotations also validate the effectiveness of the extracted summary tweets with respect to specific news articles.

Keywords: Tweet summarization · Summarization · Twitter · News articles · Relevance · Diversity

1 Introduction

The popularity of Twitter as a news media platform encourages a large fraction of the news readers to post tweets on news articles so that the same can be widely discussed by a large user base [1]. A comprehensive summary of the users' opinions extracted from these tweets would help the readers to obtain an insight of the larger debate centered around a news published in the article

© Springer International Publishing AG 2017
A. Bouguettaya et al. (Eds.): WISE 2017, Part I, LNCS 10569, pp. 33–48, 2017.
DOI: 10.1007/978-3-319-68783-4_3

and at the same time it would also help the news agencies to understand the key aspects that are likely to draw the interest of the readers. However, the summary must be relevant to the news article and at the same time it should represent a holistic view (diverse in nature) of the users' opinions. Ensuring above phenomenon in summary is a fundamentally challenging job. These challenges need to be collectively addressed, otherwise, would yield poor accuracy due to propagation of errors occurring at each step of the process.

Extracting relevant tweets based on keyword similarity is not effective because of the large difference in vocabularies of the news articles and the corresponding tweets [2]. There are few supervised learning approaches that mainly dealt with the problem of extracting relevant tweets for specific news articles. Like recent works in [2–4] have proposed mechanisms for mapping tweets to articles using supervised learning approaches based on language as well as topic models. Further, certain tweet specific features like presence of hashtags and user mentions, language independent features like non-lexical markers as well as lexical features like presence of question words and sentiment words have also been used in classification. However, the major drawback of such supervised classification approaches is the requirement of large manually annotated dataset to train the classifiers. Moreover, annotation is also error-prone as inconsistencies may arise depending upon the perception and knowledge of the annotators.

On the other hand, unsupervised summarization techniques like LexRank [5] and Latent Semantic Analysis (LSA) [6] that are based on keyword frequency works well for multi-document summarization but are inefficient for summarizing opinion tweets with respect to specific news articles. This is primarily because, in contrast to text documents, tweets are inherently noisy with low word count and high variance in word frequencies. Further they also feature high redundancy (high number of similar tweets) as well as large diversity in the vocabulary. These existing techniques are not suitable to handle these features and hence fail to capture the diverse opinions in the tweet summary. Hence to produce an effective tweet summary, techniques must be developed for extracting tweets with respect to news articles that satisfy the following objectives:

1. *Relevance:* The tweets generated in the summary must reflect relevant opinions or feelings of the readers with respect to the news article. They should not be mere facts or quotes from the news article. Metric related to relevance is introduced shortly.
2. *Diversity:* The tweets generated should capture the diverse opinions of the readers and should not be similar in nature. Metric related to diversity is introduced shortly.

In this paper we propose an unsupervised approach for summarizing diverse opinion tweets related to specific news articles that satisfy both the above stated criteria. To handle word diversity and vocabulary gap of the tweets (with respect to the news article), the relevant tweets are captured by considering both the keyword similarity with respect to the news article as well as contextual similarity captured by features like presence of relevant co-occurring hashtags. Subsequently, a weighted tweet network is formed by linking relevant tweets based on their relative keyword similarity as well as relevance of those words with the published article. Finally, to capture the diverse opinions, we group tweets by applying a community identification algorithm on the weighted tweet network. Communities and the representative tweets from those communities are selected for summary based on certain relevance and diversity metrics. Validation on empirical data set of around 800 articles shows that tweets extracted using our proposed approach are not only relevant with respect to the news article but maintains high diversity as compared to tweets extracted by using existing approaches like LexRank and LSA.

The organization of the paper is as follows: We discuss the related works in Sect. 2. In Sect. 3, we present a formal definition of the problem, a brief description of certain preliminary approaches that we use along with an overview of the dataset that we use for our experiments. In Sect. 4, we detail our proposed approach. We discuss the experiments and observations in Sect. 5 and finally conclude our study in Sect. 6.

2   Related Works

There is a plethora of available research related to the various aspects of our proposed work like (i) extracting relevant posts (ii) summarization and (iii) opinion mining. A few of these works are highlighted next.

Extracting Relevant Posts: Although features like word similarity [7,8] can extract blogs relevant to news articles, it fails in micro-blogs like Twitter due to the vocabulary difference between tweets and the original news article [2].

Krestel et al. [4] extracted relevant and diverse tweets corresponding to news articles by applying language modeling, topic modeling and logistic regression on a set of tweets with either specific seed hashtags like *obama, snowden, merkel* or hashtags co-occurring with these hashtags. However, our experiments indicate that topic models or language models fail to derive relevant information when each tweet is treated as a document. Certain works attempt to extract relevant tweets through a search process starting from the tweets containing the URL of the article [9] or the hashtags specific to the news [10] but these systems fail to capture many tweets that are relevant to the news article. In [11] Cao et al. generated relevant comments for Chinese news by mining micro-blog posts. However, the authors did not consider diversity of the comments while selecting them.

Unsupervised topic modeling techniques like Latent Dirichlet Allocation [12] fail to extract topics from tweets due to inherent noise present in tweets [13]. Mehrotra et al. [14] pools all tweets of a hashtag in a document which improves the performance over the existing topic modeling techniques in Twitter. Tweet-Motif [15] follows an unsupervised approach to cluster messages by considering hashtags as an approximate indicator of tweet topics. However, our observations indicate that neither keywords (as used in topic model based techniques) nor hashtags can independently represent a news item uniquely. Tweets related to a specific news article may include several hashtags of different news events and several news articles covering different news may be mapped to same hashtag set. Thus, extraction of specific and appropriate hashtag set for a particular news article requires both content similarity with the specific news article and contextual similarity with the news tweets (i.e. mostly quotes from the news article.)

Summarization: Public opinion to a news article covers multiple facets; thus the extracted relevant comment tweets should be summarized in a way to provide a holistic view of the opinions.

Document summarization techniques like LexRank and LSA have also been applied for summarizing tweets. In LexRank, sentences are linked based on their keyword similarities, thus forming a weighted network. Subsequently, a sentence is selected into the summary based on the importance of it's constituent keywords and it's connections to other important sentences. However, this algorithm fails in summarizing opinion tweets that are highly redundant and have large diversity in the vocabulary. LSA uses singular value decomposition on the term sentence matrix to derive a set of concepts and further select representative sentences from different concepts to maintain *diversity* in the summary. However, the major problem in applying LSA for summarizing news-specific tweets is the large variance in the frequency of the terms in the term-sentence matrix.

For multi-tweet summarization, researchers have proposed using both contextual features of the tweets as well as social influence of the users to extract predetermined length summary tweets of certain events [16,17]. Entity based summarization [18] algorithms use *affinity propagation* algorithm to group relevant hashtags into coherent topics and further summarize tweets from these groups into most relevant, insightful and diverse opinions of an entity. However, unlike the single dimensional search space of an entity the search space of news specific tweets is inherently multi-dimensional that is difficult to explore.

Opinion Mining: Opinion mining is a process of extracting and summarizing important opinions about a particular product or entity [19]. Previous works accurately identify the semantic orientations [20], syntactic relations [21] and sentiments [22–24] between opinion words and target to provide a holistic view of the existent expressions. Recent studies have also been made to extract relevant comment tweets (expressing opinions on news articles) specific to an article using supervised learning approaches [3]. However, creating such a universal dataset for all kind of news articles is difficult. Further, a common set of features might not be able to model the variance in characteristics of the comment tweets which differs across the entities involved and the region.

We next provide a formal definition of the problem and outline our approach with respect to the drawbacks in existing mechanisms.

3   Problem Definition and Preliminaries

In this section, we initially present a formal description of the problem. Subsequently, we provide a brief overview of the proposed approach to address the problem.

3.1   Problem Definition

Given a news article $\mathcal{N}$, let $\mathcal{T} = \{t_1, t_2, \cdots\}$ be the set of opinion tweets relevant to $\mathcal{N}$. The objective is to select at most n tweets in our final summary ($\mathcal{O}_n$) in such a way that it maximizes both the pairwise diversity of the tweets selected and relevance of the selected tweets with the news article. Further, we can impose additional constraints on the minimum value of the total relevance score (say R_{min}) of the selected tweets in $\mathcal{O}_n$. We can thus formally define the problem as:

$$f = \arg\max(\alpha Div(O) +$$
$$(1-\alpha)Sim(O, \mathcal{T} - O)) \quad O \subseteq \mathcal{T}, \alpha \in [0,1] \tag{1}$$
$$\text{subject to} \quad Rel(\mathcal{N}, O) \geq R_{min}$$
$$||O|| \leq n,$$

where $Div(O)$ is a function representing the pairwise diversity of the tweets in subset O and $Sim(O, \mathcal{T} - O))$ represents the similarity score of the tweets in O with respect to the tweets not selected in O (i.e. $\mathcal{T} - O$). The similarity function ensures that the representative tweets in the summary captures the holistic opinions expressed through the tweets. The objective is to select a subset O such that f is maximized subject to the relevance of selected tweets with news article (represented as $Rel(\mathcal{N}, O)$) and size constraints. $\mathcal{O}_n$ is that subset O for which f is maximized. Since the problem defined is a 0–1 Integer Programming Problem which is known to be NP-complete [25], existing methods of summarization are mainly approximation to the optimality problem. However, the major challenge for any summarization technique is to handle the redundancy of the tweets. The large diversity of the tweet vocabulary as well as the noise further adds to the complexity in tweet summarization.

3.2   Outline of the Proposed Approach

We address these issues in the proposed approach by initially finding the set of relevant opinion tweets for a news article based on their keyword similarity as well as their contextual similarity. Subsequently, we create a weighted network of the relevant tweets, where the weight is determined based on the keyword similarity in tweets and relevance of those overlapping keywords with the news article. The tweet network is analyzed further to identify the different closely connected tweets (communities), while each community represents a set of similar opinions. We then sample representative tweets from each of these communities using a *maximum marginal relevance* measure that satisfy the relevance constraint and also ensures relative diversity among the selected tweets, thus satisfying Eq. 1. The steps of our proposed approach are highlighted in Fig. 1. Steps are described in more detail in Sect. 4, where we describe our proposed approach.

3.3   Dataset

The dataset consists of both news articles as well as tweet sets.

News Article Dataset: We have considered only major (front page) political news from *New York Post*. We have crawled 800 articles for 2 months duration starting from 1st *July*, 2016 to 31st *August*, 2016.

Twitter Dataset: Twitter allows free access to approximately 1% random tweets using the Streaming API. We have used this API to crawl the tweets for 2 months duration starting from 1st *July*, 2016 to 31st *August*, 2016. We collected around 2.1 billion tweets with 77 million hashtags and 36 million unique users.

3.4 Preprocessing

We next highlight the major steps of preprocessing followed for both the articles as well as the tweets.

Preprocessing of Articles. We initially extract the nouns and verbs (that are important to differentiate articles with similar entities) from the headline and the body of every article, using the POS tagger of the NLTK toolkit. These keywords are further ranked by their *tf-idf* score computed based on the documents published in the same day. We then select the top 10 keywords of every article as its representative keywords [26].

Preprocessing of Tweets. The raw tweets collected are inherently noisy and hence needs to be cleaned for further analysis. We briefly outline the preprocessing steps that we perform on the tweets and then describe the extraction of related tweets of an article.

Types of Tweet: Based on the nature of tweet contents with respect to a specific news article, we categorize tweets into 3 different types. We provide one example of each type using tweets related to the news article entitled *Newly released Clinton emails show favors for foundation donors* published in New York Post on 22nd August, 2016.

Definition 1. *News Tweet* is a tweet with no sentiment polarity that states a fact specific to the news article or to a related news. Most often these are direct quotes, for example, the tweet, *new emails show Clinton foundation sought access to state department on donors behalf* is a news tweet based on the above mentioned article.

Definition 2. *Opinion Tweet* is a tweet that represents an opinion regarding a news article and has some associated sentiment polarity. For example, the tweet *Hillary Clinton is guilty of sin, is a traitor to our country, and is unfit to be president* indicates a negative sentiment related to the same article.

Definition 3. *Irrelevant Tweet* is a tweet that does not provide any relevant information with respect to the news article. For example, the tweet *rt twitter users call out pbs for using stock footage of dc fireworks show* is not related to this article.

As the definition of the different types of tweets suggest, each type of tweet highlights different kind of information in relation to an article. We are only interested in the *opinion tweets* with respect to a news article because it captures the reaction of readers. We identify an opinion tweet by it's sentiment score that we calculate using Vader Sentiment Analyzer [27]. The steps of further preprocessing used to filter out the opinion tweets with respect to a news article is described next.

1. *Keyword Extraction.* We remove the duplicate tweets and subsequently remove user mentions, retweet tags and URLs as well as the stop words from the tweet text. From the remaining words, we extract nouns and verbs using a POS tagger that forms the representative keywords of the tweet.
2. *Seed Tweet Set Creation.* We create an initial *seed tweet set* of an article that includes tweets having maximum keyword overlap with the article's representative keywords. The *seed tweets* are used to extract a set of most frequent hashtags(*seed hashtags*) that are present in the *seed tweets set* and the *co-occurring hashtags*, that appear in more than a threshold percentage of tweets of any *seed hashtag*. The collective set of seed as well as co-occurring hashtags is termed as a *related hashtag set* of the article.
3. *Extended Tweet Set.* We extend the seed tweet set by including relevant opinion tweets that have at least one of the related hashtags or one level of indirection in word overlap. Due to short length of tweets and vocabulary gap, we might not find direct word overlap of tweets with news article. However, with high probability we can find word overlap of such tweets with other tweets which does have direct word overlap with news. This step helps us to extend our relevant tweet set as well as to get rid off irrelevant tweets.

We next discuss our proposed approach that is applied on the filtered tweets.

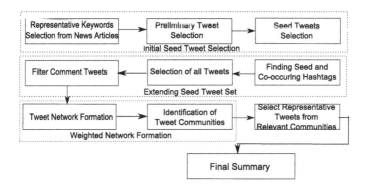

Fig. 1. Block diagram representing the proposed approach

4   Proposed Approach

In this section, we detail our methodology to extract relevant and diverse opinion tweets specific to the targeted article. We first describe the steps for creating a tweet network from the extended opinion tweet set and identifying the different communities in the network. Subsequently, we highlight the method of selecting the summary of opinion tweets from these communities.

4.1 Tweet Network Formation and Community Identification

We consider extended tweet pool for forming network and community analysis.

Tweet Network Formation: To form the tweet network, we consider every tweet as a node. Two nodes are connected by an edge if the edge weight between them exceeds a predefined threshold. Edge weight is computed based on a parameter that we term as *significance score*. *Significance score* of a keyword represents the importance (based on tf-idf) of a keyword with respect to the news article. If any keyword, say w, belongs to the representative keywords of the article then its significance score is the relative importance of w with respect to the least important keywords in representative keyword set, otherwise it is 1.

$$\mathcal{S}(w) = \frac{\text{tf-idf}(w)}{min(\text{tf-idf}(u) : \forall u \in \mathcal{K})} \quad \text{if } w \in \mathcal{K}$$
$$= 1 \quad \text{otherwise} \tag{2}$$

The weight W_{ij} of the edge connecting nodes i and j is calculated as the sum of the significance score of intersecting keywords of the two tweet nodes, normalized by the total significance score of all the unique keywords. Thus if K_i and K_j represent the keywords in nodes i and j respectively, then

$$W_{ij} = \frac{\sum_k \mathcal{S}(k)}{\sum_l \mathcal{S}(l)} \quad k \in K_i \cap K_j, \quad l \in K_i \cup K_j \tag{3}$$

The edge weight between two tweet nodes will be high if they have significant keyword overlap; however if the non-overlapping keywords have high significance score (implying tweets carry different opinions and should be part of different communities), then the edge weights will be significantly lower.

Extracting Tweet Communities: The communities in a weighted network are identified by logically partitioning the network into groups of nodes where the nodes within a partition are highly interconnected as compared to interconnections across the partitions. The quality of partitioning is measured by a term called *modularity* that compares the actual weight of the edge between two nodes within a community to the possible weight in a scenario if the connections between them were purely random.

By identifying the communities in the network, the group of tweets representing a similar class of opinions are identified, whereas each group represents a different opinion class. We show later in Sect. 5, partitioning of tweets into communities tends to make the frequency of keywords uniform with less diversity. This eases the process of selecting representative tweets from the community for summarization. We next discuss the steps taken to extract the final set of representative summary tweets.

4.2   Extracting Final Summary of Relevant and Diverse Tweets

We describe the steps followed to extract the final summary of relevant and diverse tweets based on the identified communities in the tweet network.

Ranking Communities Based on Relevance: For each community detected using the Louvain algorithm, we initially compute a community relevance score (*CommRel*) based on the cosine similarity of the top ten keywords of a community (the highest frequency keywords in all tweets of the community) and that of the top ten keywords of the seed tweets. We consider a community as relevant to the news article if the *CommRel* value is above a predefined threshold. The irrelevant communities are discarded and not used for further consideration.

Selecting a Diverse Tweet Set from Relevant Communities: After filtering out the irrelevant communities, we select a diverse set of tweets from relevant communities. The final summary tweets are selected from each relevant community C_j such that *maximum marginal relevance* [28] is achieved. *Maximum marginal relevance* ensures selection of such tweets which are highly relevant with seed tweet set, however, having minimum similarity with already selected tweet set.

$$MMR_S = max_{T_i \in C_j}[\beta Sim(K_{tweet}, K_{seed}) - (1 - \beta)Sim(K_{tweet}, K_{selected})] \quad (4)$$

where, $0 < \beta < 1$, T_i are the tweets of the relevant community C_j and $Sim(K_a, K_b)$ denotes the similarity of two keyword sets K_a and K_b. Finally, we merge the summaries from all the communities to get the aggregated summary. We next highlight the evaluation technique to observe the efficiency of the proposed method and the observed experimental results.

5   Results and Discussion

In this section, we initially describe the performance metrics and the existing summarization approaches and then compare the performance of the proposed approach with the existing summarization algorithms like LSA and LexRank. Further, we also perform a manual annotation of the extracted tweets to validate the accuracy of the proposed relevance measure. Finally, the specificity of the extracted tweets with respect to the news article is observed using a case study.

5.1   Performance Metrics

We use four different performance metrics to evaluate the quality of the summary (tweet sets).

Mean Relevance Score: The relevance score of a tweet with respect to an article is defined by the cosine similarity of the tweet keywords with the representative set of words of the news article. If a_i and b_i denotes the frequency of

i^{th} term in two documents a and b respectively, then the cosine similarity S_{ab} of the two documents will be represented as

$$S_{ab} = \frac{\sum_i a_i b_i}{\sqrt{\sum_j a_j^2} \sqrt{\sum_j b_j^2}} \qquad (5)$$

A value near to 1 indicates high similarity between a and b. If T represents a tweet set and S_{id} denotes the relevance score (cosine similarity) of the ith tweet and the news article d then the mean relevance score, $\bar{R}_T$, is given as $\bar{R}_T = \frac{\sum_{i=1}^{k} S_{id}}{k}$, where k denotes the total number of tweets in T.

Mean Diversity Score: We calculate diversity score of tweets based on two different similarity measures.

1. *Cosine Similarity:* If S_{ij} denotes the cosine similarity (as mentioned in Eq. 5) of the ith and j^{th} tweets in T, then the mean diversity score, D_T, of the set T is given as $D_T = \frac{\sum_{i,j}(1-S_{ij})}{m}$, where m denotes the total number of tweet pairs in T.
2. *Jensen Shannon Divergence: Jensen-Shannon* (JS) divergence [29] provides a measure of similarity between two probability distributions. It's based on *Kullback–Leibler* (KL) divergence. For two discrete probability distributions P and Q of the keywords in two tweets, the JS divergence of P and Q is given as

$$JSD(P||Q) = \frac{1}{2}D_{KL}(P||M) + \frac{1}{2}D_{KL}(Q||M), \qquad (6)$$

where $M = \frac{1}{2}(P+Q)$ and $D_{KL}(P||Q)$ is the KL divergence from Q to P is defined as

$$D_{KL}(P||Q) = \sum_i P(i) log \frac{P(i)}{Q(i)}, \qquad (7)$$

and represents the information gain achieved in using P instead of Q. If J_{ij} denotes the Jensen Shannon divergence of the ith and j^{th} tweets in T, then the mean diversity score, D_T, of the set T is given as $D_T = \frac{\sum_{i,j}(J_{ij})}{m}$, where m denotes the total number of tweet pairs in T. Both the divergence scores vary between 0 (lowest diversity) and 1 (highest diversity).

As tweets are highly redundant and often contains duplicate information, summary tweets of a news article must ensure both *relevance (i.e., coherence to the main article)* as well as *diversity (i.e., different from each other)*. Thus, we devised the following metrics to measure the quality of summary of any article.

1. *Coverage:* Coverage, C_T, of the tweet set T with respect to a news article by a weighted sum of it's mean relevance score $(\bar{R}_T)$ and mean diversity score, D_T, i.e., $C_T = \alpha R_T + (1-\alpha)D_T$, where $0 < \alpha < 1$. We select α as 0.5 to give equal importance to both relevance and diversity.

Though high coverage indicates a good quality summary, however, which factor (relevance or divergence) is contributing how much is not clear from coverage.

2. *Diversity-Relevance Balance Factor:* It is to understand how balanced relevance and divergence are and computed as $S_T = \frac{min(R_T, D_T)}{max(R_T, D_T)}$. A higher score ensures uniformity in both *mean relevance score* and *mean diversity score* which is the objective of summarization.

Comparison with Existing Techniques. We compare our proposed method with the existing techniques of summarization. Each of these approaches are described as follows:

1. *LexRank:* LexRank is one of the most popular document summarization algorithms that selects *representative sentences* from a document by their *page rank values*. We apply *LexRank* to summarize the extracted opinion tweets of a news article.
2. *LSA:* Latent Semantic Allocation produces the underlying concepts from documents by analyzing the existent term relationship in that document and further select sentences from each concept thus representing the whole document. We apply *LSA* on the extracted opinion tweets of a news article to give a summary of the major opinions of the article.

5.2   Results

In this section, we compare the performance of the proposed approach with all existing methods described above.

Comparing Performance Metrics. Initially, for every article, we compute the mean relevance score, mean diversity score, diversity-relevance balance factor and the coverage of the summary tweets extracted using the proposed approach and the existing methods. We then calculate the minimum, first quartile, average, third quartile and maximum of both the coverage score and diversity-relevance balance factor score of all these articles to compare the results of different algorithms.

Figure 2 highlights the performance of the proposed and existing approaches in terms of the different metrics and summarization techniques. Figure 2(a) and (b) shows the performance when Jensen-Shannon diversity score (first three) and cosine diversity score (last three) is used for summarization. Results shown in Fig. 2(a) indicate that the *Proposed* approach ensures better balance factor compared to existing methodologies. While Fig. 2(b) ensures that the proposed method also maintains better coverage score. The final coverage score of the proposed approach shows improvement up to 25% and 16% with respect to *LexRank* and *LSA* respectively.

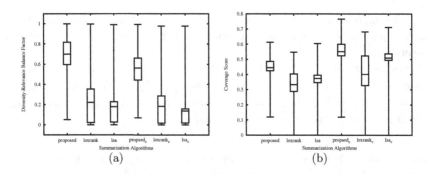

Fig. 2. (a) Shows diversity-relevance balance factor and (b) shows coverage score.

Table 1. Fraction of articles for which final extracted tweet sets are marked as relevant or irrelevant based on manual annotation as well as mean relevance score. $\mathbf{X}$ = manual annotation based and $\mathbf{Y}$ = mean relevance score based

		Y	
		Relevant	Irrelevant
X	Relevant	0.798	0.028
	Irrelevant	0.09	0.082
	Summary	**Match = 89%**	**Mismatch = 11%**

Comparison Using Manual Annotation. To evaluate the accuracy of the proposed relevance measure, we also verified the relevance of the final extracted tweets manually. To validate our results, we generated the final extracted tweets of a random 200 news articles. We provided the set of final tweets and the corresponding articles to the three manual annotators (who does not have knowledge about the details of our algorithm). Each of them annotated all the articles. An annotator marked the tweet relevant or irrelevant based on his knowledge of the article. We strongly encouraged them to try their best to understand the comment before labeling (as tweets are sometimes confusing with informal expressions and sarcasms). To compute inter-annotator agreement, we took the majority of the three for the set of tweets for an article. The final extracted tweet set was classified as relevant to the news article, if a threshold percentage (90%) of its total tweet set was marked as relevant by the majority of the manual annotator. For comparison, we also classify the final extracted tweet set as relevant based on the mean relevance score, if the value is greater than threshold (0.26). We summarize in Table 1, the fraction of articles for which the final extracted tweets are classified as relevant or irrelevant using both the techniques. We had repeated the same experiment thrice for different random set of 200 news articles. It can be observed that for 89% of the total articles, the classification made using both the approaches match, thus highlighting the accuracy of the proposed relevance measure.

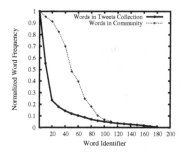

Fig. 3. Comparing the normalized word frequency (**Y** axis) of all the words (**X** axis) in relevant tweets with respect to six articles and the corresponding communities. Number of articles considered is 200.

Effectiveness of Communities. If we rank words in selected relevant tweet set in terms of importance computed based on word frequency (normalized respect to the highest frequency), then we find very small number of words are very important. These words may be related to main theme, however, a summarization based on this importance would lose diversity. However, when we group similar tweets using community analysis and do the same experiment (normalized frequency separately in each community) there would be important keywords in every community and this would help to select a diverse set of tweets in final summary. Effects of normalization is shown in Fig. 3.

Specificity of the Tweets to News Article. To establish that proposed approach is capable of extracting tweets specific to news article we do the following case study. We considered *four* related articles published in the same day. We show that tweets extracted using the proposed approach are relevant to the corresponding article while it is irrelevant when compared with other articles. In Table 2 we show the mean relevance score of extracted tweets of different news articles related to the same news event, *Dallas Shooting*. As can be observed, the mean relevance scores at the diagonals are much higher than the rest of the values indicating high specificity of the extracted tweets with respect to the corresponding news article.

Table 2. Mean relevance score of extracted tweet set against different articles related to Dallas Shooting. A value in cell ij indicates the mean relevance score of article number i with the tweets extracted for article number j

Article titles	Extracted tweet set for articles			
	1	2	3	4
1. *Dallas cop killed in attack survived three tours in Iraq*	0.28	0.04	0.1	0.1
2. *Trump, Clinton postpone campaign events after Dallas attacks*	0	0.28	0	0
3. *Dallas PD first to use a robot to kill suspect*	0.175	0.04	0.43	0.09
4. *Arsenal found in Dallas sniper's suburban home*	0.04	0.04	0.2	0.608

6   Conclusion

In this paper, we have proposed an unsupervised approach to summarize users'
opinion on a specific news article. Proposed mechanism also ensures presence of
diverse and relevant views in final summary. It can effectively handle the issues
like large vocabulary gap and diversity of the tweets with respect to specific
news articles while existing summarization techniques like LexRank and LSA
fail to do so. We have introduced *coverage* and balance factor as metrics that
capture both the relevance and diversity of the summary tweet sets with respect
to an article. The summary tweets extracted using our proposed approach have
very high coverage as compared to existing summarization approaches. Result
shows that our proposed approach produces 16–25% improvement over existing
schemes. Further it is also observed that even in the presence of several related
news articles and their corresponding tweets, our approach can easily identify
the tweets that are relevant and specific to the targeted news article. Although
the proposed method has been verified on political news stories, however, it
can also be used for extracting summary tweets for other news categories as
well. The accuracy of the proposed method can possibly be improved further
by exploring techniques that establishes the contextual relationship among the
keywords extracted from the articles and tweets.

References

1. Kwak, H., Lee, C., Park, H., Moon, S.: What is Twitter, a social network or a news
   media? In: Proceedings of the 19th International Conference on World Wide Web,
   WWW 2010, pp. 591–600. ACM, New York (2010)
2. Tsagkias, M., De Rijke, M., Weerkamp, W.: Linking online news and social media.
   In: Proceedings of the Fourth ACM International Conference on Web Search and
   Data Mining, pp. 565–574. ACM (2011)
3. Kothari, A., Magdy, W., Darwish, K., Mourad, A., Taei, A.: Detecting comments
   on news articles in microblogs. In: ICWSM, vol. 2013 (2013)
4. Krestel, R., Werkmeister, T., Wiradarma, T.P., Kasneci, G.: Tweet-recommender:
   finding relevant tweets for news articles. In: Proceedings of the 24th International
   Conference on World Wide Web, pp. 53–54. ACM (2015)
5. Erkan, G., Radev, D.R.: Lexrank: graph-based lexical centrality as salience in text
   summarization. J. Artif. Int. Res. **22**(1), 457–479 (2004)
6. Gong, Y., Liu, X.: Generic text summarization using relevance measure and latent
   semantic analysis. In: Proceedings of the 24th Annual International ACM SIGIR
   Conference on Research and Development in Information Retrieval, SIGIR 2001,
   pp. 19–25. ACM, New York (2001)
7. Ikeda, D., Fujiki, T., Okumura, M.: Automatically linking news articles to blog
   entries. In: AAAI Spring Symposium: Computational Approaches to Analyzing
   Weblogs, pp. 78–82. AAAI (2006)
8. Takama, Y., Matsumura, A., Kajinami, T.: Visualization of news distribution in
   blog space. In: Proceedings of the 2006 IEEE/WIC/ACM International Conference
   on Web Intelligence and Intelligent Agent Technology, WI-IATW 2006, pp. 413–
   416. IEEE Computer Society, Washington, D.C. (2006)

9. Štajner, T., Thomee, B., Popescu, A.-M., Pennacchiotti, M., Jaimes, A.: Automatic selection of social media responses to news. In: Proceedings of the 19th ACM SIGKDD International Conference on Knowledge Discovery and Data Mining, pp. 50–58. ACM (2013)
10. Shi, B., Ifrim, G., Hurley, N.: Be in the know: connecting news articles to relevant Twitter conversations. arXiv preprint arXiv:1405.3117 (2014)
11. Cao, X., Chen, K., Long, R., Zheng, G., Yu, Y.: News comments generation via mining microblogs. In: Proceedings of the 21st International Conference on World Wide Web, pp. 471–472. ACM (2012)
12. Blei, D.M., Ng, A.Y., Jordan, M.I.: Latent dirichlet allocation. J. Mach. Learn. Res. **3**(Jan), 993–1022 (2003)
13. Zhao, W.X., Jiang, J., Weng, J., He, J., Lim, E.-P., Yan, H., Li, X.: Comparing Twitter and traditional media using topic models. In: Clough, P., Foley, C., Gurrin, C., Jones, G.J.F., Kraaij, W., Lee, H., Mudoch, V. (eds.) ECIR 2011. LNCS, vol. 6611, pp. 338–349. Springer, Heidelberg (2011). doi:10.1007/978-3-642-20161-5_34
14. Mehrotra, R., Sanner, S., Buntine, W., Xie, L.: Improving lda topic models for microblogs via tweet pooling and automatic labeling. In: Proceedings of the 36th International ACM SIGIR Conference on Research and Development in Information Retrieval, pp. 889–892. ACM (2013)
15. O'Connor, B., Krieger, M., Ahn, D.: TweetMotif: exploratory search and topic summarization for Twitter. In: ICWSM, pp. 384–385 (2010)
16. Becker, H., Naaman, M., Gravano, L.: Selecting quality Twitter content for events. In: Adamic, L.A., Baeza-Yates, R.A., Counts, S. (eds.) Proceedings of International Conference on Weblogs and Social Media, ICWSM 2011. The AAAI Press (2011)
17. Chang, Y., Wang, X., Mei, Q., Liu, Y.: Towards Twitter context summarization with user influence models. In: Proceedings of the Sixth ACM International Conference on Web Search and Data Mining, WSDM 2013, pp. 527–536. ACM, New York (2013)
18. Meng, X., Wei, F., Liu, X., Zhou, M., Li, S., Wang, H.: Entity-centric topic-oriented opinion summarization in Twitter. In: Proceedings of the 18th ACM SIGKDD International Conference on Knowledge Discovery and Data Mining, pp. 379–387. ACM (2012)
19. Hu, M., Liu, B.: Mining and summarizing customer reviews. In: Proceedings of the Tenth ACM SIGKDD International Conference on Knowledge Discovery and Data Mining, pp. 168–177. ACM (2004)
20. Ding, X., Liu, B., Yu, P.S.: A holistic lexicon-based approach to opinion mining. In: Proceedings of the 2008 International Conference on Web Search and Data Mining, pp. 231–240. ACM (2008)
21. Qiu, G., Liu, B., Bu, J., Chen, C.: Opinion word expansion and target extraction through double propagation. Comput. Linguist. **37**(1), 9–27 (2011)
22. Ortega, R., Fonseca, A., Montoyo, A.: SSA-UO: unsupervised Twitter sentiment analysis. In: Second Joint Conference on Lexical and Computational Semantics (*SEM), vol. 2, pp. 501–507 (2013)
23. Luo, Z., Osborne, M., Wang, T.: An effective approach to tweets opinion retrieval. World Wide Web **18**(3), 545–566 (2015)
24. Bravo-Marquez, F., Mendoza, M., Poblete, B.: Combining strengths, emotions and polarities for boosting Twitter sentiment analysis. In: Proceedings of the Second International Workshop on Issues of Sentiment Discovery and Opinion Mining, p. 2. ACM (2013)
25. Sahni, S., Gonzalez, T.: P-complete approximation problems. J. ACM **23**(3), 555–565 (1976)

26. Ramos, J., et al.: Using TF-IDF to determine word relevance in document queries. in Proceedings of the First Instructional Conference on Machine Learning (2003)

27. Hutto, C.J., Gilbert, E.: Vader: a parsimonious rule-based model for sentiment analysis of social media text. In: Eighth International AAAI Conference on Weblogs and Social Media (2014)

28. Carbonell, J., Goldstein, J.: The use of MMR, diversity-based reranking for reordering documents and producing summaries. In: Proceedings of the 21st Annual International ACM SIGIR Conference on Research and Development in Information Retrieval, pp. 335–336. ACM (1998)

29. Lin, J.: Divergence measures based on the Shannon entropy. IEEE Trans. Inf. Theory **37**(1), 145–151 (1991)

Interpreting Reputation Through Frequent Named Entities in Twitter

Nacéra Bennacer, Francesca Bugiotti[✉], Moditha Hewasinghage,
Suela Isaj, and Gianluca Quercini

LRI, CentraleSupélec, Paris-Saclay University,
91190 Gif-sur-Yvette, France
{nacera.bennacer,francesca.bugiotti,moditha.hewasinghage,
suela.isaj,gianluca.quercini}@lri.fr

Abstract. Twitter is a social network that provides a powerful source
of data. The analysis of those data offers many challenges among those
stands out the opportunity to find the reputation of a product, of a person, or of any other entity of interest. Several tools for sentiment analysis
have been built in order to calculate the general opinion of an entity using
a static analysis of the sentiments expressed in tweets. However, entities
are not static; they collaborate with other entities and get involved in
events. A simple aggregation of sentiments is then not sufficient to represent this dynamism. In this paper, we present a new approach that
identifies the reputation of an entity on the basis of the set of events
it is involved into by providing a transparent and self explanatory way
for *interpreting reputation*. In order to perform this analysis we define
a new sampling method based on a tweet *weighting* to retrieve relevant
information. In our experiments we show that the 90% of the reputation
of the entity originates from the events it is involved into, especially in
the case of entities that represent public figures.

Keywords: Reputation · Frequent itemsets · Sampling · Opinion
mining

1 Introduction

Twitter is one of the most popular social media platforms. A user of Twitter can
follow any other user in the social network and the higher number of followers
he has, the more reachability his tweets have. Twitter also has mechanism to
spread information by means of retweets and favorites. The more retweets and
the more favorites a tweet gets, the more it spreads, as it gets more audience.

Given that any kind of information can be posted and shared, it is possible to
filter out tweets related to people, products, organizations or any other entity of
interest. Data crawling depends on the APIs provided by Twitter and retrieving
relevant data is a challenge, due to the noise. The opinion about an entity held
by the public is widely known as *reputation*. Natural language processing tools
are quite efficient for extracting sentiments expressed in a single tweet and the

A. Bouguettaya et al. (Eds.): WISE 2017, Part I, LNCS 10569, pp. 49–56, 2017.
DOI: 10.1007/978-3-319-68783-4_4

overall reputation of the entity can be only calculated by a simple sum of the sentiments of the individual tweets it is involved into.

However entities are characterized by their dynamism: they collaborate and create events. Public figures offer the typical example of reputation influenced by events. In this paper we show how the reputation of a person or a product can be construct as a combination of the events it is involved into. In our experiments show that 90% of the reputation of an entity comes from these events. We also propose a sampling method that is based on the retweets, the number of followers, and the favorites. We finally show that the weighted sample technique yields richer information and it is able to discover more events.

The remainder of the paper is organized as follows: in Sect. 2 we discuss the related work, we provide a detailed description of our approach in Sect. 3, we present our experimental results in Sect. 4 and we conclude in Sect. 5.

2   Related Work

Twitter has been used broadly for gathering information about an entity of interest. Finding the attitude (positive, negative, or neutral) towards a topic is known as sentiment analysis. Most of contribution in the field focuses in finding sentiments in the tweet-level [1,4,10], some of them suggest aggregating the sentiments as a simple sum [12,15,16], while the problem of the reputation of an entity has not been specifically addressed. Moreover, the usage of emoticons contributes to the sentiment [11]. In [10] they have used Internet specific acronyms, emoticons, and domain specific text processing to successfully detect the sentiments.

Machine learning techniques prove to be effective with sentiment analysis [15,17]. In order to augment the accuracy of the classifier, Semantic Sentiment Analysis is used in [13]. Emoticons, repeated letters or acronyms have been used in [3]. Domain-dependent sentiment analysis has been studied in [18] and the effect of hashtags in assigning sentiment scores to tweets in [16]. Sentistrength has been developed using emoticons, repeated letters, phrasal verbs and everyday expressions, exclamation marks and repeated punctuation and it shows a higher accuracy compared to several other learning methods [14]. [5] uses pattern discovery and mining of comparative sentences inside blogs, forums, and product reviews. [12] uses a different approach where the entities are further classified into topics and the overall opinion is summarized on the different topics.

3   Approach

In order to extract reputation of people and products from Twitter, we describe the dataset in terms of the set of Named Entities that are defined in it. As first step we select an entity of interest and we collect data related to this entity by querying Twitter. As second step we enrich the information in the tweets using a sampling technique. On top of the sampled data, as third step, we apply frequent itemset mining technique to extract the Frequent Named Entities related to the entity of interest E. As final step, we extract the reputation of the entity E analyzing the collaboration of this entity with the other related entities.

Our goal is to interpret the reputation of an entity E, having as input the Twitter data, and as output defined in Eq. 1 the reputation of an entity E as a collection of reputations of the events it is involved:

$$R = \{(i_k, R_k)|1 \leq k \leq n\} \tag{1}$$

where i_k is a frequent set of e_i and R_k is their reputation.

3.1   Weighted Sampling

The common problem that has been widely addressed is the data extraction. Twitter has provided a REST API[1] which allows running queries against the data to retrieve a sample of the actual content in Twitter. Retrieving tweets related to a specific entity is done by querying Twitter with a keyword. However this does not guarantee the relevance of the information. The most important question that we should pose is: "Do we went a statistically representative sample that aligns with the real large Twitter dataset or do we want a filtered sample that focuses on the relevant tweets?". Several papers have contributed to find a statistically representative sample [6,8], while others highlight focused on crawling ([9]) or Expert Sampling ([7]). Inspired by the latest work, we consider three main parameters that influence the quality of the tweet: the number of times the tweet is retweeted, the favorite count of the tweet, and the number of followers of the user that has tweeted.

We sample the tweets according to these parameters. All the tweets are ranked taking into account: the number of retweets, the favorite count and the number of followers of the user. Then, a weight (w_i) is assigned to tweet t_i that comes from the average of all three rankings. We sample tweets according to their weight; we generate a random number (ϱ) of each t_i, if $w_i > \varrho$ than the tweet is selected to the weighted sample.

3.2   Reputation of Frequent Named Entities

The weighted sample S will be used in order to find the reputation of the entity, aided from the frequent entities in S. Named entities (NE) carry valuable information as they represent people, location, time, and monetary values. Considering t_i as a transaction containing a set of entities e_i as items, in addition to the traditional concepts of [2]: we consider a Frequent Named Entity (FNE) as a set of e_i that is maximal for a predefined support s in S. A FNE describes an event, which is usually associated by a reputation.

In this context the *reputation* of e_i, derived from S of k tweets will be the ratio between the sum of all positive sentiments and the negative sentiments. The sum of all underlying positive sentiments of tweets, as well as the sum of the

---

[1] https://dev.twitter.com/rest/public.

negative ones can be transformed into normalized proportions that indicate the reputation of an entity $(\frac{\sum_{i=1}^{k} pos_k}{\sum_{i=1}^{k} pos_k + \sum_{i=1}^{k} neg_k}, \frac{\sum_{i=1}^{k} neg_k}{\sum_{i=1}^{k} pos_k + \sum_{i=1}^{k} neg_k})$. We propose finding FNEs and interpret the reputation of the entity of interest by the FNEs and their reputation. This approach is described in Algorithm 1. We intend to find the tweets that contain the FNEs and aggregate their sentiment (line 5, 6 and 7). In this way, the reputation of an entity can be explored analyzing its relations with other entities. This procedure is transparent to the user.

Algorithm 1. Calculating aggregated reputation of the FNEs

Input: A collection of tweets from S with their sentiment and set of entities $\{< t_i, E_i, pos_i, neg_i >\}$ where $E_i \leftarrow \bigcup_{j=1}^{n}, e_j \exists he_{ij}$, A set of FNEs $< i_k >$ explored in S list

Output: Reputation R expressed as combination of $\{(i_k, R_k) | 1 \leq k \leq n\}$

1: **for each** FNE i_k **do**
2:     $pos_k \leftarrow 0$
3:     $neg_k \leftarrow 0$
4:     **for each** $< t_i, E_i, pos_i, neg_i >$ **do**
5:         **if** $i_k \subseteq E_i$ **then**
6:             $pos_k \leftarrow pos_k + pos_i$
7:             $neg_k \leftarrow neg_k + neg_i$
8:         **end if**
9:         $R_k \leftarrow (\frac{pos_k}{pos_k + neg_k}, \frac{neg_k}{pos_k + neg_k})$
10:        add R_k to R
11:    **end for**
12: **end for**
13: **return** R

---

4   Experiments

In this section we provide a description of the experiments we performed in order to evaluate our approach: we analyzed the richness of the samples, the effectiveness of frequent entity mining, and we compared the ranking of the sample to the population. To collect data about a certain topic we run a keyword query having as parameter a single string (such as '*Obama*'). We analyzed then the text of each retrieved tweet and we improved the quality of the text by separating merged words inside the hashtag. For example #iamsohappy and #iam#sohappy will be handled by our cleaning algorithm to produce "I am so happy." We used the corpus of words of Sentistrength[2] for word identification and then different techniques for organizing the sentence and discarding not relevant words. We have used Stanford NLP[3] to identify the Named Entities from the retrieved tweets and Sentistrength as the sentiment analysis tool. Sentistrength scores a given text with a positive (from 1 to 5) and negative (from −5 to −1) values.

4.1   The Datasets

We collected four datasets of tweets: *Obama* dataset, *Trump* dataset, *La La Land* dataset and, *The Voice* dataset by querying Twitter with respective strings.

---

[2] http://sentistrength.wlv.ac.uk.
[3] https://nlp.stanford.edu.

In the context of describing the characteristics of our datasets, we define two notions: *(i)* **Density of NE** - Expresses the average number of Named Entities linked to a tweet *(ii)* **Coverage of NE** - Represents the percentage of the tweets in the dataset that contain at least one Named Entity. The datasets regarding public figures have a high density of Named Entities, as well as a high coverage (Table 1). When it comes to the coverage and the density of Named Entities, *The Voice* is inferior to all other three datasets. For instance, compared to the public figures, it has half the density and half the coverage. It is interesting for our evaluation to take into consideration datasets with different characteristics.

Table 1. Dataset characteristics related to NE

	Obama	Trump	La La Land	The Voice
Density of NE	1.818	1.888	1.382	1.061
Coverage of NE	0.916	0.897	0.630	0.547

4.2   The Richness of Weighted Sample

In our approach, we propose using weighted sampling for reputation discovery. We extract a random sample and a weighted sample from all datasets. We compare the richness of the information in terms of number of hashtags, number of URLs and number of NEs for 10 random samples and 10 weighted samples.

The average of the indicators are presented in Table 2. According to Table 2, the weighted sample is significantly richer in terms of the aforementioned indicators. Nevertheless, in terms of entities in *La La Land* and in terms of hashtags in *The Voice*, weighted sample has not been able to perform better. Since one of our parameters of interest is retweet count, sometimes for the movies and TV shows promotional tweets are retrieved, which might not be richer in information.

Table 2. Average indicators of the samples

	Random			Weighted		
	Hashtags	Entities	URLs	Hashtags	Entities	URLs
Obama	14048.6	1828.5	5007.9	14256.1	1839.8	5230.2
Trump	8450.38	1609	2981.75	8655.25	1666.12	3094.5
La La Land	7986.9	1198.9	3102.9	9799.2	1081.6	3230.1
The Voice	1047.2	2856.7	1353	668.7	3368.2	1658

4.3   Frequent Named Entity Mining in Weighted Sample

Frequent Named Entities are discovered through itemset mining techniques [2]. The tweets are considered as transactions and the NEs as itemsets. We used R to perform these experiments, arules package and eclat algorithm.

We used 50 random samples and 50 weighted samples to get an average of number of FNEs for each support value. For *Trump* dataset in Fig. 1 the weighted sample performs better for each of the support values, providing more FNEs than the random sample. *Obama* dataset in Fig. 1 shows a similar behavior as *Trump* dataset. For the same support, the weighted sample performs better, sometimes significantly better; in the low support values the weighted sample provides 20–40 more FNEs than the random sample. The weighted sample in *La La Land* (Fig. 2), in general, extracts more FNEs than the random sample. In the case of *The Voice* dataset (Fig. 2) the weighted sample is always superior to the random sample.

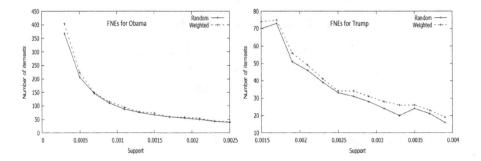

Fig. 1. Average number of itemsets for *Trump* and *Obama* datasets

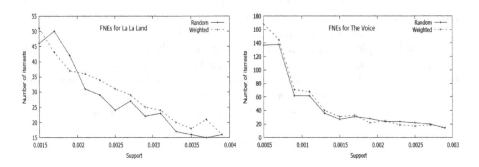

Fig. 2. Average number of itemsets for *La La Land* and *The Voice* datasets

4.4   Comparing the Ranking of the FNEs

Since we are exploring FNEs through samples, we want to guarantee that the discovered FNEs are similar to the FNEs of the population. We ran the eclat algorithm on the whole datasets to discover the FNEs. We matched and ranked the FNEs in the population and in the sample. Then we calculated the Kendall coefficient and the Spearman rank order for both rankings. We repeated the experiment for 10 samples from *Obama*, *Trump*, *La La Land* and *The Voice* dataset. The average values of 10 samples of each dataset showed a considerable similarity between the sample and the whole population in terms of ranking of itemsets; for *Obama*(0.79 and 0.89), *Trump*(0.76 and 0.60), *La La Land*(0.65 and 0.78) and *The Voice*(0.79 and 0.61) for Spearman and Kendall respectively.

4.5 Reputation Through Frequent Named Entities

In this experiment we used Algorithm 1 and for each FNE, we found the sentiment and calculated its reputation. In order to respect the frequency of the FNE in the sample, we weighted the reputation by the support of the FNE. In the end, we calculated an overall reputation as $\sum_{k=1}^{n} R_k * s_k$, where R_k is the reputation of the itemset i_k and s_k is the support of the i_k in S.

We implemented this approach for all dataset, repeating the experiment 10 times for the weighted sample. The average accuracy is 90%. Both datasets related to public figures showed a precise alignment of the reputation explored through FNEs after weighted sampling with the reputation of the whole population. Nevertheless, in the case of the movie *La La Land*, we can distinguish a difference between both results, which comes from the fact that movies are not as dynamic as public figures, therefore, the reputation of a movie is enriched by FNEs, but not defined by them. In the case of *La La Land*, through Frequent Named Entities it is possible to discover viral events; in all of our 10 samples, the first FNE was related Emma Stone and JAEBUM and had a reputation of $(+100, -0)$. It is important to note that our contribution does not focus on finding a reputation, but in enriching the interpretation of reputation by the means of Frequent Named Entities. This self-explanatory approach gives the user the possibility to interpret the information and since it breaks down the reputation of an entity into the reputation of the groups of entities it belongs to, the user has the freedom to use the pieces of reputation in a meaningful way (Table 3).

Table 3. Reputation extraction through FNEs

	Obama	Trump	La La Land	The Voice
Whole population	$(40.79, -59.20)$	$(32.04, -67.96)$	$(74.42, -25.57)$	$(56.06, -43.93)$
Weighted sample	$(40.91, -59.08)$	$(38.22, -61.77)$	$(90.87, -09.12)$	$(55.28, -44.71)$

5 Conclusions

We addressed the problem of reputation discovery and aggregation of sentiments by exploring the underlying entities that co-exist in the data. We introduced a weighted sampling technique to improve the richness of the dataset. We tested the power of Frequent Named Entity Mining on reputation discovery and we showed that FNEs contribute to around the 90% of the reputation of the entity, especially in cases of public figures, who are highly dynamic in their collaborations with other entities. In this paper we used a ranking algorithm based on properties of interest to weight the tweets. Further studies on weighting techniques or choosing and transforming the properties of interest, could improve the quality of the sample.

References

1. Agarwal, A., Xie, B., Vovsha, I., Rambow, O., Passonneau, R.: Sentiment analysis of Twitter data . In: Proceedings of the Workshop on Languages in Social Media, Association for Computational Linguistics, pp. 30–38 (2011)
2. Agrawal, R., Imieliński, T., Swami, A.: Mining association rules between sets of items in large databases. Acm Sigmod Rec. **22**, 207–216 (1993)
3. Bizhanova, A., Uchida, O.: Product reputation trend extraction from Twitter. Soc. Netw., Scientific Research Publishing (2014)
4. Bollen, J., Mao, H., Pepe, A.: Modeling public mood and emotion: Twitter sentiment and socio-economic phenomena. ICWSM Conf. **11**, 450–453 (2011)
5. Ding, X., Liu, B., Zhang, L.: Entity discovery and assignment for opinion mining applications. In: KDD Conference, pp. 1125–1134 (2009)
6. Gabielkov, M., Rao, A., Legout, A.: Sampling online social networks: an experimental study of Twitter. In: ACM SIGCOMM Conference, pp. 127–128 (2014)
7. Ghosh, S., Zafar, M.B., Bhattacharya, P., Sharma, N., Ganguly, N., Gummadi, K.: On sampling the wisdom of crowds: random vs. expert sampling of the Twitter stream. In: CKIM Conference, pp. 1739–1744 (2013)
8. Gjoka, M., Kurant, M., Butts, C.T., Markopoulou, A.: Walking in facebook: a case study of unbiased sampling of OSNs. In: Infocom, pp. 1–9 (2010)
9. Gouriten, G., Maniu, S., Senellart, P.: Scalable, generic, and adaptive systems for focused crawling. In: HT ACM Conference, pp. 35–45 (2014)
10. Hangya, V., Berend, G., Farkas, R.: SZTE-NLP: sentiment detection on Twitter messages. SEM Conf. **2**, 549–553 (2013)
11. Hogenboom, A., Bal, D., Frasincar, F., Bal, M., de Jong, F., Kaymak, U.: Exploiting emoticons in sentiment analysis. In: ACM Symposium on Applied, Computing, pp. 703–710 (2013)
12. Meng, X., Wei, F., Liu, X., Zhou, M., Li, S., Wang, H.: Entity-centric topic-oriented opinion summarization in Twitter. In: KDD Conference, pp. 379–387 (2012)
13. Saif, H., He, Y., Alani, H.: Semantic sentiment analysis of Twitter. In: Cudré-Mauroux, P., Heflin, J., Sirin, E., Tudorache, T., Euzenat, J., Hauswirth, M., Parreira, J.X., Hendler, J., Schreiber, G., Bernstein, A., Blomqvist, E. (eds.) ISWC 2012. LNCS, vol. 7649, pp. 508–524. Springer, Heidelberg (2012). doi:10.1007/978-3-642-35176-1_32
14. Thelwall, M., Buckley, K., Paltoglou, G.: Sentiment in Twitter events. ISI J. **62**(2), 406–418 (2011). Wiley
15. Van Canneyt, S., Claeys, N., Dhoedt, B.: Topic-dependent sentiment classification on Twitter. In: Hanbury, A., Kazai, G., Rauber, A., Fuhr, N. (eds.) ECIR 2015. LNCS, vol. 9022, pp. 441–446. Springer, Cham (2015). doi:10.1007/978-3-319-16354-3_48
16. Wang, X., Wei, F., Liu, X., Zhou, M., Zhang, M.: Topic sentiment analysis in Twitter: a graph-based hashtag sentiment classification approach. In: CKIM Conference, pp. 1031–1040 (2011)
17. Xiang, B., Zhou, L., Reuters, T.: Improving Twitter sentiment analysis with topic-based mixture modeling and semi-supervised training. In: ACL Conference, pp. 434–439 (2014)
18. Zhou, Z., Zhang, X., Sanderson, M.: Sentiment analysis on Twitter through topic-based lexicon expansion. In: Wang, H., Sharaf, M.A. (eds.) ADC 2014. LNCS, vol. 8506, pp. 98–109. Springer, Cham (2014). doi:10.1007/978-3-319-08608-8_9

Social Network Data Analysis

Discovering and Tracking Active Online Social Groups

Md Musfique Anwar[1]([⊠]), Chengfei Liu[1], Jianxin Li[2], and Tarique Anwar[1]

[1] Swinburne University of Technology, Melbourne, Australia
{manwar,cliu,tanwar}@swin.edu.au
[2] University of Western Australia, Perth, Australia
jianxin.li@uwa.edu.au

Abstract. Most existing works on detection of social groups or communities in online social networks consider only the common topical interest of users as the basis for grouping. The temporal evolution of user activities and interests have not been thoroughly studied to identify their effects on the formation of groups. In this paper, we investigate the problem of discovering and tracking time-sensitive activity driven user groups in dynamic social networks. The users in these groups have the tendency to be temporally similar in terms of their *activities* on the topics of interest. To this end, we develop two baseline solutions to discover effective social groups. The first solution uses the network structure, whereas the second one uses the topics of common interest. We further propose an index-based method to incrementally track the evolution of groups with a lower computational cost. Our main idea is based on the observation that the degree of user *activeness* often degrades or upgrades widely over a period of time. The temporal tendency of user activities is modelled as the *freshness* of recent activities by tracking the social streams with a fading time window. We conduct extensive experiments on two real data sets to demonstrate the effectiveness and performance of the proposed methods. We also report some interesting observations on the temporal evolution of the discovered social groups.

Keywords: Active social groups · Dynamic social networks · Group evolution

1 Introduction

The popularity of online social networks (OSNs) has attracted hundreds of millions of users. OSNs have become one of the mainstream mediums for communication. These platforms provide multiple modes of communication, enable diverse interaction types and allow sharing of user-generated contents (e.g. blogs, tweets, and videos). A social network can be modeled as a graph, where each individual is represented by a node in the graph. An edge between two nodes represents the existence of an interaction or relationship that the individuals

© Springer International Publishing AG 2017
A. Bouguettaya et al. (Eds.): WISE 2017, Part I, LNCS 10569, pp. 59–74, 2017.
DOI: 10.1007/978-3-319-68783-4_5

involved in, during the observation time. On social media platforms, the tendency of users with similar interests, choices, and preferences to get associated in OSNs leads to the formation of online (virtual) user groups or communities. Identification of such groups is an important problem in the area of social network analysis. An *online social group* or *community* can be defined as a group of users in an OSN who are closely connected to each other, interact with each other more frequently than with those outside the group, and may have interests on common topics. Many existing works have studied this problem. For instance, the linkage behavior information were explored for community prediction and clustering in [3, 8, 11]. Another direction is to explore the content of the interactions among users in social networks, e.g., [2, 4] improve the quality of discovered communities by considering user interests on different topics. However, the common aspect of the above works is that they did not study the dynamics of the structure and the content in users' online interaction. That is to say, these methods cannot identify the temporal information associated with the network, and thus cannot detect how it evolves over a period of time. Consequently, the discovered groups cannot capture the *active* relationships of social users when an emerging news or event occurs in OSNs.

In this work, we introduce a novel concept of user's *activeness*. It refers to the level of participation of a user towards certain topics, and emphasizes her degree of interest on those topics at a time period. According to our observation, the interest of users on different topics is associated with a corresponding degree of activeness, which keeps changing with time based on several external factors. Out of interest or duty, some user may continuously pay attention to certain topics and remain active (like, posting messages on those topics) as the topic continues and evolves over time. For example, two users (A and B) engaging in conversations related to politics (posting messages, replies to each other) are more likely to be grouped together than someone (C) who occasionally broadcasts posts on politics. However, the user's degree of interest and activeness may vary greatly over time. For instance, all users (A, B, C) related to politics would like to show their inclination during elections in a country. Our goal is to discover and track the active groups in online social networks (e.g., Twitter) called *active online social groups*, in which the users are closely linked to each other and have at-least a certain level of activeness in the topics of common interest.

An active online social group is depicted as a connected induced subgraph in which each node has a degree of at least k. The parameter k measures the structure cohesiveness of the social group. We apply time-based forgetting factor to the users' past activities in order to emphasize the *freshness* of users' recent activities. A sliding time window based model is imposed on social streams (e.g. tweets, posting images) in order to discover and track the evolution of social groups at different time intervals by a set of primitive evolution operations. The main contributions of this work are summarized below.

- We investigate the users' activeness for social groups detection and evolution by using the time decay strategy;

- We develop two baseline solutions and one index-based efficient solution to incrementally discover and track the evolution of user groups in an efficient manner in dynamic OSNs;
- We conduct extensive experiments to show the efficiency and effectiveness of our proposed approach using real data sets.

In the rest of the paper, we present the related works in Sect. 2, and formulate the problem of discovering and tracking active online social groups in Sect. 3. In Sect. 4, we show our approach of modelling the users' recent interests and activities using a fading time window scheme. Sections 5 and 6 present the baselines and the proposed indexed based solution, respectively. Our experimental results are shown in Sect. 7, and finally, we conclude the paper in Sect. 8.

2   Related Work

Social network analysis methods have traditionally focused on the representation of graphs as static networks. The commonly used methods for community detection are based on modularity [8], edge betweenness [3] and neighborhood concepts [11]. However, these methods focus only on the information regarding the linkage behavior (connection) for the purposes of community detection. As a result, they fail to account for other node properties and user interactions in the social graph, such as user interests. Various topic-based approaches have also been investigated in the literature, which employ textual content published by the users jointly with social connections to detect like-minded users. A link-content model is proposed by Natarajan et al. [1] to discover topic-based communities. Qi et al. [2] proposed a method based on matrix factorization by integrating the structural and content aspects of OSNs to greatly improve the effectiveness of community detection. Yang et al. [4] introduced an alternative discriminative probabilistic model (PCL-DC) to incorporate content information in the conditional link model and estimate the community membership directly. However, the above methods are not applied to detect the active social groups in a dynamic environment.

Motivated by the temporally evolving nature of social networks, some works focus on mining dynamic information networks. Palla et al. in [9] extend the popular clique percolation method (CPM) for community detection. They run the CPM on successive graph snapshots to track the changes in the groups. Kim et al. [7] identifies the correspondence between communities at different times by imposing temporal constraints on successive graph snapshots. Cuzzocrea and Folino [10] introduce graph-based model-theoretic approach from structural point of view to detect community evolution in time-evolving network. Chen et al. [6] propose a convex relaxation problem to find the overlapping community structure that maximizes a quality function associated with each snapshot subject to a temporal smoothness constraint. However, none of these approaches consider the user's degree of interest or activeness towards the topics over time.

3   Preliminary and Problem Definition

We define some basic concepts before formally introducing our problem.

Social Network: A social network is defined as a graph $G = (U, E)$, where U is the set of nodes (representing users), and E is the set of links between the nodes (representing connections between users). If two users $u, v \in U$ are connected in G, then there exists an edge $e_{uv} \in E$ in G.

k-CORE: Given an integer $k(k \geq 0)$, the k-core of a graph G, denoted by C^k, is the largest subgraph of G, such that $\forall u \in C^k, deg_{C^k}(u) \geq k$, where $deg_{C^k}(u)$ refers to the degree of node u in C^k.

Node Core Number: The core number of a node u in G is the maximum k for which u belongs in the k-core of G.

Topic: In Twitter, many tweets are accompanied by hashtags (e.g.,#election2012, #earthquake), short textual tags that get widely adopted by the Twitter users. These hashtags represent specific pieces of information that we can track as they get adopted across the network. In our work, we consider these hashtags as topics.

Activity: An activity refers to an action that a user performs at a time point. For example, when a user u_i mentions a topic T_x at time t_j, this activity is recorded as an activity tuple $\langle u_i, T_x, t_j \rangle$. An activity stream S is a continuous and temporal sequence of activities i.e. $S = \{s_1, s_2, \ldots, s_r, \ldots\}$ such that each object (s_i) corresponds to an activity tuple.

Sliding Time Window: A window of a predefined length len is moved over the activity stream S and specifies the intervals to analyze. Let $\Gamma = <t_1, t_2, \ldots, t_n>$ be a sequence of points in time, I_m an interval $[t_{i-len}, t_i]$ of length len, where $0 < len \leq i$. We partition Γ into set of equal-length intervals denoted as $\mathcal{I} = \{I_1, \ldots, I_m\}$. We consider an *overlapping window* that partially overlaps with the prior window. The degree of overlap is controlled by the parameter Δt.

Recency Score: Generally, not all the past activities of a user are equally important. We assign greater importance to user's most recent activities by a measure called *recency score*, denoted by $\mu \in [0, 1]$. Its value is computed by an exponential time-decay function in Eq. 1, which assigns lower importance to user's older activities, as they are less likely to match the user's current interest. The parameter a is an external factor to control the speed of decay and $age_{\langle u_i, T_x, t_j \rangle}$ denotes the amount of time passed since the activity occurred.

$$\mu_{\langle u_i, T_x, t_j \rangle} = exp(-a \times age_{\langle u_i, T_x, t_j \rangle}) \tag{1}$$

Active User: A user u_i is said to be active towards a topic T_x in the time interval I_m, if u_i has performed at least $\alpha(\geq 1)$ activities related to T_x within I_m, i.e., $|\{\langle u_i, T_x, t_j \rangle\}| \geq \alpha$. The set of such active users is denoted by $\mathcal{A}_{(T_x, I_m)}$.

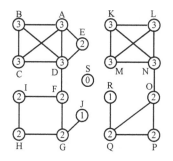

Fig. 1. Social graph (the number denotes the node core number)

Table 1. Social groups (Bold node indicates new active node at I_m)

Time	Active users	Social group ($\mathcal{G}$)
I_1	**A, B, C, D**	$\mathcal{G}_1 = \{A, B, C, D\}$
I_2	A, C, D, **K, L, M,** **N, O, P, Q**	$\mathcal{G}_1 = \{A, C, D\}$ $\mathcal{G}_2 = \{K, L, M, N, O, P, Q\}$
I_3	A, B, C, D, **E**, K, L, M, O, P, Q,	$\mathcal{G}_1 = \{A, B, C, D, E\}$ $\mathcal{G}_3 = \{K, L, M\}$ $\mathcal{G}_4 = \{O, P, Q\}$
I_4	A, C, D, **F, G, H, I,** K, L	$\mathcal{G}_5 = \{A, C, D, F, G, H, I\}$

Activeness Score: For each active user $u_i \in \mathcal{A}_{(T_x, I_m)}$, we compute an activeness score to measure the involvement of u_i towards a topic T_x in a time interval I_m, using Eqs. 2 and 3, where $t_j \in I_m$, and $\psi_{(u_i, T_x, I_m)}$ is the user's degree of activeness compared to the total number of activities performed by all the active users.

$$\psi_{(u_i, T_x, I_m)} = \frac{\sum \mu_{(u_i, T_x, t_j)} \times |\{\langle u_i, T_x, t_j \rangle\}|}{\sum_{u_z \in \mathcal{A}_{(T_x, I_m)}} |\{\langle u_z, T_x, t_j \rangle\}|} \tag{2}$$

$$\sigma_{(u_i, T_x, I_m)} = \frac{\psi_{(u_i, T_x, I_m)}}{max_{u_z \in \mathcal{A}_{(T_x, I_m)}} \{\psi_{(u_z, T_x, I_m)}\}} \tag{3}$$

Definition 1 (Activity-driven Social Group Detection). *Given a social network $G = (U, E)$, a topic T_x, an activity stream S, a positive integer k and a time-interval set $\mathcal{I}$, a sliding window slides over S with Δt amount at regular intervals, and produces a sequence of observations $\mathcal{P} = (P_{I_1}, P_{I_2}, \ldots, P_{I_\tau})$. Each P_{I_m}, which is observed at time interval I_m, is a collection of non-empty and non-overlapping set of active groups $\Phi_{I_m} = \{\mathcal{G}_1, \mathcal{G}_2, \ldots, \mathcal{G}_n\}$, such that $\bigcup_{i=1}^{n} \mathcal{G}_i \subseteq \mathcal{A}_{(T_x, I_m)}$ and $\forall i \neq j \, \mathcal{G}_i \cap \mathcal{G}_j = \emptyset$, where $\mathcal{A}_{(T_x, I_m)}$ denotes the set of active users at I_m for topic T_x and $\mathcal{G}_i$ denotes an active online social group of users having similar social activities. For each $\mathcal{G}_i \in \Phi_{I_m}$, the following properties hold.*

1. **Connectivity.** *$\forall_i \mathcal{G}_i \subset G$ is connected;*
2. **Structure cohesiveness.** *$\forall u \in \mathcal{G}_i, deg_{\mathcal{G}_i}(u) \geq k$;*
3. **Topic cohesiveness.** *$\forall u \in \mathcal{G}_i$, activeness score of a user u is $\sigma_{(u, T_x, I_m)} \geq \theta$ at time interval I_m and $\theta \in [0, 1]$ is a threshold.*

Figure 1 shows a social graph G with the core number for each node, e.g., the nodes with $3 - core$ are $\{A, B, C, D, K, L, M, N\}$. Table 1 shows the active users for a topic T_x at different time intervals. Suppose we want to find active social groups at different time intervals for $k = 2$ and topic T_x. We get $\mathcal{G}_1 = \{A, B, C, D\}$ at I_1. At I_2, one member (B) is removed from $\mathcal{G}_1$ but it still satisfies the group criteria and a new group $\mathcal{G}_2$ is formed. Then $\mathcal{G}_2$ is split into two groups

($\mathcal{G}_3 = \{K, L, M\}$ and $\mathcal{G}_4 = \{O, P, Q\}$) as a removal of inactive user N and new active users B and E joined to $\mathcal{G}_1$ at I_3. A new group $\{F, G, H, I\}$ is formed and merged with updated $\mathcal{G}_1$ to form $\mathcal{G}_5 = \{A, C, D, F, G, H, I\}$ at I_4.

4 Fading Time Window for Evolving Social Groups

A sliding window manager is generally used to keep track of social streams arriving in the system, where the time interval of incoming streams defines the size of the sliding window. The two most common variants of sliding window are - (i) count based windows, which contain the $|M|$ most recent data objects (tweets in our work); and (ii) time-based windows, which contain the objects whose time-stamps are within certain time interval. Again, there are two variants of time-based sliding window- the *disjoint* window (windows do not overlap) and the *overlapping* widow (allows windows to partially overlap).

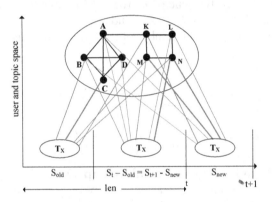

Fig. 2. An illustration of the fading time window from time t to $t + 1$.

The main objective of our work is to extract active social groups based on common actions of the users. As the user's actions (posting of messages) appear in the form of a continuous ordered stream, we adopt the time-based overlapping sliding window model. From a given time point, the system consumes the incoming tweets over a time window. It has a fixed size in terms of the selected measure and hence in order to accommodate new data, the old data must be dropped/forgotten. An important parameter of the time window approach is the step at which the window grows or slides. We introduce a fading time window scheme (illustrated in Fig. 2) by combining the sliding time window and the decaying (or Fading) function for the time-evolving graphs like Twitter. The sliding time window scheme (posts are first-in, first-out) provides a scope within which the actions can be monitored to track the evolution. Fading scheme puts a higher emphasis on users' newer activities. Since social groups evolve quickly, even within a given time window, it is important to highlight the users' recent activities and degrade the old activities using the fading scheme.

The time window length and the time interval are denoted by Len and Δt respectively. When the time window slides from time t to $t+1$, the social stream S_t from time t is updated to S_{t+1} at $(t+1)$. S_{old} is the set of old messages (tweets) from S_t that lapses at time $t+1$, and S_{new} is the set of new messages from $S_t + 1$ that appears. Clearly,

$$S_{t+1} = S_t - S_{old} + S_{new} \tag{4}$$

Figure 2 shows a social graph where users are active on a topic T_x. The weight of the line connecting a user and the topic indicates user's degree of activeness towards topic T_x for a specific time interval. Each time the sliding window shifts, the least recent tweets are replaced by the most recent tweets.

5   Baseline Solutions

In this section we present two baseline solutions, both of which are based on a naive idea. The social groups from the graph G can be identified in three steps at each time interval. The first step is to find the k-cores from G. Then for all the users from each k-core, we need to calculate their activeness scores for a given topic T_x. A user is considered as active user if her activeness score is greater than some threshold value θ for that particular time interval. Finally, we find all the connected active k-cores (considered as active social groups) from G.

Algorithm 1. basic-G

Input: $G = (U, E), T_x, S, k, \mathcal{I}$
Output: $\mathcal{P} = (P_{I_1}, P_{I_2}, ..., P_{I_\tau})$ ▷ set of active social groups Φ_{I_t} at each I_t
 1: compute the maximal k-core $C^k(G)$ of G
 2: $t \leftarrow 1$
 3: **for** each connected component H_j^k from $C^K(G)$ **do**
 4:      **for** each $u_i \in H_j^k$ **do**
 5:           compute $\sigma_{(u_i, T_x, I_t)}$
 6:           **if** $\sigma_{(u_i, T_x, I_t)} < \theta$ **then**
 7:                DFS(u_i)
 8: Output the set of active connected components Φ_{I_t} from $C^k(G)$
 9: $t \leftarrow t + 1$
10: **Procedure** DFS(u)
11: **for** each $v \in N(u, C^k(G))$ **do**   ▷ $N(u, C^k(G))$ is the list of u's neighborhood
12:      remove edge (u, v) from $C^k(G)$
13:      **if** $deg_{C^k(G)}(v) < k$ **then**
14:           DFS(v)
15: remove node u from $C^k(G)$

Intuitively, there are two possible approaches to find such groups. The first approach (called basic-G) is to explore the social network by firstly considering the degree constraint and then refining the groups further with the topic constraint, whereas the second approach (called `basic-T`) is to consider the topic constraint first, followed by the degree constraint.

In `basic-G` (shown in Algorithm 1), we first compute the maximal k-core of the graph G denoted by $C^k(G)$ for a given k. Then, we iteratively invoke the following procedure to compute the connected active k-core social groups. For each connected components H_j^k from $C_k(G)$, we remove the inactive nodes (whose $\sigma_{(u_i, T_x, I_m)} < \theta$) from H_j^k (lines 3–7). Removal of an inactive node u requires to recursively deletes all the nodes that violate the cohesiveness constraint using DFS procedure (lines 10–15). This is because the degree of u's neighbor nodes decrease by 1 due to the removal of inactive node u. This may result into violation of the cohesiveness constraint by some of u's neighbors. Due to this, they cannot be included in the subsequent social groups, and thereby we need to delete them. Similarly, we also need to verify the other hop (e.g., 2-hop, 3-hop, etc.) neighbors that they satisfy the cohesiveness constraint. The drawback of this approach is that it has to compute the k-core decomposition on the entire graph as well as has to perform DFS operation numerous times.

Algorithm 2. `basic-T`

Input: $G = (U, E), T_x, S, k, \mathcal{I}$
Output: $\mathcal{P} = (P_{I_1}, P_{I_2}, ..., P_{I_\tau})$ ▷ set of active social groups Φ_{I_t} at each I_t
1: $t \leftarrow 1$
2: **for** each $u_i \in G$ **do**
3:     compute $\sigma_{(u_i, T_x, I_m)}$
4:     **if** $\sigma_{(u_i, T_x, I_m)} < \theta$ **then**
5:         $\mathcal{A}_{(T_x, I_t)}.add(u_i)$
6: compute the induced graph G_t on $\mathcal{A}_{(T_x, I_t)}$
7: compute the maximal k-core $C^k(G_t)$ of G_t
8: Output the set of active connected components Φ_{I_t} from $C^k(G_t)$
9: $t \leftarrow t + 1$

In `basic-T` (shown in Algorithm 2), at first we compute the set of active nodes from G (lines 2–5). Then we identify the maximal k-core of $C^k(G_t)$ from the induced graph G_t formed by the set of active nodes $\mathcal{A}_{(T_x, I_t)}$ (lines 6–7). It then outputs the set of active connected components (groups) Φ_{I_t} from C^k. The above procedure is applied at each time interval. The major drawback of this approach is that it has to calculate the activeness score of all the nodes at the beginning, although many nodes may not meet the cohesiveness constraint. Again we need to compute the maximal k-core at each time interval.

6   Index-Based Method

The major drawback of `basic-G` and `basic-T` approaches is that these algorithms require much time to compute the k-core decomposition at every time interval. To efficiently discover and track the evolution of social groups, we follow an indexing mechanism called `CL-tree` (Core Label tree) [12], which organizes the k-cores into a tree structure.

6.1 Index Overview

The CL-tree index is built based on the observations that cores are nested, i.e., a $(k + 1)$ core must be contained in a k-core. A subgraph containing nodes with a minimum degree of $(k + 1)$ already satisfy the condition of a minimum degree of k. Therefore after computing the k-core of G, we assign each connected components induced by k-core as a child node of a tree. Each node in the tree contains three elements: *coreNum* (the core number of the k-core), *vertexSet* (set of graph vertices), and *childList* (list of child nodes). Using the tree structure, for a given k, it is very efficient to find all the nodes having a core of at least k by traversing the CL-tree.

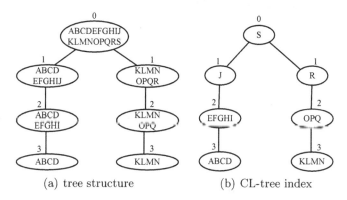

(a) tree structure   (b) CL-tree index

Fig. 3. CL-tree index for the graph in Fig. 1

6.2 Incremental Group Evolution

In the existing literature on dynamic social groups detection, there is a broad consensus (e.g. [5, 13]) on the fundamental events (e.g. growth, contraction, merging, splitting, birth and death) to characterise the evolution of dynamic social groups. In order to keep track of the evolution of social groups, we use the following primitive group evolution operations.

1. $+\mathcal{G}$: create a new group $\mathcal{G}$;
2. $-\mathcal{G}$: remove an old group $\mathcal{G}$;
3. $\uparrow (\mathcal{G}; \varphi)$: increase the size of $\mathcal{G}$ by adding set of users φ;
4. $\downarrow (\mathcal{G}; \varphi)$: decrease the size of $\mathcal{G}$ by removing set of users φ.
5. Merge: merge a set of groups into a new single group $\mathcal{G}$;
6. Split: split a single group $\mathcal{G}$ into a set of new groups and remove $\mathcal{G}$.

Our index-based incremental group maintenance algorithm IGM (Algorithm 3) starts with computing the set of active nodes having degree of at least k (lines 2–4). Then we invoke the procedure SGI (Social Group Identification) which computes the k-core connected subgraphs from the active nodes (line 5). For each existing group, we remove inactive nodes which may results violation of cohesiveness constraint for some other nodes of the group. We recursively removes those nodes using DFS procedure (lines 6–10). Again, if removing one or more nodes breaks the connectivity property of the group, we then invoke SGI (line 12) to find the active connected k-core components from the remaining active

---

Algorithm 3. IGM: Incremental Group Maintenance

Input: Φ_{I_t}, *root of* CL-tree
Output: $\mathcal{P} = (P_{I_1}, P_{I_2}, ..., P_{I_\tau})$ ▷ set of active social groups Φ_{I_t} at each I_t
1: $\Phi_{I_{t+1}} = \Phi_{I_t}$
2: **for each** $u_i \in G$ and $deg(u_i) \geq k$ **do**
3:        **if** $\sigma_{(u_i, T_x, I_{t+1})} \geq \theta$ and $u_i \notin \Phi_{I_t}$ **then**
4:              $\mathcal{A}_{(T_x, I_{t+1})}.add(u_i)$
5: $SGI(\mathcal{A}_{(T_x, I_{t+1})})$
6: **for each** $\mathcal{G}_j \in \Phi_{I_t}$ **do**
7:        **for each** $u_i \in \mathcal{G}_j$ **do**
8:              **if** $\sigma_{(u_i, T_x, I_{t+1})} < \theta$ **then**
9:                    $\Phi_{I_{t+1}}(\mathcal{G}_j).delete(u_i)$;
10:                   DFS(u_i) ▷ Procedure DFS from Algorithm 1
11:        **if** connectivity$(\Phi_{I_{t+1}}(\mathcal{G}_j))$ is false **then** ▷ check whether $\mathcal{G}_j$ is connected or not
12:              $SGI(\Phi_{I_{t+1}})$
13:              $\Phi_{I_{t+1}}.delete(\mathcal{G}_j)$
14: **for each** $\mathcal{G}_j \in \Phi_{I_{t+1}}$ **do**
15:       **for each** $u_i \in \mathcal{A}_{(T_x, I_{t+1})}$ **do**
16:             **if** edge$(\mathcal{G}_j, u_i)$ **then**
17:                   $\mathcal{G}_j.add(u_i)$
18:       find $\mathcal{G}_k$ from $\mathcal{G}_j$
19: **for each** $\mathcal{G}_j \in \Phi_{I_{t+1}}$ **do**
20:       **for each** $\mathcal{G}_o \in \Phi_{I_{t+1}}$ **do**
21:             **if** $(j \neq o)$ and connectivity$(\mathcal{G}_j, \mathcal{G}_o))$ is true **then**   ▷ check whether $\mathcal{G}_j$ and $\mathcal{G}_o$ is connected or not
22:                   $\mathcal{G}_n = \Phi_{I_{t+1}}(\mathcal{G}_j) \cup \Phi_{I_{t+1}}(\mathcal{G}_o)$
23:                   $\Phi_{I_{t+1}}.add(\mathcal{G}_n)$;
24:                   $\Phi_{I_{t+1}}.delete(\mathcal{G}_j)$; $\Phi_{I_{t+1}}.delete(\mathcal{G}_0)$;
25:                   break;
26: **return** $\Phi_{I_{t+1}}$
27: **Procedure** SGI$(\mathcal{A}_{(T_x, I_{t+1})})$
28: compute the k-core connected components from $\mathcal{A}_{(T_x, I_{t+1})}$
29: **for each** connected components $\mathcal{G}_j$ **do**
30:       **for each** $u_i \in \mathcal{G}_j$ **do**
31:             $\mathcal{A}_{(T_x, I_{t+1})}.delete(u_i)$
32:       $\Phi_{I_{t+1}}.add(\mathcal{G}_j)$
33: **return** $\Phi_{I_{t+1}}$

---

nodes of the group. We further validate the cohesiveness constraint of an existing group by adding one or more new active nodes to it. We then increase the size of that group by the number of those newly joined active nodes (lines 14–18) which are able to fulfill the requirement for cohesiveness constraint. Two groups are merged into a single group (lines 19–25) if connectivity exists between these groups. Finally, the algorithm returns the set of social groups $\Phi_{I_{t+1}}$ at I_{t+1}.

Our group evolution tracking algorithm GET (Algorithm 4) tracks the evolution of the social groups over a period of time. Given a stream of tweets, from the start time i, it tracks the primitive group evolution operations at each time, working on top of IGM algorithm (line 2). Basically, the algorithm monitors the changes of groups effected by IGM at each time interval. If a group is changed, the algorithm determines the corresponding case and outputs the group evolution patterns (lines 3–13). Note that in lines 4–7, if a group $\mathcal{G}_x$ in I_t has GroupId id, we use the convention that $\Phi_{I_t}(id) = \mathcal{G}_x$ to access $\mathcal{G}_x$ by id, and $\Phi_{I_t}(id) = \phi$; means there is no group in I_t with GroupId id. Especially, lines 6–7 mean a group in I_t evolves into a group in I_{t+1} by removing the inactive nodes in $(\Phi_{I_t}(id) - \Phi_{I_{t+1}}(id))$ first, and then adding new active nodes $(\Phi_{I_{t+1}}(id) - \Phi_{I_t}(id))$.

Algorithm 4. GET: Group Evolution Tracking

Input: $G = (U, E), T_x, S, k, \mathcal{I}$, *root* of CL-tree
Output: Primitive social group evolution operations
 1: **for** t from i to j **do**
 2:     $\Phi_{I_{t+1}} = $ IGM$(\Phi_{I_t}, root)$
 3:     **for each** $\mathcal{G}_x \in \Phi_{I_{t+1}}$ **do**
 4:         $id = GroupId(\mathcal{G}_x)$
 5:         **if** $\Phi_{I_t}(id) \neq \phi$ **then**
 6:             output $\downarrow (\mathcal{G}_x, \Phi_{I_t}(id) - \Phi_{I_{t+1}}(id))$
 7:             output $\uparrow (\mathcal{G}_x, \Phi_{I_{t+1}}(id) - \Phi_{I_t}(id))$
 8:         **else**
 9:             $+\mathcal{G}_x$
10:     **for each** $\mathcal{G}_x \in \Phi_{I_t}$ **do**
11:         $id = GroupId(\mathcal{G}_x)$
12:         **if** $\Phi_{I_{t+1}}(id) = \phi$ **then**
13:             $-\mathcal{G}_x$

To illustrate the Algorithms 3 and 4, let us consider the example graph in Fig. 1 and the Table 1. As we indexed the graph G in tree like structure (Fig. 3), so it is easy to locate nodes having $deg(u_i) \geq k$. For example, if $k = 2$, at time interval I_1 we update the activeness score of all the nodes and get the active nodes set $\mathcal{A}_{(T_x,I_1)} = \{A, B, C, D\}$ (IGM:lines 2–4). Then by passing $\mathcal{A}_{(T_x,I_1)}$ to the procedure SGI, we get active connected k-core components (i.e. active social group) of $\mathcal{G}_1 = \{A, B, C, D\}$. In the next time interval at I_2, we get a new active social group $\mathcal{G}_2 = \{K, L, M, N, O, P, Q\}$. We remove the inactive node B from $\mathcal{G}_1$ (using DFS) and find that $\mathcal{G}_1 = \{A, C, D\}$ still maintains all the constraints and survive (lines 6–10). Removing inactive node N at I_3 splits group $\mathcal{G}_2$ into two groups (lines 6–13) i.e. $\mathcal{G}_3 = \{K, L, M\}$ and $\mathcal{G}_4 = \{O, P, Q\}$. The set of new active nodes $\mathcal{A}_{(T_x, I_3)} = \{B, E\}$ at I_3 joined with the existing $\mathcal{G}_1$ (lines 14–18). At I_4, $\mathcal{G}_3$ fails to survive due to the removal of inactive node M. $\mathcal{G}_4$ is also removed as none of the nodes are active. The new active nodes $\{F, G, H, I\}$ formed a new group $\mathcal{G}_5$ and eventually merged with the updated $\mathcal{G}_1$ (as B and E are no longer active) to form a new group $\mathcal{G}_6 = \{A, C, D, F, G, H, I\}$ (lines 19–25).

The advantages of these algorithms over the basic algorithms are that it is easy to locate vertices having $deg(u_i) \geq k$ using the index structure. Again, as we maintain the groups that are computed in each time interval, it is easy to keep track of the evolution of groups over time. Furthermore, it requires to perform less DFS operations compared to the basic algorithm (basic-G).

7   Experiments

We conduct our experiments on two Twitter datasets: a small dataset CRAWL [14] and a very large dataset SNAP[1]. In the CRAWL dataset, we consider the user tweets from February to April of 2012. SNAP contains 467 million Twitter posts from 20 million users covering a 7 month period from June 1, 2009 to December 31, 2009. We randomly choose 1,00,000 users and consider their tweets from June 11, 2009 to July 12, 2009. Table 2 shows the statistics of our experimental data. Our experiments are performed on an Intel(R) Core(TM) i7-4500U 2.4 GHz Windows 7 PC with 16 GB RAM.

Table 2. Datasets

Dataset	# of nodes	# of edges	Stream size
CRAWL	9,468	1,474,510	6,919,200
SNAP	100,000	3,938,007	573,832

In the CRAWL dataset, we consider the topics of *politics* (#Obama, #election2012, #politicalcorruption etc.) and *entertainment* (#nowplaying, #bb14[2], #itunes etc.) for the experiments. Similarly, in the SNAP dataset we consider the topics *election in Iran* (#Iranelection, #helpiranelection, #Iran etc.), *Follow Friday*[3] (#FollowFriday, #FFD etc.) and the death of pop star *Michael Jackson* (#MichaelJackson, #thankyouMichael, #MJ etc.). The baseline solutions (basic-G and basic-T) and the index-based proposed method IGM (Algorithm 3) can only discover the social groups. The extended method GET (Algorithm 4) captures the evolution of groups by tracking them. The efficiency and group evolution results are shown in Sects. 7.1 and 7.2 respectively.

---

[1] http://snap.stanford.edu/data/twitter7.html.

[2] Big Brother 14 was the 14th season of the American reality television series.

[3] A tradition in which the users can recommend their followers to follow more people.

7.1 Efficiency

We present the efficiency results by varying k, threshold θ and Δt to see their effects on overall running time of all the four algorithms for *politics* in CRAWL, and *Iran election* and *Follow Friday* in SNAP. In this regard, we consider user posts for the month of February, 2012 in CRAWL.

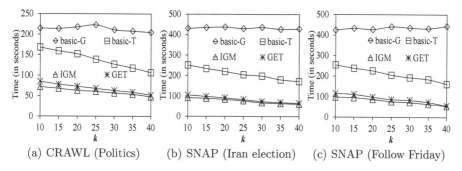

(a) CRAWL (Politics)   (b) SNAP (Iran election)   (c) SNAP (Follow Friday)

Fig. 4. Run-time for different values of k (in all cases, Time window = 5 days, $\Delta t = 1$ day, $\theta = 0.5, \alpha = 5, a = 0.02$)

Figure 4 shows the running time at different values of k. Observe that basic-G takes more time compared to the other methods, because it needs to perform the k-core decomposition on the entire graph G at the beginning and the DFS operations numerous times. The computation for basic-G depends on the size of G as we see that it takes more time for large SNAP dataset. A lower k renders more subgraphs, due to which it takes more time generally for all algorithms except basic-G. The reason is that basic-G has to remove many inactive users through numerous DFS operations for higher values of k and the number of DFS operations varies for different values of k. The index-based methods IGM and GET perform faster than basic-T because basic-T needs to compute the maximal k-core in each time interval.

Figure 5 shows the running time at different values of θ. We observe that the computation times for basic-T and indexed methods are reduced as θ increases, because it results less active users in the network. As basic-G computes the k-core decomposition on the entire G at first, it has to remove numerous inactive users through DFS for higher values of θ. We see that both IGM and GET outperform the baseline solutions.

Figure 6 shows the running time at different values of Δt. As higher values of Δt result into lesser number of time windows, the computation time decreases as Δt grows higher. We also observe that the gap between basic-G and basic-T gets smaller with higher Δt values especially for CRAWL. This is because basic-G requires to perform less number of k-core decomposition on entire G, and for smaller network like CRAWL, it requires less computation time for k-core decomposition.

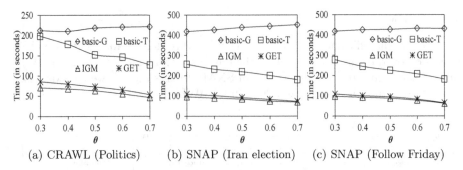

Fig. 5. Run-time for different values of θ (in all cases, Time window = 5 days, $\Delta t = 1$ day, $k = 20, \alpha = 5, a = 0.02$)

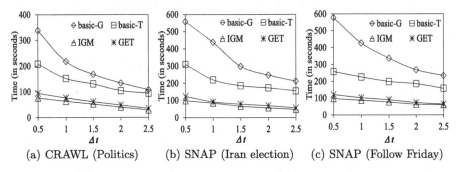

Fig. 6. Run-time for different values of Δt (in all cases, Time window = 5 days, $k = 20, \theta = 0.5, \alpha = 5, a = 0.02$)

We observe that GET takes slightly more time than IGM. This is due to the reason that GET also performs the tracking of evolution of the discovered groups in each time window, whereas IGM only discovers the groups.

7.2 Group Evolution

Figure 7(a) shows the average activeness score of a sample group from CRAWL dataset for *politics* and *entertainment* during the first half of the months of February, March and April in 2012. We observe that users in the group are mostly active on *entertainment* in February and March, and then their activeness shifts towards *politics* in April as the date of US election (in 2012) gets closer. Figure 7(b) shows the number of different evolution event types at different values of k for *politics*. We observe that low values of k lead to formation and survival of a large number groups, while high values of k results in more dissolution.

Figure 8 shows the evolution (group size transition) of a sample group over time for each of the three topics (*Iran election, Follow Friday* and the death of *Michael Jackson*) of SNAP dataset separately. We see that the members of the group are active about *Iran election* (was held on June 12, 2009) at the

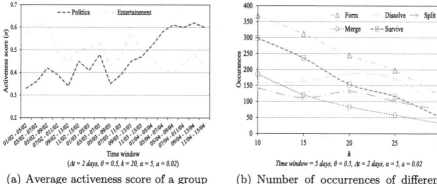

(a) Average activeness score of a group

(b) Number of occurrences of different evolutionary events

Fig. 7. Evolution results on CRAWL dataset

beginning. Afterwards, the size of the same group increases suddenly due to the death of *Michael Jackson* (died at June 25, 2009). The #FollowFriday trend is a customary Friday activity. Users are generally active during Friday as we see that the group does not exist in the time intervals that do not include Friday (for example, time interval from 13/06/2009 (Saturday)–18/06/2009 (Thursday)).

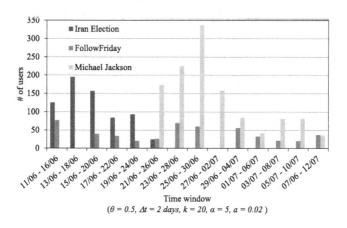

Fig. 8. Example of group evolution (SNAP dataset)

8   Conclusion

In this paper, we studied the problem of discovering and tracking user groups in dynamic online social networks. We developed two baseline solutions to discover effective social groups, and further proposed an index-based method to incrementally track the evolution of groups in an efficient manner. Our method is based on the observation that the activeness of a user towards different topics of interest keeps changing temporally. This behaviour is modelled by tracking the social streams with a fading time window. Our extensive experiments on real data sets demonstrated the efficacy of the proposed methods.

Acknowledgment. This work is supported by the ARC Discovery Projects DP160102412 and DP140103499.

References

1. Natarajan, N., Sen, P., Chaoji, V.: Community detection in content-sharing social networks. In: ASONAM, pp. 82–89 (2013)
2. Qi, G., Aggarwal, C., Huang, T.: Community detection with edge content in social media networks. In: ICDE, pp. 534–545 (2012)
3. Newman, M.E.J., Park, J.: Why social networks are different from other types of networks. Phys. Rev. E **68**, 036122 (2003)
4. Yang, T., Jin, R., Chi, Y., Zhu, S.: Combining link and content for community detection: a discriminative approach. In: KDD, pp. 927–936 (2009)
5. Greene, D., Doyle, D., Cunningham, P.: Tracking the evolution of communities in dynamic social networks. In: ASONAM, pp. 176–183 (2010)
6. Chen, Y., Kawadia, V., Urgaonkar, R.: Detecting overlapping temporal community structure in time-evolving networks. e-print: arXiv:1303.7226 (2013)
7. Kim, M., Han, J.: A particle-and-density based evolutionary clustering method for dynamic networks. In: VLDB, pp. 622–633 (2009)
8. Meo, P.D., Ferrara, E., Fiumara, G., Provetti, A.: Generalized Louvain method for community detection in large networks. In: ISDA, pp. 88–93 (2011)
9. Palla, G., Barabasi, A.L., Vicsek, T.: Quantifying social group evolution. Nature **446**(7136), 664–667 (2007)
10. Cuzzocrea, A., Folino, F.: Community evolution detection in time-evolving information networks. In: EDBT, pp. 93–96 (2013)
11. Cohen, J.: Trusses: cohesive subgraphs for social network analysis. Technical report, National Security Agency (2008)
12. Fang, Y., Cheng, R., Luo, S., Hu, J.: Effective community search for large attributed graphs. In: VLDB, pp. 1233–1244 (2016)
13. Asur, S., Parthasarathy, S., Ucar, D.: An event-based framework for characterizing the evolutionary behavior of interaction graphs. In: KDD, pp. 913–921 (2007)
14. Bogdanov, P., Busch, M., Moehli, J., Singh, A.K., Szymanski, B.K.: The social media genome: modeling individual topic-specific behavior in social media. In: ASONAM, pp. 236–242 (2013)

Dynamic Relationship Building: Exploitation Versus Exploration on a Social Network

Bo Yan[1], Yang Chen[2], and Jiamou Liu[2(✉)]

[1] Beijing Institute of Technology, Beijing, China
yanbo@bit.edu.cn
[2] University of Auckland, Auckland, New Zealand
yche767@aucklanduni.ac.nz, jiamou.liu@auckland.ac.nz

Abstract. Interpersonal relations facilitate information flow and give rise to positional advantage of individuals in a social network. We ask the question: *How would an individual build relations with members of a dynamic social network in order to arrive at a central position in the network?* We formalize this question using the dynamic network building problem. Two strategies stand out to solve this problem: The first directs the individual to exploit their social proximity by linking to nodes that are close-by, while the second tries its best to explore distant regions of the network. We evaluate and contrast these two strategies with respect to edge- and distance-based cost metrics, as well as other structural properties such as embeddedness and clustering coefficient. Experiments are performed on models of dynamic random graphs and real-world data sets. We then discuss and test ways that combine these two strategies.

Keywords: Dynamic social networks · Interpersonal ties · Network evolution · Centrality · Exploitation-exploration tradeoff

1   Introduction

The network of social relations entails important properties of individuals. Take, as an example, the structural construct of centrality [33]. Much has been revealed about the correlation between centrality and social statues [9,10,23,25]. By occupying a more central position in the social network, an individual may exercise more control over the flow of information, accessing diverse knowledge and skills, and hence gaining a higher positional advantage [38]. Exploiting this principle, individuals may cultivate relationships with others towards improving their social statues [11]. One famous example is the House of Medici, who rose to prominence in 15th century Florence through intermarriage with other noble families [35]. Another example is Moscows growing statues in 12–13th century Russia thanks to trade relationships with other towns [36]. A third example is the case of Paul Revere who successfully raised a militia during the American Revolution by strategically creating social ties [38].

Imagine that an individual tries to embed herself at the center of a social network through forming new ties. From a structural perspective, this individual

© Springer International Publishing AG 2017
A. Bouguettaya et al. (Eds.): WISE 2017, Part I, LNCS 10569, pp. 75–90, 2017.
DOI: 10.1007/978-3-319-68783-4_6

needs to choose a set of members to build links with. Here we put aside issues such as attitude, personality, and individual preferences, and focus on a structural perspective of network building. To this end, the individual may adopt an *exploitative* or an *exploratory* strategy: The former ensures that the individual exploits existing interpersonal ties and links to those that share a common social proximity; On the contrary, the latter allows the individual to explore far and bridge diverse parts of the network. A natural question arises as to which strategy is more suitable. Moreover, social networks in real life are rarely static, but rather, they constantly evolve with time. Thus the question has an extra layer of complexity: *How to incrementally build relationships in a network to gain positional advantage while the network is evolving?*

To attempt this question, we should settle several issues: Firstly, we need a notion that reasonably reflects positional advantage; here centrality metrics may be of use. Secondly, relation building costs time and effort; one needs to quantify such costs. Thirdly, one needs models on how a social network evolves.

Contribution. We list the main contributions of the paper:

1. In this paper, we propose the problem of *dynamic network building (DNB)*. The input to the problem consists of a connected graph G that undergoes a sequence of updates. The problem asks for a plan that builds edges incrementally between a node v and other nodes so that v gains centrality as G evolves. (See Sect. 2).
2. To solve this problem, we define exploitative and exploratory strategies and present heuristics to realize each strategy. (See Sect. 3).
3. We compare the heuristics over various evolution models of social networks and real-world networks. Exploration often builds less number of new links, while the exploitative strategy produces better results when other factors, such as distance and embeddedness is considered (See Sect. 4).
4. Lastly, we propose and evaluate ways that combine the exploitative and exploratory strategies (See Sect. 5).

This work is meaningful in the following ways: Firstly, the process of socialization has been studied intensively in social sciences [21,28,32]. Through formalizing and analyzing mechanisms of network building with respect to distance, embeddedness and clustering, the work quantitatively reveals fundamental insights in this otherwise rather qualitative problem domain. Secondly, the exploration-exploitation tradeoff has been a recurring theme in artificial intelligence and management science [2,6,34]. This work discovers an instance of this tradeoff in the context of social networks. Thirdly, the work opens the door to many novel applications from engineering information channels on career-based online social networks to enhancing workplace communication and collaboration through enterprise management systems.

Related Works. The establishment of interpersonal ties has been a major problem in social network analysis. Granovetter's pioneering work contrasts ties having high embeddedness (strong ties) with ties that bridge two otherwise disjoint social circles (weak ties); while embeddedness reflects important dimensions such as trust, commitment and solidarity [18], bridges are important to the exchange of knowledge and ideas [16]. We extends this discussion to study strategies for building different types of ties. Network building (NB) has been studied in [29–31]; The problem studied in this paper has crucial differences to these works: (a) While NB only operates on static networks, here we focus on evolving networks, which demand the node to be strategic towards future changes. (b) While NB focuses on smallest eccentricity, DNB aims for optimal rank on centrality. (c) DNB considers costs incurred from the distance between the two nodes when forming an edge. A large literature on *strategic network formation* explains tie establishment between rational agents using game theory; these works do not consider stochastic models of network evolution [20]. The general view that interpersonal ties bring social support and cohesion has been discussed in [24,26]. Lastly, exploratory-exploitative strategies discussed in this paper parallel the two modes of network formation in [21]; there, "meeting strangers" means exploratory encounters in the network, and "meeting friends-of-friends" means exploiting existing social circles.

2   The Dynamic Network Building Problem

Following standard convention, we view a *social network* as a graph $G = (V, E)$ where V is a set of nodes and E is a set of undirected edges on V of the form uv where $u \neq v \in V$. Here the undirected edges are abstracted models of channels of information or transactions. The set $\Gamma(u) = \{v \mid uv \in E\}$ denotes the *neighborhood* of u, consisting of all nodes that are *adjacent* to u. A *path* (of *length k*) is a sequence of nodes $u_0, u_1, \ldots, u_k$ where $u_i u_{i+1} \in E$ for any $0 \leq i < k$. The (geodesic) *distance* between u and v, denoted by $\mathsf{dist}_G(u, v)$, is the length of a shortest path between u and v. We omit the subscript G writing simply $\mathsf{dist}(u, v)$ when the underlying graph is clear. We also need the following formalism:

- For a node $s \in V$ and $v \neq s$, denote by $G \oplus_s v$ the expanded network $(V \cup \{v\}, E \cup \{sv\})$ obtained by adding sv to G.
- We assume that the social network G evolves by some (discrete-time) stochastic mechanism, which we define below:

Definition 1. *An evolution mechanism M is a function that maps a social network G to a probability distribution of social networks $M(G)$. Starting at G, the network evolves to a sample outcome of $M(G)$ in the next time step.*

Imagine v is a node who wants to build relationships in G; Let's call v the *newcomer* in this paper. We assume that (1) v is a node with few connections in G; (2) v may create edges from itself to nodes in V by paying costs (see below); and (3) v has no knowledge regarding how G may evolve.

Abstractly, one can view the interactions between v and the social network G as a two-player game: v is one player; the other player represents the evolution mechanism of the network. At each round, v creates an edge with a node in G (keeping all existing edges). For simplicity, we assume each round allows v to create at most one new edge. One may easily generalize this setting to allow v to create several edges in a single round. The evolution mechanism then modifies the updated network. Through multiple rounds, v aims to get increasingly integrated into the network. Note that we assume that v functions independently from the evolution mechanism to highlight that v builds edges without prior knowledge of the network evolution mechanism.

Definition 2. *An (ℓ-round) network building (NB) process between v and G consists of a sequence of networks $G_0 = (V_0, E_0), G_1 = (V_1, E_1), \ldots, G_\ell = (V_\ell, E_\ell)$ and a sequence of nodes $s_0 \in V_0, s_1 \in V_1, \ldots, s_{\ell-1} \in V_{\ell-1}$ such that $G_0 = G$ and each network G_{i+1} is a sample output of $M(G_i \oplus_{s_i} v)$.*

Definition 3. *An NB strategy is a function φ that outputs a node $\varphi(G)$ in a given network G. Any NB process $(G_0, \ldots, G_\ell, s_0, \ldots, s_{\ell-1})$ is said to be consistent with strategy φ if $\forall 0 \leq i < \ell \colon s_i = \varphi(G_i)$.*

Closeness centrality amounts to an important index of social capital that captures a node's ease in accessing information, social support and other resources [1,13,37]. Thus we use closeness centrality here to indicate the positional advantage of nodes. For any connected $G = (V, E)$ and $v \in V$, define

$$C_{\mathsf{Cls}}(v) = \frac{|V| - 1}{\sum_{u \in V \setminus \{v\}} \mathsf{dist}(u, v)}.$$

A higher value of $C_{\mathsf{Cls}}(v)$ implies that v is in general closer to other nodes, thus it occupies a better network position. The Cls-*rank* of v is the percentage of nodes whose closeness centrality are higher or equal to $C_{\mathsf{Cls}}(v)$:

$$\mathsf{rank}_{\mathsf{Cls}}(v) = |\{u \in V \mid C_{\mathsf{Cls}}(u) \geq C_{\mathsf{Cls}}(v)\}| / |V|.$$

We assume that the goal of v is to gain a higher closeness centrality (or a low $\mathsf{rank}_{\mathsf{Cls}}$). One way to achieve this is to build a tie between v and all nodes in the network. However, establishing new relationships requires time, efforts and resources. To identify realistic solutions, one needs to define *costs* of relationship building. Here we consider temporal and establishment costs. Temporal cost is the number of rounds in the NB process and coincides with the number of edges created for v. The *proximity principle* states that ties are generally more difficult to establish between nodes that are further apart (e.g. reciprocal of distance is a score for link prediction [27]). We thus define establishment cost as the sum of distance between v and its linked nodes (prior to edge creation).

Definition 4. *Let $(G_0, G_1, \ldots, G_\ell, s_0, s_1, \ldots, s_{\ell-1})$ be an NB process. We define the following costs:*

1. *The temporal cost is ℓ.*
2. *The establishment cost is $\sum_{i=1}^{\ell-1} \sum_{u \in S_i} \mathsf{dist}_{G_i}(v, u)$.*

We are now ready to present the *dynamic network building (DNB) problem*: Given a connected social network G and newcomer v, the problem asks for an NB strategy φ such that any NB process consistent with φ will have high $C_{\mathsf{Cls}}(v)$ (or small $\mathsf{rank}_{\mathsf{Cls}}(v)$) value, and low temporal and and establishment costs. Note that due to restriction to unary NB strategies, the temporal cost coincides with the number of edges added to the newcomer v. Figure 1 displays a simple example where the graph evolves with the dynamic BA mechanism (see below); the newcomer gains a high centrality in three rounds.

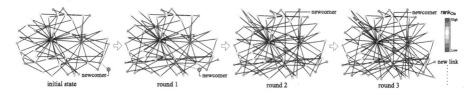

Fig. 1. A newcomer adds one edge at each round and achieves high centrality after 3 rounds, while the network is evolving.

The DNB problem differs from building relations in a static networks, which has been discussed in [29–31]: (1) As the network evolves, the NB process may last indefinitely where v tries to improves and maintains its centrality; (2) Network evolution forces v to balance between current knowledge with future predicted outcome. For example, linking to a central node will improve v's centrality quickly, but also incurs a high cost; on the other hand, linking to a low-centrality node may seem undesirable in the current network, but this link may improve the newcomer's centrality in the future. In this way, the evolution mechanism significantly impacts the newcomer's strategy.

3   Exploratory and Exploitative Strategies

Exploitative Strategy. This strategy utilizes existing social proximity of the node v and searches for the most promising node that lies within a pre-defined distance from v. Fix as parameter a centrality index $C_* : V \to \mathbb{R}$ for nodes. Let $d \geq 2$ be a *proximity threshold*. When creating an edge, the strategy traverses through nodes with distance $\leq d$ from v, and picks a node found with maximum C_* value. Procedure 1 defines one round of the exploitative strategy. For the choice of C_*, we use standard centrality metrics that reflect aspects of social capital. The variety of centrality metrics below allow us to examine different potential heuristics, which may not always correlate [7].

1. *Degree*: $C_{\mathsf{Deg}}(u) = \{w \mid uw \in E\}$.
2. *Betweenness*: $C_{\mathsf{Btw}}(u) = \sum_{s \neq u \neq t \in V} |P_{st}(u)|/|P_{st}|$ where P_{st} is the set of shortest paths between s and t, $P_{st}(u) \subseteq P_{st}$ is those shortest paths that contain u.

3. *Closeness*: $C_{Cls}(u)$.

We denote using Ld-Deg, Ld-Btw and Ld-Cls the local heuristics with centrality C_{Deg}, C_{Btw}, C_{Cls}, resp. When $d = 2$, the newcomer always links to a "friend-of-friends", a strategy studied in [21].

Procedure 1. Ld-*(G, v): Given $G = (V, E)$, $v \in V$

$U \leftarrow \{u \mid \mathsf{dist}(u, v) \le d\}$
return a node $u \in U$ with maximum $C_*(u)$

Exploratory Strategy. This strategy explores beyond the social proximity of v and links v with promising nodes in potentially distant parts of the network: The strategy takes a centrality index $C_* : V \to \mathbb{R}$ and a *distance threshold* $\gamma \in \mathbb{N}$ as parameters. Call all nodes within distance γ from v *covered*; at each round, the strategy will pick an uncovered node that has maximum C_*-value. Procedure 2 describes a single round of this strategy. We use Gγ-Deg, Gγ-Btw and Gγ-Cls to denote the exploratory heuristic that use C_{Deg}, C_{Btw} and C_{Cls}, respectively.

Since exploration creates edges to nodes that may be far away from v, this strategy by definition bridges different parts of the network more quickly than exploitation. Indeed, this can be verified using a simple example: Fix a large natural number n. Consider the *path graph* L_{2n+1} with $2n + 1$ nodes (with nodes $v_1, v_2, \ldots, v_{2n+1}$ and edges $v_1v_2, v_2v_3, \ldots, v_{2n}v_{2n+1}$). Say v_1 is the newcomer. Assuming the network is static, G2-Cls builds $O(1)$ edges from v_1 (e.g. to $v_{n+1}, v_{n-2}, v_{n+2}$) and gives v_1 the highest closeness centrality. On the contrary, the exploitative strategy will create $\Omega(n)$ new edges to have the highest closeness centrality. In the next section, we compare the two strategies above through experiments on standard network evolution mechanisms and real-world data.

Procedure 2. Gd-*(G, v): Given $G = (V, E)$, $v \in V$

Find maximum $\gamma' \le \gamma$ with $V \ne \{u \mid \mathsf{dist}(v, u) \le \gamma'\}$
$U \leftarrow V \setminus \{u \in V \mid \mathsf{dist}(v, u) \le \gamma'\}$
return a node $u \in U$ with maximum $C_*(u)$

---

4   Contrasting Exploitative and Exploratory Strategies

4.1   Network Evolution Mechanisms

We consider three standard *network formation models*. Originally, each of these model were used to generated static networks. Here we extend them so that they entail mechanisms for network generation and evolution.

(a) **Dynamic ER Model.** The Erdös-Renyi (ER) random graph model adds edges between nodes as Bernoulli random variables with probability p. The degree distribution in the resulting graph thus follows a binomial distribution $B(n-1,p)$ [8]. We extend the model to a *death-birth* evolution model: Start from an ER random graph and introduce parameter $r \in [0,1]$. At each round, first remove a randomly chosen set of nodes of size rn (i.e., death); then, add rn new nodes and link them with nodes in the graph with probability p (i.e., birth). It is clear that the operation preserves the binomial degree distribution $B(n-1,p)$.

(b) **Dynamic BA Model.** The Barabási-Albert model generates scale-free networks through a preferential attachment mechanism [4]. To define an evolving network model, we follow Barabási's dynamic extension. The key ideas include *growth, link establishment* and *node deletion* [3]. (a) Growth takes a rate $g \in [0,1]$ and adds gn new nodes; adds m edges from each new node to an existing node with probability $k_i / \sum_{v_j \in V} k_j$ where k_j is the degree of v_j, $\forall v_j \in V$. (b) Link establishment takes a parameter $\lambda \in [0,1]$; selects λn pairs of nodes in the current graph and randomly creates edges between these pairs; the probability of creating edge $v_i v_j$ is proportional to $k_i k_j$. (c) Node deletion takes a rate $d \in [0,1]$; picks dn nodes, and remove each of them (say, v_i) with probability $\frac{1/k_i}{\sum_j 1/k_j}$.

(c) **Dynamic WS-Model.** The Watts-Strogatz model starts from a regular lattice, and performs random edge rewirings (with probability β) to obtain small-world networks, which have high levels of clustering and low average path length [39]. We extend the process to an evolution mechanism: after initialization, the network evolves in each round by rewiring those previously-rewired edges to a random node. This dynamic network preserves the small-world property.

Table 1 summarizes the parameters used in our experiments. We choose these values either because they are standard choices used by others (e.g. $m = 2$ for dynamic BA [4]), or they ensure a gradual and smooth change at each round (e.g. for dynamic ER and WS models).

Table 1. Parameters for evolving network models.

Evolving Model	Parameters
Dynamic ER	$r = 5\%, np = 4$
Dynamic BA	$m = 2, \lambda = 4\%, g = 5\%, d = 2\%$
Dynamic WS	$\beta = 0.2$

Experiment 1 (Costs). Through simulating DNB processes, we compare the temporal and establishment costs between the exploitative and exploratory strategies. DNB processes are generated by applying the heuristics in conjunction with the ER, BA and WS models. As a DNB process may have indefinite length, we need a termination condition to specify when the simulation stops.

A natural method is to set a (high) threshold on centrality $C_{\mathsf{Cls}}(v)$, or set a (small) threshold on $\mathsf{rank}_{\mathsf{Cls}}(v)$, such that the process terminates once the threshold is met. There are problems with this approach: (1) It is difficult to determine a desired $C_{\mathsf{Cls}}(v)$ that facilitates fair comparisons across all evolving models. (2) In certain cases (e.g. WS model), closeness centrality of nodes are distributed within a small range; Hence, a node with low centrality may still have a small $\mathsf{rank}_{\mathsf{Cls}}$. These concerns motivate us to set a termination condition based on the ratio $g(v) = C_{\mathsf{Cls}}(v)/\mathsf{rank}_{\mathsf{Cls}}(v)$; we introduce a threshold ζ such that the process terminates at the first round when $g(v) \geq \zeta$ is satisfied.

We generate 10 networks of each size $n = 100, 200, 500, 1000$ using any network models above. We compare the exploitative with the exploratory strategies by running L2-$*$, L3-$*$ and G2-$*$ heuristic on each graph. Note that L2-$*$ only links v to her "friends-of-friends" which amounts to the most "local" exploitative strategy; L3-$*$ reaches beyond this local proximity and results in a different performance (see below); G2-$*$ is an exploratory strategy which tries its to link v to nodes outside of her local proximity. We do not include results for G3-$*$ as they are very similar to G2-$*$. For any generated graph G and heuristic, we do the following: (1) Build an edge between the newcomer v and a randomly chosen node in G. (Here we run the same experiment while linking v with initial nodes of different Cls-rank (10%–90%). The resulting temporal and establishment costs are very similar. This shows that the rank of the initial node does not significantly affect the performance of the strategies.) (2) Apply the heuristic to the evolution mechanism (corresponding to the network formation model) to generate 100 DNB processes; the threshold ζ is 33. Here, the value $\zeta = 33$ reflects the fact that the desired $C_{\mathsf{Cls}}(v) \approx 1/3$ and $\mathsf{rank}_{\mathsf{Cls}}(v) \approx 1\%$. Other outcomes with $g(v) = 33$ that have either considerably lower centrality and Cls-rank, or considerably higher centrality and Cls-rank (e.g. $(C_{\mathsf{Cls}}(v), \mathsf{rank}_{\mathsf{Cls}}(v))$ is $(1/6, 0.5\%)$ or $(1, 3\%)$) have been empirically shown to be unlikely. (3) After the DNB process terminates, measure the resulting temporal and establishment costs. (4) Finally, record the average costs among all DNB processes of the same evolving network model, initial network size, and heuristic.

A few facts stand out from the results in Fig. 2: (i) Temporal costs are mostly below 10, suggesting that a small number of edges are built by the strategies, even when the size of the network becomes 1000. (ii) For ER and WS, exploration results in a lower temporal cost compared to the exploitation. (iii) For the scale-free model BA, the number of edges built decreases as n increases. This may be due to the expanding nature of the dynamic BA model and the skewed degree distribution. As a result, exploitation creates less or similar numbers of edges than exploration. (iv) In general, exploration results in a much higher establishment costs. This is easy to understand: the strategy links to distant nodes from v. The L3 heuristics builds less edges than L2 due to the ability to traverse to a wider part of the graph. (v) The effects of the centrality metrics C_* vary with graph models: For ER, closeness centrality is in general preferred, while for BA, degree centrality is slightly more preferred. For WS, C_{Btw} is better for exploration, while C_{Cls} is better for exploitation as the graph becomes large.

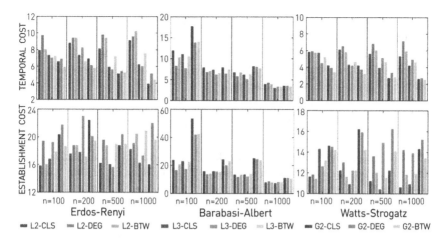

Fig. 2. Average temporal and establishment costs of simulated DNB processes by different heuristics on networks of different sizes.

We then plot v's Cls-rank as new edges are created during the DNB process; See Fig. 3. $\mathrm{rank}_{\mathsf{Cls}}(v)$ reaches $<1\%$ after 10 rounds under all heuristics. G2 improves Cls-rank faster than L2 heuristics. This is most evident for L2-Deg and L2-Cls in WS-graphs: the high level of clustering may make it hard for exploitation to get out of a dense cluster, but clustering does not seem to pose a problem for L2-Btw.

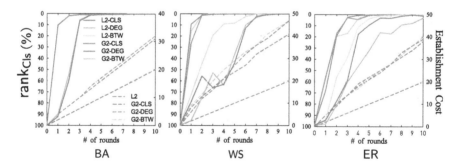

Fig. 3. Changes to $\mathrm{rank}_{\mathsf{Cls}}(v)$ (solid lines) and the establishment costs (dashed lines) during 10 rounds of the DNB process. The network starts with 1000 nodes and v initially connects to a node with the lowest centrality.

Experiment 2 (Embeddedness and clustering). Experiment 1 implies that, to some extend, exploitation and exploration perform on par with each other from a centrality perspective. In building relationship, trust, tie strength and role integrity are other important dimensions not captured by centrality alone [17]. These notions are closely affected by two concepts:

1. *Embeddedness* in a social network refers to the degree to which an individual is constrained by social relationships and is often viewed as a platform for trust [18]. The embeddedness of an edge between x, y is defined as the size of their shared neighborhoods $D(x) \cap D(y)$ [12]. We define $\mathsf{embed}(v)$ as a normalized sum of embeddedness:

$$\mathsf{embed}(v) = \frac{\sum_{vu \in E} |\{w \in V \mid wu, wv \in E\}|}{(|V| - 1)(|V| - 2)}.$$

Note that the highest $\mathsf{embed}(v)$ is 1 when v is a node in a complete graph.

2. *Clustering coefficient* of a node measures the probability of two randomly chosen friends of the node are also friends and relates to the self-identify of individuals [5]:

$$\mathsf{cc}(v) = \frac{2 \cdot |\{uw \in E \mid uv \in E, vw \in E\}|}{\deg(v)(\deg(v) - 1)}.$$

We measure $\mathsf{embed}(v)$ and $\mathsf{cc}(v)$ during DNB processes; See Fig. 4. Remarkably, $\mathsf{embed}(v)$ increases drastically with exploitation, while staying close to 0 with exploration. The clustering coefficient $\mathsf{cc}(v)$ also stays close to 0 with exploration, while with exploitation, it quickly rises to a very high level in the first few rounds, and drops down to below 0.1 after 10–15 rounds. This highlights the newcomer's ability to cut across different clusters. Overall, this experiment demonstrates the crucial difference between the strategies: while both strategies improve v's closeness centrality, exploitation enables a higher embeddedness and clustering coefficient which positively correlates with tie strength and trusts on its social relations.

4.2   Real-World Evolving Networks

We next take real-world evolving network data as case studies.

Experiment 3 (Contact networks). We take two evolving physical contact networks: The first data set records face-to-face contacts among roughly 110 attendees of ACM Hypertext 2009 conference during a 2.5-day period [19]. The second records contacts among roughly 100 employees in a French workplace June 24 to July 3, 2013 [14]. In both data sets, contacts are updated every 20 seconds. We ask the question: If a newcomer attends the conference, or joins the workplace, how does she utilize face-to-face contacts to reach a central position of the network?

We simulate DNB processes on the *accumulated network*, i.e., networks constructed by accumulating edges in previous times, using the G2 and L2 heuristics. For the first data set, an edge is built every 15 min, and for the second data set, an edge is built very hour. Figure 5 plots the changes on the newcomer's Cls-rank. All heuristics have similar results in terms of improving the Cls-rank as time progresses. As the networks are relatively small and edges are accumulated, the heuristics somehow fail to improve newcomer's rank in the last three days of

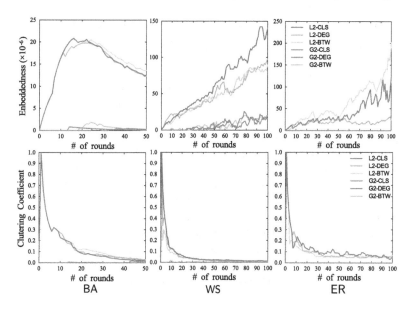

Fig. 4. Changes to v's embeddedness and cc as edges are added to v. The network starts with 1000 nodes where v is connected to a node with the lowest centrality.

the second network beyond 20%. The exploitative strategy, however, gives much smaller establishment costs. On the other hand, exploitation leads to considerably higher embeddedness and clustering coefficient than exploration.

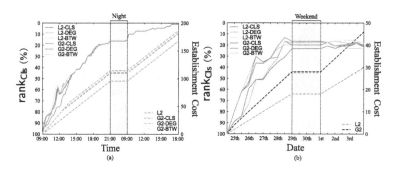

Fig. 5. The Cls-rank and establishment cost resulted from running the heuristics on (a) ACM Hypertext 2009 contact network; (b) French workplace contact network.

5   Balancing Exploitation with Exploration with UBC

As demonstrated above, exploration usually brings faster improvement to the newcomer v's centrality, while exploitation by definition incurs less establishment cost. The trade-off between the two strategies is further complicated by the evolving nature of the network (e.g. exploitation exhibits lower temporal costs than exploration in general on scale-free networks). As in many scenarios of reinforcement learning, to obtain an optimal solution one needs to strike a balance between exploitation and exploration. In this section, we adopt *upper confidence bound* (UBC), a well-known reinforcement learning method for resolving the exploitation-exploration dilemma to build relations by integrating the two strategies.

We adapt the UCB1 algorithm proposed in [2] which has a guaranteed logarithm regret uniformly over the number of rounds. Here, exploitative and exploratory strategies are regarded as two actions, 0 and 1, respectively and dynamic network building is viewed as a *2-armed bandit problem*. In each round of the DNB process, v evaluates plausible Cls-rank achieved through performing each action $i \in \{0,1\}$ using past experience, and then selects the strategy that seems to be the best. To estimate the plausible mean Cls-rank of a strategy, we introduce the following notations: Let $(G_0, G_1, \ldots, G_t, s_0, s_1, \ldots, s_{t-1})$ be a t-round DNB process. For $m < \ell$, we use $\mathsf{rank}(m)$ to denote the Cls-rank of v in the graph G_{m+1}. The estimate on the expected Cls-rank resulted by choosing action $i \in \{0,1\}$ at round $\ell+1$ is defined by

$$\Upsilon(i) = \underbrace{\frac{\sum_{j=1}^{n_i} \mathsf{rank}(r_{ij}-1) - \mathsf{rank}(r_{ij})}{n_i}}_{average\ function} + \underbrace{\sqrt{\frac{2\ln t}{n_i}}}_{padding\ function} \qquad (1)$$

where t is the number of rounds passed, n_i is the times that strategy i is selected, r_{ij} is the jth round that action i is selected. We use the difference of v's Cls-rank between two contiguous rounds to define the reward that v receives after each round. Then *Average function* denotes average reward that v has got so far by choosing action i. *Padding function* denotes an estimated uncertainty on i; the more times that i is selected, the less uncertainty it has.

At the beginning of a DNB process, we apply exploration and exploitation in the first and the second round, resp. Then, in subsequent rounds, we select the strategy that achieves the highest $\Upsilon(i)$ value (as (1)). See Procedure 3, which we call DNB_UCB. Note that the algorithm again requires fixing a centrality measure C_* and a distance threshold $\gamma \in \mathbb{N}$ as parameters as for L and G heuristics.

Procedure 3. DNB_UCBγ-*(G,v): Given $G = (V,E)$, $v \in V$

  Initialization: Select Lγ-* and Gγ-* at round 0 and round 1 resp.
  if $\Upsilon(\mathsf{L}\gamma\text{-}*(G,v)) \geq \Upsilon(\mathsf{G}\gamma\text{-}*(G,v))$ **then**
      Find maximum $\gamma' \leq \gamma$ with $V \neq \{u \in V \mid \mathsf{dist}(v,u) \leq \gamma'\}$
      $U \leftarrow V \setminus \{u \in V \mid \mathsf{dist}(v,u) \leq \gamma'\}$
      return a node $u \in U$ with maximum $C_*(u)$
  else
      $U \leftarrow \{u \in V \mid \mathsf{dist}(u,v) \leq \gamma\}$
      return a node $u \in U$ with maximum $C_*(u)$
  end if

We implement the algorithm and evaluate its performance over dynamic network models. At the beginning, all networks have 1000 nodes, and the parameters used by the evolution mechanisms are the same as in Sect. 4. Figure 6 plots the change in Cls-rank as well as the resulting establishment costs after running both the pure (exploratory or exploitative) strategies, and the mixed strategy during a fixed number of rounds. In all three types of random networks, the speed that DNB_UCB improves Cls-rank sits between the exploitative and exploratory strategies. The improvement is most evident in WS where DNB_UCB performs similarly well to exploration. On the other hand, DNB_UCB also results in a significant reduction of the establishment costs than exploratory strategy.

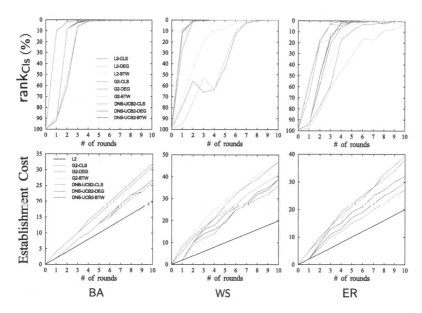

Fig. 6. The Cls-rank and establishment cost resulted from running the pure and mixed strategies on the dynamic BA, WS and ER networks. The speed that DNB_UCB improves Cls-rank sits between the exploitative and exploratory strategies. On the other hand, DNB_UCB significantly reduces the establishment costs than exploration alone.

6   Conclusion and Future Work

The paper proposes dynamic network building problem and concentrates on exploratory-exploitative strategies and their combinations. The focus is on introducing the algorithmic framework: Many methods exist for the multi-armed bandit problem; A natural future work is to compare these methods with DNB_UCB. Furthermore, despite closeness centrality's importance in capturing social support and access to resources, other measures of social capital, e.g. betweenness, eigenvector centrality, can be used as a goal for the newcomer v as well.

Other future works that we will consider include: (1) designing strategies that build relations while respecting bounded capacity of individuals, i.e., v has only a bounded number of links so severing ties may be needed. (2) investigating network building on attributed networks where pairs of nodes have different likelihood of being connected due to features such as personality, common interests, etc. (3) Studying information diffusion and influence maximization through network building. (4) A crucial future work is on applications of the presented framework. In particular, the work develops a foundation of new functionality on the Web that engineers optimal links among OSN users to enhance communication and capability. Here, a possible scenario is to recommend information agencies to a user on a career-based OSN to boost their access to resources and enhance career-prospect. Another possible scenario is to provide social support to those that are in need through social network building [40].

References

1. Agneessens, F., Borgatti, S.P., Everett, M.G.: Geodesic based centrality: unifying the local and the global. Soc. Netw. **49**, 12–26 (2017)
2. Auer, P., Cesa-Bianchi, N., Fischer, P.: Finite-time analysis of the multiarmed bandit problem. Mach. Learn. **47**(2–3), 235–256 (2002)
3. Barabási, A.-L.: Network Science. Cambridge University Press, Cambridge (2016)
4. Barabási, A.-L., Albert, R.: Emergence of scaling in random networks. Science **286**(5439), 509–512 (1999)
5. Bearman, P.S., Moody, J.: Suicide and friendships among American adolescents. Am. J. Public Health **94**(1), 89–95 (2004)
6. Benner, M., Tushman, M.: Exploitation, exploration, and process management: the productivity dilemma revisited. Acad. Manag. Rev. **28**(2), 238–256 (2003)
7. Bolland, J.: Sorting out centrality: an analysis of the performance of four centrality models in real and simulated networks. Soc. Netw. **10**, 233–253 (1988)
8. Bollobás, B.: Random Graphs, 2nd edn. Cambridge University Press, Cambridge (2001)
9. Bonacich, P.: Power and centrality: a family of measures. Am. J. Sociol. **92**, 1170–1182 (1987)
10. Borgatti, S.: Identifying sets of key players in a social network. Comput. Math. Organ. Theory **12**(1), 21–34 (2006)
11. Burt, R.: Brokerage and Closure: An Introduction to Social Capital. Oxford University Press, Oxford (2005)
12. Easley, D., Kleinberg, J.: Networks, Crowds, and Markets: Reasoning About a Highly Connected World. Cambridge University Press, Cambridge (2010)
13. Freeman, L.: Centrality in social networks: conceptual clarification. Soc. Netw. **1**, 215–239 (1979)
14. Génois, M., Vestergaard, C., Fournet, J., Panisson, A., Bonmarin, I., Barrat, A.: Data on face-to-face contacts in an office building suggest a low-cost vaccination strategy based on community linkers. Netw. Sci. **3**(3), 326–347 (2015)
15. Gleditsch, K.S.: Expanded trade and GDP data. J. Conflict Resolut. **46**(5), 712–724 (2002)
16. Granovetter, M.: The strength of weak ties. Am. J. Sociol. **78**(6), 1360–1380 (1973)

17. Granovetter, M.: Economic action and social structure: the problem of embeddedness. Am. J. Sociol. **91**(3), 481–510 (1985)
18. Gulati, R.: Alliances and networks. Strateg. Manag. J. **19**, 293–317 (1998)
19. Isella, L., Stehlé, J., Barrat, A., Cattuto, C., Pinton, J.-F., den Broeck, W.V.: What in a crowd? Analysis of face-to-face behavioral networks. J. Theoret. Biol. **271**, 166–180 (2011)
20. Jackson, M.: A Survey of Models of Network Formation: Stability and Efficiency. Group Formation in Economics. Cambridge University Press, Cambridge (2005). pp. 11–57
21. Jackson, M., Rogers, B.: Meeting strangers and friends of friends: how random are social networks? Am. Econ. Rev. **97**(3), 890–915 (2007)
22. Jackson, M.O., Nei, S.: Networks of military alliances, wars, and international trade. Proc. Nat. Acad. Sci. USA **112**(50), 15277–15284 (2015)
23. Lee, S.H., Cotte, J., Noseworthy, T.J.: The role of network centrality in the flow of consumer influence. J. Consum. Psychol. **20**(1), 66–77 (2010)
24. Liu, J., Li, L., Russell, K.: What becomes of the broken hearted? An agent-based approach to self-evaluation, interpersonal loss, and suicide ideation. In: Proceedings of the 16th Conference on Autonomous Agents and MultiAgent Systems, pp. 436–445. International Foundation for Autonomous Agents and Multiagent Systems (2017)
25. Liu, J., Moskvina, A.: Hierarchies, ties and power in organizational networks: model and analysis. Soci. Netw. Anal. Min. **6**(1), 106:1–106:26 (2016)
26. Liu, J., Wei, Z.: Network, popularity and social cohesion: a game-theoretic approach. In: AAAI, pp. 600–606 (2017)
27. Lü, L., Zhou, T.: Link prediction in complex networks: a survey. Physica A **390**(6), 1150–1170 (2011)
28. Morrison, E.: Newcomers' relationships: the role of social network ties during socialization. Acad. Manag. J. **45**(6), 1149–1160 (2002)
29. Moskvina, A., Liu, J.: How to build your network? A structural analysis. In: Proceedings of the IJCAI 2016, pp. 2597–2603. AAAI (2016)
30. Moskvina, A., Liu, J.: Integrating networks of equipotent nodes. In: Nguyen, H.T.T., Snasel, V. (eds.) CSoNet 2016. LNCS, vol. 9795, pp. 39–50. Springer, Cham (2016). doi:10.1007/978-3-319-42345-6_4
31. Moskvina, A., Liu, J.: Togetherness: an algorithmic approach to network integration. In: 2016 IEEE/ACM International Conference on Advances in Social Networks Analysis and Mining (ASONAM), pp. 223–230. IEEE (2016)
32. Newman, M.: The structure of scientific collaboration networks. Proc. Natl. Acad. Sci. USA **98**, 404–409 (2001)
33. Newman, M.: Networks: An Introduction. Oxford University Press, Oxford (2010)
34. Osugi, T., Deng, K., Scott, S.: Balancing exploration and exploitation: a new algorithm for active machine learning. In: Proceedinsg of the ICDM 2005, pp. 330–337 (2005)
35. Padgett, J.F., Ansell, C.K.: Robust action and the rise of the medici, 1400–1434. Am. J. Sociol. **98**(6), 1259–1319 (1993)
36. Pitts, F.R.: The medieval river trade network of Russia revisited. Soc. Netw. **1**(3), 285–292 (1978)
37. Sparrowe, R.T., Liden, R.C., Wayne, S.J., Kraimer, M.L.: Social networks and the performance of individuals and groups. Acad. Manag. J. **44**, 316–325 (2001)
38. Uzzi, B., Dunlap, S.: How to build your network. Harvard Bus. Rev. **83**(12), 53–60 (2005)

39. Watts, D.J., Strogatz, S.H.: Collective dynamics of 'small-world' networks. Nature **393**(6684), 440–442 (1998)
40. Yadav, A., Chan, H., Jiang, A.X., Xu, H., Rice, E., Tambe, M.: Using social networks to aid homeless shelters: dynamic influence maximization under uncertainty. In: Proceedings of the AAMAS 2016, pp. 740–748 (2016)

Social Personalized Ranking Embedding for Next POI Recommendation

Yan Long[1], Pengpeng Zhao[1]([✉]), Victor S. Sheng[2], Guanfeng Liu[1], Jiajie Xu[1], Jian Wu[1], and Zhiming Cui[1]

[1] Department of Computer Science and Technology,
Soochow University, Suzhou, China
ppzhao@suda.edu.cn
[2] Computer Science Department, University of Central Arkansas, Conway, USA

Abstract. As the increasing popularity of the applications of location-based services, points-of-interest (POI) recommendation has become a great value part to help users explore their surrounding living environment and improve the quality of life. Recently, some researchers proposed next POI recommendation, which not only exploiting the users personal interests but also considers the sequential information of users check-ins. There are some next POI recommendation models exploit Metric Embedding method to improve recommendation performance and efficiency. However, these approaches not consider social relations in next POI recommendation, which is challenging due to social relations are noisy and sparse. To this end, in this paper, we proposed a Social Personalized Ranking Embedding (SPRE) model, which integrates user personalization and social relations into consideration, to learn the social relations by social embedding for next POI recommendation. Our experiments on a real-world large-scale dataset (Foursquare) results show that our model outperforms the state-of-the-art next POI recommendation methods.

Keywords: Next POI recommendation · Metric embedding · Social relations influence

1 Introduction

As the rapid increasing popularity of the applications of location-based services, location-based social networks (LBSNs), such as Foursquare, Facebook Places, Gowalla and Yelp, have attracted the large amount of users to check in at points-of-interest (POIs), e.g., bars, restaurants and sighting sites, and share their experience of visiting these POIs with friends. POI recommendation is has a great value to help users explore their surrounding living environment and improve the quality of life, which has attracted a significant amount of research interest in developing recommendation techniques [2,4]. Recently, there are a lot of recommendation models that were put forward for POI recommendation by developing and integrating geographical influence [21], social influence [6],

© Springer International Publishing AG 2017
A. Bouguettaya et al. (Eds.): WISE 2017, Part I, LNCS 10569, pp. 91–105, 2017.
DOI: 10.1007/978-3-319-68783-4_7

context influence [19], the effect of temporal cyclic information [7,25] and their joint effect [13,22,24].

Some researchers proposed a natural extension of general POI recommendation, i.e., next POI recommendation [3]. Comparing to POI recommendation, next POI recommendation has more challenges and fewer studies are on this new problem. In addition to the user's personal interests, the next POI recommendation additionally considers the sequential information of users' check-ins. On the one hand, there are some researchers proposed Markov chain-based recommender model [26] to capture the sequential patterns of POIs for next POI recommendation. On the other hand, some researchers proposed a hybrid model [3] for next POI recommendation by combining temporal cyclic effect, social relations influence and others. However, because of the sparseness of the data, it is difficult for Markov chain-based models and others to accurately and effectively estimate the probability of visiting next POI for users. Recently, many various recommender models utilize the social relations influence to improve the recommended accuracy. However, social links are also sparse and noisy [27]. The traditional recommendation methods (such as MF and Markov chain) of recommended accuracy will be hurt since the sparsity and noise.

Recently, embedding the item in the low-dimensional Euclidean space has been widely used in various fields, especially in natural language processing and data mining. Embedding model is often used to deal with some sparse data and mine the data that has not been observed. Tang et al. [18] proposed a Large-scale Information Network Embedding model (LINE), which suits arbitrary types of information networks and easily scales to millions of nodes by using the metric embedding. Feng et al. [5] proposed a Personalized Ranking Metric Embedding (PRME) model to map each user and POI to some point in a latent Euclidean space for next new POI recommendation. However, they did not take into account some influences, such as social relations influence, context influence and temporal cyclic information, into the model.

In this paper, we proposed a new embedding model by embedding the social relations and user preference for next POI recommendation. In the real-world social relations graph, some users have no direct links, but their social network structure is similar, such as users 5 and 6 in Fig. 1. But the implicit relationships are ignored by the existing methods. In our model, we map the per-user to an object in a low-dimensional Euclidean latent space, and use the metric embedding algorithm to effectively calculate the social relationships. By social embedding method, we can be the implicit social relations in the European space. Intuitively speaking, the distance between two objects measures the intensity of the similar relationship. Since embedding model is often used to deal with sparse data and mines the data which has not been observed [5], social embedding can solve the sparse social relational data. That is, we can use the social embedding model to find the user's more similar friends even their social relations graph is sparse or unobserved, so we can utilize the social embedding model to more accurate and effective calculate the user relations, which can better to do next POI recommend. In the other words, our embedding method respectively encodes the personal preference and social dimensions in low-dimension latent space to effectively address the issues of data sparsity. According to our model, we are efficient and accurate to calculate user's preference and social relation parameters. As far as we know, we are the first of using social embedding in the next POI recommendation. Experimental results show that our Social Personalized Ranking Embedding (SPRE) model outperforms competitive baselines in terms of both effectiveness and efficiency. The primary contributions of our research are summarized as follows.

- We propose a Social Personalized Ranking Embedding (SPRE) model to joint personalized embedding and social embedding for next POI recommendation. To best our knowledge, this is first work that uses social embedding for the next POI recommendation.
- We evaluate our method with a real-world dataset. Our extensive experimental results show that our method outperforms baselines in terms of different metrics.

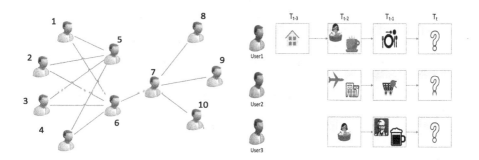

Fig. 1. Social graph   Fig. 2. Users check-in sequences

The rest of the paper is organized as follows. We introduce the related work in Sect. 2. In Sect. 3, we list notations used in the following paper and provide the formal definition of next POI recommendation with considering user personalization and social relations. Some related models contributing to our model are explained in detail in Sect. 4. Our proposed method and the model parameter Learning is discussed in Sect. 5. Our experimental results are shown in Sect. 6. Finally, we conclude our work in Sect. 7.

2 Related Work

POI recommendation has attracted many researchers and many different recommendation technologies have been developed [17,21,23,25]. Recently, some researchers pay attention to next POI recommendation, which is a natural extension of general POI recommendation. It needs timely to provide satisfactory advice based on the POIs that users access recently and their personal preferences. Most of the research is based on the Markov chain-based method to capture the POI sequential pattern and to predict the next check-ins. Zhang et al. [26] predicted the next location probability through an additive Markov chain, and assumed that the POIs that a user recently visited have more impact than the POIs that the users visited a long time ago. Liu et al. [15] exploited the transition pattern of POI categories to predict future check-ins. However, their method fails to describe the unobserved transition probability or highly depends on the category information for calculating accuracy. In this paper, we estimate the European distance between users and POIs in latent European space, so we can through the distance in space to show their transition probability.

Recommendation with social dimensions focuses on using social networks to improve the accuracy and the efficiency of recommendation [8,9,14,16]. Most existing methods are based on Metric Factorization with regularization. For example, Golbeck [8] and Massa and Avesani [16] assumed that a user's preference is similar to or influenced by their directly connected friends. experimentally, this assumption is rough. Their estimation of user's preference is not accurate because of the highly sparse directly linked lines in social networks. To overcome the sparsity of directly connected data, Krohn-Grimberghe et al. [11] recently employed a twice matrix factorization approach, and used potential social relations as regularization for users. However, their method just used *first-order proximity* to construct a Laplacian matrix for regularizing the preference of users. Our work differs from the aforementioned studies in that we can incorporate *first-order proximity* and *second-order proximity* in a unified embedding for calculating social relations.

Embedding methods have been studied for a long time and their wide applications showed that they can effectively capture latent semantic interactions. Tang et al. [18] learned words embedding to make document classification, and showed that their method is efficient with a high accuracy. Recently, some music recommendation research adopted metric embedding into optimization recommendation. Chen et al. [1] proposed a Logistic Markov embedding (LME) for generating the playlists by using metric embedding in the music playlist prediction. And then, there is some research take advantage of metric embedding in the field of next POI recommendation. GE [20] used graph-based metric embedding, and joined four embedding models (i.e., POI-POI, POI-Time, POI-Region, and POI-Word) for POI recommendation. And the PRME [5] is using metric embedding for next new POI recommendation, which is most related to ours. The research proposed a model embedded user preferences and sequential transition into two different spaces. However, this research did not use some useful characteristics, such as social relations of users. Our work differs from the studies mentioned above in that we exploit social embedding scheme to learn the user social relations and adapt Metric Embedding for the next POI recommendation by incorporating user preference and social relations influence.

3   Problem Definition and Notations

In this section, we formulate several definitions for next POI recommendation, and provide notations used in this paper.

Let us assume that there is a user u_i in a M users set $U = \{u_1, u_2, u_3, \ldots, u_M\}$, who just visited a POI l_j in the N POIs set $L = \{l_1, l_2, l_3, \ldots, l_N\}$ and checked-in at this POI at time t. And we defined the social relations set $G = (U, E)$., and U is the set of users and E is the set of edges between users. In general, if the user u_i have a friend u_j, there is an edge between u_i and u_j, the e_ij is defined as the set value of 1 to represent the relationship between them. Then, the user wants to know where she/he like to go next. Based on the user's sequential activities within a short time $\triangle T$ and the social relations G, we need to recommend u_i a POI to go next, which u_i may be interested in. According to this scenario, we formally define the next POI recommendation problem as follows.

Definition 1 (*Next POI Recommendation*). *For a user set U and a POI set L, l^c is the current POI of the user u, the next POI recommendation goal is to recommend a list of POIs that u would be interested in the next, denoted as S^{u,l^c}, which is defined as follows.*

$$S^{u,l^c} = \{l \in L\} \tag{1}$$

For example, the next POI recommendation may tell where a user goes to play after dinner, or suggest which place can be affordable for a lunch near shops where he/she is shopping, as illustrated in Fig. 2. Note that our next POI recommendation model uses the visited POIs and the social relation of a user to predict his/her next POIs. Therefore, we provide the definitions of social embedding for next POI recommendation as follows.

Definition 2 (*First-Order Proximity*). *The **first-order** proximity in a social network is the local pairwise proximity between two users (i.e., two vertices in the social network). Each pair of users is linked by an edge (u, v) with a weight $w_u v$. The weight indicates the first order proximity between u and v. The first-order proximity is 0, if no edge is observed between u and v.*

In a real-world network, the *first-order* proximity often represents the similarity of two nodes. Such as, between each other are friends who want to share a similar interest in the social network, such as users 6, 7 in Fig. 1. However, the links observed are only a small proportion in a real-world information network, and with many others missing [12]. Therefore, we define the *second-order* proximity to complement the first-order proximity and save it in the network structure.

Definition 3 (*Second-Order Proximity*). *The **second-order** proximity between two users (u, v) in a social network is the similarity between their neighborhood network structures. The first-order proximity of u with all other users* is denoted $p_u = (w_{u,1}, \ldots w_{U,|v|})$, *and the second-order proximity between u and* *v is resolved by the similarity between p_u and p_v. If no vertex is linked from/to* *both u and v, the second-order proximity between u and v is 0.*

In the social graph, some users have no direct links, but their social network structure is similar, such as users 5 and 6 in Fig. 1. Their *second-order* proximity is high. In our paper, we utilize both *first-order* and *second-order* proximity for social embedding. We will introduce our social embedding model in detail in next section. Notations used in this article are shown in Table 1.

Table 1. Notations

Symbols	Interpretation
U, L	The set of users and POIs
O_{ij}	Distance of point i and j in the latent European space
$X(i)$	Location of i in the latent European space
G	The graph of the user social information
K	The number of dimensions of the latent space
$\triangle T$	The time period threshold

4   Preliminary Models

4.1   Metric Embedding Technology

The Metric Embedding (ME) model is often used to deal with some sparse data and mine the data which has not been observed [5]. To help understand the specific meaning of metric embedding, we use POI embedding as an example to explain metric embedding. In the POI embedding model, we present each POI as one point in a latent Euclidean space. We also use the Euclidean distance of POIs in the latent Euclidean space to estimate the transition probability, no matter the transition information have been unobserved.

In the POI Embedding model, each POI l has a position $X(l)$ in the latent space. Given a pair of POIs l_i and l_j, we can use the Euclidean distance of the pair of POIs l_i and l_j to estimate their transition probability. The larger the distance, the lower the transition probability is. That is, the transition probability is defined as follows.

$$\hat{P}(l_j|l_i) = \sigma(\|X(l_i) - X(l_j)\|^2) \tag{2}$$

where $\|X(l_i) - X(l_j)\|^2 = \sum_{k=1}^{K}(X_K(l_i) - X_K(l_j))^2$, K is the number of dimensions of the latent space and the $\sigma(z) = 1/(1 + exp(-z))$ is the logistic function as the method for normalization, which is in accordance with our hypothesis about the relationship between the distance and the transition probability.

4.2 Personalized Embedding Model

Here we will present the PRME [5], the state-of-the-art Personalized Ranking Embedding Model to utilize two latent spaces, i.e., the *sequential transition space* and the *user preference* space.

In the Personalized Embedding Model, each POI l has one latent position $X^S(l)$ in the sequential transition space, This is similar to the ME, and $O^S_{l_i,l_j} = \|X(l_i)^S - X(l_j)^S\|^2$ on behalf of the Euclidean distance of the POIs i and j. In the user preference space, each POI l has one latent position $X^U(l)$ and each user u has one latent position $X^U(u)$, and the Euclidean distance of user u and POI l in the user preference space is defined as $O^U_{u,l} = \|X(l_i)^U - X(l_j)^U\|^2$.

In the above spaces only the observed check-ins are exploited to learn the latent position of each POI and each user. Since the observed data is very sparse, we fit the rankings for the POI transition for learning the latent position. Consequently, we can additionally make use of the unobserved data to learn the parameters. We assume that the observed under a POI compared with current POI more relevant than was not observed. For example, POI l_i is observed, and next POI l_j is not. We can compare the Euclidean distance with the user current location $X(u)$. In other words, we use Euclidean distance to rank the POIs instead of utilizing the exponential function, which is used in previous studies. The ranking can be defined as follows.

$$\hat{P}(l_i|u) > \hat{P}(l_j|u) = \sigma(\|X(u) - X(l_j)\|^2) > \sigma(\|X(u) - X(l_i)\|^2)$$
$$\Rightarrow \|X(u) - X(l_j)\|^2 < \|Xl_u) - X(l_i)\|^2 \tag{3}$$
$$\Rightarrow O_{u,l_i} - O_{u,l_j} > 0$$

We model personalized sequential transition by integrating two kinds of metrics for a candidate POI l. Given current location l_c of user u in sequential transition space, we can use a linear interpolation to weight the two metrics.

$$O_{u,l_c,l} = \alpha O^U_{u,l} + (1 - \alpha) O^S_{l_c,l} \tag{4}$$

where $\alpha \in [0, 1]$ controls the weight of different kinds of spaces.

5 Social Personalized Ranking Embedding (SPRE)

In this section, we will present the social embedding model in Sect. 5.1. And in Sect. 5.2, we propose our model: Social Personalized Ranking Embedding (SPRE) model, which jointly incorporates personalized embedding and social embedding. In the end, we will be detailed introduce our model parameter learning.

5.1 Social Embedding Model

Based on this fundamental assumption, the goal of social embedding is to describe relations between users, regardless whether a link between two users is observed or not. To achieve the goal, formally, given the user-user graph $G = (U, E)$, where U is the set of users and E is the set of edges between users. Each edge $e_{i,j}$ in the graph has a source node (u_i) and a target node (u_j). In this section, we propose a novel solution as a variant of LINE [18] to combine both first-order and second-order proximities in a unified framework for depicting the social relations.

The first-order proximity refers to the local pairwise proximity between the vertices in the network. Thus, we define the joint probabilities of any pair of nodes as the models of the first-order proximity.

$$p_1(u_i, u_j) = \frac{1}{1 + exp(-q_i^T \cdot q_j)} \tag{5}$$

where q_i is the embedding vector of vertex u_i, and q_j is the embedding vector of vertex u_j. Equation 5 defines a distribution $p(\cdot, \cdot)$ over the space $U \times U$ and its empirical probability, which can be defined as $\hat{p}(i, j) = \frac{w_{ij}}{W}$, where $W = \Sigma_{(i,j) \in E}$. The way to preserve the first-order proximity is to minimize the following objective function.

$$O_1 = d(\hat{p_1}(\cdot|\cdot), p_1(\cdot|\cdot)) \tag{6}$$

where $d(\cdot, \cdot)$ is the distance between two distributions. By omitting some constants and replacing $d(\cdot, \cdot)$ with KL-divergence, we can have:

$$O_1 = - \sum_{(i,j) \in E} w_{ij} log \, p_1(U_i|U_j) \tag{7}$$

We can present each vertex in the K-dimensional space by finding $\{q_i\}_{1...|U|}$ that minimizes Eq. 7.

In the second-order proximity, we suppose similar context nodes tend to have similar meanings. This supposition leads to a more reliable way to measure the relations between users, which could solve the social link sparsity well. Based on the above assumptions, we define the conditional probability of vertex u_j generates vertex u_i as follows.

$$p_2(u_j|u_i) = \frac{exp(q_j^T \cdot q_i)}{\sum_{k=1}^{|v|} exp(q_k^T \cdot q_i)} \tag{8}$$

where q_i and q_j are the embedding vector of vertex U_i and U_j, Respectively. Equation 8 defines a conditional distribution $p(\cdot|u_i)$ over all the vertices.

We make the conditional distribution $p(\cdot|u_i)$ close to its empirical distribution $\hat{p}(\cdot|u_i)$ for preserving the weight w_{ij} on edge e_{ij}. The empirical distribution can be defined as $\hat{p}(u_j|u_i) = \frac{w_{ij}}{deg_i}$. Then, we minimize the following objective function.

$$O_2 = \sum_{i \in U} \lambda_i d(\hat{p_2}(\cdot|u_i), p_2(\cdot|u_i)) \tag{9}$$

where $d(\cdot, \cdot)$ is the distance between two distributions, and λ_i is the prestige of vertex u_i in the network, which can be measured by the degree $deg_i = \sum_j w_{ij}$. Omitting some constants, the objective function Eq. 9 can be calculated by replacing $d(\cdot, \cdot)$ with KL-divergence as follows.

$$O_2 = - \sum_{(i,j) \in E} w_{ij} log \ p_2(\cdot|u_i) \tag{10}$$

By learning $\{u_i\}_{i=1...|U|}$ to minimize Eq. 10, we are able to present every user u_i with a K-dimensional space.

5.2 Social Personalized Ranking Embedding (SPRE) Model

In Personalized embedding model we assume that if the time interval between two adjacent check-ins is larger than the threshold $\triangle T$, we just consider the user preference. Then we recount the distance metric $O_{u,l_c,l}$ as follows.

$$O_P = O_{u,l_c,l} = \begin{cases} O^U_{u,l} & \text{if } \triangle(l, l_c) > \triangle T, \\ \alpha O^U_{u,l} + (1-\alpha) O^S_{l_c,l} & \text{otherwise.} \end{cases} \tag{11}$$

where $\triangle(l, l_c)$ is the time difference of the successive POIs (l and l_c).

And in social model we linearly combine the first-order and second-order proximities together to preserve both proximities with interest propagation for embedding the social networks.

$$O_S = \beta O_1 + (1 - \beta) O_2 \tag{12}$$

where $\beta \in [0, 1]$ is the strength weight; O_1 and O_2 are the first-order objective and second-order objective respectively. According to Eq. 12, we can obtain a better result to represent the distance between users.

In the above two embedding model, we have a detailed description of the personalized embedding and the social embedding. In this section, we normalize O_S and O_P by using Min-Max scaling to avoid the impacts of excessive value, which could lead to inaccurate results.

Given a current location l^c of user u and the user social relation links, we define the distance metric O as follows.

$$O = \mu O_S \cdot O_P + (1 - \mu)O_P \tag{13}$$

where $\mu \in [0,1]$ is the coefficient to control the proportion of the personalized embedding and social embedding. With the primary assumption in Sect. 4.1, (i.e., the larger the distance, the lower the transition probability is), we can rank the TOP-n next POI recommendation for user u.

5.3 Optimization Learn

By using maximum a posterior (MAP) based Bayesian Personalized Ranking (BPR) [17] approach for personalized embedding, and maximizing a criterion for social embedding, we develop the SPRE Model as follows.

$$\Theta = \underset{\Theta}{\arg\max} \prod_{u \in U} \prod_{l^c \in L} \prod_{l_i \in L} \prod_{l_j \in L} \prod_{(i,j) \in E} P(>_{u,l^c} |\Theta)P(G|\Theta)P(\Theta) \tag{14}$$

where the major parameters of personalized embedding are $\Theta = \{X_L^S, X_L^U, X_U^U, X_U^G\}$ and $P(>_{u,l^c} |\Theta)$. The social embedding is denoted $P(G|\Theta)$.

By using the logistic function to calculate the above two probabilities, the two probabilities can be further estimated as follows.

$$P(>_{u,l^c} |\Theta) = \sigma(O_{u,l^c,l_j} - O_{u,l^c,l_i}) \tag{15}$$

$$P(G|\Theta) = \sigma(O_S) \tag{16}$$

Then, we have the final objective function in Eq. 17 with Gaussian priors on the parameters Θ. ω is a regularization term parameter.

$$
\begin{aligned}
\Theta &= \underset{\Theta}{\arg\max} \ln \prod_{u \in U} \prod_{l^c \in L} \prod_{l_i \in L} \prod_{l_j \in L} \prod_{(i,j) \in E} \sigma(O_{u,l^c,l_j} - O_{u,l^c,l_i})\sigma(O_S)P(\Theta) \\
&= \underset{\Theta}{\arg\max} \sum_{u \in U} \sum_{l^c \in L} \sum_{l_i \in L} \sum_{l_j \in L} \sum_{(i,j) \in E} (\ln(\sigma(O_{u,l^c,l_j} - O_{u,l^c,l_i})) \\
&\quad + \ln(\sigma(O_S)) - \omega\|\Theta\|^2
\end{aligned}
\tag{17}
$$

We adopt a widely used stochastic gradient descent (SGD) algorithm to optimize the objective function in Eq. 18. Based on the previous check-in records and the social graph, we can construct the training tuples. And the updating procedure is defined as follows, where η is the learning rate.

$$\Theta \leftarrow \Theta + \eta \frac{\partial}{\partial \Theta}(\ln(\sigma(O_{u,l^c,l_j} - O_{u,l^c,l_i})) + \ln(\sigma(O_S)) - \omega\|\Theta\|^2) \tag{18}$$

6   Experiments

6.1   Experimental Settings

Datasets. In the experiments, we use a publicly available real-world large-scale LBSNs dataset Foursquare to evaluate our proposed SPRE model. The statistics of the dataset is shown in Table 2. The whole dataset covers 483,813 check-in histories and 121,142 POIs in the local scope. There are 4,163 users in this dataset, who live in the California, USA. There are 32,512 friendship links, and the average number of friends per user is 6.7.

Table 2. Description of our dataset Foursquare

Statistics	Counts
Number of users	4,163
Number of POIs	121,142
Number of the whole check-ins	483,813
Friendship links	32,512
Average friends per user	6.7

Baselines. We compare our SPRE model with the following three methods.

- **SocialMF:** This system combines both a user-item matrix and positive links for recommendation [10], which is a special case of the proposed framework with only positive links. It utilizes the user-item matrix and a user-user matrix.
- **MESocial:** This method joints the user preference embedding and the social embedding together as a baseline. It does not include the POI sequential transition, and is like traditional POI recommendation. However, it is an embedded method.
- **PRME:** PRME [5] is a personalized ranking metric embedding algorithm by integrating the sequential transition of POIs and user's preference. It maps the POI-POI and the POI-user to two different spaces. One is the sequential transition space, and the other is the user preferences space. It also exploits the metric embedding method for the next POI recommendation.

Evaluation Metrics. The performance of our SPRE model and the three comparison methods is measured in terms of two well known measure metrics, namely Recall@N and Precision@N, which are defined as follows.

$$Recall@N = \frac{\#hit@N}{N_v} \tag{19}$$

$$Prescision@N = \frac{\#hit@N}{N} \qquad (20)$$

where $\#hit@N$ denotes the number of POIs that user u visited after the time t in Top-n recommended, and N_v is the total number of POIs that user u have visited after the time t.

6.2   Experimental Results

The experimental results of our SPRE model and the three baselines are shown in Fig. 3. All the methods are running with well-tuned parameters for their effectiveness. Note that we usually ignore a large N in top-n recommendation tasks. We only show the performance for $N = \{5, 10, 20, 30\}$ respectively. In our proposed methods, we set $k = 80, \triangle T = 6\,h$. Besides, we set personalized embedding component weight α as 0.2, and the social embedding component weight β as 0.6. The learning rate η is set as 0.0025, and the SPRE component weight μ is set as 0.3.

In accordance with the time each user visits, we use the first ten months of data as a training set when comparing with the other three baselines. From the results, as shown in Fig. 3, we observe that our method generally outperforms two baselines. SocialMF has a lower precision and recall. This indicates that the traditional MF method is not suitable for the next POI recommendation since it does not make use of sequential information and can't solve the sparsity issue of both POI check-ins and social links. By comparing SocialMF and MEsocial in Fig. 3, we can find out the metric embedding technology can effectively solve the problem of data sparseness, both in POI check-ins and social links. Sequential transition has an important effect in next POI recommendation by observing the differences between ME-social and SPRE. Figure 3 also show that our proposed SPRE model obviously outperforms PRME. This is because our proposed SPRE model joints the social influence to the model. This shows that the social influence indeed improves the accuracy of next POI recommendation.

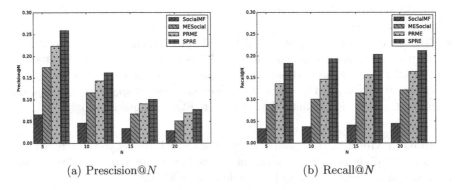

(a) Prescision@N (b) Recall@N

Fig. 3. The result of methods.

6.3 Impact of Different Parameters

Impact of the Component Weight μ. Figure 4 shows the impact of the component weight μ. The performance at $\mu = 0$ (only personalized embedding) is much better than $\mu = 1$ (only social embedding). That is, user preference is more important than social influence in the next POI recommendation. The best results are obtained at $\mu = 0.3$. Hence, we set the $\mu = 0.3$ in our following experiments. We also notice that there is a sharp drop when we change the weight μ from 0.3 to 0.8. This shows that too much social dimension factors can degrade the performance of the next POI recommendation.

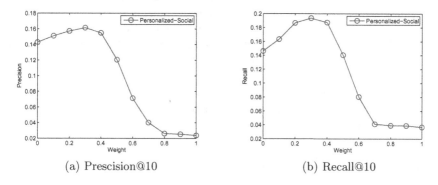

(a) Prescision@10   (b) Recall@10

Fig. 4. Effect of component weight μ

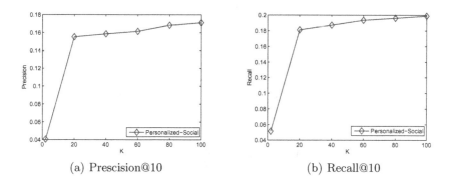

(a) Prescision@10   (b) Recall@10

Fig. 5. Effect of the number of dimension K.

Impact of Metric Dimensions K. Figure 5 shows the experimental results under different metric dimensions K (including the personalized embedding and social embedding metric dimensions). When $K > 20$, the recommendation quality growth becomes smooth. The performance increases with the increment of K.

This is because a high dimension can reflect the potential measurement better. Figure 5 also shows that it is less helpful to improve the performance by increasing K, when the number of dimensions is larger than 80. Empirically, we chose $K = 80$ in our experiments because we simultaneously consider both the model quality and the running time.

7   Conclusion

In this paper, we studied the next POI recommendation problem. We proposed a Social Personalized Ranking Embedding (SPRE) model to joint personalized embedding and social embedding together, which can learn the user preference and social relations in a low-dimension latent space. As far as we know, we are the first to integrate personalized embedding and social embedding in the next POI recommendation. Experimental results on the real-world LBSNs data Foursquare validated the performance of our proposed SPRE method. Our extensive experimental results also show that our method outperforms baselines regarding Top-n recommendation. In the future, we will integrate other dimensions into the model for next POI recommendation, such as semantic information, context information, and temporal cyclic information.

Acknowledgments. This work was partially supported by Chinese NSFC project (61472263, 61402312, 61402311, 61472268).

References

1. Chen, S., Moore, J.L., Turnbull, D., Joachims, T.: Playlist prediction via metric embedding. In: ACM Knowledge Discovery and Data Mining, pp. 714–722 (2012)
2. Cheng, C., Yang, H., King, I., Lyu, M.R.: Fused matrix factorization with geographical and social influence in location-based social networks. In: AAAI Conference on Artificial Intelligence (2012)
3. Cheng, C., Yang, H., Lyu, M.R., King, I.: Where you like to go next: successive point-of-interest recommendation
4. Cho, E., Myers, S.A., Leskovec, J.: Friendship and mobility: user movement in location-based social networks. In: ACM SIGKDD International Conference on Knowledge Discovery and Data Mining, pp. 1082–1090 (2011)
5. Feng, S., Li, X., Zeng, Y., Cong, G., Chee, Y.M., Yuan, Q.: Personalized ranking metric embedding for next new POI recommendation. In: International Conference on Artificial Intelligence, pp. 2069–2075 (2015)
6. Ference, G., Ye, M., Lee, W.C.: Location recommendation for out-of-town users in location-based social networks. In: ACM International on Conference on Information and Knowledge Management, pp. 721–726 (2013)
7. Gao, H., Tang, J., Hu, X., Liu, H.: Exploring temporal effects for location recommendation on location-based social networks. In: ACM Conference on Recommender Systems, pp. 93–100 (2013)
8. Golbeck, J.: Generating predictive movie recommendations from trust in social networks. In: Stølen, K., Winsborough, W.H., Martinelli, F., Massacci, F. (eds.) iTrust 2006. LNCS, vol. 3986, pp. 93–104. Springer, Heidelberg (2006). doi:10.1007/11755593_8

9. Hu, B., Ester, M.: Spatial topic modeling in online social media for location recommendation. ACM (2013)
10. Jamali, M., Ester, M.: Trustwalker: a random walk model for combining trust-based and item-based recommendation. In: ACM SIGKDD, pp. 397–406 (2009)
11. Krohn-Grimberghe, A., Drumond, L., Freudenthaler, C.: Multi-relational matrix factorization using bayesian personalized ranking for social network data. In: ACM International Conference on Web Search and Data Mining, pp. 173–182 (2012)
12. Liben-Nowell, D., Kleinberg, J.: The link prediction problem for social networks. In: Twelfth International Conference on Information and Knowledge Management, pp. 556–559 (2003)
13. Liu, B., Fu, Y., Yao, Z., Xiong, H.: Learning geographical preferences for point-of-interest recommendation. In: ACM SIGKDD International Conference on Knowledge Discovery and Data Mining, pp. 1043–1051 (2013)
14. Liu, B., Xiong, H.: Point-of-interest recommendation in location based social networks with topic and location awareness (2013)
15. Liu, X., Liu, Y., Aberer, K., Miao, C.: Personalized point-of-interest recommendation by mining users' preference transition. In: ACM International Conference on Conference on Information Knowledge Management, pp. 733–738 (2013)
16. Massa, P., Avesani, P.: Trust-aware recommender systems. In: ACM Conference on Recommender Systems, pp. 17–24 (2007)
17. Rendle, S., Freudenthaler, C.: BPR: bayesian personalized ranking from implicit feedback. In: Conference on Uncertainty in Artificial Intelligence, pp. 452–461 (2009)
18. Tang, J., Qu, M., Wang, M., Zhang, M., Yan, J., Mei, Q.: Line: large-scale information network embedding, pp. 1067–1077
19. Wang, W., Yin, H., Chen, L., Sun, Y., Sadiq, S., Zhou, X.: Geo-sage: a geographical sparse additive generative model for spatial item recommendation. In: ACM SIGKDD International Conference on Knowledge Discovery and Data Mining, pp. 1255–1264 (2015)
20. Xie, M., Yin, H., Wang, H., Xu, F., Chen, W., Wang, S.: Learning graph-based POI embedding for location-based recommendation. In: ACM International on Conference on Information and Knowledge Management, pp. 15–24 (2016)
21. Ye, M., Yin, P., Lee, W. C., Lee, D.L.: Exploiting geographical influence for collaborative point-of-interest recommendation. In: SIGIR 2011, Beijing, China, pp. 325–334, July 2011
22. Yin, H., Cui, B., Huang, Z., Wang, W., Wu, X., Zhou, X.: Joint modeling of users' interests and mobility patterns for point-of-interest recommendation. In: ACM International Conference on Multimedia, pp. 819–822 (2015)
23. Yin, H., Sun, Y., Cui, B., Hu, Z., Chen, L.: Lcars: a location-content-aware recommender system. In: ACM SIGKDD, pp. 221–229 (2013)
24. Yin, H., Zhou, X., Shao, Y., Wang, H., Sadiq, S.: Joint modeling of user check-in behaviors for point-of-interest recommendation. In: ACM International on Conference on Information and Knowledge Management, pp. 1631–1640 (2016)
25. Yuan, Q., Cong, G., Ma, Z., Sun, A., Thalmann, N.M.: Time-aware point-of-interest recommendation. In: International ACM SIGIR Conference on Research and Development in Information Retrieval, pp. 363–372 (2013)
26. Zhang, J.D., Chow, C.Y., Li, Y.: Lore: exploiting sequential influence for location recommendations, pp. 103–112
27. Zhang, Q., Wang, H.: Not all links are created equal: an adaptive embedding approach for social personalized ranking. In: The International ACM SIGIR Conference, pp. 917–920 (2016)

Assessment of Prediction Techniques: The Impact of Human Uncertainty

Kevin Jasberg and Sergej Sizov[✉]

Web Science Group, University of Duesseldorf,
Universitaetsstr. 1, 40225 Duesseldorf, Germany
{kevin.jasberg,sergej.sizov}@uni-duesseldorf.de
https://jasbergk.wixsite.com/research

Abstract. Many data mining approaches aim at modelling and predicting human behaviour. An important quantity of interest is the quality of model-based predictions, e.g. for comparative analysis and finding a competition winner with best prediction performance. In real life, human beings meet their decisions with considerable uncertainty. Its assessment and resulting implications for the statistically evident evaluation of predictive models are in the main focus of this contribution. We identify relevant sources of uncertainty as well as the limited ability of its accurate measurement, propose an uncertainty-aware methodology for more evident evaluations of data mining approaches, and discuss its implications for existing quality assessment strategies. Specifically, our approach switches from common point-paradigm to more appropriate distribution-paradigm. The proposed methodology is exemplified in the context of recommender systems and their established metrics of prediction quality. The discussion is substantiated by comprehensive experiments with real users and large-scale simulations.

Keywords: Human uncertainty · User noise · RMSE · Magic barrier · Distribution-paradigm

Track: Empirical evaluation, Exploratory

1 Introduction

A broad range of algorithms and approaches in data mining aims at modelling and predicting aspects of human behaviour. These efforts are motivated by many practically relevant applications, including various recommender systems, content personalisation, targeted advertising, along with many others. The comparative assessment of methods usually involves implicit or explicit knowledge about user behaviour, either by observing user interactions or by asking users explicitly. In many situations, particular individuals may meet their decisions with considerable uncertainty. In other words, they would not exactly reproduce their decisions when asked twice or multiple times. Consequently, observed decisions must be seen as single draws from individual "feeling"-distributions,

© Springer International Publishing AG 2017
A. Bouguettaya et al. (Eds.): WISE 2017, Part I, LNCS 10569, pp. 106–120, 2017.
DOI: 10.1007/978-3-319-68783-4_8

resulting from complex cognition processes, and influenced by multiple factors (e.g. mood, media literacy, etc.). Moreover, and even more important, our knowledge about such distributions may be very limited due to natural restrictions of human behaviour, i.e. it is practically not possible to require the necessary amount of repeated trials for precise location of the underlying distribution parameters. The presence of human uncertainty and our incomplete knowledge about its properties naturally raise the question of assessment validity and reliability. If some approach R_1 shows better results than approach R_2 in the sense of a certain quality metric (prediction accuracy, user satisfaction, etc.) given reference data, can we consider this as a statistically evident proof that approach R_1 *is indeed* better? In the common sense of statistical hypothesis testing, the confident conclusion can be made if the opposite case has a very low probability (type I error) to happen. Under appropriate accounting for human uncertainty, such evidence is often hard to reach.

As a motivating example, we consider the task of rating prediction (common to recommender systems research), along with the Root Mean Square Error (RMSE) [9] as a widely used metric for prediction quality. In a systematic experiment with real users (described in more detail in the forthcoming sections), individuals rated certain media items (movie trailers) multiple times. Only 27% of users have shown constant rating behaviour; 73% of them have given at least two different ratings to the same item; 49% of users have given three or more different responses. Based on the observations made so far, we constructed individual uncertainty models for every user and thus, the considered quality metric (in our case, RMSE) became a random quantity, characterized by a certain probability density function (PDF).

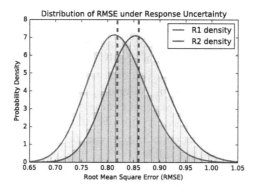

Fig. 1. RMSE as random quantity (Color figure online)

Figure 1 shows corresponding results for two sample recommenders: best possible prediction R_1 (the mean of observed user responses) (red chart) vs. random predictions around the mean R_2 (blue chart); here R_1 is supposed to be the better system by design. As can be seen, there is a large overlap between both PDFs inducing a probability of $P(\text{"}R_2 \text{ better than } R_1\text{"}) \approx 0.33$. Insofar, the comparison of point-wise calculated quality metrics in a particular experiment is not necessarily evident for a statistically sound proof of method advantages. Without any loss of generality, the observations made so far can be considered as an indicative motivation for a more careful analysis of the following **research questions:**

Q1: How well is human uncertainty measurable and what are the implications of its incomplete assessment onto possible model comparisons?

Q2: How well can distinguishability be reached under the human uncertainty assumption, specifically
  (a) What is a natural metric for the distinguishability of two different models?
  (b) What kind of statistical evidence indicates that a model can still be improved?
  (c) What makes a difference between two models statistically significant?

2   Related Work

In the context of this paper, we exemplify our approach by scenarios from the field of recommender systems as summarised in [14] and focus specifically on comparative evaluation metrics. Recommender systems were initially based on demographic, content-based and collaborative filtering. An overview of these techniques is given in [4]. As collaborative filtering recently turned out to be one of the most successful techniques, they rapidly got into the centre of further research. A roadmap to collaborative filtering as well as a profound discussion on its predictive performance is provided by [17].

Due to the importance of evaluating those recommender systems in terms of their model-based prediction quality, different metrics have been introduced, such as the root mean squared error (RMSE), mean absolute error (MAE), mean average precision (MAP) and normalized discounted cumulative gain (NDCG) (see [1]). Further possible quality-related dimensions of interest in recommender assessment (user satisfaction, precision/recall, etc.) are summarised in [9].

All mentioned quantities have in common the need for human input, either by asking the users explicitly or by observing their interactions. In both the cases, human responses may show a considerable degree of uncertainty, resulting from complex cognition processes and multiple influential factors. Consequently, the main results shown in this contribution can be easily adopted for general cases without substantial loss of validity.

The idea of uncertainty is not only related to predictive data mining but also to measuring sciences such as physics or biology. In this area, a science called metrology has been developed, which is about accurate and precise measurement. Recently, a paradigm shift was initiated on the basis of a so far incomplete theory of error (see [7]), so that variables are currently modelled by probability density functions, and quantities of interest are obtained by means of a convolution of these densities. This model is described in [12]. A feasible framework for computing these convolutions via Monte-Carlo-simulation is given by [13].

We employ this model as a basis for our modelling of uncertainty for addressing similar issues in the field of computer science.

The complexity of human perception and cognition can be addressed by means of latent distributions (see [18]), resulting in varying observations. This idea is widely used in cognitive science and in statistical models for ordinal data. For example, so-called CUB models for ordinal data [10] assume the Gaussian as a latent response model underlying the observations. We adopt the idea of modelling user uncertainty by means of individual Gaussians following the argumentation in [10] for constructing our own response models.

The human impact on the prediction quality was noticed in 2009 when [2] stated, that users are inconsistent in giving feedback and therefore establish an unknown amount of noise that challenges the validity of collaborative filtering. In consequence, [15] has shown that quality metrics cannot exceed certain barriers, grounded in the collective uncertainty of observed user decisions. In order to collect information about human uncertainty, we follow [3] by using repeated rating scenarios for same users and items within conducted experiments designed in accordance with experimental psychology [6,11]. On the basis of the information gathered by using this approach, the authors of [3] were able to develop a preprocessing in order to de-noise the underlying data set of ratings and therefore yield better prediction accuracy. In contrast, we distinguish between non-significant deviations (natural human noise) and significant ones (model induced noise). In this paper, we use the same measuring instrument to collect uncertainty information as in [3] but in this contribution, we also focus on the influence of this uncertainty on the accuracy of recommender systems under the view of metrology. We also take the idea of a pre-processing to reduce the impact of human uncertainty on RMSE under this different perspective.

3   Modelling Human Uncertainty

For evaluating the quality of model-based predictions exemplified by recommender system accuracy, we compare internally computed predictors against real user ratings. Let $\mathcal{I} = \{1, \ldots, I\}$ be the index set of I items and $\mathcal{U} = \{1, \ldots, U\}$ the index set of U users. When several users have rated several items, we obtain $n \leq U \cdot I$ pairs (π_ν, r_ν) of predictors π_ν and ratings r_ν that can be matched against each other where $\nu \in \mathcal{U} \times \mathcal{I}$ is a multi-index. These quantities allow computing single scores of accuracy metrics (e.g. RMSE) which corresponds to the commonly used point-paradigm. By using the metrologic distribution-paradigm, we explicitly account for human uncertainty and its resulting rating uncertainty.

We consider all the given ratings to be a family of random variables $R_\nu \sim \mathcal{N}(\mu_\nu, \sigma_\nu)$ which are assumed to be normally distributed as also done in [10]. From this point of view, a given rating r_ν can be seen as the output of a random experiment that is somehow related to human cognition. Hereunder, human uncertainty is strongly related to statistical randomness and the standard deviation σ_ν becomes a natural measure of human uncertainty. In this case, the RMSE becomes a random variable itself, since it is a composition of continuous maps of random variables. The distribution emerges as a convolution of n density functions under the given mathematical model

$$\text{RMSE} = \sqrt{\sum_{\nu \in \mathcal{U} \times \mathcal{I}} \frac{(\pi_\nu - R_\nu)^2}{n}}. \tag{1}$$

As an example, we consider all n rating distributions to be i.i.d. with $R_\nu \sim \mathcal{N}(\pi_\nu, 1)$ that is, the predictors of our recommender systems perfectly match with the mean of our rating distributions. With these distributions, we want to derive the RMSE's density gradually by specifying the densities for every step of computation that has to be done for calculating the entire RMSE. First, we consider the initial step $S_\nu^1 := \pi_\nu - R_\nu$ which is a random variable distributed by $\mathcal{N}(0,1)$. Then as sum of n standard normal distributed random variables, the second step $S_\nu^2 := \sum_\nu (S_\nu^1)^2$ yields a $\chi^2(n)$-distribution with n degrees of freedom. Hence, a scaling by $1/n$ will lead to a gamma distribution $S_\nu^3 := \frac{1}{n} \cdot S_\nu^2 \sim \Gamma(\frac{n}{2}, \frac{2}{n})$ and finally for the last step, $S_\nu^4 := \sqrt{Z_\nu^2} \sim \text{Nakagami}(\frac{n}{2}, 1)$ yields the Nakagami-distribution since it is the square root of a gamma-distributed random variable. Under all these conditions, we yield the RMSE not to be a single point but rather to be a Nakagami-distributed random variable with density function

$$f(x) = \frac{2m^m}{\Gamma(m)} x^{2m-1} \exp\left(-mx^2\right) \quad \text{where} \quad m = n/2. \tag{2}$$

whose expectation

$$\mathbb{E}(\text{RMSE}) = \frac{\Gamma(\frac{n+1}{2})}{\Gamma(\frac{n}{2})} \sqrt{\frac{2}{n}} \tag{3}$$

is the average RMSE score according to the point paradigm when repeating the rating scenario infinitely. The advantage of this approach is, that it additionally provides a non-vanishing variance

$$\mathbb{V}(\text{RMSE}) = 1 - \frac{2}{n} \cdot \left(\frac{\Gamma(\frac{n+1}{2})}{\Gamma(\frac{n}{2})}\right)^2 \tag{4}$$

as a measure of the uncertainty that is related to the RMSE. The fact that a different RMSE score is achieved each time the rating scenario is repeated, corresponds to drawing a random number from a given RMSE distribution within the distribution-paradigm. Considering a data set of uncertain ratings, two different recommender systems would obtain different RMSEs on this dataset, denoted X_1 and X_2. Let $f_{X_1}(x)$ and $f_{X_2}(x)$ the probability density functions of X_1 and X_2. If those densities overlap, then there is also a non-vanishing possibility of error when building a ranking order by evaluating single scores only (point-paradigm). Let x_1 and x_2 denote two realisations of the RMSEs X_1 and X_2 and let $x_1 < x_2$ be the ranking order by using the point-paradigm, then the probability P_ε of error for this decision is given by $P_\varepsilon := P(X_1 > X_2)$ with

$$P(X_1 > X_2) := \int_{-\infty}^{\infty} f_{X_2}(x)\left(1 - F_{X_1}(x)\right) dx \leq 0.5 \tag{5}$$

where $F_{X_1}(x) := \int_{-\infty}^{x} f_{X_1}(t)\,dt$ denotes the cumulative distribution function of f_{X_1}. Later, it will be shown that a ranking built by using the point-paradigm is associated with considerable errors caused by human uncertainty. However, this can virtually be subtracted out in a pre-processing step.

From the view of the distribution-paradigm, each time a given rating is compared with a model-based prediction, we must examine whether the observed deviations are significant or just in nature of contingency, i.e. the influence of human uncertainty. In doing so, we divide the set of all deviations into two subsets. One subset contains all the deviations around the predictor π_ν that can be considered as human uncertainty and the other subset contains all deviations whose extent cannot be explained by this uncertainty and thus seems to be induced by the prediction model. In this case, it seems viable to calculate the RMSE by taking into account only those deviations that are related to the algorithm rather than to human uncertainty. Similarly to the classic RMSE, we refer to this more natural metric as the significant RMSE (**sRMSE**). Following this approach, we have to use statistical hypothesis testing to decide whether a realisation r_ν of the rating distribution R_ν is equal to a model-based prediction π_ν or not. In mathematical notation, we have to test

$$H_0: r_\nu = \pi_\nu \quad \textbf{vs.} \quad H_1: r_\nu \neq \pi_\nu \tag{6}$$

for every multi-index ν at a given significance level α. For known density functions f_{R_ν} of the rating distributions R_ν the critical region can be constructed as the complement of $I_\nu = [\pi_\nu - a;\, \pi_\nu + a]$ where a is chosen such that

$$\int_{\pi_\nu - a}^{\pi_\nu + a} f_{R_\nu}(x)\,dx = 1 - \alpha. \tag{7}$$

We now yield the probability density function of the sRMSE by a convolution of the pseudo-restrictions $f_{R_\nu}|_{I_{95}^c}(x) := \mathbb{I}_{I_{95}^c}(x) \cdot f_{R_\nu}(x)$ where $\mathbb{I}$ is the indicator function. Due to this definition, the sRMSE grants assessment of different recommender systems with much lower probabilities of error. This can be explained by not taking into account the stabilising centre of all the rating-distributions and as the RMSE amplifies the remaining extreme values by its quadratic term (see Eq. 1), the distributions rapidly differ under increase of false predictions. Having in mind this mathematical model of human uncertainty in terms of the novel metrologic distribution-paradigm, we elaborate on our research questions by examination of real life scenarios.

4   User Study and Simulations

In practice, the application of the previously described model is technically challenging. Let the rating distributions $R_\nu \sim \mathcal{N}(\mu_\nu, \sigma_\nu)$ be not necessarily equal for every ν. As it has been shown in [5], the sum of squared deviations receives the density of a non-central χ^2-distribution. At this point, it is quite hard to find a closed form for the RMSE density. It turns out that efficient dealing with the RMSE's distribution can only be maintained by using statistical simulations when general cases are taken into account. In this paper we use **Monte-Carlo-Simulations** (MC) as described in [13]: For every input variable $R_\nu \sim \mathcal{N}(\mu_\nu, \sigma_\nu)$ we take a sample $\mathcal{S}(R_\nu) := \{r_\nu^1, \ldots, r_\nu^\tau\}$ of τ pseudo-random numbers (trials) that are drawn from this specific distribution. Due to the randomness, further computations may fluctuate slightly, but his effect diminishes for a high number of trials. In our analyses, we reached stable results by setting $\tau = 10^6$. With these samples we compute $\mathcal{S}(\text{RMSE})$ by

$$\mathcal{S}(\text{RMSE}) = \left\{ y_j = \sqrt{\sum_\nu \frac{(\pi_\nu - r_\nu^j)^2}{n}} : j = 1, \ldots, \tau \right\}. \tag{8}$$

Post hoc illustration of this sample by a normalised relative histogram with b bins lead to an approximation of the RMSE's density. Our analyses often focus on the error probability P_ε as described in Eq. 5. In the following numerical simulations this probability is efficiently computed by

$$P_\varepsilon = P(\text{RMSE1} > \text{RMSE2}) = |A|/\tau \tag{9}$$

where A is the set of all $(r_i, s_j) \in \mathcal{S}(\text{RMSE1}) \times \mathcal{S}(\text{RMSE2})$ holding the condition $r_i > s_i$ for $i = 1, \ldots, \tau$. For modelling human uncertainty, we assume a set of known rating distributions, based on perceptions about real user behaviour from comprehensive user experiments.

User Experiments

Our experiment is set up with Unipark's[1] survey engine whilst our participants were committed from the crowdsourcing platform Clickworker[2]. During the experiment, participants watched theatrical trailers of popular movies and television shows and provided ratings on a 5-star scale multiple times in random order. The submitted ratings have been recorded for five out of ten fixed trailers so that the remaining trailers act as distractors triggering the misinformation effect, i.e. memory is becoming less accurate because of interference from post-event information. Altogether, we received a rating tensor $R_{u,i,t}$ with $\dim(R) = (67, 5, 5)$, having $N = 1\,675$ data points in total, where the coordinates (u, i, t) encode the rating that has been given to item i by user u in the t-th trial. From this dataset we derive a unique rating distribution for every user-item-pair by considering tensor-slices in trial-dimension $R_{u,i} := R_{u,i,\bullet} = \{R_{u,i,t}|t = 1, \ldots, 5\}$ which can be easily depicted in a relative histogram and modelled by a certain rating distribution. In our experiment, only few tensor slices contain constant ratings and hence lead to a vanishing variance. Performing an item-wise analysis, the fraction of tensor slices with non-zero variance ranges from 50 to 90% that is, only every second participant is able to reproduce its own decisions for the best case.

---

[1] http://www.unipark.com/de/.
[2] https://www.clickworker.de/.

For the worst case, only one out of ten participants is able to precisely reproduce a rating. All tensor slices containing a non-vanishing variance are checked for normality by a KS-test at $\alpha = 0.05$. The null hypothesis was never rejected, allowing to keep the Gaussian distribution as a possible model (rationally, it exhibits maximum entropy among all distributions with finite mean/variance and support on $\mathbb{R}$).

Research Question Q1: Measurability of Human Uncertainty

Description: Based on our user study, we assume $R_\nu \sim \mathcal{N}(\mu_\nu, \sigma_\nu)$. Since this study only surveyed a sample rather than an entire population, point estimates for the distribution parameters would be inappropriate. Instead, confidence intervals have to be specified. Following [8], the confidence interval for the parameter μ_ν can be received by

$$\mu_\nu \in \left[\bar{x}_\nu - t_{(1-\frac{\alpha}{2};n-1)}\frac{s_\nu}{\sqrt{n}} \; ; \; \bar{x}_\nu + t_{(1-\frac{\alpha}{2};n-1)}\frac{s_\nu}{\sqrt{n}} \right] \qquad (10)$$

where $\bar{x}$ and s are the point estimates for the mean and bessel-corrected standard deviation and $t_{(p;k)}$ represents the p-quantile of the t-distribution with k degrees of freedom. Following [16], the confidence interval of σ_ν is given by

$$\sigma \in \left[s\sqrt{(n-1)/\chi^2_{(1-\frac{\alpha}{2};n-1)}} \; ; \; s\sqrt{(n-1)/\chi^2_{(\frac{\alpha}{2};n-1)}} \right] \qquad (11)$$

where $\chi^2_{(p;k)}$ is the p-quantile of the χ^2-distribution with k degrees of freedom. This means that we can not simply determine a single rating distribution for each data set. Instead, a variety of rating distributions needs to be computed for each user-item-pair where the associated parameters are drawn from the corresponding confidence interval. Even for large-scale computations, the resulting RMSE does not possess a stable density function. However, we can consider borderline cases which reveal the maximum span in which we can expect results for the density function of the RMSE. On this basis we run three simulations:

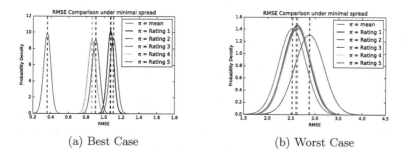

(a) Best Case   (b) Worst Case

Fig. 2. Borderline cases of RSME for different recommender systems

Simulation 1: In Simulation 1 we compute these borderline cases by assigning the parameters μ_ν and σ_ν as the lower limits of the corresponding confidence interval and the upper limits respectively. In doing so, we first build six recommender systems by defining their predictors via where k denotes the k-th recommender systems. Then, for every recommender systems we compute a sample $\mathcal{S}(\mathrm{RMSE}(R\,k))$ for all borderline cases as described in Eq. 8 and generate the ML-density functions. In this simulation we use $\tau = 10^6$ MC-trials for steadiness of histograms as well as $b = 55$ bins for accurate display of densities. Figure 2 shows the impact of the uncertainty of the re-rating-proceeding. Whilst we can recognise a good resolution for three groups of RMSEs in the minimum case, this is virtually no longer possible for the maximum case. The true distributions of the individual RMSEs can vary between these two thresholds but remain unknown to us on the basis of the information collected. In short, with only five re-ratings it is not possible to get high-quality uncertainty information, but it must be said that this phenomenon is not grounded within the point-paradigm itself. In practice, we have to distinguish between two different types of uncertainty: On the one hand, there is the human uncertainty (leading from scores to distributions) which is in the main focus of this contribution. But on the other hand, there is also a kind of measurement error which we call the method uncertainty. The variability for the RMSE distributions in Fig. 2 is completely explained by the impact of this method uncertainty.

Simulation 2: The method uncertainty can be reduced by decreasing the width of the confidence intervals that scale with $1/n^q$ for some $q \in \mathbb{R}$. To this end, it is necessary to increase the number of re-ratings. Accordingly, the borderline cases of the RMSE converge to a stationary state for large n. In this Simulation we estimate the number of re-ratings in order to get stable results, so we can speak of the *true* RMSE. As a measure of this convergence, we calculate the intersection area of the minimum and maximum density for each recommender system. As can be seen from Fig. 3(a), we need about 1000–2000 re-ratings, so that both distributions converge to a steady state by more than 90%. This means that users in a real rating scenario would have to re-evaluate the same item at least 1000 times in order to locate the RMSE-distribution accurately.

Simulation 3: If it is not feasible to calculate the stationary state with the re-rating-proceeding, then it might be sufficient to only gather samples as large as to exclude the high error probabilities of the maximum case. This is simulated by fixing the point estimates $\bar{x}$ and s and artificially increasing the sample size n to calculate the boundary points of our confidence intervals in Eqs. 10 and 11. With those, we determine the error probabilities for a point-paradigm ranking of recommender system 1 to all the other recommender systems for each of the borderline cases. Figure 3(b) depicts the error probabilities $P_\varepsilon = P(\mathrm{RMSE}(R1) > \mathrm{RMSE}(R3))$ for the minimum and the maximum case. All the other cases of $P_\varepsilon = P(\mathrm{RMSE}(R1) > \mathrm{RMSE}(R\,k))$ lead to equivalent results for $k \neq 1$. As we can see, we would need about 500 re-ratings to regard the RMSE approximation to be satisfactory, if we accept a maximum of $P_\varepsilon \approx 0.10$.

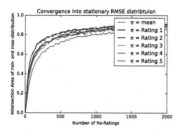

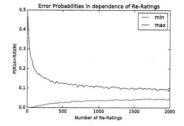

(a) Convergence of minimum and maximum RMSE

(b) Example of convergence of error probabilities

Fig. 3. Convergence into the stationary state

Research Question Q2b: Statistical Evidence for Improvements

Here, we examine the conditions under which a single recommender system can not be distinguished from a theoretically optimal recommender system by means of the RMSE. The idea of this investigation is to create a copy of a given recommender system and to distort this copy by artificial uniform-noise. This is done by resampling its predictors $\pi_1 \in [(1-p)\pi_0 \, ; \, (1+p)\pi_0]$ assuming a uniform distribution. In this case, a noise fraction of p means that those new predictors deviate from the originals by $100p\%$. The RMSE thereby receives a shift on the x-axis so that it's possible to build a ranking along with its associated error probability. We can apply these as a function of the noise component. Noise is, in this context, a specific quantity for inducing differences in recommender system quality in a controlled manner.

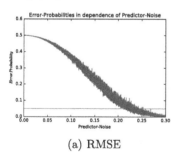

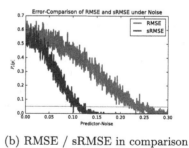

(a) RMSE

(b) RMSE / sRMSE in comparison

Fig. 4. Error probabilities as a function of artificial predictor noise

Simulation 4: The expected value of a random variable is the value which is obtained on average in the case of infinite repetitions of the random experiment and thus has the smallest sum of squared deviations. Theoretically, this property makes the arithmetic mean $\bar{x}_{u,i}$ of the data series $R_{u,i}$ the optimal predictor. Hence, we define the optimal recommender system by setting $\pi_{u,i} := \bar{x}_{u,i}$, so statements can be generated which are correct for very large investigations on the average. To this optimum, we additionally create a copy which we distort by artificial uniform-noise as described and specify that two recommender systems can be distinguished significantly if the error probability is less than 5%. In this simulation we again use $\tau = 10^6$ MC-trials for each of the 10^6 data points (p, P_ε), having 10^{12} trials in total. Figure 4(a) shows the curve of the error probability where the width of this graph is an artefact of the uniform-noise. We can see that the error probability drops below the 5% mark in a range of 21% to 24%, i.e. only then distinctions to the optimum can be reliably detected. This proves the existence of a certain barrier of prediction quality so that any superior recommender system can not be differentiated from the best possible recommender system anymore.

Research Question Q2c: Significant Differences of two Models

In real life, assessments compare several recommender systems among each other. This is taken into account in the following simulations.

Simulation 5: We generate two copies of an optimal recommender, with different proportions of added noise in such a way that the relative noise difference of both copies remains constant. Then, we compute the resulting RMSEs for both copies together with an error probability for the point-paradigm ranking. By increasing the noise for both copies whilst keeping their relative difference constant, we generate an offset (deviation from the optimum or prediction quality) and can thus plot the error probabilities against this offset for different noise ratios. This simulation was performed with 10^{12} data points. Figure 5 depicts the family of curves mapping the noise offset to the corresponding error probabilities. The offset represents background noise and is a measure of the deviation from the best possible recommender system, i.e. the larger the offset, the worse the prediction quality of the recommender system. The colours encode the relative difference Δ of two recommender systems among each other. For the green curve (representing 10% noise of difference), an x-value of 0.15 means that RS1 has a noise of 15% whereas RS2 has a noise of 25%. The corresponding y-value indicates the error probability for ranking both of these recommender systems using the point paradigm. It is apparent from this Figure, that two systems can not be brought into a ranking order without considerable error probability if their relative difference is below 15%, regardless of their basic prediction quality. Figure 5 also reveals that only for noise differences of more than 20%, two different systems can be distinguished starting from a certain quality. As a result, we recognise the following: The better a system becomes, the more improvement does a revision need in order to be detected with statistical evidence.

Simulation 6: In order to make our results more tangible and comparable to current competitions (e.g. the Netflix Prize), we define the RMSE difference as the relative difference in the expectation values of both distributions for this difference uses to be the best estimation for an infinitely repeated rating scenario. We rerun the last simulation, but now determine the RMSE distances by using adaptive noise: We only add so much noise until we reach the desired

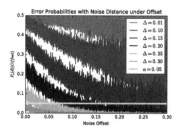

(a) artificial noise as a measure for recommender difference

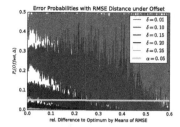

(b) relative RMSE-differences as a measure for recommender difference

Fig. 5. Error Probabilities for two suboptimal recommender systems

RMSE difference. Then we compare the error probabilities by means of those RMSE distances. For the RMSE distances, a similar result is obtained. Two systems differing by 10% in terms of RMSE must deviate more than 40% from the optimum to be distinguished significantly. In reverse, if the closeness of two systems to the theoretical optimum (i.e. the offset) remains unknown - which is probably always the case in real life assessment - then both systems would only be distinguishable with statistical evidence, if they differ at least 20% in terms of the RMSE (since only the 20%-curve is below the 5%-mark for any offset).

Human Accuracy Metrics

At this point, we investigate the resolution properties of two recommender systems by means of the sRMSE. This is performed by a hypothesis test as described in Sect. 3 and considering only significant deviations from the rejection range to compute an RMSE. As a result, the sRMSE could theoretically distinguish between two recommender systems even with fewer deviations.

Simulation 7: In practice, the hypothesis test is performed by constructing a symmetric interval around the predictor π_ν within the rating distribution of R_ν until the density's area over this interval sums up to 0.95. All values in this interval do not represent any significant deviations and are not taken into account in the sRMSE. We hence generate pseudo-random numbers according to R_ν until we have $\tau = 10^6$ values in the rejection range and use these to compute the sRMSE distribution. For these density functions, we now repeat the procedure from simulation 4. The results are depicted in Fig. 4(b). Here we see error curves under noise in the form of a comparison of RMSE and sRMSE. It can be seen that the sRMSE grants substantially faster distinguishability from an optimum with statistical evidence than the traditional RMSE. Using this metric, a recommender system can already be distinguished from a theoretical optimum with 10% of noise whereas the RMSE would probably need more than 20%. A repetition of simulation 5 and 6 leads to equivalent results. This proves the better distinguishing features of the sRMSE as predicted by theory.

5   Discussion

The lessons learned so far can be summarised as follows:

1. Due to the blur of the RMSE, an ordering relation is sometimes very difficult to define; we can only give probabilities for the existence of a particular order relation: The probability $P_\varepsilon := P(R1 > R2 | \mathbb{E}(R1) < \mathbb{E}(R2))$ for making an error when following the point-paradigm has proven to be an intuitive and very good metric. It correlates positively with the overlap of two RMSE distributions and is hence a good measure for the distinguishability of two recommender systems and also serves as a p-value for hypothesis testing.
2. A recommender system is only to be significantly distinguished from an optimum if it differs by more than 21 to 24% in terms of noise. Below this limit, it cannot be distinguished with evidence.
3. The distinguishability of two systems is not dependent solely on its (noise) difference, but also on their basic quality, that is, from their distance to a theoretical optimum. The worse two recommender systems predict, the less they have to differ in order to be distinguished evidently and vice versa.
4. Methods for collecting uncertainty information are yet to imprecise; the parameters of the rating distributions have such wide confidence intervals, that specifying RMSE densities is not reliable. We need between 500 and 1000 re-ratings to exclude the worst case and about 2000 re-ratings for stable results. The method of re-rating-proceeding must, therefore, be improved.

The most notable results are 2 and 3, since they show a natural limit for the resolution of evaluation metrics (which is also always present in the point paradigm but can not be made accessible). Result (2) implies the existence of an equivalence class of optimal recommenders because all recommender systems below a certain RMSE value are no longer to be distinguished from the optimum. Result (3) generalises this fact and raises the fundamental question of assessment evidence. On the basis of our results, the suggested solution of using the sRMSE has proven to be quite fruitful for evaluating prediction quality. In the our simulations, the sRMSE outperformed the traditional RMSE by far, i.e. the resolution capability for two recommender systems was doubled.

6   Conclusion and Future Work

It has been shown that accounting for natural human uncertainty is essential for objective and statistically evident interpretation of ratings and their predictions. In this contribution, we considered recommender systems and their assessment by means of the RMSE, as a characteristic evaluation scenario. It can be assumed that similar influences might be observed for other metrics accounting for uncertain inputs, such as ratings and browsing behaviour. For example, the results presented here could be reproduced in an equivalent form for the metrics average absolute deviation and mean signed deviation. Similar influences might be found not only in recommender systems but also anywhere in predictive data mining where human behaviour is to be analysed. We were, therefore, able to provide initial indications that human uncertainty may have a striking influence on the predictive data mining and thus on all the areas that build upon it. On this basis, further research may lead into various directions: For **theoretical research**, the overall goal is to develop a complete mathematical model of human uncertainty providing large connectivity for practical applications. For **practical research**, it would be quite profitable to assimilate technical approaches and sensitising them for human uncertainty. This could be done by developing Bayesian prediction models with informative priors based on advanced experiments.

References

1. ACM: Workshop on Recommendation Utitlity Evaluation: Beyond RMSE, vol. 9, Dublin, Ireland, September 2012
2. Amatriain, X., Pujol, J.M., Oliver, N.: I like It.. I like it not: evaluating user ratings noise in recommender systems. In: Houben, G.-J., McCalla, G., Pianesi, F., Zancanaro, M. (eds.) UMAP 2009. LNCS, vol. 5535, pp. 247–258. Springer, Heidelberg (2009). doi:10.1007/978-3-642-02247-0_24
3. Amatriain, X., Pujol, J.: Rate it again: increasing recommendation accuracy by user re-rating. In: Proceedings of the Third ACM Conference on Recommender Systems, pp. 173–180. ACM (2009)
4. Bobadilla, J., Ortega, F., Hernando, A., Gutiérrez, A.: Recommender systems survey. Knowl.-Based Syst. **46**, 109–132 (2013)
5. Chan, F.K.: Miss distance-generalized variance non control chi distribution. In: AAS/AIAA Space Flight Mechanics Meeting, pp. 11–175 (2011)
6. Döring, N., Bortz, J.: Forschungsmethoden und Evaluation in den Sozial- und Humanwissenschaften. Springer-Lehrbuch, 5th edn. Springer, Heidelberg (2016). doi:10.1007/978-3-642-41089-5
7. Grabe, M.: Grundriss der Generalisierten Gauß'schen Fehlerrechnung. Springer, Heidelberg (2011). doi:10.1007/978-3-642-17822-1
8. Henze, N.: Stochastik für Einsteiger. Springer Spektrum, Heidelberg (2013)
9. Herlocker, J.L.: Evaluating collaborative filtering recommender systems. ACM Trans. Inf. Syst. **22**(1), 5–53 (2004)
10. Iannario, M.: Modelling uncertainty and overdispersion in ordinal data. Commun. Stat. - Theory Methods **43**, 771–786 (2014)
11. Intraub, H.: Presentation rate and the representation of briefly glimpsed pictures in memory. J. Exp. Psychol. **6**, 1–11 (1990)
12. JCGM: Guide to the expression of uncertainty in measurement. Technical report, BIPM (2008)
13. JCGM: Supplement 1 to the GUM - propagation of distributions using a Monte Carlo method. Technical report, BIPM (2008)
14. Ricci, F., Rokach, L., Shapira, B. (eds.): Recommender Systems Handbook. Springer, Boston (2015). doi:10.1007/978-1-4899-7637-6
15. Said, A., Jain, B.J., Narr, S., Plumbaum, T.: Users and noise: the magic barrier of recommender systems. In: Masthoff, J., Mobasher, B., Desmarais, M.C., Nkambou, R. (eds.) UMAP 2012. LNCS, vol. 7379, pp. 237–248. Springer, Heidelberg (2012). doi:10.1007/978-3-642-31454-4_20
16. Sauerbier, T.: Statistik für Wirtschaftswissenschaftler. Oldenbourg (2003)

17. Su, X., Khoshgoftaar, T.M.: A survey of collaborative filtering techniques. Adv. Artif. Intell., 01 (2009)
18. Vandekerckhove, J.: A cognitive latent variable model for the simultaneous analysis of behavioral and personality data. J. Math. Psychol. **60**, 58–71 (2014)

Data Mining

Incremental Structural Clustering for Dynamic Networks

Yazhong Chen[1], Rong-Hua Li[1], Qiangqiang Dai[1], Zhenjun Li[1(✉)],
Shaojie Qiao[2], and Rui Mao[1]

[1] College of Computer Science and Software Engineering,
Shenzhen University, Shenzhen, China
874908722@qq.com, rhli@szu.edu.cn, 1134133685@qq.com,
15323940@qq.com, mao@szu.edu.cn
[2] Chengdu University of Information Technology, Chengdu, China
qiaoshaojie@gmail.com

Abstract. Graph clustering is a fundamental tool for revealing cohesive structures in networks. The structural clustering algorithm for networks (SCAN) is an important approach for this task, which has attracted much attention in recent years. The SCAN algorithm can not only use to identify cohesive structures, but it is also able to detect outliers and hubs in a static network. Most real-life networks, however, frequently evolve over time. Unfortunately, the SCAN algorithm is very costly to handle such dynamic networks. In this paper, we propose an efficient incremental structural clustering algorithm for dynamic networks, called ISCAN. The ISCAN algorithm can efficiently maintain the clustering structures without recomputing the clusters from scratch. We conduct extensive experiments in eight large real-world networks. The results show that our algorithm is at least three orders of magnitude faster than the baseline algorithm.

1 Introduction

Network data are ubiquitous. Most real-world networks such as social networks, communication networks, and biological networks contain community structures. Discovering the community structures from a network is very useful for a number of applications. For example, in the biological network, a community may represent the molecule with common properties. In the communication network, a community may denote a close group which frequently communicate with each other.

Graph clustering is a fundamental tool to identify such community structures. In the last decade, there are a huge number of models and algorithms that have been proposed for graph clustering. A comprehensive survey on graph clustering and community detection algorithms can be found in [8]. Among all these algorithms, the structural graph clustering algorithm SCAN proposed in [23] is an notable algorithm which has been successfully used in many network analysis tasks [23]. Unlike many other graph clustering algorithms, the streaking

© Springer International Publishing AG 2017
A. Bouguettaya et al. (Eds.): WISE 2017, Part I, LNCS 10569, pp. 123–134, 2017.
DOI: 10.1007/978-3-319-68783-4_9

feature of SCAN is that it is not only able to detect the clusters of a network, but it can also be identify hubs and outliers.

The idea of the SCAN algorithm is similar to a density-based clustering algorithm DBSCAN, which has been widely used for clustering spatial data. Specifically, the SCAN algorithm first defines the structural similarity between two end vertices of each edge in the graph. If the structural similarity for an edge is no less than a given threshold ε, then this edge will be preserved. Otherwise, the algorithm can delete that edge. After this processing, the vertex in the remaining graph that has at least k neighbors is called a core vertex. Then, the algorithm uses the core vertices as seeds, and expands the clusters from the seeds by following the structural similarity edges (more details can be found in Sect. 2).

Unfortunately, the SCAN algorithm is tailored for static graph data. However, real-world networks typically evolve over time. The naive structural clustering algorithm to handle the dynamic networks is to recompute all clusters from scratch using the SCAN algorithm. Clearly, such a naive solution is very costly, as the time complexity of SCAN algorithm is $O(m^{1.5})$ (m denotes the number of edges of the graph), which is nonlinear with respective to the graph size [2].

To overcome this problem, we propose an efficient incremental structural clustering algorithm for dynamic networks, called ISCAN. The ISCAN algorithm can efficiently maintain the clusters generated by the SCAN algorithm without recomputing all the clusters. Specifically, when an edge updating (insertion or deletion), the ISCAN algorithm only works on a small number of edges (i.e., the edges that their structural similarity may update). The structural similarity of the edges may decrease and increase when an edge updating (see Sect. 3). When the structural similarity of an edge decreases, we may need to split the clusters. On the other hand, when the structural similarity of an edge increases, we may merge the clusters. In ISCAN, we propose a BFS-forest structure to maintain the clusters. Each BFS-tree represents a cluster. We also use a set Φ to maintain the set non-tree edges such that the structural similarity of these edges are larger than the threshold ε. When the algorithm splits a BFS-tree, we need to scan the set Φ to check whether the split tree can be merged again by an edge in Φ. We conduct extensive experiments in eight large real-world networks. The results show that the ISCAN algorithm is at least three orders of magnitude faster than the baseline algorithm.

The rest of this paper is organized as follow. In Sect. 2, we briefly introduce the SCAN algorithm. We propose the ISCAN algorithm in Sect. 3. The experimental results are reported in Sect. 4. We survey the related work and conclude this paper in Sects. 5 and 6 respectively.

2   Preliminaries

In this section, we briefly introduce several key concepts used in the SCAN algorithm [23]. Let $G = (V, E)$ be a graph, where V and E denote the set of vertices and edges respectively. The vertex neighborhood of a vertex $v \in V$ is defined as $\Gamma(v) \triangleq \{w \in V | (v, w) \in E\} \cup \{v\}$. The structural similarity between two end vertices of an edge (u, v) is defined as

$$\sigma(u, v) \triangleq \frac{|\Gamma(u) \cap \Gamma(v)|}{\sqrt{|\Gamma(u)||\Gamma(v)|}}. \tag{1}$$

If u and v are not end vertices of an edge, we define $\sigma(u, v) = 0$. In the SCAN algorithm, if $\sigma(u, v)$ is no less than a given parameter ε, the vertices u and v will be assigned into the same cluster. The ε-neighborhood of a node v is defined as

$$N_\varepsilon(v) \triangleq \{w \in \Gamma(v) | \sigma(w, v) \geq \varepsilon\}. \tag{2}$$

A vertex v is called a core vertex if and only if $|N_\varepsilon(v)| \geq \mu$, i.e., $CORE_{\varepsilon,\mu}(v) \Leftrightarrow |N_\varepsilon(v)| \geq \mu$. In the SCAN algorithm, if v is a core vertex and $u \in N_\varepsilon(v)$, u will be assigned to the cluster where v belongs to, and we call u is directly structural reachable from v (denoted by $DirREACH_{\varepsilon,\mu}(v, u)$). Formally, we define direct structure reachability as

$$DirREACH_{\varepsilon,\mu}(v, u) \Leftrightarrow CORE_{\varepsilon,\mu}(v) \wedge w \in N_\varepsilon(v) \tag{3}$$

If $DirREACH_{\varepsilon,\mu}(v, u)$ and $DirREACH_{\varepsilon,\mu}(u, w)$ hold, we call w is structural reachable from v (denoted by $REACH_{\varepsilon,\mu}(v, w)$). Formally, it is defined by

$$REACH_{\varepsilon,\mu}(v, w) \Leftrightarrow \exists v_1, ..., v_n \in V : v_1 = v \wedge v_n = w \wedge \forall i \in \{1, ..., n-1\} : DirREACH_{\varepsilon,\mu}(v_i, v_{i+1}). \tag{4}$$

If there exists a vertex $v \in V$ such that $REACH_{\varepsilon,\mu}(v, u)$ and $REACH_{\varepsilon,\mu}(v, w)$ hold, we call u and w meeting structure connectivity, denoted by $CONNECT_{\varepsilon,\mu}(u, w)$. Based on the above definitions, the cluster C in SCAN is defined as

Definition 1. $CLUSTER_{\varepsilon,\mu}(C) \Leftrightarrow$

(1)   Connectivity : $\forall u, w \in C : CONNECT_{\varepsilon,\mu}(u, w)$
(2)   Maximality : $\forall u, w \in V : u \in C \wedge REACH_{\varepsilon,\mu}(u, w) \Rightarrow w \in C$

The SCAN algorithm aims to find all clusters defined in Definition 1. Note that there may exist some vertices that do not belong to any cluster. Those vertices are considered as hubs if they bridge different clusters, otherwise they will be classified as outliers [23]. The SCAN algorithm first finds a core vertex, and then creates a new cluster for that core vertex. Then, the algorithm traverses the ε-neighborhood of the core vertex in a BFS (Breadth-first search) manner to add vertices into the cluster. When all the vertices are visited, the algorithm terminates. Note that the SCAN algorithm is tailored for static graphs, and it is nontrivial to maintain the clusters when the graphs evolve over time. In this paper, we focus on such a cluster maintenance problem when the graph is updated by an edge insertion and deletion.

3   Incremental Structure Clustering Algorithm

To maintain the clusters, a naive algorithm is to recompute all clusters by invoking SCAN when inserting or deleting an edge. Clearly, such a naive algorithm is inefficient. Below, we propose the ISCAN algorithm to maintain the clusters without recomputing all clusters. Our algorithm is based on the following key observations.

Observation 1. Consider an edge $e = (u, v)$. Let $N(e_{uv}) \triangleq \Gamma(u) \cup \Gamma(v)$, $R(e_{uv}) \subseteq E$ be the set of edges with two end vertices in $N(e_{uv})$. When insert or delete an edge $e = (u, v)$, we only need to update the structural similarity between the two end vertices of an edge in $R(e_{uv})$. There is no need to update the structural similarity between the two end vertices of an edge in $E \backslash R(e_{uv})$. When adding or removing an edge $e = (u, v)$, the structural similarity may increase or decrease for different edges in $R(e_{uv})$. Below, we focus mainly on the edge insertion case, and similar results also hold for the edge deletion case. When inserting an edge $e = (u, v)$, we have three different cases.

Algorithm 1. SCAN [23]

Input: $G=(V,E)$, parameters ε and μ
Output: The BFS-forest and the non-tree-edge set Φ
1: **for all** Unclassified vertex $v \in V$ **do**
2:     **if** $CORE_{\varepsilon,\mu}(v)$ **then**
3:         Generate new clusterID for v; Insert all $x \in N_\varepsilon(v)$ into queue Q;
4:         $x.parent \leftarrow v$; {// create the tree strucuture}
5:         **while** $Q \neq \emptyset$ **do**
6:             $y \leftarrow Q.pop()$; $R=\{x \in V \mid DirREACH_{\varepsilon,\mu}(y, x)\}$;
7:             **for all** $x \in R$ **do**
8:                 **if** x is unclassified **then**
9:                     Assign the current clusterID to x; Insert x into queue Q;
10:                     $x.parent \leftarrow y$;
11:                 **else**
12:                     **if** $x.parent \neq y \wedge y.parent \neq x$ **then** Insert the edge (y, x) into Φ;
13:             Remove y from Q;

First, the structural similarity between (u, v), i.e., $\sigma(u, v)$ increases to $\frac{|\Gamma(u) \cap \Gamma(v)|}{\sqrt{(|\Gamma(u)|+1)(|\Gamma(v)|+1)}}$ after inserting (u, v). Here $\Gamma(v)$ denotes the vertex neighborhood of v before inserting (u, v). This is because there is no edge between (u, v) before inserting (u, v), thus $\sigma(u, v) = 0$ before adding (u, v) by definition. Second, if (w, u, v) forms a triangle after inserting (u, v), $\sigma(w, v)$ will increase to $\frac{|\Gamma(w) \cap \Gamma(v)|+1}{\sqrt{|\Gamma(w)|(|\Gamma(v)|+1)}}$ based on the following lemma.

Lemma 1. $\frac{|\Gamma(w) \cap \Gamma(v)|}{\sqrt{|\Gamma(w)||\Gamma(v)|}} < \frac{|\Gamma(w) \cap \Gamma(v)|+1}{\sqrt{|\Gamma(w)|(|\Gamma(v)|+1)}}$

Proof. First , we have $\frac{|\Gamma(w) \cap \Gamma(v)|^2}{|\Gamma(w)||\Gamma(v)|} / \frac{(|\Gamma(w) \cap \Gamma(v)|+1)^2}{|\Gamma(w)|(|\Gamma(v)|+1)} = |\Gamma(w) \cap \Gamma(v)|^2 (\sqrt{|\Gamma(v)|} + 1)/\sqrt{|\Gamma(v)|}(|\Gamma(w) \cap \Gamma(v)| + 1)^2$. Then, we have $|\Gamma(w) \cap \Gamma(v)|^2 (\sqrt{|\Gamma(v)|} +$

$1)/\sqrt{|\Gamma(v)|}(|\Gamma(w) \cap \Gamma(v)| + 1)^2 \leq |\Gamma(w) \cap \Gamma(v)|(\sqrt{|\Gamma(v)|} + 1)/\sqrt{|\Gamma(v)|}(|\Gamma(w) \cap \Gamma(v)| + 1)$. Since $|\Gamma(w) \cap \Gamma(v)| \leq |\Gamma(v)|$, we have $|\Gamma(w) \cap \Gamma(v)|(\sqrt{|\Gamma(v)|} + 1)/\sqrt{|\Gamma(v)|}(|\Gamma(w) \cap \Gamma(v)| + 1) \leq 1$. This completes the proof.

Third, if the vertices (w, u, v) do not form a triangle after adding (u, v), $\sigma(w, v)$ decreases to $\frac{|\Gamma(w) \cap \Gamma(v)|}{\sqrt{|\Gamma(w)|(|\Gamma(v)| + 1)}}$. Based on this observation, when the structural similarity of (w, v) increases, we may merge two clusters. On the other hand, when the structural similarity decreases, we may need to split a cluster.

Observation 2. A crucial observation is that the clustering procedure of SCAN will generate a BFS-forest where each BFS-tree is a cluster [23]. Note that all the non-leaf nodes in a BFS-tree are the core vertices. Based on this, we can use the BFS-forest to maintain the clusters when the graph changes. In Algorithm 1, we give a modified SCAN algorithm to generate the BFS-forest (see lines 4 and 10).

3.1 The ISCAN Algorithm

As shown in Observation 1, each edge updating (inserting or deleting) can lead to the structural similarity decreasing or increasing. When the structural similarity of an edge (u, v) increases, the algorithm may need to merge the clusters of u and v if $u(v)$ is directly structural reachable from $v(u)$. Moreover, the vertices u and v may become core vertices, if they are not core before updating. On the other hand, if the structural similarity of (u, v) decreases, the algorithm may need to split the cluster, because $\sigma(u, v)$ may be smaller than the threshold ε. Also, the vertices u and v may become non-core vertices if they are core vertices before updating. The challenge is how can we maintain the BFS-forest structure to handle all these cases.

Algorithm 2. ISCAN

Input: $G=(V,E)$, ε and μ, the updated edge $e = (u, v)$, the non-tree-edge set Φ
Output: The updated clusters
1: Let $N(e_{uv}) \triangleq \Gamma(u) \cup \Gamma(v)$, $R(e_{uv}) \subseteq E$ be the set of edges with two end vertices in $N(e_{uv})$;
2: Recompute the structural similarity for all edges in $R(e_{uv})$;
3: **for all** $w \in N(e_{uv})$ **do**
4:     **if** w is a core vertex **then** MergeCluster(w,Φ);
5: **for all** $e = (\tilde{u}, \tilde{v}) \in R(e_{uv})$ **do**
6:     **if** $\sigma(\tilde{u}, \tilde{v}) \geq \varepsilon \wedge \sigma'(\tilde{u}, \tilde{v}) < \varepsilon$ **then** SplitCluster(e,Φ);
7: **for all** $(\tilde{u}, \tilde{v}) \in \Phi$ **do**
8:     **if** $\tilde{u}$.ClusterID $\neq \tilde{v}$.ClusterID **then** Merge($\tilde{u},\tilde{v}$); {// merge the trees that contain w and u}

To tackle this challenge, we additionally maintain a set Φ which stores all the non-tree edges (u, v) such that $v(u)$ is directly structural reachable from u (v). Recall that by the SCAN algorithm, there may exist an edge (u, v) meeting the

$DirREACH$ relationship, i.e., $v(u)$ is directly structural reachable from $u(v)$ and (u, v) is not in any BFS-tree. We make use of the set Φ to keep all these edges. In other words, we classify the edges that satisfy the $DirREACH$ relationship into two classes: tree edge which is stored in the BFS-forest, and non-tree edge which is kept in Φ. When we split a BFS-tree into two sub-treess, we need to scan Φ to check whether these sub-trees can be merged again by an edge in Φ. The ISCAN algorithm maintains both the BFS-forest structure and the set Φ. Initially, we can obtain Φ using the modified SCAN algorithm as shown in Algorithm 1 (see line 12).

The ISCAN algorithm is outlined in Algorithm 2. It consists of three steps to maintain the clusters after an edge (u, v) updating. In the first step, the algorithm considers the case of structural similarity increasing. In this case, the algorithm scans the core vertices to maintain the BFS-forest and Φ. The algorithm recomputes the structural similarity for each edge in $R(e_{uv})$, because the structural similarity for these edges may be updated. For each core vertex in $N(e_{uv})$, the algorithm invokes Algorithm 3 to maintain the set Φ and merge the clusters (lines 1–4).

Algorithm 3. MergeCluster(w,Φ)

1: **if** w is unclassified **then** Create a new clusterID for w;
2: **for all** $u \in N_\varepsilon(w)$ **do**
3:     **if** u is classified **then**
4:         **if** u is non-core vertex **then**
5:             **if** $(u.clusterID = w.clusterID \wedge u.parent \neq w) \vee$
6:                 $(u.clusterID \neq w.clusterID)$ **then**
7:                 Insert edge (w, u) into Φ;
8:         **else**
9:             **if** $u.clusterID = w.clusterID \wedge u.parent \neq w \wedge w.parent \neq u$ **then**
10:                 Insert edge (w, u) into Φ;
11:             **if** $u.clusterID \neq w.clusterID$ **then**
12:                 Merge(w,u); {// merge the trees that contain w and u}
13:     **else**
14:         $u.clusterID \leftarrow w.clusterID$; $u.parent = w$;

In Algorithm 3, the algorithm first checks whether the core vertex w is classified or not. If it is unclassified (i.e., w does not belong to any cluster), we create a cluster ID for w. Then, the algorithm traverses the ε-neighborhood of w. For each neighbor u in $N_\varepsilon(w)$, if u is unclassified, then we add u into the same cluster as w, and set w as the parent for u (line 13). Otherwise, the algorithm checks whether u is a core vertex. If that is the case, the algorithm verify whether (w, u) is a tree edge. If it is not a tree edge and w and u have the same cluster ID, we insert (w, u) into Φ (lines 8–9). If w and u have different cluster IDs, we merge the two trees (i.e., clusters) of w and u (line 10–11). On the other hand, if u is not a core vertex, we consider two cases. First, if (w, u) is not a tree edge and w, u have the same cluster ID, we insert (w, u) into Φ. Second, if (w, u) have different cluster IDs, we also add (w, u) into Φ (lines 4–6). For this case, we will add u into the cluster of w in the third step.

Algorithm 4. SplitCluster($e = (u, v)$,Φ)

1: **if** both u and v are core vertices **then**
2:    **if** $u.parent \neq v.parent \vee v.parent \neq u.parent$ **then**
3:       Delete (u, v) from Φ;
4:    **else**
5:       Remove (u, v) from the BFS-tree.
6: **if** u is core and v is not core **then**
7:    **if** $v.parent = u$ **then**
8:       Remove (u, v) from the BFS-tree.
9:    **else**
10:       Remove (u, v) from Φ;
11: **if** v is core and u is not core **then**
12:    Similar processing as the case of "u is core and v is not core";
13: **if** both u and v are not core vertices **then**
14:    **if** u is core before updating **then**
15:       **if** $v.parent = u$ **then**
16:          Remove (u, v) from the BFS-tree.
17:       **else**
18:          Remove (u, v) from Φ;
19:    **if** v is core before updating **then**
20:       Similar processing as the case of "u is core before updating";

In the second step, Algorithm 2 considers the case of when the structural similarity decreases. To this end, Algorithm 2 scans all the edges in $R(e_{uv})$. For an edge $e = (\tilde{u}, \tilde{v})$, if the structural similarity for e before updating (denoted by $\sigma(\tilde{u}, \tilde{v})$) is no less than ε and the structural similarity for e after updating (denoted by $\sigma'(\tilde{u}, \tilde{v})$) is smaller than ε, the algorithm invokes Algorithm 4 to split the BFS-trees and also maintain the set Φ.

In Algorithm 4, we consider four different cases for the input edge (u, v). First, both u and v are core vertices after updating. In this case, if (u, v) is not a tree edge, we delete (u, v) from Φ (lines 2–3). Otherwise, we remove (u, v) from the corresponding BFS-tree (line 5). Second, u is a core vertex and v is not. In this case, if u is a parent of v before updating, we remove (u, v) from the corresponding BFS-tree (lines 7–8). Otherwise, we remove it from Φ (line 10). Third, v is a core vertex, but u is not. This case is similar to the second case, thus we omit the details. Fourth, both u and v are not core vertices. In this case, we need to consider whether $u(v)$ is core before updating. If both u and v are not core vertices before updating, we do nothing. If u (v) is core and (u, v) is a tree edge, we delete (u, v) from the BFS-tree (lines 14–16). Otherwise, delete (u, v) from Φ.

In the third step, Algorithm 2 scans each edge $(\tilde{u}, \tilde{v})$ in Φ, and merge two clusters by the edge $(\tilde{u}, \tilde{v})$ if $\tilde{u}$ and $\tilde{v}$ have different cluster IDs. Since the ISCAN algorithm enumerates all the possible cases for updating both the BFS-forest and Φ, it is correct. Below, we analyze the time and space complexity of the algorithm.

Complexity Analysis. We first analyze the time complexity of the ISCAN algorithm. Let m and n be the number of edges and vertices of the graph G respectively. Let $\tilde{m} = |\Phi|$ be the size of Φ. Clearly, $\tilde{m}$ is much smaller than m in real-world graphs. In our experiments, we show that in the Youtube social network $m = 2,987,624$ whereas $\tilde{m} = 3,210$. Initially, the algorithm recomputes the structural similarity for all edge in $R(e_{uv})$. Let $O(T)$ be the time spent in this initial step. Since $|R(e_{uv})|$ is very small, $O(T)$ typically can be dominated by $O(m)$ in real-world graphs. In the first step, the cluster merging procedure can be done in $O(n)$ time, because in the worst case, we merge at most $O(n)$ trees. In the second step, we also at most split $O(n)$ clusters, thus the time spent in this step can be bounded by $O(n)$. In the last step, the algorithm takes $O(\tilde{m})$ time to scan Φ and merge the clusters. Putting it all together, we can conclude that the time complexity is $O(m + n)$. In the experiments, we will show that the time usage of our algorithm is much less than such a worst case bound. For the space complexity, our algorithm only need to maintain the BFS-forest and Φ which is dominated by $O(m + n)$.

4   Performance Studies

In this section, we conduct extensive experiments to evaluate the performance of the proposed algorithm. We implement two algorithms: ISCAN and Basic. The ISCAN algorithm is the proposed algorithm, while the Basic algorithm recomputes the clustering results using the SCAN algorithm when the graph changes. We implement these algorithms in C++. All the experiments are conducted in a Linux Server with 2 CPUs and 32 GB main memory.

Dataset. We use four real-world large datasets in the experiments. The detailed statistics of the datasets are summarized in Table 1. All these datasets are downloaded from (http://konect.uni-koblenz.de/networks/). The first three datasets (Youtube, Pokec, and Flixster) are social networks, and the following three datasets (WebGoogle, WebBerkStan, and TREC) are web graphs. The Skitter dataset is a computer network and the RoadNetPA dataset is a road network.

Parameter Setting. There are two parameters in our algorithm: ε, and μ. As recommended in [23], we set the default values of ε and μ by 0.5 and 2, respectively. We vary ε from 0.3 to 0.8, and vary μ from 2 to 7. In all experiments, when varying a parameter, we set the default value for the other parameter. In all experiments, we randomly insert and delete 1000 edges from the original network. For each edge update, we invoke the ISCAN and Basic algorithm to update the clustering results. We record the total time for each algorithm to handle the 1000 edge insertions and deletions.

Efficiency Testing (vary ε). In this experiment, we evaluate the efficiency of our algorithm when varying ε. The results are shown in Fig. 1. As can be seen, the ISCAN algorithm is at least three orders of magnitude faster than the Basic algorithm over all the datasets. For example, in Youtube dataset, when $\varepsilon = 0.5$, our algorithm takes only 10 s to update 1000 edges, whereas the

Table 1. Datasets

Datasets	Number of nodes	Number of edges
Youtube	1,134,000	2,987,000
Pokec	1,632,000	22,301,000
Flixster	2,523,000	7,918,000
WebGoogle	875,000	4,322,000
WebBerkStan	685,000	6,649,000
TREC	1,601,000	6,679,000
Skitter	1,696,000	11,095,000
RoadNetPA	1,088,000	1,541,000

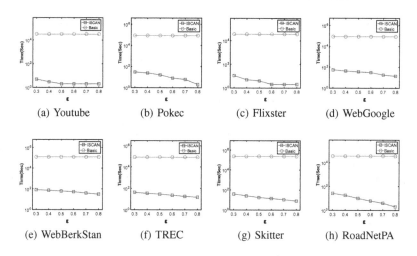

(a) Youtube   (b) Pokec   (c) Flixster   (d) WebGoogle

(e) WebBerkStan   (f) TREC   (g) Skitter   (h) RoadNetPA

Fig. 1. Comparison between ISCAN and Basic (vary ε)

Basic algorithm consumes more than 10000 s. Moreover, the running time of our algorithm generally decreases with increasing ε, while the running time of Basic keeps stable with varying ε. The reason is as follows. When ε is large, the clusters obtained by the SCAN algorithm are relatively stable with respect to an edge updating. As a result, our algorithm may only need to update a small amount of edges. For the Basic algorithm, the algorithm always invoke SCAN to recompute the clusters, thus its running time is insensitive to an edge updating.

Efficiency Testing (vary μ). In this experiment, we compare the efficiency between ISCAN and Basic when varying μ. The results are reported in Fig. 2. From Fig. 2, we can see that the ISCAN algorithm is at least three orders of magnitude faster than the Basic algorithm with different μ values in all datasets. Furthermore, the running time of ISCAN decreases as μ increase. The rationale is as follow. When the graph updating, the larger value of μ, the less influence for the original clusters. Therefore, our algorithm is more efficient when μ is large.

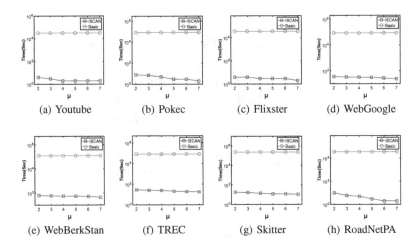

Fig. 2. Comparison between ISCAN and Basic (vary μ)

Similarly, for the Basic algorithm, it is robust with respect to the parameter μ, as it always recompute the clusters using the SCAN algorithm.

To summarize, we can conclude that the ISCAN algorithm is very efficient in practice. As shown in Figs. 1 and 2, under the default parameter setting, the ISCAN algorithm takes only a few seconds to update the clusters in a large graph (e.g., in Pokec dataset, it has more than 22 million edges) with 1000 edge updates. These results demonstrate the high efficiency of the proposed algorithm.

5   Related Work

Structural Graph Clustering. The original structural graph clustering algorithm (SCAN) was proposed by Xu et al. in [23]. Recently, Shiokawa et al. [20] proposed an improved SCAN algorithm called SCAN ++. The SCAN ++ algorithm is based on a new data structure, called directly two-hop-away reachable node set (DTAR). Specifically, DTAR maintains the set of two-hop-away nodes from a given node which are likely to be in the same cluster as the given node. To further reduce the running time of the SCAN algorithm, Chang et al. [2] developed a two-step algorithm called pSCAN. The pSCAN algorithm first clusters the core nodes, and then clusters the border nodes. They also proposed an efficient technique to cluster the core nodes based on a union-find structure. All those SCAN algorithms are tailored for the static graphs, and they are costly to handle the dynamic graphs.

Cohesive Subgraph and Community Detection. Our work is closely related to the cohesive subgraph detection problem which aims to find the densely connected subgraphs from a graph. There are a number of cohesive subgraph models proposed in the literature. Notable examples consist of the maximal clique [4], k-core [12,15,24], k-truss [5,21], maximal k-edge connected subgraph (MkCS) [1,3,25], locally dense subgraph [14], influential community [10,11], and so on.

All those methods can be used to find the non-overlapped communities, and a comprehensive survey on the other community detection algorithms can be found in [8]. Another line of studies focus on finding overlapped communities. For example, Cui et al. [6] proposed an α-adjacency γ-quasi-k-clique model to study the problem of overlapped community search. More recently, Huang et al. [9] introduce a k-truss community model to detect overlapped communities. An excellent survey on overlapped community detection can be found in [22].

Community Maintenance in Dynamic Networks. The community maintenance problem in dynamic networks is an important task in social network analysis [7]. Our work is also closely related to this issue. For the community maintenance problem, it is very often not necessary to recompute the communities when the graph changes. One can only need to detect the affected edges or nodes in a community after the the graph updating. Clearly, different community models have different community updating strategies. Notable community updating algorithms are listed as follows. For the maximal clique model, Cheng et al. [4] introduced an algorithm for dynamically updating the maximal cliques in massive networks. For the k-core model, Li [12] proposed an efficient core maintenance in large dynamic graphs. Similarly, for the k-truss model, Huang [9] proposed an efficient truss maintenance algorithm for dynamic networks. Different from all the existing algorithm, in this paper, we study the problem of dynamically updating the clustering results generated by the SCAN algorithm. Our algorithms may also work on location-based social networks [?], spatial networks [17,19] and trajectory data [16,18]. In the next step, we will study dynamic algorithms in the metric space [13].

6 Conclusion

In this paper, we study the incremental structural clustering problem for dynamic network data. We propose a new algorithm called ISCAN to efficiently maintain the clusters generated by the SCAN algorithm. In the ISCAN algorithm, we use a BFS-forest and a non-tree edge set structure to maintain the clusters. We conduct comprehensive experiments over eight large real-world networks, and the results demonstrate the high efficiency of our algorithm.

Acknowledgement. We thank anonymous reviewers for their insightful comments. The work was supported in part by NSFC Grants (61402292, U1301252, 61033009), NSF-Shenzhen Grants (JCYJ20150324140036826, JCYJ20140418095735561), and Startup Grant of Shenzhen Kongque Program (827/000065).

References

1. Akiba, T., Iwata, Y., Yoshida, Y.: Linear-time enumeration of maximal k-edge-connected subgraphs in large networks by random contraction. In: CIKM (2013)
2. Chang, L., Li, W., Lin, X., Qin, L., Zhang, W.: pSCAN: Fast and exact structural graph clustering. In: ICDE, pp. 253–264 (2016)

3. Chang, L., Yu, J.X., Qin, L., Lin, X., Liu, C., Liang, W.: Efficiently computing k-edge connected components via graph decomposition. In: SIGMOD (2013)
4. Cheng, J., Ke, Y., Fu, A.W.C., Yu, J.X., Zhu, L.: Finding maximal cliques in massive networks. ACM Trans. Database Syst. **36**(4), 21 (2011)
5. Cohen, J.: Trusses: cohesive subgraphs for social network analysis. Technique report (2005)
6. Cui, W., Xiao, Y., Wang, H., Lu, Y., Wang, W.: Online search of overlapping communities. In: SIGMOD (2013)
7. Eppstein, D., Galil, Z., Italiano, G.F.: Dynamic graph algorithms. In: Algorithms and theory of computation handbook, p. 9 (1996)
8. Fortunato, S.: Community detection in graphs. Phys. Rep. **486**(3–5), 75–174 (2010)
9. Huang, X., Cheng, H., Qin, L., Tian, W., Yu, J.X.: Querying k-truss community in large and dynamic graphs. In: SIGMOD (2014)
10. Li, R., Qin, L., Yu, J.X., Mao, R.: Influential community search in large networks. PVLDB **8**(5), 509–520 (2015)
11. Li, R., Qin, L., Yu, J.X., Mao, R.: Finding influential communities in massive networks. VLDB J. (2017)
12. Li, R., Yu, J.X., Mao, R.: Efficient core maintenance in large dynamic graphs. IEEE Trans. Knowl. Data Eng. **26**(10), 2453–2465 (2014)
13. Mao, R., Zhang, P., Li, X., Liu, X., Lu, M.: Pivot selection for metric-space indexing. Int. J. Mach. Learn. Cybern. **7**(2), 311–323 (2016)
14. Qin, L., Li, R., Chang, L., Zhang, C.: Locally densest subgraph discovery. In: KDD (2015)
15. Seidman, S.B.: Network structure and minimum degree. Soc. Netw. **5**(3), 269–287 (1983)
16. Shang, S., Chen, L., Jensen, C.S., Wen, J.R., Kalnis, P.: Searching trajectories by regions of interest. IEEE Trans. Knowl. Data Eng. **99**, 1–1 (2017)
17. Shang, S., Chen, L., Wei, Z., Jensen, C.S., Wen, J.R., Kalnis, P.: Collective travel planning in spatial networks. In: IEEE International Conference on Data Engineering, pp. 59–60 (2017)
18. Shang, S., Ding, R., Yuan, B., Xie, K., Zheng, K., Kalnis, P.: User oriented trajectory search for trip recommendation. In: EDBT, pp. 156–167 (2012)
19. Shang, S., Ding, R., Zheng, K., Jensen, C.S., Kalnis, P., Zhou, X.: Personalized trajectory matching in spatial networks. VLDBJ **23**(3), 449–468 (2014)
20. Shiokawa, H., Fujiwara, Y., Onizuka, M.: Scan++: efficient algorithm for finding clusters, hubs and outliers on large-scale graphs. PVLDB **8**(11), 1178–1189 (2015)
21. Wang, J., Cheng, J.: Truss decomposition in massive networks. PVLDB **5**(9), 812–823 (2012)
22. Xie, J., Kelley, S., Szymanski, B.K.: Overlapping community detection in networks: the state-of-the-art and comparative study. Acm Comput. Surv. **45**(4), 43 (2011)
23. Xu, X., Yuruk, N., Feng, Z., Schweiger, T.A.J.: Scan: a structural clustering algorithm for networks. In: KDD, pp. 824–833 (2007)
24. Zheng, D., Liu, J., Li, R., Aslay, Ç., Chen, Y., Huang, X.: Querying intimate-core groups in weighted graphs. In: 11th IEEE International Conference on Semantic Computing, ICSC (2017)
25. Zhou, R., Liu, C., Yu, J.X., Liang, W., Chen, B., Li, J.: Finding maximal k-edge-connected subgraphs from a large graph. In: EDBT (2012)

Extractive Summarization via Overlap-Based Optimized Picking

Gaokun Dai$^{(\boxtimes)}$ and Zhendong Niu

Beijing Institute of Technology, Beijing, China
dgkmao2340465@gmail.com, zniu@bit.edu.cn

Abstract. Optimization-based methods regard summarization as a combinatorial optimization problem and formulate it as weighted linear combination of criteria metrics. However due to inconsistent criteria metrics, it is hard to set proper weights. Subjectivity problem also arises since most of them summarize original texts. In this paper, we propose overlap based greedy picking (OGP) algorithm for citation-based extractive summarization. In the algorithm, overlap is defined as a sentence containing several topics. Since including overlaps into summaires indirectly impacts on salience, summary size and content redundancy, OGP effectively avoids the problem of inconsistent metric while dynamically involving criteria into optimization. Despite of greedy method, OGP proves above $(1 - 1/e)$ of optimal solution. Since citation context is composed of objective evaluations, OGP also solves subjectivity problem. Our experiment results show that OGP outperforms other baseline methods. And various criteria proves effectively involved under the control of single parameter β.

Keywords: Overlap-based optimization · Non-decreasing submodular objective function · Citation-based extractive summarization

1 Introduction

Moving to a new area is always painful for researchers, especially when the knowledge within the field becomes boosted and complicated. Researchers need to spend great amount of efforts on reading various papers for a deeper understanding. So the tool which summarizes papers with only several sentences could help researchers to study in a more efficient way.

To generate a good summary, various criteria should be taken into consideration. Typically summaries should contain the most salient contents while avoiding repetition or redundancy. Meanwhile, summaries should be concise and contain as many topics as possible. One of solutions is to employ centroid-based methods [8,9] where each topic represents a knowledge aspect of a paper. Then the summary is generated by selecting the most salient sentence per topic, which efficiently avoids redundancy while remaining salience. However, centroid-based methods would incredibly increase summary size if topics are numerous to cover.

© Springer International Publishing AG 2017
A. Bouguettaya et al. (Eds.): WISE 2017, Part I, LNCS 10569, pp. 135–149, 2017.
DOI: 10.1007/978-3-319-68783-4_10

To solve this problem, a simple but effective solution is to select top-N sentences, sorted according to salience [29]. In this way, summary size is reduced while remaining content salience as much as possible and still guaranteeing low redundancy. Nevertheless, top-N strategy would ignore less salient topics, which leads to information loss. Since full topic coverage becomes urgent especially when the space of output is limited [20], top-N strategy is not always preferred.

Optimization-based methods try to involve all criteria into optimization process and pursue balance instead of inclination. Generally, the objective function is formulated as the weighted linear combination of criteria measurements. Then integer linear program(ILP)[27,28] is employed to get optimal solution. Nevertheless since obtaining optimal solution of the optimization problem with cardinality constraint is NP-hard, fast and performance guaranteed greedy algorithms [15,16,38] is preferred. Obviously weight setting is extremely important when measuring performance of greedy methods. However since criteria metrics are usually inconsistent, it is hard to set proper weight for each criterion in the linear combination.

In this paper, we propose Overlap based Greedy Picking(OGP) algorithm which formulates optimization process based on trade-off between overlaps and non-overlaps for citation-based extractive summarization tasks. In this method, overlap is defined as a sentence containing several topics while non-overlap is a sentence containing single topic. And we claim that including overlaps into summaries have following three impacts in detail.

1. Under the constraint of full topic coverage, less sentences are required if more overlaps included.
2. Since reiterating same topics is disgusting, overlaps are preferred especially when they can avoid topic repetitive coverage, which implies reducing content redundancy.
3. More overlaps might cause less salient summary since overlap is not always salient enough to represent every contained topic due to limited sentence length.

Thus formulated as the trade-off process between overlaps and non-overlaps, OGP indirectly involve various criteria and effectively avoids the problem of inconsistent metrics. Despite of a greedy method, OGP proves performance guaranteed due to the design of objective function mentioned in Sect. 3.

Generally there are two main contributions. Firstly we propose an effective optimization methodology that formulates summarization based on the trade-off between overlaps and non-overlaps, which indirectly involves criteria and avoids the problem of inconsistent criteria metrics. Secondly we design an objective function, which guarantees the performance of greedy solution. Further via a single parameter β in the objective function, balance status of criteria can be easily controlled.

The rest of this paper is organized as follows. We introduce related work in Sect. 2. The formulation of the constrained optimization problems are discussed in Sect. 3. Section 4 shows the detail of OGP. In Sect. 5, details about datasets, baseline approaches and evaluation metrics are presented. Then experiments results and relative discussion are described in Sect. 6. Section 7 concludes the main work in this paper and suggestions on future work.

2   Related Work

2.1   Citation-Based Summarization

Citances refer the text spans around the citations and contain information about the important facts in the cited paper [24]. [12,30] extend the definition of citances to including non-explicit text spans which talk about the cited paper but not explicitly expressed. Thus citation context, which is composed of all citances, provides a rich context about the important knowledge aspects of the cited paper and is suitable to be used for summarization [24]. [21] produce the impact-based summary of a single research paper based on the language modeling method, where citances are extracted to describe the impact. [29] finish citation-based single document summarization on scientific articles by applying clustering firstly to find different contributions of a target paper and then utilizing LexRank [5] for sentence selection. [25] prove the possibility of citation-based survey, where citation categorizations are proposed to obtain a survey. Further [22] compare surveys originated from abstracts, full papers and citation context, whose results indicate that multi-document technical survey creation benefits much from citations. Besides of containing enough information for summaries, citation context also help to solve the subjectivity problem, since citances are composed of objective evaluations from other scholars. Thus in this paper, we propose summarization method purely based on citation context.

2.2   Optimization-Based Approaches

Various summary criteria should be considered to generate qualified summaries. Salience is one of the most popular metric, which is often measured by sentence-level features such as position, TFIDF. Relying on graph theory, graph model attracts attention, where Random walk [4,5] is applied to assign centrality score to measure salience. Besides other graph theory technologies such as minimum dominating set model [34] and graph cut [31] are later applied to score salience. Nevertheless, most of these centroid-based methods pay little attention on limiting redundancy. Consider a scenario where a text contains one central topic and several other related topics. In order to gain salience, sentences containing the same central topics in this case are preferred, which definitely cause redundancy. A methodology for pruning redundancy is via clustering, where each group of similar sentences represents a single topic [7,19,29,35]. Picking the most central sentence per cluster could effectively avoid redundancy while remaining salience. We regard this cluster-based methodology deserving attention for its simple and effective way of restricting redundancy, even though most hard clustering methods ignore the existence of overlap as [36] states.

Optimization-based methods regard summarization as combinational problem, where objective function is formed as a weighted linear combination of various criteria metrics. Maximum Marginal Relevance(MMR) [2] cast redundancy penalty on the centrality score to reduce redundancy. [18] describe MMR as a knapsack problem and propose to utilize an integer linear program(ILP) solver to find the optimal solution. Complying with knapsack constraint, maximum coverage model [6,37] is proposed where sentences are picked to maximize information units coverage. To tackle coherence problem, dependency-based discourse tree [10] and graph model [27,28] is also employed, where ILP is used to get optimal solution. Since it often takes much time to get optimal solution, several performance-guaranteed fast approaches are proposed. [26] add redundancy-penalty constraint over the objective function and find feasible approximate solution based on Lagrange heuristics. [14] formalize the problem as submodular function maximization, which is solved via a simple greedy algorithm near-optimally. [15] extend the cardinality constraint to a general budget constraint and [16] points out monotone nondecreasing submodular function is an ideal objective function for optimization-based summarization. Then [17] introduces steps to design a more complicated submodular objective function using submodular shells. Later [23,38] proves submodular objective function also works well in the context of discourse structure. However we find most of current optimization-based methods mentioned above pay little attention on involving summary size into optimization. And naive top-N strategy is usually employed to comply with length constraint, which is indeed a post-processing style. Thus while top-n strategy violates the principle of optimization, we believe methods which dynamically involves various criteria including summary size is better.

3   Problem Formulation

In order to take various criteria dynamically involved as mentioned above, we formulate the citation-based extractive summarization task as the constrained optimization problem. And the problem is formally shown in following Eq. (1).

$$Maximize. \sum_n^N Rep_n = \sum_n^N \{\frac{\sum_c^C \Phi_n(c)S_{nc}}{\Phi_n} + \beta[\sum_c^C \Phi_n(c)]^2\}$$

$$ST. \sum_n^N \Phi_n(c) = 1, \forall c \in C \tag{1}$$

Equation (1) shows the objective function of the optimization problem. n denotes a candidate sentences and N is the set of all sentences. c denotes a topic of articles and C represent all topics, which are explored by the method in Sect. 4.1. $\Phi_n(c)$ equals 1 when candidate sentence n is selected and covers c, otherwise 0. Φ_n refer the amount of topics that sentence n contains. Rep_n is the metric to measure sentence n, which is introduced further in Sect. 4.2. The first part of Rep_n is to measure the salience density where S_{nc} is the metric to measure salience of sentence n to cover topic c. The second part measures the amount of topics that sentence n covers. Note that under the constraint in Eq. (1), all topics should be covered, which guarantees full topic coverage. Thus the Rep_n of sentence n could be affected when some topics which n could cover are covered by others, because each topic should be covered by a single candidate sentence according to the constraint.

According to Eq. (1), the objective function is designed to maximize sum of Rep scores for a summary. And Rep reflects the performance of salience density and amount of covering topics while β is the relative importance of these two metrics. Generally compared to non-overlaps, overlaps contains more topics and have relatively lower salience density, as mentioned above. Thus when Rep focuses more on salience density, non-overlaps are preferred. Conversely when Rep focuses more on the amount of contained topics, overlaps are preferred. In fact, we claim that content redundancy is also implicitly involved in Rep. And the level of content redundancy is lower when overlaps are preferred, which is explained in Sect. 4.2. Thus based on the trade-off process between overlaps and non-overlaps under the control of β, Eq. (1) transforms extractive summarization task into the trade-off between criteria of salience, summary size and content redundancy.

4 Proposed Method

To solve the constrained optimization problem mentioned in Sect. 3, Overlap-based Greedy Pick(OGP) is proposed. Generally OGP could be defined as a three-step process. Firstly, a fast single-linkage hierarchical method [1] is employed to explore topics of the citation context. Sentences containing more than one topic are defined as overlaps. Secondly representativeness metric is proposed to enable the trade-off process between overlaps and non-overlaps. Finally, a performance guaranteed greedy algorithm is designed to solve the optimization problem and generate summaries.

4.1 Overlap Discovery

Topic exploration is the main task for overlap discovery, where topics are defined as sentences containing more than one topic. Then cluster-based method is employed to explore topics, where each cluster of sentences is defined as one topic. Since each candidate sentence could contain several paper topics, traditional cluster-based method which enforces one cluster per sentence is not suitable. Further, efficiency of the clustering method is very important especially when there is still much subsequent work to do. Then a fast single-linkage hierarchical clustering method [1] is employed in this paper.

[1] is a graph-based clustering method. So in order to apply this method, graph model is built firstly where each node represents candidate sentence and the edge denotes the corresponding sentence similarity based on typical TF.IDF metric. [1] treats links or edges as the cluster target instead of nodes and pursues

maximal links within each cluster. Then the overlap is the node with links belonging to different clusters, which indirectly solve the problem of being enforced to belong to only one cluster. Specifically, Jaccard Index [11] is employed to measure the similarity of link pairs as Eq. (2) shows, where e_{ik}, e_{jk} represent the edge between k,i and k,j respectively. Neighbors of node i are indicated by ni.

$$S(e_{ik}, e_{jk}) = \frac{|n(i) \bigcap n(j)|}{|n(i) \bigcup n(j)|} \qquad (2)$$

Then single-linkage hierarchical clustering process is applied. Initially, individual cluster is assigned to each link. Cluster pairs containing link pairs with currently largest similarity, merged together at each step until all separate clusters merged into the single one. Partition density in Eq. (4) is recorded at each step to measure current clustering result. And finally the result turn to be the one with the highest partition density. Obviously this hierarchical merging mechanism indeed consumes very little time, which would not be increased sharply with the boosted data scale.

$$D_c = \frac{m_c - (n_c - 1)}{n_c(n_c - 1)/2 - (n_c - 1)} \qquad (3)$$

$$D = \frac{2}{M} \sum_c m_c \frac{m_c - (n_c - 1)}{(n_c - 2)(n_c - 1)} \qquad (4)$$

In Eq. (3) D_c indicates the link density of cluster c. The overall partition density D in Eq. (4) is then the average of D_c, weighted by the fraction of present links. m_c, n_c represent the number of links and number of nodes in this cluster. Thus D_c is in essence the number of links in cluster c normalized by the minimum and maximum numbers of links between those connected nodes. For example, cluster results corresponding to maximum D in Eq. (4), would be near-complete subgraph and the link amount is close to $n(n - 1)/2$, where n is the number of member within the cluster.

4.2 Representativeness Metric

After revealing the knowledge structure or exploring topics, overlaps and non-overlaps are easily recognized. And we propose a new metric, representativeness to measure these candidate sentences. As mentioned in Sect. 3, the goal of optimization over criteria is indirectly realized via the trade-off process between overlaps and non-overlaps. Thus representativeness metric should reflect all of sentence salience, length and content redundancy.

To measure sentence salience, LexRank [5] is employed based on the graph model, where eigenvector of the largest eigenvalue is calculated via power method and corresponding eigenvector entries are cast as salience score. And some modifications is applied to LexRank in this paper to make salience score globally meaningful, which is shown in Eq. (5). The basic principle is that due to the effect of normalization of LexRank, the scores of nodes in larger cluster is relatively lower than those in smaller cluster. So the reverse value of highest salience score in a cluster is set as the bench.

$$S_{nc} = \frac{L_c(n)}{L_c(M_c)} \cdot \frac{1}{L_c(M_c)} \tag{5}$$

In Eq. (5), $L_c(n)$ is the LexRank score of node n in cluster c. $L_c(M_c)$ refers the highest LexRank value in cluster c. Set C refers to the clusters node n represent in reality. Then S_{nc} is the global salience score of node n covering cluster c. It is easily recognized that the sentence with higher salience score is likely the core in a large cluster, which means it containing most important topic.

Besides of salience, the factor of summary size should also get involved dynamically. As mentioned above, the goal is indirectly achieved by containing more overlaps, which helps to fulfill the constraint of full topic coverage with less sentences. Then we add coverage bonus metric to reflect overlaps ability to cover topics, which is simply set as the square of the amount of topics the sentence covers. Note that only the sentence with highest salience for every topic would be selected if coverage bonus is simply set as the amount of topics the sentence covers. Because in that case, the sum of the coverage bonus for all selected sentences always equals N since every topic should be covered by single sentence according to the constraint in Eq. (1), which means non-sense in terms of the objective function. In fact with the increase of the power, the ability to cover topics become more and more important. And we simply set square to avoid overwhelming effect of high power. Then representativeness is formally proposed in following Eq. (6),

$$Rep_n = \frac{\sum_c^C \Phi_n(c) S_{nc}}{\Phi_n} + \beta [\sum_c^C \Phi_n(c)]^2 \tag{6}$$

where n is the candidate node and c is the explored topic. $\Phi_n(c)$ equals 1 if sentence n is covering topic c, and 0 otherwise. Generally, there are two parts to form the representativeness. Φ_n refer the amount of topics that sentence n could cover. And the first part of Eq. (6) could be viewed as salience density score of a candidate sentence. And the second part is the corresponding coverage bonus, which is simply set as the square of the amount of topics candidate sentence n really covers. Limited to sentence length, overlaps might have lower salience density than non-overlaps. Nevertheless overlaps would enjoy higher coverage bonus, which helps to compete with non-overlaps in terms of the metric in Eq. (6). Most of time, these two factors are incompatible where broader coverage might be the result of including more low-salience-score overlaps while pursing higher salience density usually means including more non-overlaps with high salience score. β is the coefficient to control this optimization, ranging from 0 to infinite. When β is small, salience density dominates while coverage casts little influence. In this case, the optimization process resembles a salience-based method. As β increases, the topic coverage factor is more and more important, which leads to including more overlaps in the summary.

We also claim that content redundancy is implicitly involved in representativeness metric. When β increases, the performance that the amount of topics a sentence really covers is the key for the metric in Eq. (6). In this case, sentences would be preferred when they can cover multiple topics and more importantly avoid repetitive topic coverage. Since the constraint that a topic should be covered by single sentence, repetitive topic coverage would contribute nothing to representativeness score. Thus, content redundancy is reduced with the increasing β. Conversely when β decrease, salience density of a candidate is the key. In this case, the scenario of repetitive coverage is ignored. And in fact, many sentences contain multiple topics instead of pure one. So this mechanism would definitely increase the possibility of redundant content.

4.3   Overlap-Based Greedy Pick

Since this constrained optimization problem is complicated and NP-hard, we propose Overlap-based Greedy Picking(OGP) algorithm for solution. OGP generates summaries by recursively selecting the candidate sentence with currently highest representativeness scores until achieving full topic coverage. The pseudo code is shown in Algorithm 1.

In each recursion in Algorithm 1, the main purpose is to select the candidate sentence with currently highest representativeness score. Firstly for all available sentences, $GetCurrentAvailableTopics$ is the function to find all topics which could be covered by sentence n but not yet covered by other sentences picked before. $GetContainTopics$ is the function to calculate the amount of topics sentence n contains. Then the corresponding representativeness score is calculated as Eq. (6). After traversing all sentences, the one with highest score is picked out and included into the summary. Then, $RemovePick$ is designed to remove the picked sentence and covering topics out of corresponding sets. At last next recursion begins until all topics are covered.

Although a greedy meghod, we claim that OGP shown in Algorithm 1 is guaranteed near-optimal, above $(1 - 1/e)$ of the optimal solution. To start with, we show some related theorems. Theorem 1 shows that a function is submodular if it has a diminishing return property. And Theorem 2 shows if the objective function is submodular and monotonic, the performance of the greedy solution is above $(1 - 1/e)$ of the optimal solution. Next we prove that the designed objective function in Eq. (1) is a monotonic submodular function. Obviously Eq. (1) is monotone increasing, because representativeness score in Eq. (6) is always non-negative, which leads to non-decreasing objective function. Then we prove the submodularity of Eq. (1), which is formally presented in proof. N is the current generated summary containing all previous selected sentences. OF is the objective function in Eq. (1). Then the diminishing property of submodularity in Theorem 1 is directly proved by comparing increasing level of OF when including a sentence into the summary. Note that $\Phi_n^N(c)$ under the smaller background N is not less than $\Phi_n^{N'}(c)$ under a larger N', since some available topics might be covered by others when the amount of sentences in current summary is increased. Finally as OF is monotonic and submodular, the performance of OGP shown in

Algorithm 1 is guaranteed above $(1 - 1/e)$ of the optimal solution according to Theorem 2.

Theorem 1. *A function $F : 2^V \to R$ is submodular if for all $s \in V$ and every $S \subseteq S' \subseteq V$, it satisfies $F(S \cup s) - F(S) \geq F(S' \cup s) - F(S')$*

Theorem 2. *Suppose F is monotonic and submodular. Then greedy algorithm gives constant factor approximation: $F(A_{greedy}) \geq (1 - 1/e)maxF(A)$*

Proof. Submodularity of Objective Function (OF)

Suppose $N \subseteq N'$ and $f(\tilde{N}) = OF(\tilde{N} \cup n) - OF(\tilde{N}) = Rep_n = \frac{\sum_c^C \Phi_n(c)S_{nc}}{\Phi_n} + \beta[\sum_c^C \Phi_n(c)]^2$

$f(N') - f(N) = \frac{\sum_c^C [\Phi_n^{N'}(c) - \Phi_n^N(c)]S_{nc}}{\Phi_n} + \beta \sum_c^C [\Phi_n^{N'}(c) - \Phi_n^N(c)] \sum_c^C [\Phi_n^{N'}(c) + \Phi_n^N(c)]$

Since $\Phi_n(c), \Phi_n, S_{nc} \geq 0$ and $\Phi_n^{N'}(c) \leq \Phi_n^N(c), \forall n$

Then $f(N') - f(N) \leq 0$ and $OF(N' \cup n) - OF(N') \leq OF(N \cup n) - OF(N)$

So the Objective Function or OF is submodular.

Algorithm 1. Overlap-based Greedy Pick

INPUT: C_R is the set of topics currently not covered; N_R is the set of overlaps whose potential covering topics are not yet completely covered by picked ones.

1: **Function:** OGP(C_R, N_R)
2: **if** $Length(C_R) = 0$ **or** $Length(N_R) = 0$ **then**
3:     **return** 0
4: **end if**
5: $Repre \leftarrow 0$
6: $N_{max} \leftarrow NULL$
7: **for** n in N_R **do**
8:     $C_r \leftarrow GetCurrentAvailableTopics(C_R, n)$
9:     $Salience \leftarrow 0$
10:     $CoverTopics \leftarrow 0$
11:     **for** c in C_r **do**
12:         $Salience \leftarrow Salience + L_c(O)/L_c(M_c)^2$
13:         $CoverTopics \leftarrow CoverTopics + 1$
14:     **end for**
15:     $Repre_n \leftarrow Salience/GetContainTopics(n) + \beta * CoverTopics^2$
16:     **if** $Repre < Repre_n$ **then**
17:         $N_{max} \leftarrow n$
18:     **end if**
19: **end for**
20: $C_R, N_R \leftarrow RemovePick(C_R, N_R, N_{max})$
21: **return** OGP(C_R, N_R) + $Repre$
22:

---

5   Experiment

5.1   DataSets

CL-Scisumm 2014 is the subtrack of TAC 2014 Biomedical Summarization Track. The dataset contains 10 references papers, each of which has up to 10 Citing Papers. Three annotators are employed to manually annotate corresponding citing text spans and reference text spans, which leads to three little different training datasets. To avoid the difference between three annotators, all outcomes are the average of results generated based on these three datasets. Then for this extractive summarization task, the gold summary is generated by manually selecting important citances from the citation context. In this paper, we simply focus on the Task 2, that is generating a faceted summary of up to 250 words based on the reference spans and citing spans. And in order to comply with the length constraint, we simply select top 250 words in case of length exceeding.

5.2   Evaluation Method

Based on the given gold summaries, officially ROUGE-L is employed to measure summaries. ROUGE [13] stands for Recall-Oriented Understudy for Gist Evaluation, which serves as a metric to determine the quality of a summary by comparing it to the ideal one. It is proved highly correlated to human evaluation [13] and chosen to be the standard measure in DUC 2004 summarization tasks. ROUGE-L is a sentence-level metric, focusing on the similarity in terms of longest common subsequence. Three detailed metrics are presented, those are average recall, average precision and average f1-measure. Since the size of a summary would itself cast influence on the precision and recall, average f1-measure value is served as the ultimate criteria.

5.3   Baseline Approaches

In order to verify the performance of the proposed optimization approaches, CLexRank, MEAD and ACL Anthology online toolkit are employed to generate summaries above CL-Scisumm 2014 mentioned in Sect. 5.1. Also the best-recorded system in the competition or MQ is included for comparison.

C-LexRank [29] also employ a graph-based model. It first employs a hierarchial agglomeration algorithm [3] to explore clusters from the graph model. And each cluster of sentences represent a topic. Then it employs LexRank to score sentence within each topic and picks the most salient sentence per topic to form the summary. However, the graph clustering method it employs [3] simply assumes that one sentence contains only one topic and pays no attention to the potential overlap structure.

MEAD [32] is a multi-document summarizer, which generate summaries by picking top-n sentences sorted by scores. Assigned by a feature-based classifier, the score is calculated by linear combination of centroid and position.

ACL Anthology [33] contains all papers published by ACL and other related organizations. It employs string-based heuristics approaches to extract all the citation sentences for an article and generates a citation-based summary, which contains five sentences.

6   Results and Discussion

6.1   Overall Performance

Table 1 shows result comparison between OGP and other baseline approaches. In practice, $OGP_{\beta=0.027}$ is chosen as the representation. And it is intuitive to find that $OGP_{\beta=0.027}$ is better than CLexRank and MEAD in terms of ROUGE-L F-score. Further $OGP_{\beta=0.027}$ is even shorter than both of CLexRank and MEAD, which definitely increases reliability. ACL toolkit is restricted to only five sentences, whose ROUGE-L F-score is low. In order to avoid the influence of summary size, OGP_{TOP5} is to pick top five sentences sorted by representativeness score. And the result of the comparison between OGP_{TOP5} and ACL toolkit shows OGP surpassing ACL default summarization method. As for MQ, all we know is the ROUGE-L F-score, which is still lower than the proposed method. Figure 1 shows the change trend of summary quality generated by OGP with different β. Obviously the above conclusion still works regardless of β.

Table 1. Comparison results of OGP and other baseline approaches.

ROUGE_METHOD		$OGP_{\beta=0.027}$	OGP_{TOP5}	CLexRank	MEAD	ACL	MQ
ROUGE-L	AVG_R	0.40982	0.32776	0.37516	0.37466	0.13453	
	AVG_P	0.68594	0.63628	0.63313	0.56310	0.25407	
	AVG_F	0.49795	0.42477	0.45975	0.44146	0.17197	0.260
AVG_SIZE		195.26	119.71	195.93	221.17	115.7	

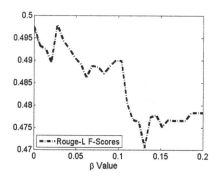

Fig. 1. ROUGE-L F-scores of OGP with various β

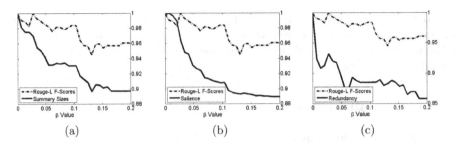

Fig. 2. β's controlling effect over summary size, salience and content redundancy. All of three Figures also reflect the change of ROUGE-L F-score with different β. Figure 2a shows β effect on summary size. Figure 2a shows the influence of β on summary salience score. Figure 2c shows the β effect on content redundancy.

6.2   Effectiveness of β

β is the parameter designed to control the trade-off process between various criteria according to OGP shown in Algorithm 1. When β increases, summary size is reduced since more overlaps are included. Meanwhile content redundancy is also improved since in this case overlaps are preferred when they can avoid topic repetitive coverage. However limited to sentence length, more overlaps might reduce summary salience. To verify the conclusion, summaries are generated by OGP with different β and the result is shown in Fig. 2. In order to make the result more intuitive, all data shown in Fig. 2 is normalized by dividing maximum value, which makes y value of each figure range from 0 to 1. Figure 2a specifically shows the change of summary size with different β. The result turns out that summary size is reduced up to 10% with increasing β. Figure 2b shows how salience score shown in Eq. (5) changes with different β. From the Fig. 2b, salience is nearly monotone decreasing and reduced up to 11%. Figure 2c shows the change of content redundancy with different β. Content redundancy is defined as number of times that topics are repetitive covered. Note that ideally every topic is covered once where redundancy should be zero. Thus redundancy should be normalized via a two-step process. Firstly redundancy is subtracted by the amount of topics. Then normalized redundancy is calculated by dividing current maximum redundancy value. According to the results from Fig. 2c, content redundancy generally decreases with the increasing β, whose rate is up to 14%. Generally, the results shown in Fig. 2 comply with the conclusion of β effect mentioned in Sect. 4.1.

ROUGE-L F-score is used to measure the summary quality. Although fluctuations exists and the rate is relatively small, summary quality generally decreases with increasing β from Fig. 1. Note that longer summary size is definitely beneficial for improving summary quality, since topics might be explained better with more sentences. Thus the scenario of summary quality decreasing could be easily explained by the synthetic effect by summary size, salience and content redundancy. Since the sum of the drop rate of both summary salience and salience is greater than the improve rate of redundancy, the synthetic effect reflected to the general summary quality becomes negative with increasing β, which could be easily recognized from Fig. 2. Also since the gap between these rate is small and the gap is not always negative, the drop of ROUGE-L F-score is slow and unstable.

7 Conclusion and Future Work

In this paper, we mainly propose a new optimization methodology to generate citation-based summarization. Criteria such as summary size, salience and content redundancy is indirectly involved via the optimization between overlaps and non-overlaps, which successfully solve the problem of inconsistent metrics. To solve the optimization problem, a performance-guaranteed greedy algorithm OGP is proposed. Further via a single parameter β, balance status of criteria is effectively controlled.

There are still several tasks for future work. Firstly, the value of β in the experiment is limited to a certain range in advance. And the task to find proper β could be studied in future. Secondly and most importantly, the change of summary quality is strongly related to the synthetic effect of summary size, salience and content redundancy. So the task to explore such relationship is extremely important for better understanding the contribution of each criterion.

References

1. Ahn, Y.Y., Bagrow, J.P., Lehmann, S.: Link communities reveal multiscale complexity in networks. Nature, pp. 761–764 (2010)
2. Carbonell, J., Goldstein, J.: The use of MMR, diversity-based reranking for reordering documents and producing summaries. In: Proceedings of the 21st Annual International ACM SIGIR Conference on Research and development in information retrieval, pp. 335–336 (1998)
3. Clauset, A., Newman, M.E., Moore, C.: Finding community structure in very large networks. Phys. Rev. E **70**, 1–6 (2004)
4. Erkan, G., Radev, D.R.: Lexpagerank: Prestige in multi-document text summarization. In: Conference on Empirical Methods in Natural Language Processing, pp. 365–371 (2004)
5. Erkan, G., Radev, D.R.: Lexrank: Graph-based lexical centrality as salience in text summarization. J. Artif. Intell. Res. **22**, 457–479 (2004)
6. Filatova, E., Hatzivassiloglou, V.: A formal model for information selection in multi-sentence text extraction. In: Proceedings of the 20th International Conference on Computational Linguistics, pp. 397–403 (2004)
7. Fung, P., Ngai, G., Cheung, C.S.: Combining optimal clustering and hidden Markov models for extractive summarization. In: Proceedings of the ACL 2003 Workshop on Multilingual Summarization and Question Answering, pp. 21–28 (2003)
8. Harabagiu, S., Lacatusu, F.: Topic themes for multi-document summarization. In: Proceedings of the 28th Annual International ACM SIGIR Conference on Research and Development in Information Retrieval, pp. 202–209 (2005)

9. Hardy, H., Shimizu, N., Strzalkowski, T., Ting, L., Zhang, X., Wise, G.B.: Cross-document summarization by concept classification. In: Proceedings of the 25th Annual International ACM SIGIR Conference on Research and Development in Information Retrieval, pp. 121–128 (2002)

10. Hirao, T., Yoshida, Y., Nishino, M., Yasuda, N., Nagata, M.: Single-document summarization as a tree knapsack problem. In: Conference on Empirical Methods in Natural Language Processing, pp. 1515–1520 (2013)

11. Jaccard, P.: Etude comparative de la distribution florale dans une portion des Alpes et du Jura. Impr. Corbaz (1901)

12. Kaplan, D., Iida, R., Tokunaga, T.: Automatic extraction of citation contexts for research paper summarization: a coreference-chain based approach. In: Proceedings of the 2009 Workshop on Text and Citation Analysis for Scholarly Digital Libraries, pp. 88–95 (2009)

13. Lin, C.Y.: Rouge: a package for automatic evaluation of summaries. In: Text Summarization Branches Out: Proceedings of the ACL-04 Workshop, pp. 74–81 (2004)

14. Lin, H., Bilmes, J., Xie, S.: Graph-based submodular selection for extractive summarization. In: IEEE Workshop on Automatic Speech Recognition & Understanding, ASRU 2009, pp. 381–386 (2009)

15. Lin, H., Bilmes, J.: Multi-document summarization via budgeted maximization of submodular functions. In: Human Language Technologies: The 2010 Annual Conference of the North American Chapter of the Association for Computational Linguistics, pp. 912–920 (2010)

16. Lin, H., Bilmes, J.: A class of submodular functions for document summarization. In: Proceedings of the 49th Annual Meeting of the Association for Computational Linguistics: Human Language Technologies, pp. 510–520 (2011)

17. Lin, H., Bilmes, J.: Learning mixtures of submodular shells with application to document summarization. In: Proceedings of the Twenty-Eighth Conference on Uncertainty in Artificial Intelligence, pp. 479–490 (2012)

18. McDonald, R.: A study of global inference algorithms in multi-document summarization. In: Amati, G., Carpineto, C., Romano, G. (eds.) ECIR 2007. LNCS, vol. 4425, pp. 557–564. Springer, Heidelberg (2007). doi:10.1007/978-3-540-71496-5_51

19. McKeown, K., Klavans, J., Hatzivassiloglou, V., Barzilay, R., Eskin, E.: Towards multidocument summarization by reformulation: progress and prospects. In: Sixteenth National Conference on Artificial Intelligence and the Eleventh Innovative Applications of Artificial Intelligence Conference Innovative Applications of Artificial Intelligence, pp. 453–460 (1999)

20. Mei, Q., Guo, J., Radev, D.: Divrank: the interplay of prestige and diversity in information networks. In: Proceedings of the 16th ACM SIGKDD International Conference on Knowledge Discovery and Data Mining, pp. 1009–1018 (2010)

21. Mei, Q., Zhai, C.: Generating impact-based summaries for scientific literature. In: Proceedings of the Meeting of the Association for Computational Linguistics, pp. 816–824 (2008)

22. Mohammad, S., Dorr, B., Egan, M., Hassan, A., Muthukrishan, P., Qazvinian, V., Radev, D., Zajic, D.: Using citations to generate surveys of scientific paradigms. In: Proceedings of Human Language Technologies: The 2009 Annual Conference of the North American Chapter of the Association for Computational Linguistics, pp. 584–592 (2009)

23. Morita, H., Sasano, R., Takamura, H., Okumura, M.: Subtree extractive summarization via submodular maximization. In: Proceedings of the 51st Annual Meeting of the Association for Computational Linguistics, pp. 1023–1032 (2013)

24. Nakov, P.I., Schwartz, A.S., Hearst, M.: Citances: citation sentences for semantic analysis of bioscience text. In: Proceedings of the SIGIR 2004 workshop on Search and Discovery in Bioinformatics, pp. 81–88 (2004)
25. Nanba, H., Okumura, M.: Towards multi-paper summarization using reference information. In: International Joint Conference on Artificial Intelligence, pp. 926–931 (1999)
26. Nishikawa, H., Hirao, T., Makino, T., Matsuo, Y.: Text summarization model based on redundancy-constrained knapsack problem. In: Proceedings of COLING 2012: Posters, pp. 893–902 (2012)
27. Parveen, D., Mesgar, M., Strube, M.: Generating coherent summaries of scientific articles using coherence patterns. In: Conference on Empirical Methods in Natural Language Processing, pp. 772–783 (2016)
28. Parveen, D., Ramsl, H.M., Strube, M.: Topical coherence for graph-based extractive summarization, pp. 1949–1954 (2015)
29. Qazvinian, V., Radev, D.R.: Scientific paper summarization using citation summary networks. In: Proceedings of the 22nd International Conference on Computational Linguistics, pp. 689–696 (2008)
30. Qazvinian, V., Radev, D.R.: Identifying non-explicit citing sentences for citation-based summarization. In: Proceedings of the 48th annual meeting of the association for computational linguistics, pp. 555–564 (2010)
31. Qian, X., Liu, Y.: Fast joint compression and summarization via graph cuts. In: Conference on Empirical Methods in Natural Language Processing, pp. 1492–1502 (2013)
32. Radev, D., Allison, T., Blair-Goldensohn, S., Blitzer, J., Celebi, A., Dimitrov, S., Drabek, E., Hakim, A., Lam, W., Liu, D., et al.: Mead-a platform for multidocument multilingual text summarization (2004)
33. Radev, D.R., Muthukrishnan, P., Qazvinian, V., Abu-Jbara, A.: The ACL anthology network corpus. Lang. Resour. Eval. **47**, 919–944 (2013)
34. Shen, C., Li, T.: Multi-document summarization via the minimum dominating set. In: Proceedings of the 23rd International Conference on Computational Linguistics, pp. 984–992 (2010)
35. Siddharthan, A., Nenkova, A., McKeown, K.: Syntactic simplification for improving content selection in multi-document summarization. In: Proceedings of the 20th international conference on Computational Linguistics, pp. 896–902 (2004)
36. Skabar, A., Abdalgader, K.: Clustering sentence-level text using a novel fuzzy relational clustering algorithm. IEEE Trans. Knowl. Data Eng. **25**, 62–75 (2013)
37. Takamura, H., Okumura, M.: Text summarization model based on maximum coverage problem and its variant. In: Conference of the European Chapter of the Association for Computational Linguistics, pp. 505–513 (2009)
38. Vigneshwaran, L.J.K.P.M., Sharma, M.V.V.D.M.: Non-decreasing sub-modular function for comprehensible summarization. In: Proceedings of NAACL-HLT, pp. 94–101 (2016)

Spatial Information Recognition in Web Documents Using a Semi-supervised Machine Learning Method

Hendi Lie[(⊠)], Richi Nayak, and Gordon Wyeth

School of Electrical Engineering and Computer Science,
Science and Engineering Faculty, Queensland University of Technology,
Brisbane, Australia
{h2.lie,r.nayak,gordon.wyeth}@qut.edu.au

Abstract. Web documents are a promising source of spatial information. With information recognition and extraction, this information can be used in various applications such as building semantic maps and indoor robotic navigation. In this paper, we present a novel methodology to identify spatial information in web documents using semi-supervised trained machine learning classifiers. The semi-supervised models trained with the half amount of data available yield only the F-score of 4% and 9% inferior to the supervised models trained with complete data on classifying spatial entities and relationships respectively.

Keywords: Spatial role labelling · Web mining · Named entity recognition · Relation extraction

1 Introduction

Web documents obtained from the Internet are a promising source of symbolic, natural language spatial information. Given an object of interest, a simple Google search (or browsing on websites dedicated to navigation purpose) normally would return abundant descriptive information of location of the object of interest. There are a number of applications that can utilize the processed spatial information such as for building a semantic map [1] and navigating robots through indoor environment [2]. These systems require a precise and accurate input; however, a general search would return a large amount of natural language description of the environment. Manual generation is infeasible and might not be available when the target environment is unknown, restricting their potential use cases. An automated method to generate spatial information of an arbitrary target location from the web would solve this problem and allow wider applications of these consumer systems.

In this paper, we specify "spatial information" as a combination of "spatial entities" and "spatial relations". A spatial entity is a spatial object of interest such as room, building or lecture theatre. Given a spatial entity, a spatial relation is identified as its direction and relative positioning with another spatial entity. For example, in the given sentence "Lecture theatre B103 is located beside computer lab B105"; "Lecture theatre B103" and "computer lab B105" will be identified as spatial entities connected through the spatial relation of "beside".

A. Bouguettaya et al. (Eds.): WISE 2017, Part I, LNCS 10569, pp. 150–164, 2017.
DOI: 10.1007/978-3-319-68783-4_11

With the vast amount of information available on the web, implementing a machine learning-based method for spatial information recognition is a natural choice. Given the labelled spatial information, supervised machine learning models will allow extraction of patterns and utilize them to find any new spatial information related to an object. Unfortunately, our research discovered that not much annotated web-based spatial information data set is available. A manual annotation process is infeasible for this scale of data. A novel method to build an effective spatial information recognition system using only a small amount of annotated data needs to be devised.

A spatial information recognition (SIR) system requires three components to achieve its purpose. Firstly, spatial entities mentioned in web documents need to be accurately identified. Next, from several identified spatial entities in a sentence, we must determine possible spatial relations and their description between a pair of spatial entities. These two objectives must be achieved using only a small amount of annotated training data. Lastly, this vast amount of spatial information should be represented in ranking order based on the relevancy to the original object of interest.

The problem of identifying spatial entities in a sentence comes closest to the task of named entity recognition (NER). NER is a well-studied topic in Information Retrieval (IR) with development of multiple algorithms and approaches [3]. Spatial entity is a well-studied class in NER. Most NER approaches focus on recognizing high level spatial entities e.g. countries, cities, suburbs and landmarks [3], while we seek to find low-level spatial entities e.g. campus, building, office room and floor number. Different to the proposed work, in our prior work, we introduced a supervised spatial entity recognition task using some very specific patterns to identify explicitly mentioned spatial entities without regards to their relations [4].

The task of finding spatial relations between spatial entities falls into the area of Spatial Role Labelling (SpRL) [5]. For each word in a sequence, a SpRL method aims to assign a role as trajector, landmark or spatial indicator. Trajector and landmark can be identified as part of spatial entity recognition, while a spatial indicator is identified as a spatial relation. Multiple machine learning approaches have been proposed to solve this task and to classify the type of spatial indicator and relations [6, 7]. However, all previous works in SpRL have been concentrated on general type of spatial entity (i.e. a trajector/landmark could be a person or animal), while we aim to specifically identify buildings and rooms.

These previous methods in spatial entity and relation recognition rely on abundant annotated data sets using supervised learning. These approaches are not suitable for situations where labelled training data is scarce. Semi-supervised learning is often used in several applications when faced with the lack of annotated data sets [8]. In this paper, we propose to employ the semi-supervised approach in the SIR task.

In response to a query, not every information found can be highly relevant, even some partially relevant results can be returned by the state-of-the-art search engines [9]. Relevancy ranking is a well-studied subject in IR in order to return top relevant documents only, with approaches ranging from simple scoring functions to machine learning-based methods [9]. As runtime performance is a concern, we need to choose a method that can run fast yet produce accurate results. To our knowledge, no SIR system has employed result ranking as part of the approach.

In this paper, we propose a novel methodology to identify and rank spatial information present in web documents in response to a user query. This novel approach contains four critical components: the query bot, spatial entity tagger, spatial relation tagger and result ranker. For both spatial entity and relation tagger, we explored semi-supervised learning to enable accurate identification of spatial information without expensive manual annotation of training documents. To the best of our knowledge, the proposed framework is the first attempt to solve the spatial entity and spatial role labelling problem on the domain of web documents using semi-supervised learning. Extensive empirical analyses with both the supervised and semi-supervised models reveal that the proposed framework is able to perform SIR with high F1 score. The final component, result ranker, has been able to score highly on $NDCG_k$ metric.

2   The Proposed Method of Spatial Information Recognition

The proposed SIR method consists of 4 main modules as illustrated in Fig. 1.

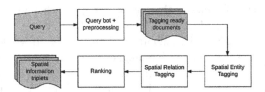

Fig. 1. The proposed spatial information recognition (SIR) system.

Web Documents Processing: Given a query, the query bot will run a search on a search engine to find relevant web documents. Each retrieved document is pre-processed to remove unwanted characters, extract individual tokens/sentences and produce training features. For example, the sentence "Mary © had a little lamb... ☺" will be returned as "Mary had a little lamb." After removing special symbols.

Spatial Entity Tagging: For spatial entity tagging, we generate 12 feature groups containing 33 features for each word in a sentence, ranging from the token to syntactic features such as prefix and suffix of the token. Table 1 details them, with examples on the sentence "Mary had a little lamb" and word "Mary" as the first word of the sentence (at index 0).

The tagging ready documents are then passed to the spatial entity tagger to extract spatial entities. As each spatial entity can be represented with multiple words, the IOB2 labelling standard [10] is used to label them. An example of spatial entities extracted from a sentence applying the proposed model is given in Text 1.

[**Computer lab S510**] is located in front of [**tutorial room S509**].

Text 1. Sample spatial entities from a sentence. The two spatial entities are in bold.

As shown in Fig. 2, we propose a semi-supervised learning approach using the bootstrapping technique to identify the spatial entity in a sentence/document. We first

Table 1. Features generated for spatial entity tagging.

Feature	Explanation	Example
Token	A single word as part of a sentence	$w_0 = Mary$
Token lower	Token in lower case	$wl_0 = mary$
POS	Part of speech (POS) label of the token	$pos_0 = NNP$
Chunk	Shallow parsing chunk of a token, uses the IOB2 labelling standard [10]	$chunk_0 = B - NP$
Word shape	The shape of the token, determined by characters composing it	$w\_shape_0 = ULLL$ (uppercase and 3 lowercase letters)
Word shape simplified	Simplified version of token shape by removing consecutively repeated shape characters	$w\_shape\_simple_0 = UL$
Word type	Type of token such as 'AllDigits', 'AllUpper' or 'AllSymbol'	$w\_type_0 = AllLetter$
Prefix	The first characters of a token	$p0_0 = M$, $p1_0 = Ma$ (first two characters)
Suffix	The last characters of a token	$s0_0 = y$, $s1_0 = ry$ (last two characters)
Digit and alpha-numeric	Boolean feature indicating whether a token fulfils specific alphanumeric patterns	$2d_0 = no$, $4d_0 = no$ (token is not consisted of 2 digits or 4 digits characters)
Is uppercase/ lowercase	Boolean features indicating a token fulfilling specific lowercase and uppercased character patterns	$lu_0 = yes$, $au_0 = no$ (Mary does not start with an uppercase letter and is not all uppercase letters)
Contain character	Boolean features indicating whether a token has a specific character	$cu_0 = yes$, $cd_0 = no$ (token has an uppercase letter and no digit character)

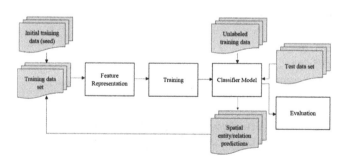

Fig. 2. The proposed semi-supervised learning approach for SIR

train a classifier using a small annotated subset of the data set. The rest of the data set is then split into 9 parts and treated as unlabeled data. On each iteration of the training process, the model is used to label one part of this data and the predictions are then added into the training data set. This approach allows us to reduce the amount of annotated data set to be used in training the classifiers and work with the limited annotated datasets. We choose linear-chain conditional random field (CRF) [11] as the learning algorithm. CRF has been widely used to label sequential tokens in sentences and it is the algorithm of choice in previous attempt to solve the general SpRL problem [12]. In addition, CRF allows label probability output, allowing us to control confidence bound used to filter prediction labels before re-added into training data. The proposed method can be implemented with any learning algorithm such as support vector machine (SVM) or hidden Markov model (HMM) or combination.

Let T be number of tokens in a sequence, K be number of features, x be the vector of feature values generated for a token from a sequence, θ_k be feature weight associated with k-th feature and y be a spatial entity label of the token. With f_k as feature function, the linear-chain CRF computes probability p of y over x as in [11]:

$$p(y|x) = \frac{1}{Z(x)} \prod_{t=1}^{T} \exp\left\{ \sum_{k=1}^{K} \theta_k f_k(y_t, y_{t-1}, x_t) \right\} \tag{2.1}$$

$$Z(x) = \sum_{y} \prod_{t=1}^{T} \exp\left\{ \sum_{k=1}^{K} \theta_k f_k(y_t, y_{t-1}, x_t) \right\} \tag{2.2}$$

To optimize the performance of the model, a greedy forward feature selection algorithm (as detailed in Fig. 3) is proposed to find the best feature set to represent the data set. Given the candidate features, the process starts by initiating an empty set of features. On each iteration, the algorithm adds the best performing candidate feature into the feature set. Mean F1 score is used to measure the performance of each additional feature, validated on the 5-fold setting. This process will be repeated until all candidate features are exhausted.

```
F := empty feature set; X := set of candidate features
Z := best performing feature set

While X not empty do
  For candidate_feature in X do
    Z := F + candidate_feature
    If Z performs the best do
      best_feature := candidate_feature
  Remove best_feature from X, Append best_feature to F
  If F performs better than Z do Z := F   End.
```

Fig. 3. Greedy forward feature selection algorithm used to find the best feature set

Spatial Relation Tagging: Next, the spatial relation tagger is employed to find spatial relations amongst these spatial entities. We focus on identifying spatial information in the form of entity-relation-entity triplets, where the mentions of spatial entities are in the same sentence and the descriptor of the spatial relation between these entities are present in a sub-sentence between the spatial entities.

Given N number of labelled spatial entities in a sentence, we generated $\binom{N}{2}$ pairs of spatial entities (ENT_1, ENT_2) where:

$$ENT_1 : \{w_i, w_{\{i+1\}}, w_{\{i+2\}}, \ldots, w_{\{i+k\}}\} \tag{2.3}$$

$$ENT_2 : \{w_j, w_{\{j+1\}}, w_{\{j+2\}}, \ldots, w_{\{j+l\}}\}$$
$$i \leq i+k < j \leq j+l; k, l \geq 0 \tag{2.4}$$

Such that $ENT_1 \cap ENT_2 = \emptyset$ and w being a keyword in a sentence or in an entity. Assuming each pair of (ENT_1, ENT_2) are spatially related, we define the task of spatial relation tagging as finding a possible spatial relation descriptor in the subsequence

$$S = \{ENT_1, w_{i+k+1}, w_{i+k+2}, \ldots, w_{j-2}, w_{j-1}, ENT_2\} \tag{2.5}$$

Using this method, the entity pair is effectively used as a starting and ending anchor of the subsequence. While this entity pair generation process can be expensive, in practice, the average N is 1.466. It is possible for zero or more spatial entities to be contained in subsequence between the anchors and they should not be tagged as spatial relation. An example of extracted spatial relation is presented in Text 2.

[Computer lab S510] is located **in front** of [tutorial room S509].

Text 2. Sample spatial relation from a sentence. The spatial relation is bold and underlined.

In addition to 12 feature groups used for spatial entity tagging, we use additional 3 features for relational tagging to indicate the entity and relation positions. Table 2 details these additional features using the sentence in Text 2 and the word "front" as token (at index 6).

We use the semi-supervised learning approach and greedy forward feature selection algorithm in the same fashion as the spatial entity tagging module.

Table 2. Additional features generated for spatial relation tagging.

Feature Group	Explanation	Example
Start	Distance of a token from the last token of starting spatial entity anchor of the subsequence	$start_6 = 4$
End	Distance of a token from the first token of ending spatial entity anchor of the subsequence	$end_6 = 2$
Is entity	Boolean feature to indicate whether a token is a spatial entity or not. It was designed to reduce possible search space in sub-sequences where a non-anchor spatial entity is present	$is\_entity_6 = no$

Ranking Spatial Information: Lastly, extracted spatial information triplets are ranked based on their relevancy to the given query. There is a possibility that the classifiers in previous modules may pick up spatial information irrelevant to the query or a large amount of spatial information is presented on a query topic. Ranking allows emphasizing on the more relevant and useful information and enables the output to be easily used in any possible applications.

Let Q be the user query on which spatial information is to be obtained, QT be the TF-IDF representation of the query, $WD = \{WD_1, WD_2, WD_3, \ldots, WD_N\}$ be the set of web documents returned by a search engine and WDT_i be TF-IDF representation of WD_i. Let $SD_i = \{SD_{i1}, SD_{i2}, SD_{i3}, \ldots, SD_{im}\}$ be the set of spatial information triplets extracted from WD_i, and SDT_{ij} be the TF-IDF representation of SD_{ij}.

Let $CS(A, B)$ be cosine similarity between two non-zero vectors:

$$CS(A, B) = \frac{\vec{A} \cdot \vec{B}}{\|\vec{A}\| \cdot \|\vec{B}\|} \tag{2.6}$$

The ranking score W of a spatial information SD_{ij} obtained from the web document WD_i with query Q is calculated by:

$$W(SD_{ij}, WD_i, Q) = CS(QT, WDT_i) + CS(QT, SDT_{ij}) \tag{2.7}$$

Spatial information triplets with higher W score mean higher relevancy to the supplied query and they are ranked higher in the output.

3 Empirical Analysis

Datasets: We used two data sets sourced from the web as detailed in Table 3. The first dataset was sourced from LostOnCampus[1] (LoC), an Australian campus directory website. Another data set was constructed from self-tour documents of numerous universities in Australia, USA and Great Britain (e.g.[2,3,4]). All documents (in form of HTML or PDF) in the data sets were split into individual sentences using NLTK's punkt tokenizer[5]. Each sentence was then pre-processed using methods described in Sect. 2. Spatial entities and relations present in the pre-processed data set were then manually annotated and the data set was split into 80/20% distribution for training and testing purpose using random sampling.

Evaluation Criteria: Performance of spatial entity and spatial relation tagging models was evaluated using precision (P), recall (R) and $F1$ measure. Precision denotes the

---

[1] http://lostoncampus.com.au.

[2] http://www.columbia.edu/content/self-guided-walking-tour.html.

[3] http://www.unimelb.edu.au/campustour/self-guided.

[4] http://www.bristol.ac.uk/university/visit/walking-tour.html.

[5] http://www.nltk.org/api/nltk.tokenize.html#module-nltk.tokenize.punkt.

Table 3. Data sets used for training the classifiers.

Data set	#Docs	#Sentences	#Entities	#Relations	Type
LoC (D1)	200	3878	2131	267	HTML
University self-tour (D2)	19	2435	1939	134	PDF

proportion of predictions that are true positives and Recall shows percentage of ground truth that are positively predicted [3]. To determine whether a prediction is a true positive, we used two different evaluations, complete and partial. In the complete matching evaluation, a prediction from the model is only counted as a match (or true positive) if it exactly matches an answer label. For spatial relation, this means matching the spatial relation descriptor precisely given two anchoring spatial entities. Meanwhile for partial evaluation, a true positive is awarded if the prediction overlaps at least one token of a spatial entity or relation label.

To measure the performance of a semi-supervised model against the corresponding supervised model, we calculated the average loss of each P, R and F1 score on various experiments. We conjecture that a semi-supervised model should demonstrate inferior results to the equivalent supervised model, thus the better performing semi-supervised models should produce the least average performance loss. An average performance loss for a performance measure method $x \in \{P, R, F1\}$ is computed by formula:

$$Average\_loss_x = \frac{1}{N} \sum_{i=1}^{N} (B_{xi} - Sp_{xi}) \qquad (3.1)$$

where N is the number of samples taken, and B_{xi} and Sp_{xi} are the score on performance measure x achieved by supervised and semi-supervised models respectively. In all experiments, we employed an implementation of linear-chain CRF called CRFSuite [14].

In evaluating the performance of our ranking method, we measured NDCG$_k$ (Normalized Discounted Cumulative Gain at K) score [13], which demonstrates the accuracy of the relevancy ranking method against human given score. Let $R = \{R_1, R_2, \ldots, R_k\}$ be the relevancy score given by human to the top k ranked spatial information and Rs be the R sorted in descending order. NDCG$_k$ is computed as:

$$NDCG_k = \frac{\sum_{i=1}^{k} \frac{2^{R_i} - 1}{\log_2(i+1)}}{\sum_{i=1}^{k} \frac{2^{Rs_i} - 1}{\log_2(i+1)}} \qquad (3.2)$$

3.1 Supervised Models Performance

Before experimenting with semi-supervised trained classifiers, we trained and measured performance of supervised-trained CRF models with 100% training data as our baseline result. To achieve optimal performance from the features, we employed the greedy feature selection algorithm presented in Fig. 3 to select the optimal feature set.

Table 4. Supervised models' performance.

Task	Evaluation	Data Set	Precision	Recall	F1
Spatial entity	Complete	D1	0.846	0.775	0.809
		D2	0.791	0.588	0.675
	Partial	D1	0.927	0.818	0.869
		D2	0.872	0.628	0.730
Spatial relation	Complete	D1	0.855	0.855	0.855
		D2	0.719	0.500	0.590
	Partial	D1	0.873	0.857	0.865
		D2	0.750	0.511	0.608

Table 4 shows the results obtained from the supervised models on the two data sets for both spatial entity and relation identification.

3.2 Semi-supervised Models Performance

We started the training process with a small amount of seed data and gradually increased it. To explore the effect of the size of the seed data to performance of the model, we conducted experiments with $P \in \{10\%, 20\%, 30\%, 40\%, 50\%\}$ training data as seed data. Another parameter that was examined in experiments is the probability lower bound, used to filter out lower probability predictions to be added into the training data set on each iteration of the semi-supervised training process. This value varies from 0 (i.e. all predictions are added) to 0.9 (i.e. only predictions with probability $\geq$ 0.9 are added back to the training data set) with a regular increment of 0.1.

From Figs. 4 and 5, in the spatial entity recognition task, it is observed that increase in probability lower bound positively correlates to precision and negatively impacts the recall and F1 scores. Setting a higher threshold helps maintaining the quality of candidates and consequently the fewer candidates are identified. The severity of the impact varies according to different seed data sizes. While smaller seed sizes score relatively low without any bound, their precision score increases and F1 and recall rate decrease sharply as the probability bound value progresses. In comparison, larger seed data sizes generally perform relatively better at zero probability bound and their scores change at much slower rate. We believe this is attributed to the diversity of patterns contained in the seed data. Smaller seed data has more homogenous patterns, thus leading the model learning only few patterns repeatedly during training iterations, increasing precision score and dropping recall and F1 values. In contrast, larger seed data contains heterogeneous patterns, allowing the models to learn varied patterns, subsequently results in higher recall.

As shown in Table 5, observations on the average loss scores between the semi-supervised and baseline supervised models further confirm the positive correlation of larger seed data to the overall performance of the models. The only anomaly is found in the university self-tour precision score, which shows that the 10% sample to have the least loss. This can be attributed to the homogeneity of patterns in the seed data. As this data has about 50% and 30% less instances of sentence and spatial entity respectively compared to the LoC data set, the effect of homogeneity is more severe.

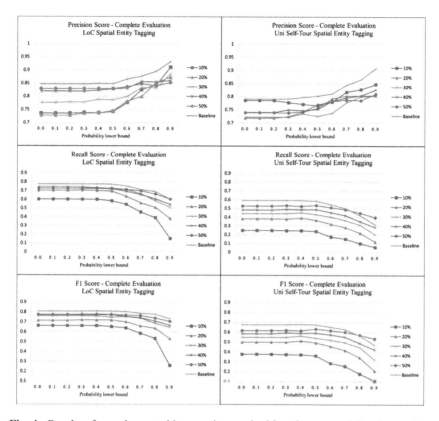

Fig. 4. Results of complete matching: semi-supervised learning on spatial entity tagging.

Performance of models on the spatial relation tagging task was found very similar to the performance on the spatial entity tagging task. Due to space constraint, we are not able to present the individual graphs. Increase in probability lower bound positively correlates to the precision rate and negatively impacts the recall and F1 score. Similar effect on the seed data size to the degree of impact is observed. Smaller seed sizes show more volatile changes, while larger seed data increase/decrease at much slower rate. On LoC dataset, the highest value of probability lower bound (0.9) on the smallest seed data resulted in no predictions being accepted in retraining steps.

Table 6 reports the average loss between the semi-supervised scores and the baseline supervised scores on the spatial relation tagging task. It confirms the positive correlation of larger seed data to the precision, recall and F1 score of the model. On both data sets, the largest seed data (50%) scored the least average loss on recall and F1 score. Similar to spatial entity tagging, homogeneity of patterns in smaller seed data results in lower precision loss. In LoC data set, however, high values of confidence bound result in the smallest data set to have the highest average loss.

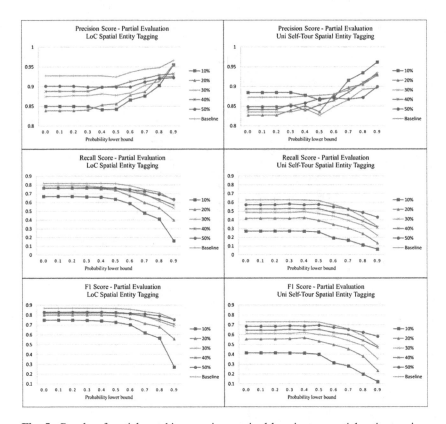

Fig. 5. Results of partial matching: semi-supervised learning on spatial entity tagging.

Table 5. Spatial entity tagging: average loss in performance of semi-supervised models against the supervised models. Lowest loss values are bolded.

Data set	Avg. loss/seed size		10%	20%	30%	40%	50%
D1	Complete	Precision	0.088	0.095	0.062	0.032	**0.030**
		Recall	0.227	0.111	0.052	0.056	**0.037**
		F1	0.198	0.113	0.059	0.047	**0.033**
	Partial	Precision	0.067	0.068	0.044	0.031	**0.030**
		Recall	0.218	0.101	0.049	0.060	**0.038**
		F1	0.186	0.096	0.050	0.049	**0.034**
D2	Complete	Precision	**0.025**	0.061	0.072	0.051	0.056
		Recall	0.335	0.209	0.139	0.090	**0.029**
		F1	0.331	0.194	0.129	0.081	**0.033**
	Partial	Precision	**−0.011**	0.024	0.034	0.019	0.022
		Recall	0.347	0.210	0.132	0.083	**0.018**
		F1	0.342	0.191	0.116	0.068	**0.015**

Table 6. Spatial relation tagging: average loss in performance of semi-supervised models against the supervised models. Lowest loss values are bolded.

Data set	Avg. loss/seed size		10%	20%	30%	40%	50%
D1	Complete	Precision	0.030	0.018	**−0.009**	−0.008	0.004
		Recall	0.350	0.180	0.173	0.128	**0.121**
		F1	0.260	0.115	0.100	0.074	**0.072**
	Partial	Precision	0.039	0.027	**−0.001**	0.001	0.013
		Recall	0.351	0.181	0.174	0.146	**0.127**
		F1	0.265	0.120	0.105	0.088	**0.081**
D2	Complete	Precision	**−0.120**	−0.099	−0.046	−0.044	−0.064
		Recall	0.336	0.274	0.231	0.223	**0.123**
		F1	0.361	0.270	0.216	0.204	**0.089**
	Partial	Precision	**−0.120**	−0.099	−0.046	−0.044	−0.064
		Recall	0.336	0.274	0.231	0.223	**0.123**
		F1	0.361	0.270	0.216	0.204	**0.089**

3.3   Benchmarking with Similar Methods

No specific method exists that can explicitly be used for comparison here. So to analyze the performance of the proposed spatial entity and relation classifiers, we compared them against two most relevant state-of-the-art methods, namely SVM-HMM [15] and Stanford NER [16]. SVM-HMM is a support vector machine model designed for the sequential tagging task and has been used in the SpRL task before [6]. In the benchmarking process, we retrained the SVM-HMM classifier using the same feature sets used as the CRF classifier used in this paper. As suggested in the manual page[6], each feature value is represented as binary feature and we performed a 5-fold cross validation grid search on its parameters to ensure best possible performance.

The well-known Stanford NER system uses a general linear-chain CRF sequence model for tagging tasks. We compared the CRF classifiers to Stanford NER in two different experiment settings. Firstly, we retrained the classifier to tag the spatial entity and spatial relation in our data sets using supervised learning. Here, all features and learning settings used by Stanford NER for the CoNLL2003 task (using the–goodconll flag[7]) were retrained. This benchmark allows us to check the effectiveness of our feature selection algorithm in improving the accuracy of our model. In addition to supervised learning, we also trained Stanford NER in semi-supervised setting. More specifically, we used the 50% seed data size with no probability bound (0.0). As shown in Table 7, on all tasks and data sets, our classifiers outperform the results from both systems. This ascertains that the CRF classifier and feature sets employed in our proposed method are optimal in tagging the spatial entities and relations.

---

[6] https://www.cs.cornell.edu/people/tj/svm_light/svm_hmm.html.

[7] http://nlp.stanford.edu/projects/project-ner.shtml.

Table 7. Benchmark results vs similar systems.

Task/learning method	Model/ data set	Complete F1		Partial F1	
		D1	D2	D1	D2
Entity/supervised	Our CRF	**0.809**	**0.675**	**0.869**	**0.730**
	Stanford	0.786	0.644	0.855	0.702
	SVM-HMM	0.775	0.665	0.849	0.717
Entity/semi-supervised	Our CRF	**0.773**	**0.615**	**0.828**	**0.684**
	Stanford	0.762	0.568	0.823	0.637
Relation/supervised	Our CRF	**0.855**	**0.590**	**0.865**	**0.608**
	Stanford	0.850	0.571	0.860	0.574
	SVM-HMM	0.816	0.571	0.816	0.574
Relation/semi-supervised	Our CRF	**0.784**	**0.455**	**0.784**	**0.455**
	Stanford	0.772	0.413	0.772	0.413

Table 8. $NDCG_k$ for various queries used in the tests.

Query	Top-k					
	1	3	5	10	15	20
S515, S Block, QUT Gardens Point	1.000	0.927	0.973	0.975	0.977	0.977
The Cube, P Block, QUT Gardens Point	0.333	0.629	0.817	0.854	0.875	0.878
Z410, S Block, QUT Gardens Point	0.333	0.516	0.596	0.694	0.811	0.816
Pitch Drop Experiment, Priestly Building, UQ St. Lucia	1.000	1.000	1.000	1.000	1.000	1.000
Dorothy Hill Library, Hawken Engineering Building, UQ St. Lucia	1.000	1.000	1.000	0.901	0.918	0.978
Postgraduate Study Area, Biological Sciences Library Building, UQ St. Lucia	1.000	1.000	1.000	0.967	0.971	0.995
PBL Room 2, Faculty of Science, UQ St. Lucia	0.000	0.235	0.447	0.664	0.728	0.728
Queensland Micro and Nanotechnology Centre, Griffith Nathan	1.000	1.000	1.000	0.946	0.932	0.981
N72 Seminar Room, Glyn Davis Building, Griffith Nathan	1.000	0.980	0.981	0.936	0.938	0.979
N12 Chaplaincy, Sewell Building, Griffith	1.000	1.000	0.854	0.663	0.790	0.900
Mean	0.767	0.829	0.867	0.86	0.894	0.923
Std. Dev.	0.387	0.273	0.197	0.135	0.09	0.092

3.4 Ranking Result: Overall Performance of the Proposed System

In response to a query, spatial entity and relation tagging recognition tasks are performed on the retrieved documents. The proposed ranking method then ranks the spatial information in relevancy order. We used spatial information extracted from LoC documents of 3 universities in Brisbane, Australia (147 triplets from 448 documents) and 10 queries of various locations in these universities. For each query, we computed

Table 9. Human scoring guideline for result ranking experiments.

Score	Explanation
4	One or both spatial entities match the query
3	One or both spatial entities are in the same level
2	One or both spatial entities identified are in the same building
1	One or both spatial entities identified are in the same campus
0	No relevant spatial information is obtained

relevancy score of each triplet using the ranking method. Each returned triplet was then given score from 0–4 by three human annotators based on relevancy to the query. The scoring guide is given in Table 9. $NDCG_k$ scores are then computed using the mean score. Table 8 presents the $NDCG_k$ values measured on spatial information ranking for the given query set. The average NDCG value exceeds 0.75 and reached its highest at $k = 20$ with 0.923. This ascertains that the proposed method is able to find the relevant and required information accurately.

4 Conclusion, Limitation and Future Work

We have presented a novel method of identifying the spatial information in the web documents with semi-supervised machine learning classifiers. CRF models were trained in the semi-supervised bootstrap fashion to identify spatial entities and spatial relation descriptions of these entities. We were able to achieve high performance models with much less manually annotated training data available. While increase on the seed data size has proven to be positively correlated to model performance, our experiments showed with only 50% of the training data as seed, the semi-supervised trained models showed just less than 9% loss on F1 metrics. To ensure the relevancy of produced spatial information for a query, we employed the query bot which applies relevancy filter on input web documents using Google Search, and ranks the spatial information triplets. The $NDCG_k$ evaluation shows that the proposed method can find highly accurate spatial information in response to the user queries.

For future work, we plan to expand the candidate feature sets. All features used in this work consist of only local syntactic knowledge, we plan to include external knowledge and more sophisticated local features such as word clustering, phrasal clustering [17]. In addition, during spatial relation recognition, we assumed that all spatial entities in a sequence are related and $\binom{N}{2}$ candidate pairs were generated from N spatial entities. This large amount of candidate spatial entity pairs could be reduced by validating the possible relation before the tagging process, effectively reducing the search space and increasing accuracy of the system. Lastly, we plan to extend the spatial relationship tagging system to identify spatial relations not located between two candidate spatial entities. Such example is the sentence "Next to Bridger Bowl is Bohart Ranch", where the spatial relation "next to" is mentioned before its spatial entities "Bridger Bowl" and "Bohart Ranch".

References

1. Walter, M.R., Hemachandra, S., Homberg, B., Tellex, S., Teller, S.: A framework for learning semantic maps from grounded natural language descriptions. Int. J. Robot. Res. **33** (9), 1167–1190 (2014)
2. Talbot, B., Schulz, R., Upcroft, B., Wyeth, G.: Reasoning about natural language phrases for semantic goal driven exploration. In: Proceedings of the Australasian Conference on Robotics and Automation 2015 (2015)
3. Nadeau, D., Sekine, S.: A survey of named entity recognition and classification. Linguisticae Investigationes **30**, 3–26 (2007)
4. Hou, J., Schulz, R., Wyeth, G., Nayak, R.: Finding within-organisation spatial information on the Web. In: Pfahringer, B., Renz, J. (eds.) AI 2015. LNCS, vol. 9457, pp. 242–248. Springer, Cham (2015). doi:10.1007/978-3-319-26350-2_21
5. Kolomiyets, O., Kordjamshidi, P., Bethard, S., Moens, M.-F.: Semeval-2013 task 3: spatial role labeling. In: Second Joint Conference on Lexical and Computational Semantics (* SEM), Volume 2: Proceedings of the Seventh International Workshop on Semantic Evaluation (SemEval 2013), pp. 255–266 (2013)
6. Bastianelli, E., Croce, D., Nardi, D., Basili, R.: UNITOR-HMM-TK: Structured kernel-based learning for spatial role labeling. In: Second Joint Conference on Lexical and Computational Semantics (* SEM), vol. 2, pp. 573–579 (2013)
7. Roberts, K., Harabagiu, S.M.: UTD-SpRL: a joint approach to spatial role labeling. In: Proceedings of the First Joint Conference on Lexical and Computational Semantics-Volume 1: Proceedings of the main conference and the shared task, and Volume 2: Proceedings of the Sixth International Workshop on Semantic Evaluation, pp. 419–424 (2012)
8. Prakash, V.J., Nithya, L.M.: A survey on semi-supervised learning techniques. Int. J. Comput. Trends Technol. **8**(1), 25–29 (2014)
9. Manning, C., Raghavan, P., Schütze, H.: Introduction to Information Retrieval. Cambridge University Press, Cambridge (2008)
10. Cho, H.-C., Okazaki, N., Miwa, M., Tsujii, J.: Named entity recognition with multiple segment representations. Inf. Process. Manag. **49**(4), 954–965 (2013)
11. Sutton, C., McCallum, A.: An introduction to conditional random fields. Found. Trends Mach. Learn. **4**(4), 267–373 (2011)
12. Mani, I., et al.: SpatialML: annotation scheme, resources, and evaluation. Lang. Resour. Eval. **44**(3), 263–280 (2010)
13. Kaggle: Normalized Discounted Cumulative Gain
14. Okazaki, N.: CRFsuite: a fast implementation of Conditional Random Fields (CRFs) (2007)
15. Joachims, T.: SVM-HMM: sequence tagging with SVMs (2008)
16. Finkel, J.R., Grenager, T., Manning, C.: Incorporating non-local information into information extraction systems by Gibbs sampling. In: Proceedings of the 43rd Annual Meeting on Association for Computational Linguistics, pp. 363–370 (2005)
17. Tkachenko, M., Simanovsky, A.: Named entity recognition: exploring features. Proc. KONVENS **2012**, 118–127 (2012)

When Will a Repost Cascade Settle Down?

Chi Chen[✉], HongLiang Tian, Jie Tang, and ChunXiao Xing

Research Institute of Information Technology, Tsinghua National Laboratory for
Information Science and Technology, Department of Computer Science and
Technology, Tsinghua University, Beijing 100084, China
chenchi14@mails.tsinghua.edu.cn

Abstract. Repost cascades play a critical role in information diffusion
on social media sites. They are developed by series of reposts and stop
eventually. Substantial previous work has studied and predicted various
aspects of repost cascades such as growth, burst and recur. However,
how or even whether it is possible to predict when a repost cascade will
settle down remains to be an open problem. Existing models cannot be
directly applied to solve the problem as the feature based models are
sensitive to features, while the point process based models assume that
the followers of all reposters are disjoint. In this paper, we propose a novel
definition settling time to model this problem. We develop a point process
based model to get rid of the restriction in previous studies and make
an accurate prediction of the settling time. We conduct an extensive set
of experiments on Sina Weibo dataset. The results show that our model
achieves over 10% performance gain than the state-of-the-art approaches
after observing the cascades for 24 h.

Keywords: Repost cascade · Point process · Prediction

1 Introduction

Information diffusion is an important topic for many application such as data
integration [21], disaster prediction [3] and epidemic transmission prediction [8].
Reposting on social media, like Weibo, Facebook, Twitter, etc., is an important
online mechanism for information diffusion: a post is reposted by one's followers,
then several of them repost it again, and a *repost cascade* can gradually develop
[10]. Substantial previous work has studied how to characterize and predict the
growth of such repost cascades [2,5,6,22]. Bursty nature in repost cascades is
also largely explored in lots of research with the aim of predicting the burst time
[17]. However, it is not clear how, or even whether it is possible, to predict when
a repost cascade will settle down. Thus we introduce *"settling time"* in a repost
cascade which represents when the repost cascade is about to settle down.

Predicting the settling time in repost cascades is vitally important for
crisis management and rumor control. For example, Hillary Clinton and Donald
Trump have emerged as the two front-runners in the 2016 U.S. presidential
campaign [18]. However, a series of explosive news, such as "Trump Sexual

© Springer International Publishing AG 2017
A. Bouguettaya et al. (Eds.): WISE 2017, Part I, LNCS 10569, pp. 165–179, 2017.
DOI: 10.1007/978-3-319-68783-4_12

Assault Allegations" and "Clinton's Emails", leads to a negative effect on their election. Facing many objections, besides hoping that the news will be flipped over as soon as possible, they are eager to find a way to deal with this situation in order to win the election. If they knew when the incidents will subside, what or even how factors influence the settling time of such cascades, they could estimate the loss and take appropriate measures in time to reduce their loss.

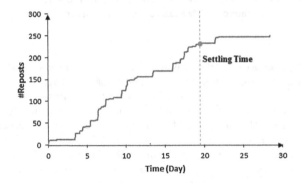

Fig. 1. An example of settling time on Sina Weibo. The Y-axis represents the cumulative number of reposts. This post has more than 200 reposts within 1 month. The speed of reposting rapidly decreases after the settling time and the cascade eventually stops.

We start our research with an example of settling time on Sina Weibo. Figure 1 shows the cumulative number of reposts over time in a repost cascade with more than 200 reposts within 1 month from Sina Weibo which is the most popular microblog in China. Many bursts following by a short dormant period are observed before the settling time, while the speed of reposting rapidly decreases after the settling time and the cascade eventually stops.

Existing models for predicting cascade related topics cannot be directly applied to predict the settling time, because repost cascades near the settling time have some special characteristics such as low incremental and long reposting interval as Fig. 1 shows. Thus previous related models have to be modified to predict the settling time. There are two major types of approaches for cascade related prediction: feature based models and point process based models. The feature based models predict the eventual growth [2,10,22] or future burst time [11,17] through utilizing a rich set of features which are extracted from historical data. However, feature extraction is expensive and it is hard to know whether there are other effective features. Point process based models model the formation of a repost cascade in a network by using point process theory. Nevertheless, existing point process based models assume that the followers of all reposters are disjoint [9,22], while there are a large number of common followers for reposters in the real network structure on social media. The assumption limits the performance of cascade related prediction because it does not realize that there are many special followers who have reposted or received the post before. In other words, the point process based models in previous work ignore the real network structure and only consider the number of followers for a reposter on social media. Due to the characteristics of settling time prediction we mentioned before, the prediction leads to a big deviation.

To address above issues, we propose a novel definition of "settling time" to model when a repost cascade will settle down and illustrate its prediction task. We develop PPST (*Point Process Based Model for Predicting the Settling Time*) where we remove the assumption of disjoint followers, introduce a new concept named "influenced followers" for a reposter who receive the post for the first time, and we develop a method to model the concept. In the end, we conduct an extensive set of experiments on Sina Weibo dataset. The results show that our model achieves over 10% performance gain than the state-of-the-art approaches after observing the cascades for 24 h. Moreover, we study the three primary factors on Sina Weibo, which is the biggest microblog website in China, including the reaction time, the infectious rate and the number of followers.

Organization. The remainder of this paper is organized as follows. Section 2 describes previous related work about repost cascades. In Sect. 3, we formally define the settling time and introduce the prediction task. We describe our method including the modeling and predicting in Sect. 4. In Sect. 5, we introduce our dataset and present the prediction result. In the end, we conclude in Sect. 6.

2   Related Work

Repost cascade has been a hot topic for a long time [14]. Many papers have analyzed and explored the formation of repost cascades. The most relevant work to our task is the research of cascade growth prediction. Recent models for predicting the growth of repost cascades are generally characterized by two types, feature based models and point process based models.

Feature based models predict the eventual growth through extracting effective features from historical data including content features, original poster features, structural features and temporal features [22]. Then various learning algorithms are applied, such as regression models [1], passive-aggressive algorithm [7], and probabilistic collaborative filtering [20]. These models are feature driven, thus, they are sensitive to quality and quantity of features. In contrast, our model requires no feature engineering and results in a formula that allows it to predict the settling time of repost cascades as they are spreading through the network.

Point process based models are different from the feature based models, which model the formation process of repost cascades by utilizing point process theory. Such models have been successfully applied to predict the eventual growth [22] and dynamic repost activity [9] in repost cascades. A major distinction between our model and existing models based on point process is that we consider the real network structure of reposters and remove the assumption that the followers of all reposters are disjoint in repost cascades. Due to special characteristics of settling time prediction, our model receives much better performance than previous approaches on Sina Weibo.

3   Problem Definition

In this section, we start with some basic definitions, then introduce the definition of the settling time. Finally, we give the prediction task in detail. To specify when a repost cascade begins to settle down, we discretize time into equal time windows, and predict the window which a repost cascade starts to settle down from. In order to define the settling time, we firstly define effective peaks whose height and width satisfy some conditions. Then the end time of the last effective peak is defined as the settling time of the repost cascade. It should be noted that all time value in our paper is the relative time to the original posting time.

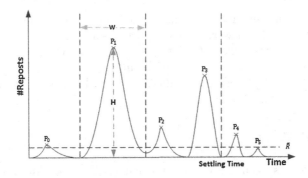

Fig. 2. Multiple peaks occur in the number of reposts over time. P_1 and P_3 are effective peaks, while P_0, P_2, P_4 and P_5 are not effective peaks.

Now we consider a reposting time sequence $S = \{T_0, T_1, T_2, T_3, \ldots T_n\}$ with n reposts, in which T_0 is the first posting time and T_i is the i-th reposting time. Repost activity sequence can be worked out by reposting time sequence S and time window size Δ_{obs}. Repost activity sequence is denoted by $C = \{R_0, R_1, R_2, R_3, \ldots R_m\}$ with m time windows. R_i represents the number of reposts in the i-th time window. The average number of reposts in a time window for C is denoted by $\overline{R}$. Similarly, $C_{(i,j)} = \{R_i, R_{i+1}, \ldots R_j\}$ is the repost activity subsequence from $i * \Delta_{obs}$ to $j * \Delta_{obs}$. In particular, if any R_k is not greater than R_{k+1} in a repost activity subsequence $C_{(a,b)}$, we call $C_{(a,b)}$ a non-descending repost activity subsequence. In the similar situation, if any R_k is not less than R_{k+1}, $C_{(a,b)}$ is a non-ascending repost activity subsequence.

Definition 1 *Effective Peak.* Two repost activity subsequence $C_{(i,j)}$ and $C_{(j,k)}$ form a peak $P_{(i,j,k)}$, if the following condition is satisfied: $C_{(i,j)}$ is a non-descending repost activity subsequence, and $C_{(j,k)}$ is a non-ascending repost activity subsequence. In particular, if $R_j \geq max(H_0, \overline{R} * m)$ and $k - i \geq W_0$, we define the peak $P_{(i,j,k)}$ as an effective peak.

Definition 2 Settling Time. If the last effective peak in repost activity sequence is $P_{(i,j,k)}$. The settling time window is defined as $STW = k$, and the settling time is $ST = k * \Delta_{obs}$.

In Definition 1, H_0 and W_0 limit the minimal height and width of effective peaks. In practice, we set $\Delta_{obs} = 6\,\text{h}$, $m = 10$, $H_0 = 10$ and $W_0 = 12\,\text{h}$. Figure 2 shows an example of repost activity. There are six peaks in the repost activity sequence. P_1 and P_3 are effective peaks because of their height and width, while the others are not effective peaks. According to the definition, the end time of the last effective peak P_3 is the settling time.

Prediction Task. Given the data from historical reposts before T, how to predict the settling time ST of a repost cascade.

To predict the settling time, we predict the repost activity sequence at first. That is to say, we firstly predict the repost activity R_i where $i * \Delta_{obs} > T$, then the settling time can be worked out by using its definition. Thus, our major task is predicting the repost activity sequence $\hat{C}$ by using the given reposts before T.

4   Modeling and Predicting

In this section, we model the formation of repost cascades by using point process theory for predicting the repost activity after observing them for a period of time. First, we illustrate our model in detail in Sect. 4.1, including the estimation and modeling of the three important components. In Sect. 4.2, we describe how to predict the repost activity in the future.

To work out the repost activity at time t, three problems should be solved in the formation process of repost cascades. They are:

P1. How many users have received the post until time t?
P2. Among the users who have received the post, how many of them will make a decision at time t, repost or not?
P3. Among the users who make a decision at time t, how many of them will repost the post?

Next, we model and predict the repost activity so as to solve the above three problems.

4.1   Modeling

To predict the repost activity, firstly we formally define the relationship between the repost intensity $\lambda(t)$ and the number of reposts δR from t to $t + \Delta$.

$$\delta R = \int_t^{t+\Delta} \lambda(s)ds \tag{1}$$

The repost activity $\lambda(t)$ at time t depends on the reposts before t. The reason is that users only receive the posts from their followees. If a user reposts the post, there must be a followee who reposted it before. Hence, the repost activity is modeled by infectious rate $p(t)$, repost time T_i, the number of influenced followers I_i, and reaction time distribution $\phi(s)$. The exact relationship is described:

$$\lambda(t) = p(t) \sum_{i:T_i<t} I_i \phi(t - T_i) \tag{2}$$

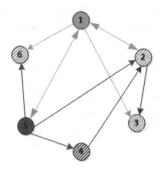

Fig. 3. An example to explain the influenced followers. If A follows B, there will be an edge from B to A. In this network, 1 reposts the post, then 5 reposts it again. 5 only influences 4, because 2, 3, 6 have been influenced after 1's reposting.

In Eq. (2), $\phi(t-T_i)$ is the *memory kernel*, which measures the probability of reposting after $t - T_i$ since the first time the followers receive the post at time T_i. $s = t - T_i$ is called reaction time whose distribution $\phi(s)$ has been studied a lot [4,19]. We discuss $\phi(s)$ on Sina Weibo later in detail. $p(t)$ is the infectious rate and we allow it to change over time for accurate prediction, although the infectious rate is the constant value in many existing models [12,13,23]. We estimate and model the infectious rate $p(t)$ later.

Once a post is reposted, there are followers who is the "first time" to receive the post. We call them influenced followers for this repost. Previous studies made the assumption that the followers of all reposters are disjoint [9,22], thus, the number of influenced followers for a repost in their work would be defined as the number of reposter's followers. In our paper, we remove the assumption, and the number of influenced followers for the i-th repost is denoted by I_i in considering the real network structure. In other words, the followers who have received the post before are excluded in the influenced followers in our method. An example of calculating I_i is showed in Fig. 3. When we use the assumption from previous work to the network in Fig. 3, the second reposter 5 influences four followers because 5 has four followers. However, after considering the real network structure, only one follower is influenced by 5's reposting for the first time. Hence, three followers are repeatedly calculated after using the assumption. Closer network structure for reposters leads to a larger deviation when the

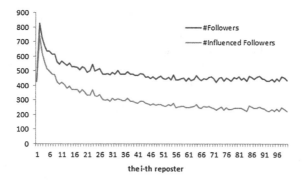

Fig. 4. The average number of influenced followers and followers for early 100 reposts of more than 10000 posts on Sina Weibo. The difference between two metrics is gradually increasing.

assumption is used. Figure 4 shows the difference between the number of followers and the number of influenced followers on Sina Weibo. Now we assume that the number of influenced followers only depends on time t. The number of influenced followers at time t is denoted by $I(t)$, and we propose a method to model $I(t)$ at the end of this section.

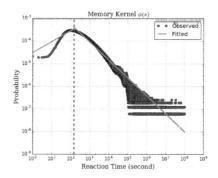

Fig. 5. Reaction time distribution and estimated memory kernel $\phi(s)$. The reaction time is plotted on logarithmic axes. The decreasing trend suggests a power law distribution.

Reaction Time $\phi(s)$. Reaction time represents how long it takes for a person to repost after receiving the post. Reaction time on Twitter has been studied [22], while the distribution of reaction time on Sina Weibo is still unexplored. Here we describe how to work out the memory kernel $\phi(s)$ which is a probability distribution of the reaction time. The reaction time is the time interval between the time, when the post is first received by a follower, and the follower's reposting time. Thus we calculate the reaction time by using reposting time sequence and the relationship of reposters in the real network structure. Figure 5 shows the distribution of reaction time and the estimated memory kernel $\phi(s)$. The distribution is showed to be heavily tailed, and divided into two parts: increasing in the first near 2.5 min, followed by a power-law decay. Then we fit the reaction time by using the following equation:

$$\phi(s) = \begin{cases} 0 & (s \leq 0), \\ a \times s^b & (0 < s \leq s_0), \\ c \times s^{-d} & (s > s_0). \end{cases} \tag{3}$$

After fitted by using the least square error, the five parameters are worked out, $s_0 = 148$, $a = 3 \times 10^{-5}$, $b = 0.479$, $c = 2.1 \times 10^{-2}$, $d = 0.79$. The figure suggests that reposting is likely occurred within 2 h, after that, the probability of reposting is rapidly decreasing. Consequently, posts have less probability to be reposted with time passing by.

Infectious Rate $p(t)$. The infectious rate is one of three primary components for repost intensity. Here we estimate it by using the method which was supposed by Kobayashi et al. [9]. They made an assumption that the infectious rate in a small time window $t \in [t_{st}, t_{en}]$ is constant. Thus it is calculated by using integration to both sides of Eq. (2).

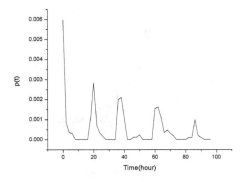

Fig. 6. An example of infectious rate $p(t)$ over time from Sina Weibo.

$$\hat{p}_t = \frac{\delta R}{\sum_i n_i \{\Phi(t_{en} - T_i) - \Phi(t_{st} - T_i)\}} \tag{4}$$

where δR is the number of reposts in the time window $[t_{st}, t_{en}]$, and $\Phi(t)$ is the integral of the memory kernel, $\Phi(t) = \int_0^t \phi(s)ds$. In Sect. 5, we discuss the window size $\Delta_p = t_{en} - t_{st}$ in detail. An example of estimated infectious rate is shown in Fig. 6.

Intuitively, the infectious rate gradually decreases with regular fluctuation. Next, we model the infectious rate depending on time t by using the following equation.

$$p(t) = p_0 \{1 - r_0 sin(\frac{2\pi}{T_m}(t + \phi_p))\}(\frac{t}{\Delta_p})^{-\alpha} \tag{5}$$

T_m is set to 24 h which reflects the period of fluctuation. The parameters p_0, r_0, ϕ_p, α correspond to the intensity of activity, the relative amplitude of the oscillation, the phase and the rate of decay respectively. These 4 parameters are figured out through fitting Eq. (5). The least square error is used in the fitting process. The average and variance of the fitted parameters in our dataset are $p_0 = 0.01 \pm 0.0002$, $\phi_p = 0.125 \pm 0.01$, $\alpha = 0.46 \pm 0.007$ and $r_0 = 0.23 \pm 0.02$.

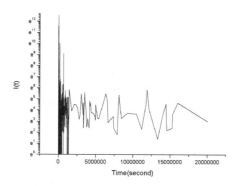

Fig. 7. An example of the number of influenced followers over time from Sina Weibo.

Number of Influenced Followers $I(t)$. According to the assumption that the number of influenced followers I only depends on time t, I is modeled by using history data I_i. Figure 7 shows an example of $I(t)$ on Sina Weibo. There are two properties for $I(t)$: decreasing amplitude and increasing period of oscillation. Decreasing amplitude suggests that early reposts have greater influence than the later. The longer period of oscillation is the result of the lower density of reposts. Based on above observation, we propose a model for the time dependence $I(t)$:

$$I(t) = e^{\gamma + \beta sin(\frac{2\pi T_m}{t}(t+\phi_I))} e^{\frac{-t}{\tau_m}} \tag{6}$$

We set T_m to 24 h. γ, β, ϕ_I, τ_m correspond to the average intensity, the amplitude of oscillation, the phase and the rate of decay respectively. These 4 parameters are figured out from the history reposts, then be used in the prediction. We also use the least square error to fit Eq. (6). The average and variance of the fitted parameters in our dataset are $\gamma = 4.59 \pm 0.38$, $\beta = 2.17 \pm 0.72$ and $\phi_I = 3.09 \pm 0.03$.

4.2 Prediction

To predict repost activity sequence $\hat{C}$, we firstly predict $\hat{\lambda}(t)$ which is the repost intensity at time t. To calculate $\hat{\lambda}(t)$, all reposts before t are needed. Unfortunately, only the reposts before T are observed. Here we consider the repost intensity into two parts: one comes from the data before T, while the other is from the reposts which are published between T and t. The first and second parts are denoted by $f(t)$ and $g(t)$. That is:

$$\hat{\lambda}(t) = f(t) + g(t) \tag{7}$$

According to Eq. (2), $f(t)$ is formally defined as:

$$f(t) = p(t) \sum_{i:T_i < T} I_i \phi(t - T_i) \tag{8}$$

We calculate the second part as follow:

$$g(t) = p(t) \int_T^t \hat{\lambda}(s) I(s) \phi(t - s) ds \tag{9}$$

In practice, we calculate the integral by using rectangle method. The time step is set to 1 min. After predicting the repost activity sequence $\hat{C}$, the settling time on $\hat{C}$ is worked out by using the definition of the settling time.

5   Experiment

In this section, we firstly introduce the dataset, the evaluation metric and the baselines used in the prediction. Second, we present the comparison result of the settling time prediction. In the end, we discuss the parameter Δ_p which is used in the estimation of the infectious rate in our model.

Dataset. Our dataset is the complete set of over 1,000,000 users and more than 80,000 posts on Sina Weibo which is the most popular microblog in China. For each post, all reposters, their reposting time and followers are included in our dataset. Thus the reaction time and the number of influenced followers for a repost is worked out from the dataset. Our experiment focuses on more than 30,000 posts which have more than 50 reposts within one month in the dataset to evaluate the performance of the settling time prediction.

Evaluation Metric. Similar to [15], we use Mean Absolute Percentage Error (MAPE) to evaluate the performance of all prediction models. The MAPE measures the average deviation between predicted and real settling time over an aggregation of posts. STW_i denotes to the settling time window of the i-th post. $\hat{STW}_i$ denotes to the predicted settling time window of the i-th post. For a dataset of M posts, the MAPE is defined as:

$$MAPE = \frac{1}{M} \sum_{i=1}^{M} \frac{|\hat{STW}_i - STW_i|}{STW_i} \tag{10}$$

Baselines. There are two typical models in previous work for predicting the repost activity. Thus, we use the two models in the prediction by using the definition of the settling time on their result. Nextly, we briefly introduce the two models below.

- **Linear regression(LR)** [16]: This method considers that the logarithm of repost activity is linear. The repost activity $R(t)$ is predicted as below after training n_{train} posts:

$$\hat{R(t)} = R(T)exp(\hat{l_t} + \hat{\sigma_t}^2/2)$$

where $\hat{R(t)}$ is the prediction result of $R(t)$, $R(T)$ is the number of reposts in the time T, $\hat{l_t}$ is obtained by minimizing the squared error

$$E_t(l_t) = \sum_{k=1}^{n_{train}} \{logR_k(t) - l_t - logR_k(T)\}^2$$

$R_k(t)$ is the number of reposts at time t in the k-th post from the training set. $\hat{\sigma_t}^2$ is determined by $\hat{\sigma_t}^2 = E_t(l_t)/n_{train}$.

- **TiDeH** [9]: The second method is based on Hawkes process with the assumption that the followers of all reposters are disjoint. This method predicts the repost activity by using:

$$\hat{\lambda(t)} = f(t) + d_p p(t) \int_T^t \hat{\lambda(s)}\phi(t-s)ds$$

$$f(t) = p(t) \sum_{i:t_i < T} d_i \phi(t - t_i)$$

where d_i is the number of followers for the i-th repost, d_p is the average number of followers for each repost before T.

Table 1. The performance of PPST

T (hour)	24	36	48
MAPE	0.4113	0.2762	0.2103
T (hour)	60	72	84
MAPE	0.1711	0.1416	0.1216

5.1   Performance Analysis

Given the reposts before T, the repost activity after T and the settling time can be predicted. We run our model for each post and compute the MAPE by using Eq. (10). In this section, we set Δ_p to 1 h in the experiment of TiDeH and PPST. Table 1 shows the MAPE of our model with different observation time T. After observing the cascades for over 60 h, the MAPE of our model is less than 20%. With longer observing time, better performance is achieved in PPST.

Figure 8 shows the performance comparison for LR, TiDeH and our model PPST. In the experiment for LR, we random select 80% posts from the dataset

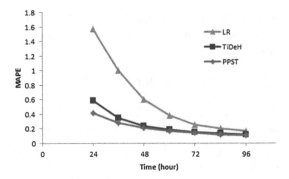

Fig. 8. The comparison of performance with baselines. The performance of PPST is over 10% better than the baselines after observing the cascades for 24 h.

to train the model. The MAPE of the three methods is decreasing along with the growth of observation time T. Compared with LR, the MAPE of TiDeH is closer to our model and achieves less than 50% after observing the cascades for 36 h. After observing them for 24 h, the MAPE of PPST is 41%, while 58% in TiDeH. Our model performs much better than TiDeH with shorter observation time. One of the reason is that our model predicts the number of influenced followers in the future through regression, while TiDeH only uses the fixed value d_p. LR and PPST show a big difference in the MAPE: LR achieves more than 100% MAPE, while the MAPE of PPST is less than 50% after observing the cascades for 24 h. It suggests that the point process based models make a more accurate prediction than the simple regression model.

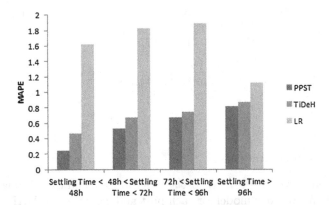

Fig. 9. The performance of LR, TiDeH and PPST with different settling time after observing the cascades for 24 h. PPST performs much better with shorter settling time.

The performance of the three models on different settling time after observing the cascades for 24 h is showed in Fig. 9. LR model has higher MAPE than the other two models in each item, while the performance of TiDeH and our model is close. For PPST and TiDeH, shorter settling time obtains better performance: the MAPE of the two models is less than 50% for the cascades with settling time which is less than 48 h, while more than 80% MAPE is achieved for the cascades with settling time which is over 96 h. When the settling time is less than 48 h, our model achieves more than 20% better performance than TiDeH. It indicates that the effect of the assumption used in previous work is more significant on the cascades with short settling time.

5.2 Discussion

In this section, we discuss the parameter Δ_p which is the window size used in the estimation of the infectious rate. Figure 10 shows the performance for repost cascades with various settling time after observing them for 24 h. PPST on the cascades with the settling time that is less than 48 h performs the best and the MAPE achieves less than 30% when Δ_p is 1 h. The cascades whose settling time is above 48 h are not predicted well because there is not enough observation time to model the infectious rate and the number of influenced followers for the cascades with long settling time. As Fig. 10 shows, $\Delta_p = 1$ h gets the best performance in each item. The difference of the MAPE between $\Delta_p = 1$ h and others is smaller and smaller along with the increasing settling time. When the settling time is over 96 h, the MAPEs of four Δ_p are close. These suggest that shorter window size Δ_p better fits the infectious rate.

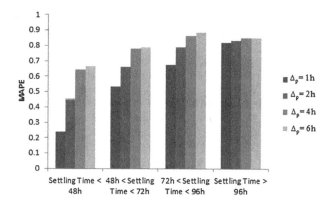

Fig. 10. The performance of the different Δ_p with varies settling time after observing the cascades for 24 h. $\Delta_p = 1$ h gets better performance when the settling time is larger than 24 h.

6   Conclusion

In this paper, we firstly propose the settling time and its prediction task, then we develop PPST for predicting the settling time as the repost cascade unfolds over the network. In our model, we remove the assumption in previous work, introduce and model a new concept "influenced followers" which plays a significant role in the settling time prediction. We conduct an extensive set of experiments on Sina Weibo dataset. The results show that our model achieves over 10% performance gain than the state-of-the-art approaches after observing the cascades for 24 h. Predicting the settling time is a novel issue and PPST is an effective model to predict the settling time. We hope that our method will be applied in the field of crisis management and rumor control.

Acknowledgment. This work was supported by NSFC (91646202), the National High-tech R&D Program of China (SS2015AA020102), Research/Project 2017YB142 supported by Ministry of Education of The People's Republic of China Research Center for Online Education Qtone Education Group Online Education Fund, the 1000-Talent program, Tsinghua University Initiative Scientific Research Program.

References

1. Agarwal, D., Chen, B.C., Elango, P.: Spatio-temporal models for estimating click-through rate. In: Proceedings of the 18th International Conference on World Wide Web, pp. 21–30. ACM (2009)
2. Cheng, J., Adamic, L., Dow, P.A., Kleinberg, J.M., Leskovec, J.: Can cascades be predicted? In: Proceedings of the 23rd International Conference on World Wide Web, pp. 925–936. ACM (2014)
3. Chongfu, H.: Principle of information diffusion. Fuzzy Sets Syst. **91**(1), 69–90 (1997)
4. Crane, R., Sornette, D.: Robust dynamic classes revealed by measuring the response function of a social system. Proc. Nat. Acad. Sci. **105**(41), 15649–15653 (2008)
5. Gao, J., Shen, H., Liu, S., Cheng, X.: Modeling and predicting retweeting dynamics via a mixture process. In: Proceedings of the 25th International Conference Companion on World Wide Web, pp. 33–34. International World Wide Web Conferences Steering Committee (2016)
6. Goel, A., Munagala, K., Sharma, A., Zhang, H.: A note on modeling retweet cascades on twitter. In: Gleich, D.F., Komjáthy, J., Litvak, N. (eds.) WAW 2015. LNCS, vol. 9479, pp. 119–131. Springer, Cham (2015). doi:10.1007/978-3-319-26784-5_10
7. Hong, L., Dan, O., Davison, B.D.: Predicting popular messages in twitter. In: Proceedings of the 20th International Conference Companion on World Wide Web, pp. 57–58. ACM (2011)
8. Khelil, A., Becker, C., Tian, J., Rothermel, K.: An epidemic model for information diffusion in manets. In: Proceedings of the 5th ACM International Workshop on Modeling Analysis and Simulation of Wireless and Mobile Systems, pp. 54–60. ACM (2002)
9. Kobayashi, R., Lambiotte, R.: TiDeH: time-dependent hawkes process for predicting retweet dynamics. arXiv preprint arXiv:1603.09449 (2016)
10. Kupavskii, A., Ostroumova, L., Umnov, A., Usachev, S., Serdyukov, P., Gusev, G., Kustarev, A.: Prediction of retweet cascade size over time. In: Proceedings of the 21st ACM International Conference on Information and Knowledge Management, pp. 2335–2338. ACM (2012)

11. Luo, Z., Wang, Y., Wu, X., Cai, W., Chen, T.: On burst detection and prediction in retweeting sequence. In: Cao, T., Lim, E.-P., Zhou, Z.-H., Ho, T.-B., Cheung, D., Motoda, H. (eds.) PAKDD 2015. LNCS, vol. 9077, pp. 96–107. Springer, Cham (2015). doi:10.1007/978-3-319-18038-0_8

12. Matsubara, Y., Sakurai, Y., Prakash, B.A., Li, L., Faloutsos, C.: Rise and fall patterns of information diffusion: model and implications. In: Proceedings of the 18th ACM SIGKDD International Conference on Knowledge Discovery and Data Mining, pp. 6–14. ACM (2012)

13. Mohler, G.O., Short, M.B., Brantingham, P.J., Schoenberg, F.P., Tita, G.E.: Self-exciting point process modeling of crime. J. Am. Stat. Assoc. **106**(493), 100–108 (2011)

14. Rogers, E.M.: Diffusion of Innovations. Simon and Schuster, New York City (2010)

15. Shen, H.W., Wang, D., Song, C., Barabási, A.L.: Modeling and predicting popularity dynamics via reinforced poisson processes. arXiv preprint arXiv:1401.0778 (2014)

16. Szabo, G., Huberman, B.A.: Predicting the popularity of online content. Commun. ACM **53**(8), 80–88 (2010)

17. Wang, S., Yan, Z., Hu, X., Philip, S.Y., Li, Z.: Burst time prediction in cascades. In: AAAI, pp. 325–331 (2015)

18. Wang, Y., Li, Y., You, Q., Zhang, X., Niemi, R., Luo, J.: Voting with feet: who are leaving hillary clinton and donald trump? arXiv preprint arXiv:1604.07103 (2016)

19. Zaman, T., Fox, E.B., Bradlow, E.T., et al.: A bayesian approach for predicting the popularity of tweets. Ann. Appl. Stat. **8**(3), 1583–1611 (2014)

20. Zaman, T.R., Herbrich, R., Van Gael, J., Stern, D.: Predicting information spreading in Twitter. In: Workshop on Computational Social Science and the Wisdom of Crowds, NIPS, vol. 104, pp. 17599–17001. Citeseer (2010)

21. Zhang, Y., Li, X., Wang, J., Zhang, Y., Xing, C., Yuan, X.: An efficient framework for exact set similarity search using tree structure indexes. In: ICDE, pp. 759–770 (2017)

22. Zhao, Q., Erdogdu, M.A., He, H.Y., Rajaraman, A., Leskovec, J.: SEISMIC: a self-exciting point process model for predicting tweet popularity. In: Proceedings of the 21th ACM SIGKDD International Conference on Knowledge Discovery and Data Mining, pp. 1513–1522. ACM (2015)

23. Zhou, K., Zha, H., Song, L.: Learning social infectivity in sparse low-rank networks using multi-dimensional hawkes processes. In: AISTATS, vol. 31, pp. 641–649 (2013)

Pattern Mining

Mining Co-location Patterns with Dominant Features

Yuan Fang, Lizhen Wang$^{(\boxtimes)}$, Xiaoxuan Wang, and Lihua Zhou

Department of Computer Science and Engineering,
School of Information Science and Engineering,
Yunnan University, Kunming 650091, China
{fangyuan, lzhwang, lhzhou}@ynu.edu.cn,
wangxiaoxuan1037@163.com

Abstract. The spatial co-location pattern mining discovers the subsets of spatial features which are located together frequently in geography. Most of the studies in this field use prevalence to measure a co-location pattern's popularity, namely the frequencies of a spatial feature set participating in a spatial database. However, in some cases, users are not only interested in identifying the prevalence of a feature set, but also the features playing the dominant role in a pattern. In this paper, we focus on mining dominant-feature co-location pattern (DFCP). We firstly propose a new measure, namely disparity, to measure the disparity of features in a pattern. Secondly, we formulate the DFCP mining problem to determine DFCP and extract dominant features. Thirdly, an efficient algorithm is proposed for mining DFCP. Finally, we offer an experimental evaluation of the proposed algorithms on both real data sets and synthetic data sets in terms of efficiency, mining results and significance. The results show that our method can effectively discover DFCPs.

Keywords: Dominant feature · Co-location pattern · Feature disparity

1 Introduction

Spatial co-location mining has been a problem of great practical importance due to its broad applications for environmental protection [15], public transportation [14], location-based service [11] and urban planning [12]. Most of the studies in this field adopt a Participation Index [1] to measure a co-location pattern's prevalence, namely the frequency of a spatial feature set which locate together in a spatial database. For example, {Restaurant, Supermarket, Coffee shop} is a prevalent co-location pattern which means the restaurant, supermarket, and coffee shop are located together frequently. However, in some cases, users are not only interested in identifying the prevalence of a pattern, but also the features playing the dominant role in a pattern.

Identifying the dominant features within a co-location pattern can not only provide indications of which features are co-located frequently but also help to reveal which certain features dominate the rest of features in a co-location pattern. Therefore, it is important to determine whether a co-location pattern has dominant features and extract the dominant features from prevalent co-location patterns in certain applications. One

© Springer International Publishing AG 2017
A. Bouguettaya et al. (Eds.): WISE 2017, Part I, LNCS 10569, pp. 183–198, 2017.
DOI: 10.1007/978-3-319-68783-4_13

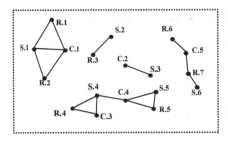

Fig. 1. Dataset of {R, S, C}   **Fig. 2.** Dataset of {H, D, F}

example of such applications is extracting dominant species from plant communities. Even though prevalent co-location mining on the vegetation data can discover the coexistence relation of plants, botanists require additional information on the dominant species among the prevalent co-located plant species. Another application is identifying the dominant facility in the facility points-of-interest of urban. In urban planning, the co-location information can be used to analyze the consistency between neighboring facilities. Identifying dominant facility can further support rational urban planning and business decision making. Therefore, it is necessary to identify the co-location patterns with dominant features.

For example, {Restaurant(R), Supermarket(S), Coffee shop(C)} and {Hospital(H), Drugstore(D), Flower store (F)} are two prevalent co-location patterns with the same participation index, their spatial distribution is shown in Figs. 1 and 2 respectively. The points represent the instances of features, and the lines between each two points illustrate the neighbor relationship between the two instances accordingly. From the perspective of pattern prevalence, the two co-location patterns seem to be very similar. However, in Fig. 1, if any feature of the pattern is deleted, the rest features will still be co-located frequently. This is to say that none of the features can dominate other features in Fig. 1, the status of all the features in this pattern are equivalent. Therefore, it indicates no dominant feature in the co-location pattern {Restaurant, Supermarket, Coffee shop}. On the contrary, in Fig. 2, the features "Flower store" and "Drug store" always appear simultaneously with the feature "Hospital". More specifically, the feature "Hospital" is always the center feature clustered around by the other two features. There are many instances of "Flower store" or "Drug store" neighboring to "Hospital" separately. Nevertheless, there is no additional proximity relation between "Flower store" and "Drug store" without "Hospital". According to this observation, it can be argued that "Hospital" is the dominant feature in the co-location pattern {Hospital, Drugstore, Flower store}. According to whether a pattern contains a dominant feature or not, different prevalent co-location patterns can be classified either as non-dominant-feature co-location patterns (Non-DFCPs) or as dominant-feature co-location patterns (DFCPs). The features within a Non-DFCP are correlated with each other while taking an equal position in the pattern. In a DFCP, some particular features dominate the rest of the features. The position of dominant features and the other features in a DFCP is nonequivalent. The aforementioned pattern {Restaurant, Supermarket, Coffee shop} is a Non-DFCP and {Hospital, Drugstore, Flower store} is a DFCP.

With above discussion as the starting point, this paper focuses on mining DFCPs from prevalent co-location patterns. The contributions of our work can be summarized as follows: (1) proposing the new concepts of dominant feature and dominant-feature co-location pattern (DFCP); (2) proposing a new measure, namely disparity, to measure the feature position in a co-location pattern; (3) formulating the problem of mining DFCP and proposing an efficient algorithm to identify the DFCPs and corresponding dominant features; (4) evaluating the proposed method with existing traditional co-location pattern mining method on both synthesized and real data sets in terms of efficiency, mining results, and significance.

The remainder of the paper is organized as follows: Sect. 1 offers an introduction on related works; Sect. 3 presents the basic concepts and describes related measures; Sect. 4 demonstrates an efficient DFCP mining algorithm; the experimental evaluation is discussed in Sect. 5; Sect. 6 ends this paper with some conclusive remarks.

2   Related Works

For over a decade, the co-location pattern mining has been attracting abundant research interests and widely used in many applications such as environmental protection [15], public transportation [14], location-based service [11] and urban planning [12], etc. Shekhar and Huang [1] first defined the concept of spatial co-location patterns. The co-location mining aims to find all subsets of spatial features, which located together frequently. In literature [1], the authors proposed to use participation index to measure the prevalence of a co-location. According to the anti-monotone property of the participation index, a Join-based co-location pattern mining algorithm was proposed. However, this original algorithm is subject to a large amount of instance join operations. Since then, various algorithms have been developed such as the Join-Less algorithm [2], the Partial-join algorithm [3], CPI-tree algorithm [4], iCPI-tree algorithm [5] and Order-Clique-Based algorithm [6] to avoid the expensive join operation and improve the efficiency. Although these developments have made significant contributions to algorithm efficiency, due to the vastness of spatial data, the typical co-location mining framework always leads to large collections of results, which make people hardly understand and identify the targeted ones. This has become one of the biggest obstacles in the studies and applications of co-location mining. In order to resolve this problem, many researchers have done many works to reduce the number of pattern results by mining the co-location pattern with a specific relationship or a specific target. Literature [6–8] focused on mining maximal co-location patterns, top-k closed co-location patterns and representative co-location patterns respectively, which effectively reduce the prevalent co-location results. In order to improve the applicability of the co-location patterns, many researchers addressed to find the co-locations with specific target based on expert knowledge such as high utility co-location pattern mining [9], ontology-based co-location pattern mining [10], co-location pattern mining under domain-constrains [13]. For discovering more interesting knowledge hidden in prevalent co-location pattern, some researchers have also committed to finding the co-location with a specific relationship such as causal relationship [17], competitive relationship [16].

Extracting the co-locations with specific relationships not only reveals the substantive connection between spatial features but also reduces the number of prevalent co-location pattern results. The dominant relation is widely appeared both in nature and in human society. There is a lot of work on dominant factor analysis in many specific applications such as public safety [18], power system [19], audio recognition [20], dominant species identification and urban planning. Therefore, mining co-location patterns with dominant relation and extracting dominant features is a significant but challengeable task.

3 Preliminary and Problem Formulation

This section firstly offers a review on the preliminary concepts of typical prevalent co-location pattern mining framework, then a formal definition of feature disparity to measure the feature disparity. Furthermore, the definitions of proposed dominant feature and dominant-feature co-location pattern (DFCP) are provided and finally, the problem of DFCP mining is formulated.

3.1 Basic Concept

Given a set of spatial features $F = \{f_1, f_2 \ldots, f_n\}$, a set of their spatial instances $S = S_1 \cup S_2 \cup \ldots \cup S_n$, where $S_i(1 \leq i \leq n)$ is a set of instance of feature f_i, and a **spatial neighbor relationship** between instances R over S, the Euclidean metric is used for the R. Each instance x of feature f_i is represented as $f_i.x$, two instances $f_i.x$ and $f_j.y$ are neighbors of each other if the Euclidean distance between them is not greater than a distance threshold d. A **k-size co-location** $c = \{f_1, \ldots, f_k\}$ is a subset of F ($c \subseteq F$). $l = \{f_1.x_1, f_2.x_2, \ldots, f_k.x_k\}$ ($l \subseteq S$) is called a **row-instance** of c while l includes all the feature types of c and forms a **clique relationship** under R. The set of all the row-instances of c is called **table-instance** $T(c)$.

Typically, the prevalence of a k-size co-location $c = \{f_1, \ldots, f_k\}$ is measured by the **participation index** and **participation ratio**. The participation ratio $PR(c,f_i)$ ($f_i \in c$) for feature type f_i in c is the fraction of feature f_i which participates in the table instance of c. The participation ratio is defined as $PR(c,f_i) = \frac{|\pi_{f_i}(T(c))|}{|T(\{f_i\})|}$, where π is the relational projection operation. The **participation index PI(c)** of c is the minimum participation ratio $PR(c,f_i)$ in all features f_i in c: $PI(c) = \min_{i=1}^{k}(PR(c,f_i))(f_i \in c)$. Given a user-specified prevalence threshold *min_prev*, a co-location c is **prevalent** if $PI(c) \geq min\_prev$.

Example 1. Figure 2 is a spatial database with 3 features: H has 5 instances, D has 8 instances and F has 7 instances. H.1 is the first instance of feature H. Co-location {H, D, F} is the super co-location of {H, D}, {D, F} and {H, F}. The co-location instances of all the co-locations in Fig. 1 are shown in Fig. 3(a) and the co-location instances of all the co-locations in Fig. 2 are shown in Fig. 3(b) respectively. In Fig. 3 (b), suppose *min_prev* = 0.4, for co-location pattern {H, D, F}, the table instance of {H, D, F} is shown in Fig. 3(a), the PI({H, D, F}) = min(PR({H, D, F}, H), PR ({H, D, F}, D), PR({H, D, F}, F)) = min(0.6, 0.5, 0.57) = 0.5 ≥ *min_prev*, so c is a prevalent co-location pattern. Similarly, the PI({R, S, C}) = 0.5.

R S		S C		R C		R S C		
R.1	S.1	S.1	C.1	R.1	C.1	R.1	S.1	C.1
R.2	S.2	S.3	C.2	R.2	C.1	R.2	S.1	C.1
R.3	S.3	S.4	C.3	R.4	C.3	R.4	S.4	C.3
R.4	S.4	S.4	C.4	R.5	C.4	R.5	S.5	C.4
R.5	S.5	S.5	C.4	R.6	C.5			
R.7	S.6			R.7	C.5			
0.86	0.86	0.86	0.8	0.67	0.8	0.57	0.5	0.6

(a) Co-location instances of {R,S,C}

H D		D F		H F		H D F		
H.1	D.1	D.1	F.1	H.1	F.1	H.1	D.1	F.1
H.1	D.2	D.5	F.5	H.2	F.2	H.4	D.5	F.5
H.2	D.3	D.7	F.6	H.3	F.3	H.5	D.7	F.6
H.3	D.4	D.8	F.7	H.3	F.4	H.5	D.8	F.7
H.4	D.5			H.4	F.5			
H.4	D.6	0.5	0.57	H.5	F.6	0.6	0.5	0.57
H.5	D.7			H.5	F.7			
H.5	D.8							
1	1			1	1			

(b) Co-location instances of {H,D,F}

Fig. 3. The Table instances of {R, S, C} and {H, D, F}

3.2 Definitions

According to the discussion in Sect. 1, if a co-location pattern is DFCP, there will be disparity between the positions of its features. In order to test the feature position to identify DFCP and further extract the dominant features, we can calculate the disparity degree as follows: Firstly, we calculate the influence of each feature to the pattern. Secondly, we propose a measure, namely feature disparity, to measure the disparity between features. Thirdly, we define the dominant relation between features.

From Fig. 3, we can easily find a fact: the instances of c_{k-1} which appears in c_{k-1} but not appears in c_k mean these instances are not dominated by the feature $c_k - c_{k-1}$. Thus, for a k-size co-location pattern c_k, we can evaluate all its features influence from its k sub patterns.

Definition 1 (Loss Ratio). Given a k-size $(k > 2)$ prevalent co-location pattern $c_k = \{f_1, f_2, \ldots, f_k\}$, and a sub pattern of c_k: $c_{k-1} = \{f_1, f_2, \ldots, f_{k-1}\}$, for feature $f_i (f_i \in c_{k-1}) (1 \leq i \leq k-1)$, the loss ratio of f_i from c_{k-1} to c_k is defined as follows:

$$\text{LR}(c_k, c_{k-1}, f_i) = \frac{|\pi_{f_i}(T(c_{k-1}))| - |\pi_{f_i}(T(c_k))|}{|T(\{f_i\})|} \quad (1)$$

Note that $\text{PR}(c_k, f_i) \leq \text{PR}(c_{k-1}, f_i)$ is ensured according to the anti-monotonicity of participation ratio which is referred in [1], it can be confirmed that $0 \leq \text{LR}(c_k, c_{k-1}, f_i) \leq \text{PR}(c_{k-1}, f_i)$. Loss ratio describes the fraction of feature's instances loss from a co-location to its sub pattern. We then aggregate these ratios to compute a minimum single value of loss index from c_{k-1} to c_k.

Definition 2 (Loss Index). The loss index from c_{k-1} to c_k is defined as:

$$\text{LI}(c_k, c_{k-1}) = \min_{i=1}^{k-1}(\text{LR}(c_k, c_{k-1}, f_i)) \quad (2)$$

Example 2. In Fig. 4, the loss ratio of feature H from {H, D, F} to {H, D} is:

LR({H, D, F}, {H, D}, H) = PR({H, D}, H) − PR({H, D, F}, H) = 1 − 0.6 = 0.4

LR({H, D, F}, {H, D}, D) = PR({H, D}, D) − PR({H, D, F}, D) = 0.5

The loss index LI({H, D, F}, {H, D}) = min(LR({H, D, F}, {H, D}, H), LR({H, D, F}, {H, D}, D)) = min(0.4,0.5) = 0.4. Similarly, LI({H, D, F}, {H, F}) = 0.4, LI({H, D, F}, {D, F}) = 0.

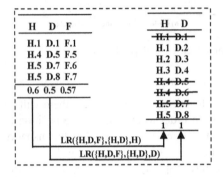 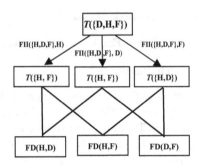

Fig. 4. Instance loss from {H, D} to {H, D, F}   **Fig. 5.** The process of finding dominant features

The loss index from c_{k-1} to c_k LI(c_k, c_{k-1}) represents the fraction of individual neighbor clique relation of c_{k-1} which is no feature $f (f = c_k - c_{k-1})$ involved, characterizes the instance loss of c_{k-1} when a new feature f join in c_{k-1} to form a higher-size pattern c_k. The higher of the loss index is, the more instances of f are irrelevant with c_{k-1}, the less dominant of feature f to the features in c_{k-1}. Therefore, we give the definition of feature influence index as follows:

Definition 3 (Feature Influence Index). Given a k-size co-location pattern c_k, the influence index of feature $f_i (f_i \in c_k)$ $(1 \leq i \leq k)$ in c_k is defined as:

$$FII(c_k, f_i) = 1 - LI(c_k, c_{k-1}) \tag{3}$$

where $c_{k-1} = c_k - \{f_i\}$.

Feature influence FII(c_k, f_i) visually represents the influence of feature f_i on other features of c_k, the higher of FII(c_k, f_i) is, the more instances of features in c_k may be dominated by f_i.

Example 3. In Fig. 3, for co-location pattern {H, D, F}, the feature influence of feature H in {H, D, F} is:

$$FII(\{H,D,F\}, H\}) = 1 - LI(\{H,D,F\}, \{F,D\}) = 1.$$
$$FII(\{H,D,F\}, D\}) = 1 - LI(\{H,D,F\}, \{H,F\}) = 0.6.$$
$$FII(\{H,D,F\}, F\}) = 1 - LI(\{H,D,F\}, \{H,D\}) = 0.5.$$

Feature influence index measures the influence of single feature on other features in the same pattern. Therefore, we can use the feature influence index to measure the feature disparity. The higher the disparity between two features is, the more likely that a dominant relation exists between them.

Definition 4 (Feature Disparity). Given a k-size $(k > 2)$ co-location pattern $c_k = \{f_1, f_2, \ldots, f_k\}$, the feature disparity between $f_i (f_i \in c_k) (1 \le i \le k)$ and $f_j (f_j \in c_k) (1 \le j \le k, i \ne j)$ in c_k is:

$$\text{FD}(f_i, f_j) = |\text{FII}(c_k, f_i) - \text{FII}(c_k, f_j)| \tag{4}$$

According to the Definition 5, it is easily proof that feature disparity is symmetrical and non-negative.

Definition 5 (Dominant Relation). Given a k-size$(k > 2)$ prevalent co-location pattern $c_k = \{f_1, f_2, \ldots, f_k\}$, a minimum feature disparity threshold min_fd $(0 \le min\_fd \le 1)$, the feature $f_i (f_i \in c_k) (1 \le i \le k)$ dominates $f_j (f_j \in c_k) (1 \le j \le k, i \ne j)$ if they meet two conditions: (1) $\text{FD}(f_i, f_j) \ge min\_fd$ and (2) $\text{FII}(c_k, f_i) > \text{FII}(c_k, f_j)$.

It can be noticed that if there exists a dominant relation between features in a prevalent co-location, then the co-location contains dominant features.

Example 4. In Fig. 5, after calculating the loss index, we can obtain the feature influence value and then calculate the feature disparity between two features in the pattern. Setting the $min\_fd = 0.4$:

$$\text{FD}(H, D) = \text{FII}(\{H, D, F\}, H) - \text{FII}(\{H, D, F\}, D) = 0.4.$$
$$\text{FD}(H, F) = 0.5, \text{FD}(D, F) = 0.1.$$

In $\{H, D, F\}$, $\text{FD}(H, F) = 0.4 \ge min\_fd$, $\text{FII}(\{H, D, F\}, H) > \text{FII}(\{H, D, F\}, D)$, so H is a dominant feature of $\{H, D, F\}$.

Figure 5 illustrates the process for finding dominated features. We introduce two different extreme value functions – a minimum function and a maximum function – to obtain the greatest influence and least influence for a co-location pattern and further help to optimize the mining process.

Definition 6 (Max-Feature Influence Index). Given a k-size co-location pattern $c_k = \{f_1, f_2, \ldots f_k\}$, the maximum feature influence index of c_k is defined as:

$$\text{max\_FII}(c_k) = \max_{i=1}^{k}(\text{FII}(c_k, f_i)) \tag{5}$$

Definition 7 (Min-Feature Influence Index). Given a k-size co-location pattern $c_k = \{f_1, f_2, \ldots, f_k\}$, the minimum feature influence index of c_k is defined as:

$$\text{min\_FII}(c_k) = \min_{i=1}^{k}(\text{FII}(c_k, f_i)) \tag{6}$$

Example 5. In Fig. 3b, for co-location pattern {H, D, F}, the max-feature influence index is:

$$\text{max\_FII}(\{H, D, F\}) = \max(\text{FII}(\{H, D, F\}, H), \text{LI}(\{H, D, F\}, F), \text{LI}(\{H, D, F\}, D)$$
$$= \max(1, 0.6, 0.5) = 1$$

The min_feature influence index is: min_FII({H, D, F}) = 0.5

Lemma 1. If $FD(f_i, f_j) \geq min\_fd$, and FII $(c_k, f_i) > $ FII (c_k, f_j), then FII $(c_k, f_i) -$ min_FII$(c_k) \geq min\_fd$;

Proof. Because of $\text{FII}(c_k, f_i) > \text{FII}(c_k, f_j) \geq \text{min\_FII}(c_k)$, then

$$\text{FII}(c_k, f_i) - min\_\text{FII}(c_k) \geq \text{FD}(f_i, f_j)$$
$$\text{FII}(c_k, f_i) - \text{FII}(c_k, f_j) \geq min\_fd$$

Lemma 2. Given a k-size($k > 2$) co-location pattern $c_k = \{f_1, f_2, ..., f_k\}$, two features $f_i(f_i \in c_k)$ $(1 \leq i \leq k)$ and $f_j(f_j \in c_k)$ $(1 \leq j \leq k)$ $(i \neq j)$, if $\text{max\_FII}(c_k) - \text{min\_FII}(c_k) \geq min\_fd$, c_k is a DFCP.

Proof. Assume $\text{FII}(c_k, f_i) > \text{FII}(c_k, f_j)$, if $\text{FD}(f_i, f_j) = \text{FII}(c_k, f_i) - \text{FII}(c_k, f_j) \geq min\_fd$, according to Definition 5, f_i is a dominant feature, then c_k is a DFCP.

Because of $\text{FII}(c_k, f_i) \leq \text{max\_FII}(c_k)$ and $\text{FII}(c_k, f_j) \geq \text{min\_FII}(c_k)$, $\text{max\_FII}(c_k) - \text{min\_FII}(c_k) \geq \text{FII}(c_k, f_i) - \text{FII}(c_k, f_j) \geq min\_fd$. Then, the feature which has $\text{max\_FII}(c_k)$ is the dominant feature in c_k, so c_k is a DFCP.

3.3 Problem Formulation

According Lemmas 1 and 2, we optimize the mining process. It can be formulated as follows:

Definition 8 (Max-Feature Disparity). Given a k-size $(k > 2)$ prevalent co-location pattern $c_k = \{f_1, f_2, ..., f_k\}$, the max_feature disparity of the feature $f_i(f_i \in c_k)$ $(1 \leq i \leq k)$ in c_k is defined as:

$$\text{max\_FD}(c_k, f_i) = |\text{FII}(c_k, f_i) - \text{min\_FII}(c_k)| \tag{7}$$

Given a minimum feature disparity threshold min_fd $(0 \leq min\_fd \leq 1)$, if $\text{max\_FD}(f_i, c_k) \geq min\_fd$, then feature f_i is a dominant feature.

Some co-location pattern recommendation applications require that interesting patterns should be both prevalent and with dominant features. On the one hand, a dominant-feature co-location means that there is more information to support specific decision-making. Thus, it guarantees the recommendation is guiding. On the other hand, identifying the dominant-feature co-location can further decrease the number of prevalent patterns. Thus, it improves the availability of patterns. Therefore, with the definition of disparity and dominant features, we formulate the problem of mining DFCP as follows.

Mining Dominant-Feature Co-location Patterns. Given a minimum prevalence threshold min_prev, a minimum feature disparity threshold min_fd, a k-size co-location pattern c_k is a DFCP if it meets the following conditions:

(1) $PI(c_k) \geq min\_prev$
(2) $\max\_FII(c_k) - \min\_FII(c_k) \geq min\_fd$

Extracting Dominant Feature. Given a minimum feature disparity threshold min_fd, a k-size co-location pattern c_k, feature $f_i (f_i \in c_k) (1 \leq i \leq k)$ is a dominant feature in DFCP if $\max\_FD(f_i, c_k) \geq min\_fd$.

Example 6. In Fig. 3, setting prevalent threshold $min\_prev = 0.4$, disparity threshold $min\_fd = 0.4$. For co-location pattern {H, D, F}, PI({H, D, F})=0.5 $\geq min\_prev$, $\max\_FII(\{H, D, F\}) - \min\_FII(\{H, D, F\}) = 0.5 \geq min\_fd$, then {H, D, F} is a DFCP.

FII({H, D, F}, H) − min_FII({H, D, F}, H) = 1 − 0.5 $\geq min\_fd$.
FII({H, D, F}, D) − min_FII({H, D, F}, H) = 0.6 − 0.5 = 0.1 $< min\_fd$.
FII({H, D, F}, F) − min_FII({H, D, F}, H) = 0 $< min\_fd$.
H is a dominant feature in {H, D, F}.
For co-location pattern{R, S, C}: PI({R, S, C}) = 0.5 $> min\_prev$
max_FII({H, D, F}) − min_FII({H, D, F}) = 0.9 − 0.71 = 0.19 $< min\_fd$
We can determine that {R, S, C}is a non-DFCP prevalent co-location pattern without calculating the feature disparity between features.

4 Algorithm

In this section, we will demonstrate a general algorithm for mining DFCPs on prevalent co-location patterns. The mining framework of DFCP consists of two stages. In stage 1, the feature influence of each feature in a prevalent co-location pattern is computed, and then the DFCP is selected by a min_fd. Stage 2 extracts the set of dominant feature of a DFCP by a min_fd. Algorithm AMDFCP is the DFCP mining framework

Algorithm: AMDFCP

Input: S: a spatial data set; F:a feature set; I: A instances set of corresponding F; d: a distance threshold; *min_prev*: a prevalence threshold; *min_fd*: a feature disparity threshold

Output: A collection of significant co-location patterns: DFCP-set

Variables: k:co-location size; C_k: k-size candidate prevalent co-location pattern set; P_k: k-size prevalent co-location pattern set; PR_c: a collection of participation ratio of prevalent co-location pattern c;

Method:

```
(1) SN=gen_star_neighborhoods(F,S,d);
(2) P₁=F, k=2, DFCP=∅;
(3) WHILE(Pₖ₋₁≠∅) DO
(4) Cₖ=gen_candidate_co-location(k, Pₖ₋₁)
(5)    FOR EACH c∈Cₖ DO
(6)       IF calculate PI(c) ≥min_prev DO
(7)          FOR EACH p∈Pₖ₋₁(c) and PR_c DO
(8)             LI(c,p)= calculate_LI(PR_p, PR_c);
(9)             FII_set(c)←{1- LI(c,p), c-p};
(10)          END DO
(11)          min_FII(c)=calculate_min_FII(FII_set(c));
(12)          max_FII(c)=calculate_max_FII(FII_set(c));
(13)          IF calculate max_FII(c)-min_FII(c)≥min_fd DO
(14)             FOR EACH fᵢ∈c DO
(15)                IF calculate max_FD(c, fᵢ)≥min_fd DO
(16)                   DFset(c)←fᵢ;
(17)                END DO
(18)             END DO
(19)             DFCP-set←{c, DFset(c)};
(20)          END DO
(21)       END DO
(22)    END DO
(23) k=k+1;
(24) END DO
```

Line 1 generates a set of star instances based on a distance threshold d. Lines 2–4 generate k-size co-location candidate pattern sets. Lines 5–22 describe DFCP mining includes: Loss index calculation, disparity test and dominant feature extraction processing. Lines 5–6 calculate participation index. Lines 7–10 compute the feature loss index and feature influence from a prevalent pattern to its sub pattern set. Lines 11–12 aggregate the feature influence to max_FII and min_FII to prune the DFCP mining and dominant feature extracting. Line 13 is a pruning strategy according to Lemma 2, which uses the full range of feature disparity, replace testing disparity of each feature pair. In line 13, if pattern c is DFCP, then lines 14–18 extract the dominant features through testing each feature in the pattern. Line15 is another pruning strategy, which uses the difference value between each feature influence and minimum feature influence to replace calculating the disparity between a feature and all the rest of features in a pattern. Line 19 stores the DFCPs with dominant features. Lines 3–23 are executed repeatedly and finally return a collection of DFCP set.

5 Experimental Study

In this section, comprehensive experiments are conducted to evaluate the proposed concepts and algorithms from multiple perspectives on both synthetic and real data sets. All the algorithms are implemented in Visual C#. All of our experiments are performed on a 2.4 GHZ, 8 GB-memory and Intel Core i3.

5.1 Experiment on Synthetic Data

5.1.1 Synthetic Data Generation

There are 4 synthetic datasets used in our experiments. We use different data generation methods to generate a synthetic dataset with different distributions. Dataset 1, dataset 3, and dataset 4 are randomly generated according to the Poisson distribution and distributed evenly in 1000×1000 space. Dataset 2 is generated for simulating the data distribution with dominant features. In order to generate the synthetic data as far as possible consistent with the dominant feature distribution characteristics, we assign different distributions of feature instances that classified as dominant features and non-dominant features respectively by controlling the specific parameters. Compared to randomly sow the data point on a plane, we increase the density of the instance of non-dominant features which around the dominant feature by setting a density parameter α and adjust the distance between the instances of same dominated feature by setting distance parameter β to ensure they are not located too closely and tightly. In our experiments, Four synthetic data sets are generated which shown in Table 1. Especially, we assign 5 dominant features and 15 non-dominant features in dataset 2.

Table 1. The experimental data set

Dataset	Instance amount	Feature amount	Data type
Dataset 1	20,000	20	Synthetic
Dataset 2	20,000	20	Synthetic
Dataset 3	40,000	25	Synthetic
Dataset 4	80,000	30	Synthetic
Dataset 5	26,546	16	Real
Dataset 6	335	32	Real

Table 2. The default values of the parameters

Parameters	Default values
Instance amount	40000
min_prev	0.3
Distance threshold	30
min_fd	0.2

We implement DFCP mining algorithm on multiple synthetic datasets and compare the efficiency of the DFCP mining algorithm and the traditional co-location Join-less [2] algorithm by considering the effect of the number of instances, the participation index threshold, the distance threshold, and the significance threshold. Table 2 shows the default parameters in the experiment.

5.1.2   Efficiency

The Effect of Prevalence Threshold. Figure 6 shows the running time of DFCP mining algorithm at five different minimum prevalence thresholds (*min_prev*) on four synthetic data sets respectively. For each data set, the run time decreases as the *min_prev* increases. For all data sets, the running time increases as the data volume increases and the *min_prev* decreases. For dataset 4, the effect of *min_prev* on algorithm performance is particularly evident, this is because low threshold and dense data lead to huge table-instances of candidates, table-instances' computation affects the performance of the algorithm. For dataset 1 and dataset 2, the running time of dataset 2 is longer than dataset 1 since the computation cost in DFCP process in dataset 2 is more than in dataset 1.

The Effect of Distance Threshold. Figure 7 depicts the running time on four synthetic data sets with respect to the variation of threshold distance. It can be observed that for each data set, the run time decreases as the distance threshold increases. For all datasets, the running time increases with the increase of the data volume and distance threshold. It can also be observed that the effect of large distance threshold on algorithm performance is especially obvious, which indicates that the performance of the algorithm is mainly affected by the data density. For dataset 1 and dataset 2, when the distance is at 10 and 20, the efficiency on dataset 1 is better than dataset 2, however, with the increase of distance, the situation has reversed. It is because when the distance is low, the computation cost of DFCP mining process affects the efficiency of dataset 2, yet when the distance is high, the auto-correlation which happened in dataset 1 is more frequent than dataset 2 so that dataset 2 performs better.

The Effect of Co-location Feature Disparity Threshold. Figure 8 demonstrates the running time on four synthetic data sets with respect to the variation of disparity threshold (*min_fd*) respectively. Comparing the running time on the four datasets, for each dataset, running time decreases more rapidly when the significance threshold increases. We note that the change of *min_fd* is particularly evident for the performance of algorithms on dense datasets. The efficiency of dataset1 is better than dataset 2 because the number of DFCP in dataset 2 is more than dataset 1 and the computation cost of dataset 2 is more than dataset 1.

5.1.3   The Mining Results of AMDFCP vs Join-Less

Figure 9 shows the mining results of AMDFCP mining algorithm and the Join-less algorithm on four synthetic data sets respectively under the default parameters. The number of DFCPs increases as the data volume increases, indicating that the denser the data, the more prominent the characteristics of the feature correlation in the prevalent co-locations. With the increase of the volume of data set, the number of DFCPs is much less than that of prevalent co-locations. Obviously, the number of DFCP in dataset 2 is far more than dataset 1, that means our algorithm can identify DFCPs correctly.

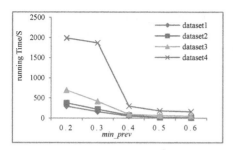

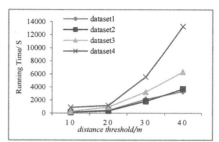

Fig. 6. Running time on different synthesized data sets (w.r.t *min_prev*)

Fig. 7. Running time on different synthesized data sets (w.r.t *distance*)

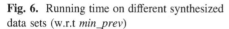

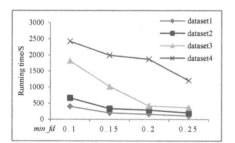

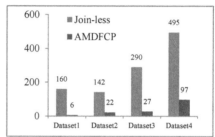

Fig. 8. Running time on different synthesized data sets (w.r.t *min_fd*)

Fig. 9. The comparison of mining results on different synthesized data sets

5.2 Experiments on Real Data

Two real-world data sets are used in our experiments. The first one is the points of interest (POI) in Beijing which consists of 26,546 spatial instances and 16 spatial feature types. The spatial distance threshold is 50 by default (meaning 50 m in the real world). The second data set is the vegetation data of the "Three Parallel Rivers Area" which consists of 335 spatial instances and 32 spatial feature types. The spatial distance threshold is 6000 by default (meaning 6 km in the real world). We set the default values as *min_prev* = 0.3 and *min_fd* = 0.2.

Compared to the synthetic data, the spatial correlation of real data is higher, thus the results have practical significance. We conduct experiments to record the number of patterns for AMDFCP algorithm compared with the number of Join-less algorithms by considering the change of prevalence threshold, the distance threshold, and the feature disparity threshold. The results of varying *min_prev* on POI dataset and vegetation dataset are presented respectively in Figs. 10 and 11. It can be observed that AMDFCP algorithm reduces the number of mining results as high as 50% because with decreasing of *min_prev*, the results of the Join-less algorithm are increased quickly, AMDFCP algorithm can filter the low correlated prevalent patterns efficiently. Figures 12 and 13 illustrate the results by varying distance threshold on POI dataset and vegetation dataset respectively. The number of DFCPs increases slower than prevalent co-location patterns as the data volume increases, it indicates that the mining result is less affected by distance thresholds. This is because the disparity metrics can avoid the result explosion for aggregation of a large number of instances under the dense data.

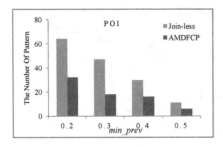

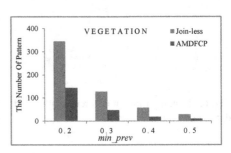

Fig. 10. The mining results with different *min_prev* on POI data set

Fig. 11. The mining results with different *min_prev* on vegetation data set

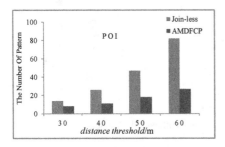

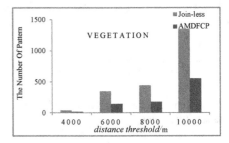

Fig. 12. The mining results with different distance threshold on POI data set

Fig. 13. The mining results with different distance threshold on vegetation data set

5.3   The Real Application of Significant Co-location Mining

We use POI datasets to present and explain the practical application of DFCP mining. Table 3 shows the results of DFCP mining, note that the dominant feature is labeled by "*". The mining results indicate that the DFCP mining can offer targeted and abundant information.

Table 3. The results of DFCP mining

DFCP	PI	FD
min_prev = 0.3; *min_fd* = 0.4; *distance threshold* = 50		
{Chinese Restaurant*, Parking Lot, clothing store}	0.42	0.4
{Chinese Restaurant, Hotel*, Parking Lot}	0.35	0.41
{Hotel*, clothing store*, Parking Lot}	0.42	0.47, 0.47
{Chinese Restaurant*, Cafe, Hotel*}	0.46	0.43, 0.4

6   Conclusions

In this study, a new approach of mining the dominant-feature co-location patterns (DFCPs) is proposed to reveal the dominant relation between features of a pattern and reduce the prevalent co-location pattern results. The DFCP mining problem consists of the DFCP determination and dominant features extracting. In order to formulate the problem of DFCP mining, we firstly propose a measure, namely disparity, to measure the feature influence disparity, then an algorithm AMDFCP is designed to mine DFCP and dominant features. Finally, experimental results demonstrated in this paper have exemplified that DFCP mining method proposed in this paper can be utilized to identify the DFCP from prevalent co-location patterns efficiently and the experimental results in POI data have also supported the proposal that DFCP mining can provide effective support to users for specific applications.

Acknowledgments. This work is supported by the National Natural Science Foundation of China (61472346, 61662086), the Natural Science Foundation of Yunnan Province (2015FB114, 2016FA026), the Project of Innovation Team of Yunnan University.

References

1. Huang, Y., Shekhar, S., Xiong, H.: Discovering co-location patterns from spatial data sets: a general approach. TKDE **16**(12), 1472–1485 (2004)
2. Yoo, J.S., Shekhar, S.: A joinless approach for mining spatial co-location patterns. TKDE **18**(10), 1323–1337 (2006)
3. Yoo, J.S., Shekhar, S., Smith, J., Kumquat, T.P.: A partial join approach for mining co-location patterns. In: Proceedings of the 12th Annual ACM International Workshop on Geographic Information Systems (GIS 2004), pp. 241–249 (2004)
4. Wang, L., Bao, Y., Lu, J., Yip, J.: A new Join-less approach for co-location pattern mining. In: 8th IEEE International Conference on Computer and Information Technology, pp. 197–202. IEEE Press, New York (2008)
5. Wang, L., Bao, Y., Lu, Z.: Efficient discovery of spatial co-location patterns using the iCPI-tree. Open Inf. Syst. J. **3**(1), 69–80 (2009)
6. Wang, L., Zhou, L., Lu, J., Yip, J.: An order-Clique based approach for mining maximal co-locations. Inf. Sci. **179**(19), 3370–3382 (2009)
7. Yoo, J.S., Bow, M.: Mining top-k closed co-location patterns. In: IEEE International Conference on Spatial Data Mining and Geographical Knowledge Services, pp. 100–105. IEEE Press, New York (2011)
8. Liu, B., Chen, L., Liu, C., Zhang, C., Qiu, W.: RCP mining: towards the summarization of spatial co-location patterns. In: Claramunt, C., Schneider, M., Wong, R.C.-W., Xiong, L., Loh, W.-K., Shahabi, C., Li, K.-J. (eds.) SSTD 2015. LNCS, vol. 9239, pp. 451–469. Springer, Cham (2015). doi:10.1007/978-3-319-22363-6_24
9. Wang, X., Wang, L., Lu, J., Zhou, L.: Effectively updating high utility co-location patterns in evolving spatial databases. In: Cui, B., Zhang, N., Xu, J., Lian, X., Liu, D. (eds.) WAIM 2016. LNCS, vol. 9658, pp. 67–81. Springer, Cham (2016). doi:10.1007/978-3-319-39937-9_6
10. Bao, X., Wang, L., Chen, H.: Ontology-based interactive post-mining of interesting co-location patterns. In: Li, F., Shim, K., Zheng, K., Liu, G. (eds.) APWeb 2016. LNCS, vol. 9932, pp. 406–409. Springer, Cham (2016). doi:10.1007/978-3-319-45817-5_35

11. Yu, W.: Spatial co-location pattern mining for location-based services in road networks. Expert Syst. Appl. **46**, 324–335 (2016)
12. Yu, W., Ai, T., He, Y.: Spatial co-location pattern mining of facility points-of-interest improved by network neighborhood and distance decay effects. Int. J. Geogr. Inf. Sci. **31**(2), 280–296 (2016)
13. Flouvat, F., Soc, J., Desmier, E.: Domain-driven co-location mining. GeoInformatica **19**(1), 147–183 (2015)
14. Shekhar, S., Huang, Y.: Discovering spatial co-location patterns: a summary of results. In: Jensen, C.S., Schneider, M., Seeger, B., Tsotras, V.J. (eds.) SSTD 2001. LNCS, vol. 2121, pp. 236–256. Springer, Heidelberg (2001). doi:10.1007/3-540-47724-1_13
15. Mohan, P., Shekhar, S., Shine, J.A.: A neighborhood graph based approach to regional co-location pattern discovery: a summary of results. In: 19th ACM SIGSPATIAL International Conference on Advances in Geographic Information Systems, pp. 122–132. ACM (2011)
16. Lu, J., Wang, L., Fang, Y., Li, M.: Mining competitive pairs hidden in co-location patterns from dynamic spatial databases. In: Kim, J., Shim, K., Cao, L., Lee, J.-G., Lin, X., Moon, Y.-S. (eds.) PAKDD 2017. LNCS, vol. 10235, pp. 467–480. Springer, Cham (2017). doi:10. 1007/978-3-319-57529-2_37
17. Lu, J., Wang, L., Fang, Y.: Mining causal rules hidden in spatial co-locations based on dynamic spatial databases. In: IEEE 2016 International Conference on Computer, Information and Telecommunication Systems, pp. 1–6. IEEE Press, New York (2016)
18. Muhaya, F.: Dominant factors in national information security policies. J. Comput. Sci. **6**(7), 808–812 (2010)
19. Peng, Y., Dong, H.: Dominant factors mining in power system based on clustering analysis. Electr. Power **39**(12), 16–19 (2006)
20. Gu, J., Lu, L., Cai, R., Zhang, H.-J., Yang, J.: Dominant feature vectors based audio similarity measure. In: Aizawa, K., Nakamura, Y., Satoh, S. (eds.) PCM 2004. LNCS, vol. 3332, pp. 890–897. Springer, Heidelberg (2004). doi:10.1007/978-3-540-30542-2_110

Maximal Sub-prevalent Co-location Patterns and Efficient Mining Algorithms

Lizhen Wang, Xuguang Bao, Lihua Zhou, and Hongmei Chen[✉]

Department of Computer Science and Engineering,
School of Information Science and Engineering, Yunnan University,
Kunming 650091, China
hmchen@ynu.edu.cn

Abstract. Spatial prevalent co-location pattern mining plays an important role to identify spatially correlated features in many applications, such as Earth science and public transportation. Observe that the existing approaches only consider the clique instances where feature instances form a clique and may neglect some important spatial correlations among features in practice, in this paper, we introduce *star participation instances* to measure the prevalence of co-location patterns such that spatially correlated instances which cannot form cliques will also be properly considered. Then we propose a new concept of *sub-prevalent co-location patterns (SCP)* based on the star participation instances. We present two efficient algorithms, the prefix-tree-based algorithm (PTBA) and the partition-based algorithm (PBA), to mine all the *maximal* sub-prevalent co-location patterns (MSCP) in a spatial data set. PTBA adopts a typical candidate generate-and-test method starting from candidates with the longest pattern-size, while PBA is performed step by step from 3-size core patterns. We demonstrate the significance of the new concepts as well as the efficiency of our algorithms through extensive experiments.

Keywords: Spatial co-location pattern mining · Sub-prevalent co-location patterns (SCP) · Star participation ratio (SPR) · Star participation index (SPI)

1 Introduction

With the development of location-based services and spatial information processing techniques, large amounts of data with spatial contexts have become available. Examples of such data include preprocessed remote sensing or medical imaging data, VLSI chip layout data, and geographic search logs comprising with associated locations [1, 2].

Spatial co-location patterns represent subsets of spatial features (or spatial objects, attributes), and co-location mining is essential to reveal the frequent co-occurrence patterns among spatial features in various applications. For example, these techniques can show that West Nile virus usually appears in areas where mosquitoes are abundant and poultry are kept; or that botanists discover that 80% of sub-humid evergreen broadleaved forests grow with orchid plants [3, 4].

© Springer International Publishing AG 2017
A. Bouguettaya et al. (Eds.): WISE 2017, Part I, LNCS 10569, pp. 199–214, 2017.
DOI: 10.1007/978-3-319-68783-4_14

The traditional model of mining prevalent co-location patterns was proposed by Shekhar and Huang [2, 3]. In this model, the prevalence measure of co-location patterns is defined based on *clique instances* under spatial neighbor relationships. In detail, the prevalence of a co-location c is the minimum participation ratio $Pr(f_i, c)$ among all features f_i in c; The participation ratio $Pr(f_i, c)$ of feature f_i in a co-location c is the fraction of the instances of f_i that participate in co-location instance (clique instance, the set of instances has clique relationship) of c. Figure 1 shows an example of the spatial data sets with three spatial features {A, B, C} where two spatial instances is connected if their corresponding distance is smaller than the given threshold and $A.i$ denotes the i-th instance of feature A. The co-location pattern {A, B, C} in Fig. 1 is not prevalent when the minimum prevalence threshold *min_prev* = 0.4, because there is only one clique instance {A.1, B.1, C.1} for pattern {A, B, C} and the prevalence of {A, B, C} is only 0.2. But we can see that there are three out of four instances of A which are neighbors to instances of B and C, and that B has two out of five instances which are neighbors to instances of A and C, and C has two out of three. That is to say, at least 40% of instances of each spatial feature in {A, B, C} are close to instances of the other features. This implies that instances from A, B and C are spatially correlated, which is neglected if we only consider the clique instances.

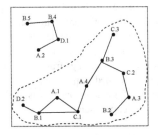

Fig. 1. An example of the spatial data sets

Discovering the patterns that are similar to {A, B, C} in Fig. 1 can help people to gain insights into the distribution of spatial features in spatial environments.

Suppose {A, B, C} is discovered from the distributional data set of vegetation in a certain region, then the presence of other vegetation in the neighborhood of growing any vegetation in {A, B, C}, i.e., {A, B, C} represents symbiotic vegetation in the region. Such patterns are useful in practical applications. For example, they can help vegetation distribution analysis, in the site selection for investigating vegetation, and in vegetation protection.

Alternatively, suppose feature A represents "hospital", B is "resident area" and C is "bus station" in Fig. 1. Let's analyze the distribution of instances of spatial feature set {"hospital", "resident area", "bus station"}. We can see that the resident area "B.3" is near to the hospital "A.4" as well as the bus stations "C.2" and "C.3", whilst the bus station "C.2" or "C.3" is not in the neighborhood of the hospital "A.4", but then "A.4" is near to another bus station "C.1". In fact, the layout of the three feature types within the dotted line has satisfied a co-located correlation.

The problem in the traditional models of mining prevalent co-location patterns is that they only choose clique instances as the measure of interest. However, the statement that "the presence of B and C in the neighborhood of an instance of the spatial feature A" obviously does not necessarily implies that instances of B and C are neighbors. To tackle the problem, this paper advocates a new concept of *sub-prevalent co-location patterns,* by replacing clique instances with star participation instances.

A traditional prevalent co-location pattern must be a sub-prevalent co-location pattern, but not vice versa. That is to say, the set of sub-prevalent co-locations will be larger, and the user may face a larger number of patterns and often not know what course to take. A key observation is that the downward inclusion property is satisfied by sub-prevalence. This enables *maximal* sub-prevalent co-location pattern (MSCP) mining to be studied in this paper.

Thus, our contributions in this paper are as follows:

- First, we define a new concept of sub-prevalent co-location patterns (SCPs) by introducing star participation instances. This concept becomes more reasonable and significant with the frequent co-occurrences of the features involved in real data sets.
- Second, we propose two novel MSCP mining algorithms, namely the prefix-tree-based algorithm (PTBA) and the partition-based algorithm (PBA). At the same time, an intersection-based method is proposed to compute the interest measures of patterns containing star participation instances.
- Third, the advantages and disadvantages of the two algorithms are analysed in depth, and we experimentally evaluate our works on synthetic and real data sets. The experimental results show that our algorithms are scalable and the mined patterns are longer than the traditional results and can capture the star-correlations of the spatial features.

Related Works: Koperski and Han [5] first proposed the problem of mining association rules based on spatial relationships. The work discovers the subsets of spatial features frequently associated with a specific *reference feature*, e.g., mines. A set of neighboring objects of each reference object is converted to a transaction. Wang et al. [6] proposed a novel method based on the partition of spatial relationships for mining multilevel spatial association rules from transactions. In their method, the introduction of an equivalence partition tree method makes the discovery of rules efficient.

Morimoto [7] discovers frequent co-located features sets using a *support count* measure. This approach uses a space partitioning and non-overlap grouping scheme for identifying co-located instances. However, the explicit space partitioning approach may miss co-location instances across partitions. Shekhar and Huang [2, 3] proposed the *minimum participation ratio* based on *clique instances* to measure the frequency of a co-location pattern. This is a statistically meaningful interest measure for co-location patterns. Based on the interest measure, many mining algorithms were proposed [2–4, 8–10]. The set of maximal prevalent co-location patterns implicitly and concisely represents all prevalent co-location patterns [11–14].

Our work on the interest measure lies between the *reference object* model of Koperski and Han [5] and the *minimum participation ratio* measure of Shekhar and

Huang [2, 3]. With no specific reference feature, directly applying the model of [5] for co-location mining may not capture the co-locations found by the techniques in this paper, whilst the measure in [2, 3] may miss some reasonable and useful co-located patterns in practical applications due to the requirement to use clique instances. In the designs of the algorithms for mining MSCPs, the idea of using the prefix-tree structure to store all maximal candidates comes from the work presented in [11]; the idea of pruning breadth-first in candidate space search tree can be traced back to [13, 14]. In the literatures we have found no mention of our core pattern method based on partitions.

The remainder of the paper is organized as follows. Section 2 defines the related concepts of SCPs mining, and proves the *anti-monotonic property* of SCPs. Two novel MSCP mining algorithms are presented in Sects. 3 and 4 respectively. We show the experimental evaluation in Sect. 5. Finally, we conclude the paper in Sect. 6.

2 Basic Concepts and Properties

We first present the basic concepts of the maximal sub-prevalent co-location pattern mining and discuss its anti-monotonic property.

In a spatial data set, let F be a set of n *features* $F = \{f_1, f_2, \ldots f_n\}$. Let S be a set of *instances* of F, where each instance is a tuple < instance-id, location, spatial features > . Let R be a *neighbor relationship* over locations of instances. R is symmetric and reflexive. In this paper, we use the Euclidean distance with a distance threshold d as the neighbor relationship.

A co-location c is a subset of spatial features, $c \subseteq F$.

We use the star neighborhoods instance, which was introduced in the join-less algorithm [8], as the method for materializing the relationship between spatial instances. The star neighborhoods instance is a set comprising a center instance and other instances in its neighborhood. Following is a formal definition.

Definition 1 *(Star Neighborhoods Instance, SNsI)*: $SNsI(o_i) = \{o_j \mid distance(o_i, o_j) \leq d$, where d is a neighbor relationship distance threshold$\}$ is the star neighborhoods instance of o_i, and o_i is the label of $SNsI(o_i)$. In other words, $SNsI(o_i)$ is the set consisting of o_i and the other spatial instances located within distance d from o_i, i.e., having neighbor relationships with o_i.

In Fig. 1, a point represents an instance and a solid line between two points represents the neighbor relationship between two instances. $X.i$ is the i-th instance of the feature X. As can be seen in Fig. 1, $SNsI(A.1) = \{A.1, B.1, C.1\}$, and $SNsI(B.3) = \{A.4, B.3, C.2, C.3\}$.

Based on the concept of star neighborhoods instance, we define the concept of star participation instances, and then we use the participation ratio and the participation index defined by Huang et al. [3] to characterize how frequently instances of different features in a co-location pattern are neighbors.

Definition 2 *(Star Participation Instance, SPIns)*: $SPIns(f_i, c) = \{o_i \mid o_i$ is an instance of feature f_i and $SNsI(o_i)$ contains instances of all features in $c\}$ is the star participation instance of feature f_i in c. In other words, $SPIns(f_i, c)$ *is* the set consisting of instances of f_i whose star neighborhoods' instances contain all features in c.

Definition 3 *(Star Participation Ratio, SPR)*: $SPR(f_i, c) = |SPIns(f_i, c)|/|S_{f_i}|$ is the star participation ratio of feature f_i in a co-location c, where S_{f_i} is the set of instances of f_i. In other words, $SPR(f_i, c)$ is the fraction of instances of f_i that occur in the star participation instance of f_i in c.

Definition 4 *(Star Participation Index, SPI)*: $SPI(c) = \min_{f \in c}\{SPR(f, c)\}$ is the star participation index of a co-location c. In other words, $SPI(c)$ is the minimum star participation ratio $SPR(f_i, c)$ among all features f_i in c.

Definition 5 *(Sub-Prevalent Co-location Patterns, SCP)*: A co-location c is a sub-prevalent co-location pattern, if its star participation index is no less than a given sub-prevalence threshold *min_sprev*, that is, $SPI(c) \geq min\_sprev$.

For example, in Fig. 1, the star participation instance $SPIns(A, \{A, B, C\})$ of feature A in $\{A, B, C\}$ is $\{A.1, A.3, A.4\}$, $SPIns(B, \{A, B, C\}) = \{B.1, B.3\}$, and $SPIns(C, \{A, B, C\}) = \{C.1, C.2\}$. Thus, the star participation ratio $SPR(A, \{A, B, C\})$ of A in $\{A, B, C\}$ is 3/4 because 3 out of 4 instances of A occur in star participation instance of A in $\{A, B, C\}$. Similarly, $SPR(B, \{A, B, C\}) = 2/5$, and $SPR(C, \{A, B, C\}) = 2/3$. Therefore, $SPI(\{A, B, C\}) = \min\{3/4, 2/5, 2/3\} = 2/5 = 0.4$. If the sub-prevalence threshold *min_sprev* is no more than 40%, $\{A, B, C\}$ is a sub-prevalent co-location pattern.

Lemma 1 *(Monotonicity of SPR and SPI).* Let c and c' be two co-locations such that $c' \subseteq c$. Then, for each feature $f \in c'$, $SPR(f, c') \geq SPR(f, c)$. Furthermore, $SPI(c') \geq SPI(c)$.

Proof. For the first claim in the lemma, we only need to show that for a spatial feature $f \in c'$, $|SPIns(f, c')| \geq |SPIns(f, c)|$.

Since $c' \subseteq c$, every star participation instance of feature f in c contains instances of all features in c'. Thus, the inequality holds.

The second claim follows from the fact that $SPI(c') = \min_{f \in c'}\{SPR(f, c')\} \geq \min_{f \in c}\{SPR(f, c)\} = SPI(c)$. □

This lemma establishes the downward inclusion property of *SPR* and *SPI*.

It is obvious that we can mine longer co-location patterns after introducing the sub-prevalent concept. But the mined results might be massive and mutually inclusive. Maximal sub-prevalent co-location pattern mining can resolve this problem.

Definition 6 *(Maximal Sub-Prevalent Co-location Patterns, MSCP)*: Given a sub-prevalent co-location $c = \{f_l, ..., f_v\}$ in a set of spatial features $F = \{f_1, ..., f_n\}$, $l, v \in \{1, 2, \cdots, n\}$, if none of c's super-patterns are sub-prevalent, the co-location c is called a MSCP.

The set of MSCPs is a compact representation of a large number of SCPs. The MSCPs form the minimal set which can lead to all the SCPs.

Based on Lemma 1, two novel mining algorithms are proposed, respectively in Sects. 3 and 4, to discover the complete set of MSCPs from a spatial data set.

3   A Prefix-Tree Based Algorithm

First, we note that star neighborhood instances are equivalent to each other, i.e., if $o_j \in SNsI(o_i)$, then $o_i \in SNsI(o_j)$. So, our algorithm firstly collects those pair of instances which are neighbors to each other, and then selects 2-size SCPs by comparison with a given sub-prevalence threshold *min_sprev*.

Then, due to Lemma 1, SCP candidates can be generated from the set of 2-size SCPs by detecting feature sets which can form cliques. For example, if there was a set of 2-size SCPs: $SCP_2 = \{\{A, B\}, \{A, C\}, \{A, D\}, \{B, C\}, \{B, D\}, \{C, D\}, \{C, E\}, \{D, E\}\}$, the feature set $\{A, B, C, D\}$ is a 4-size SCP candidate because it forms a clique under the 2-size sub-prevalent relationship, while the feature set $\{B, C, D, E\}$ does not belong to the 4-size candidate set because $\{B, E\}$ is not a 2-size SCP.

We can use a lexicographic order based method for generating all SCP candidates. In order to reduce the candidate search space dynamically in the mining process, all generated candidates are organized into a *prefix tree*. In the prefix tree, a branch is created for a candidate and new branch would share common prefixes. For example, as shown in Fig. 2, the candidate $\{A, B, C, D\}$ leads to the first branch of the candidate search space tree. The second candidate $\{A, B, C\}$ does not form a new branch due to sharing common prefixes with the first branch, whilst $\{A, B, D\}$ forms a new branch due to sharing two common prefixes. The feature set identifying each node will be referred to as the node's *head*, while possible extensions of the node are called the *tail*. For example, consider nodes Y' and Y'' in Fig. 2; their head is $\{A\}$ and $\{A, B\}$ respectively, and the tail is the set $\{B, C, D\}$ and $\{C, D\}$ respectively. The head union the tail (HUT) of Y' is $\{A, B, C, D\}$.

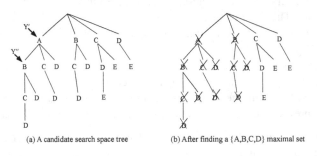

(a) A candidate search space tree   (b) After finding a {A,B,C,D} maximal set

Fig. 2.   Candidate search space tree and pruning

Our mining method starts with the longest $l = l_{max}$ co-location candidates in the search space, and determines whether they are MSCPs as specified in Definition 6. Once the algorithm determines that a candidate is maximal, the algorithm can begin a pruning process. That is, the algorithm checks breadth-first whether the HUT of each node in the candidate search space tree is a subset of a current maximal set. If the HUT is a subset of any MSCPs, the sub-tree whose root is the node is pruned. For example, in Fig. 2(a), let's assume that $\{A, B, C, D\}$ is a MSCP. Then the algorithm checks breadth-first for candidates to prune in the search space tree. In the first level, the HUT of node A is

{A, B, C, D}. Since it is a subset of {A, B, C, D}, the algorithm prunes the sub-tree whose root is A. The next node in that level is node B whose HUT is {B, C, D}. Since this candidate is a subset of {A, B, C, D}, the algorithm prunes this sub-tree. The pruning continues with the rest of the nodes in the first level. Then the algorithm starts pruning the next level until the subset tree resembles Fig. 2(b).

Finding MSCPs and running the pruning scheme happens at each l-size until there are no more candidates to process.

In the method discussed above, computing the sub-prevalence measure of a candidate c, i.e. $SPI(c)$, is a core problem. Based on the 2-size co-location instances (neighborhoods instances' pairs), we can compute the sub-prevalence measure of any k-size pattern ($k > 2$) by the following lemma 2.

Lemma 2 (*the Sub-prevalence Measure of k-size Co-locations (k > 2)*). Given the 2-size co-location instances of a spatial data set, the star participation index of a k-size co-location $c = \{f_1, \dots f_k\}$ can be calculated as follows:

$$SPI(c) = min\{|\bigcap_{j=2,\dots k} SPIns(f_1, \{f_1, f_j\})|/|S_{f_1}|, \dots| \bigcap_{j=1,\dots k-1} SPIns(f_k, \{f_k, f_j\})|/|S_{f_k}|\}$$

(1)

Proof. For $c = \{f_1, \dots f_k\}$, according to Definition 2, $SPIns(f_i, c) = \bigcap_{j=1,\dots k, j \neq i} SPIns(f_i, \{f_i, f_j\})$. From Definitions 3 and 4, the formula for calculating $SPI(c)$ is obvious. □

This method is a typical candidate generate-and-test method. We call it prefix-tree-based algorithm (PTBA) due to the fundamental nature of the method's candidate search space tree.

In PTBA, if the longest size of MSCPs is not long enough, the breadth-first candidate pruning would not be ideal. As an improvement, we propose the following pruning lemma to prune sub-trees or nodes in a candidate search space tree when a candidate is **not** a MSCP.

Lemma 3 (*Depth-first pruning*). If a candidate $c = c' \cup \{X\} \cup \{Y\}$ is not maximal sub-prevalent, then

(1) The node "Y" can be pruned;
(2) If $SPI(c' \cup \{X\} \cup \{Y\}) = SPI(c' \cup \{X\})$, then the sub-tree whose root is node "X" can be pruned;
(3) If $SPI(c') = SPI(c' \cup \{X\})$, then the node "$Y$" which is a brother of the node "X" can be pruned.

Proof. Case (1) is obvious because the candidate c is not a maximal sub-prevalent co-location; For case (2) $SPI(c' \cup \{X\} \cup \{Y\}) = SPI(c' \cup \{X\})$, because $SPI(c' \cup \{X\}) = SPI(c' \cup \{X\} \cup \{Y\}) < min\_sprev$, according to case (1), the sub-tree whose root is "X" can be pruned; For case (3) $SPI(c') = SPI(c' \cup \{X\})$, because $SPI(c' \cup \{Y\}) = SPI(c' \cup \{X\} \cup \{Y\}) < min\_sprev$, according to case (1), the node "Y" in $c' \cup \{Y\}$ can also be pruned. □

We have presented a PBTA for mining all MSCPs, but the large number of candidates in the candidate search space tree and the computational complexity of sub-prevalence measures for long patterns limit the scale of spatial data sets that can be handled. In the next section, a novel partitioning technique is proposed that can resolve these problems efficiently.

4 A Partition-Based Algorithm

4.1 PB Method

This section presents an interesting method called the partition-based algorithm (PBA), which adopts a divide-and-conquer strategy as follows. First, it divides the set of 2-size SCPs into the set of lexicographic order strings by the relation $=_{\text{head}}$. It then mines each string separately based on a *core pattern method*. We first define the related concepts as follows.

Definition 7 (partition pattern (PP) and core pattern (CP)). A 2-size pattern contained in a co-location pattern c is called a **partition pattern** of c, if its star participation index (SPI) is the biggest in all 2-size co-locations of c. The two sub-patterns of c, which are divided by the *PP*, are called **core patterns** of c.

For example, if the PP of co-location $c = \{A, B, C, D, G\}$ is $\{C, D\}$, then CPs of c are $\{A, B, C, G\}$ and $\{A, B, D, G\}$. Because the SPI of PP is the biggest in all the 2-size co-locations of c, the CPs of c may be key patterns deciding on whether c is sub-prevalent or not.

The Core Pattern Method to Identify SCPs: Given a l-size co-location $c = \{f_1,\ldots f_l\}$ and its PP $\{f_{l-1}, f_l\}$, we divide c into two $(l-1)$-size CPs $\{f_1,\ldots,f_{l-2},f_{l-1}\}$ and $\{f_1,\ldots,f_{l-2},f_l\}$ by $\{f_{l-1},f_l\}$. If two CPs $\{f_1,\ldots,f_{l-2},f_{l-1}\}$ and $\{f_1,\ldots,f_{l-2},f_l\}$ are sub-prevalent, and the two additional conditions shown below are also satisfied, then it can be determined that c is a SCP.

The additional condition (1):

$$|SPIns(f_{l-1}, \{f_1,\ldots,f_{l-2},f_{l-1}\}) \cap SPIns(f_{l-1}, \{f_{l-1},f_l\})|/|S_{f_{l-1}}| \geq min\_sprev \text{ and } |SPIns(f_l, \{f_1,\ldots,f_{l-2},f_l\}) \cap SPIns(f_l, \{f_{l-1},f_l\})|/|S_{f_l}| \geq min\_sprev$$

The additional condition (2):

$$\min\{|SPIns(f_1, \{f_1,\ldots,f_{l-2},f_{l-1}\}) \cap SPIns(f_1, \{f_1,\ldots,f_{l-2},f_l\})|/|S_{f_1}|,\ldots|SPIns(f_{l-2}, \{f_1,\ldots,f_{l-2},f_{l-1}\}) \cap SPIns(f_{l-2}, \{f_1,\ldots,f_{l-2},f_l\})|/|S_{f_{l-2}}|\} \geq min\_sprev$$

In fact, $SPI(c) = \min\{|SPIns(f_1, \{f_1,\ldots,f_{l-2},f_{l-1}\}) \cap SPIns(f_1, \{f_1,\ldots,f_{l-2},f_l\})|/|S_{f_1}|,\ldots|SPIns(f_{l-2}, \{f_1,\ldots,f_{l-2},f_{l-1}\}) \cap SPIns(f_{l-2}, \{f_1,\ldots,f_{l-2},f_l\})|/|S_{f_{l-2}}|, |SPIns(f_{l-1}, \{f_1,\ldots,f_{l-2},f_{l-1}\}) \cap SPIns(f_{l-1}, \{f_{l-1},f_l\})|/|S_{f_{l-1}}|, |SPIns(f_l, \{f_1,\ldots,f_{l-2},f_l\}) \cap SPIns(f_l, \{f_{l-1},f_l\})|/|S_{f_l}|\}$.

For identifying two CPs $\{f_1, ..., f_{l-2}, f_{l-1}\}$ and $\{f_1, ..., f_{l-2}, f_l\}$, we can deal with them recursively using the core pattern method. In fact, the core pattern method employs a bottom-up iteration strategy, which starts from identifying 3-size core patterns, then to 4-size core patterns,...until l-size pattern. With the following example, we will illustrate how the PB method works in detail.

Suppose there is the set of 2-size SCPs: $SCP_2 = \{\{A, B\}, \{A, C\}, \{A, D\}, \{A, F\},$ $\{A, G\}, \{B, C\}, \{B, D\}, \{B, G\}, \{C, D\}, \{C, F\}, \{C, G\}, \{D, E\}, \{D, F\}, \{D, G\}, \{E, F\}\}$ in a certain spatial data set whose feature set is $\{A, B, C, D, E, F, G\}$.

We first partition SCP_2 into the set of lexicographically ordered strings under the relation $=_{head}$. This resulting set is denoted L_1. Thus, we have $L_1 = \{\delta_{head}(A) =$ ABCDFG, $\delta_{head}(B) =$ BCDG, $\delta_{head}(C) =$ CDFG, $\delta_{head}(D) =$ DEFG, $\delta_{head}(E) =$ EF, $\delta_{head}(F) = $ F, $\delta_{head}(G) = $ G$\}$. We call this partition **Partition_1**.

Then, according to 2-size non-SCPs, ordered strings in L_1 are divided further. In our example, there are 2-size non-SCPs $\overline{SCP_2} = \{AE, BE, BF, CE, EG\}$. So, the ordered string $\delta_{head}(A) =$ ABCDFG is further divided into two strings ABCDG and ACDFG by BF. We continue to divide above sub-strings respectively until there exists no 2-size non-SCP in it or its size is less than 2. Thus, L_1 is replaced with $L = \{$ABCDG, ACDFG, BCDG, CDFG, DEF, DFG, EF$\}$. The non-sub-prevalent-based partition is called **Partition_2**.

Next, we deal with strings in L one by one using the core pattern method to obtain whole MSCPs. For our example, we first consider ordered string "ABCDG". Suppose $\{C, D\}$ is the PP of $c = \{A, B, C, D, G\}$, then the CPs of c are $\{A, B, C, G\}$ and $\{A, B, D, G\}$. Continuing to partition, we may get 3-size CPs $\{A, B, C\}$ and $\{A, B, G\}$ for $\{A, B, C, G\}$, and $\{A, B, D\}$ and $\{A, B, G\}$ for $\{A, B, D, G\}$. We call the partition obtaining CPs based on PP as **Partition_3**. The real line part in Fig. 3 is the result of partitioning $\{A, B, C, D, G\}$ by Partition_3.

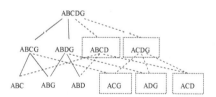

Fig. 3. $\{A, B, C, D, G\}$ and its sub-sets with head of "A"

We can directly compute the SPI values of 3-size core patterns using Lemma 2, using the results of 3-size core patterns to identify the related 4-size core patterns by the core pattern method, and then 5-size patterns. For our example:

If two 3-size core patterns $\{A, B, C\}$ and $\{A, B, G\}$ are sub-prevalent, and their additional conditions (1) and (2) are satisfied, then we can determine that 4-size co-location $\{A, B, C, G\}$ is sub-prevalent. Similarly, we can identify $\{A, B, D, G\}$, and then identify $\{A, B, C, D, G\}$.

Once the 5-size pattern $\{A, B, C, D, G\}$ is not a SCP, the rest of the 4-size patterns with the same head (beside two CPs) $\{A, B, C, D\}$ and $\{A, C, D, G\}$ (see the dotted line part in Fig. 3) need to be identified. From Fig. 3, we note that there are common patterns for identifying higher-size patterns. For example, {A, B, G} is common for identifying {A, B, C, G} and {A, B, D, G}, and {A, B, C} is common for identifying {A, B, C, G} and {A, B, C, D}. Thus it is better to store the results of lower-size patterns. In addition, if the 5-size pattern {A, B, C, D, G} is sub-prevalent, then the remaining 4-size patterns need not be identified.

When finishing the processing of judging the prevalence of ordered string "ABCDG", the next ordered string "ACDFG" in $\delta_{head}(A)$ is dealt with. Then we obtain the set $MSCP_A$ of MSCPs with the head of "A", and then prune all subsets of $MSCP_A$ in L. In the following, the ordered strings with a head of "B" in L are handled.

4.2   Comparison with the Prefix-Tree-Based Method

The core problems of the prefix-tree-based algorithm (PTBA) presented in Sect. 3 are: (1) the candidate search space tree might be too large to store and search when we confront a data set which has a large number of features; (2) the total cost to compute SPIs of candidate patterns by Lemma 2 might be too expensive due to a large number of intersection operations. The PBA is aimed to relieve these two problems.

First, we divide candidates into equivalent classes by Partition_1 and Partition_2, and deal with a pattern in an equivalent class which has the same head each time, so as to resolve the problem of the candidate search space tree being too large.

Second, differing from the PTBA method which starts with the identification of the longest patterns, the PBA method is performed step by step from 3-size core patterns. With regard to the expected cost of the intersection operations as a whole, the core pattern method is better than the PTBA method. For example, for a 6-size pattern $c = \{A, B, C, D, E, F\}$, the cost of directly computing SPI(c) by Lemma 2 is 24 intersection operations. Because computing the star participation instance SPIns(A, c) of feature A in c needs 4 times the intersection operations, if c is not sub-prevalent, then the cost of computing a 5-size sub-set of c is 15, and the cost of a 4-size is 8. But if we use the core pattern method, then a 3-size pattern costs 3. To compute a 4-size pattern based on two 3-size core patterns needs 4 times the intersection operations. Similarly, the cost of a 5-size is 5 and 6-size's is 6. Thus, the expected cost of the PTBA method for a 6-size pattern c is $(24 + 54 + 78 + 90)/4 = 61.5$, while the expected cost by the core pattern method (PBA) is $(12 + 24 + 34 + 40)/4 = 27.5$.

Third, if a pattern considered is not sub-prevalent, the core pattern method could find it sooner because Partition_3 is based on a partition pattern which SPI is the biggest in all the 2-size co-locations with respect to it. In addition, the middle computational results could either be MSCPs or be used in the computation of other corresponding patterns.

5   Experiments

In this section, we evaluate the performance of the proposed two algorithms and analyze the difference in the results from the traditional maximal prevalent co-location mining (MCP). To aid analysis, we have improved the join-less algorithm [8] so as to mine all MCPs. The improved algorithm is called the M-join-less algorithm. We have optimized implementation of M-join-less and, compared with other algorithms for mining MCPs, it seems M-join-less can deal with more data for the same run-time.

All algorithms are implemented using C# under Windows 7 and executed on a normal PC with Intel i3-3240 @3.4 GHz CPU and 4 GB of memory. In these experiments, we have manually verified that the results of all algorithms presented in this section are the expected results.

5.1 Comparison of Computational Complexity Factors

We examined the **costs** of the computational complexity factors using 3 synthetic data sets of different density. A sparse data set generated 6-size MSCPs out of 20 features while a dense data set generated 9-size MSCPs out of 20 features. In order to observe the advantage of PTBA compared with PBA, a dense* (see Table 1) data set was generated which generated 13 MSCPs out of 15 features. In Table 1, all data values except total execution time represent percent values. Overall, we can see that M-join-less is always slower than our two algorithms, because M-join-less computes prevalence measures of candidates by identifying clique instances, and the cost of computing clique instances is more expensive than that of intersection operations. Thus we discarded M-join-less and only discuss PBA and PTBA in our later comparisons.

Table 1. Comparison of computational complexity factors

Method / data type factor(%)	M-join-less		PTBA			PBA		
	sparse	dense	sparse	dense	dense*	sparse	dense	dense*
$T_{gen\ 2\ prev\ col}$	6.12	0.6	0.2	0.001	3.40	16.25	0.9	1.87
$T_{gen\ candi}$	2.58	3.85	1.96	0.29	0.44	-	-	-
$T_{gen\ candi\text{-}tree}$	-	-	6.49	0.74	0.74	-	-	-
$T_{gen\ 2\text{-}non\ prev\ col}$	-	-	-	-	-	2.52	0.001	0.001
$T_{order\ features}$	"	"	"	"	"	13.89	0.002	0.001
$T_{Part\ 1+Part\ 2}$	-	-	-	-	-	1.77	0.002	0.002
$T_{Pruning}$	-	-	3.12	39.21	61.53	8.59	4.16	0.26
$T_{gen\ mscp}$	91.29	95.64	88.19	59.76	33.88	56.89	94.94	97.88
Total execution time(Sec)	3.03	108.36	0.65	102.17	1.32	0.48	7.23	9.17

With a very dense data set like dense*, PTBA shows much better performance than PBA, but in both sparse and dense data, PBA shows better performance than PTBA. This is because PBA is a bottom-up method and, if a data set is sparse, it will stop more quickly than PTBA, while PTBA is a top-down method, and it can find long MSCPs more quickly. Also we see that in the PTBA method, the denser a data set is, the longer the percentage $T_{pruning}$ costs. Similarly, the better performance PTBA shows, the worse PBA shows. In addition, we notice that T_{gen_mscp} and $T_{pruning}$ take a much bigger portion of total costs than any other costs. Clearly, the different candidate identification methods are the core distinction between two algorithms. Note that the factor $T_{order\_features}$ is merely for selecting core patterns of candidates in PBA.

5.2   Comparison on Expected Costs in Identifying Candidates

From Table 1, we saw that no one method has the absolute advantage in the two candidate identification methods, so we compared the expected costs of identifying candidates in PBA with those in PTBA.

We used a dense data set and a sparse data set to evaluate the expected performance ratio. The spatial frame size $D \times D$ is chosen as 1000×1000 and 5000×5000 for the dense and sparse data sets respectively. We select all k-size candidates for each data set, where size k is 6, 8, 10, 12, 14, 16 and 18. Then, we set *min_sprev* as 0.2, 0.4, 0.6 and 0.8 respectively, obtained each candidate's execute time and then calculated the average execute time per size for all *min_sprev*. In Fig. 4, the y-axis represents the ratio of average execute time of PBA to the average execute time of PTBA. The results show that, as the size of candidate increases, the expected cost ratio is reduced overall, especially for the sparse data set. At the same time, we also note that the experimental results are basically consistent with the analysis in Sect. 4.2.

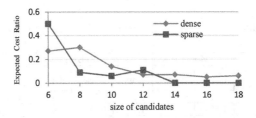

Fig. 4. Comparison of expected cost in identifying candidates

5.3   Scalability Tests

We examined the scalability of PTBA and PBA with several workloads, e.g. different numbers of instances, numbers of features, neighbor distance thresholds, and sub-prevalence thresholds. We compared the total computation time for finding all MSCPs.

Effect of the Number of Instances. First, we compared the effect of the number of instances in PTBA and PBA. As shown in Fig. 5(a), when the number of instances is fewer, PBA shows better performance than PTBA, but as the data set grows denser and denser, PBA costs more and more time on calculation while PTBA has a drop after the point of 80K (80000) instances. When the number of instances reaches 120K, the data set is dense enough that the longest candidate pattern is sub-prevalent, so PTBA and PBA stop quickly, having spent most time on generating 2-size prevalent patterns. Because the cost for creating the prefix trees in PTBA is more than the cost on Partition_1 and Partition_2 in PBA, overall PTBA costs a little more than PBA. Figure 5(a) shows that both algorithms' scale to large dense data sets.

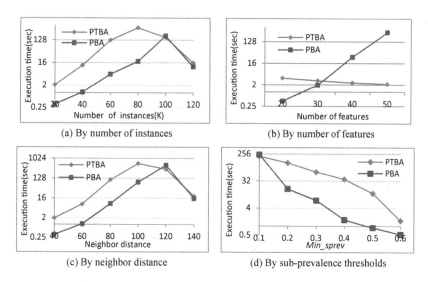

Fig. 5. Evaluation and comparison of the two algorithms

Effect of the Number of Features. In the second experiment, we compared the performance of PTBA and PBA as a function of the number of features. Figure 5(b) shows the results. PTBA is better than PBA when the number of features is over 30. As the number of features increased, the execution time of PTBA decreased, whilst that of PBA increased. The reason is that, under the same number of instances, the increase of features causes the number of instances per feature to be decreased, which in turn may lead to a decrease in the number of instances per 2-size pattern. PBA spends much more time on Partition_2 and obtains more shorter-size candidates, because the number of non-prevalent 2-size patterns becomes larger. Overall, the two algorithms both show good performance even if the number of features reaches 50.

Effect of Neighbor Distance. The third experiment examined the effect of different neighbor distances. As Fig. 5(c) shows, when the distance threshold is below 100, PTBA and PBA all increase quickly because the increase of neighbor distance makes the neighborhood areas larger and increases the number of size-2 co-location instances, although PBA performs better than PTBA. When the distance reaches 120, PTBA becomes a little better than PBA because the neighborhood area is large enough that PTBA can stop earlier than PBA.

Effect of Sub-prevalence Threshold. In the final scalability experiment we examined the performance effect of different sub-prevalence threshold *min_sprev*. Overall, the execution time decreases for both algorithms as the sub-prevalence threshold increases, as shown in Fig. 5(d). However, the PBA method reduces the computation time by a larger magnitude for lower threshold values.

5.4   Evaluation with Real Data Sets

The experimental real data sets come from the rare plant distribution data sets of the "Three Parallel Rivers of Yunnan Protected Areas". Figure 6 gives the distribution of plant data in two-dimensional space, where the X and Y coordinates represent the instances' locations relative to 0° longitude and latitude lines respectively in meters. We can see that this is a zonal plant distribution data. Because plants suffer from the effects of altitude or river formations, zonal data are common in natural ecological studies.

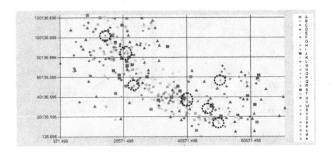

Fig. 6. The distribution of a certain plant data

Table 2. Mining Results on the Plant Data Set in Fig. 6

Size	N. of MSCPs	N. of MCPs	Size	N. of MSCPs	N. of MCPs
2	5	15	7	23	31
3	41	70	8	35	9
4	64	115	9	10	5
5	95	98	10	7	3
6	72	73	11	3	–

First, we tested the mining result difference between MSCPs under star participation instances and MCPs under clique instances. In the experiment, the rare plant distribution data set in Fig. 6 was chosen, where the number of features (plants) is 31 and the number of instances is 336. We used a rectangular spatial framework of size $130K \times 130K$, a neighbor distance threshold $d = 12000$ m, and prevalence thresholds $min\_sprev = min\_prev = 0.3$. Table 2 shows the number of MSCPs/MCPs with different lengths.

From Table 2 we can see that, with the same data set and parameter settings, the MSCP mining method can generate longer patterns than the MCP mining method. This is because, if an instance occurs in a clique instance, it must be in a star participation instance, but a star neighbor relationship may not be a clique relationship. From the table, we also found that the percentage of long patterns in MSCPs is greater than that in MCPs. There are 15.5% MSCPs whose size is over 7, compared to only 4.1% MCPs. Generally, users are more interested in long patterns, because long patterns contain more interesting information and short patterns usually are obvious.

Let us take some mined results to analyze in detail. Three mined 11-size MSCPs are {B, G, J, O, Q, R, T, V, W, c, e}, {G, J, O, Q, R, T, V, W, a, c, e} and {G, H, J, O, Q, R, V, W, a, c, e}, whilst three mined 10-size MCPs are {G, J, O, R, T, V, W, a, c, e}, {G, J, Q, R, T, V, W, a, c, e} and {J, O, Q, R, T, V, W, a, c, e}. We note that each 10-size MCPs is a sub-set of one in the 11-size MSCPs. We denote the rare plant "O" with dashed circles in Fig. 6, and then we can see that it is basically distributed in an axial line which is the focus area of growing plants. That is why it occurs in three mined 11-size MSCPs, the same cases to the plants "G" and "Q".

Second, we used a large scale vegetation distribution data set in the "Three Parallel Rivers of Yunnan Protected Area" to examine the efficiency of the PBA method. In this data set, the total number of features is 15, and the total number of instances is 487,857. We used 0.3 as min_sprev, and 1000 m, 3000 m, and 5000 m as the neighborhood distance thresholds respectively. The running time of PBA is respectively 148.26(s), 1967.84(s), and 7906.74(s).

6 Conclusions

In this paper, we analyzed the limit of the traditional clique instances based measure for spatial co-location pattern mining in some practical applications. A new concept of sub-prevalent co-location patterns based on star participation instances was defined. It was especially important to show that the proposed star participation ratio (SPR) and star participation index (SPI) obey the downward inclusion property, thus allowing maximal co-location pattern mining and interactive pruning.

The Sub-Prevalent Co-location Miner, which contains two novel algorithms for mining maximal sub-prevalent co-location patterns, was presented. The proposed algorithms were evaluated using theoretical and experimental methods. Empirical evaluation shows that the partition-based method performs much better than the prefix-tree-based method when we confront sparse data sets but it becomes slower in very dense data sets. Compared with the M-join-less algorithm for mining maximal prevalent co-locations, our Miner is more efficient.

An important future direction is to extend the maximal sub-prevalent co-location mining to other types of spatial data including spatio-temporal data and uncertain data.

Acknowledgments. This work is supported by the National Natural Science Foundation of China (61472346, 61662086), the Natural Science Foundation of Yunnan Province (2015FB114, 2016FA026), the Project of Innovation Team of Yunnan University.

References

1. Wang, X., Zhao, Y., Nie, L., Gao, Y.: Semantic-based location recommendation with multimodal venue semantics. IEEE Trans. Multimedia **17**(3), 409–419 (2015)
2. Shekhar, S., Huang, Y.: Co-location rules mining: a summary of results. In: The IEEE International Symposium on Spatial and Temporal Databases (SSTD), pp. 236–256 (2001)

3. Huang, Y., Shekhar, S., Xiong, H.: Discovering co-location patterns from spatial data sets: a general approach. IEEE Trans. Knowl. Data Eng. **16**(12), 1472–1485 (2004)
4. Wang, L., Jiang, W., Chen, H., Fang, Y.: Efficiently mining high utility co-location patterns from spatial data sets with instance-specific utilities. In: Candan, S., Chen, L., Pedersen, T. B., Chang, L., Hua, W. (eds.) DASFAA 2017. LNCS, vol. 10178, pp. 458–474. Springer, Cham (2017). doi:10.1007/978-3-319-55699-4_28
5. Koperski, K., Han, J.: Discovery of spatial association rules in geographic information databases. In: The 4th International Symposium on Large Spatial Databases (SSD 1995), pp. 47–66 (1995)
6. Wang, L., Xie, K., Chen, T., Ma, X.: Efficient discovery of multilevel spatial association rules using partitions. Inf. Softw. Technol. **47**(2005), 829–840 (2005)
7. Morimoto, Y.: Mining frequent neighboring class sets in spatial databases. In: The 7th ACM International Conference on Knowledge Discovery and Data Mining (SIGKDD), pp. 353–358 (2001)
8. Yoo, J.S., Shekhar, S.: A joinless approach for mining spatial colocation patterns. IEEE Trans. Knowl. Data Eng. (TKDE) **18**(10), 1323–1337 (2006)
9. Xiao, X., Xie, X., Luo, Q., Ma, W.: Density based co-location pattern discovery. In: The 16th ACM SIGSPATIAL International Conference on Advances in Geographic Information Systems (GIS), NY, USA, pp. 11–20 (2008)
10. Yao, X., Chen, L., Peng, L., Chi, T.: A co-location pattern-mining algorithm with a density-weighted distance thresholding consideration. Inf. Sci. **396**(2017), 144–161 (2017)
11. Wang, L., Zhou, L., Lu, J., Yip, J.: An order-clique-based approach for mining maximal co-locations. Inf. Sci. **179**(2009), 3370–3382 (2009)
12. Yoo, J.S., Bow, M.: Mining top-k closed co-location patterns. In: The IEEE International Conference on Spatial Data Mining and Geographical Knowledge Services (ICSDM), pp. 100–105 (2011)
13. Yoo, J.S., Bow, M.: Mining maximal co-located event sets. In: Huang, J.Z., Cao, L., Srivastava, J. (eds.) PAKDD 2011. LNCS, vol. 6634, pp. 351–362. Springer, Heidelberg (2011). doi:10.1007/978-3-642-20841-6_29
14. Yoo, J.S., Bow, M.: Mining spatial co-location patterns: a different framework. Data Mining Knowl. Discov. J. **24**(1), 159–194 (2012)

Overlapping Communities Meet Roles and Respective Behavioral Patterns in Networks with Node Attributes

Gianni Costa$^{(\boxtimes)}$ and Riccardo Ortale$^{(\boxtimes)}$

ICAR-CNR, Via P. Bucci 41c, 87036 Rende, CS, Italy
{costa,ortale}@icar.cnr.it

Abstract. We present a new approach to the unsupervised and joint analysis of overlapping communities, roles and respective behavioral patterns in networks with node attributes. The proposed approach relies on an innovative Bayesian probabilistic model of the statistical relationships among communities, roles, behavioral role patterns and attributes. Essentially, under the devised model, behavioral role patterns define the abstract social functions underlying roles. Also, attributes are affiliated to roles. Moreover, links are explained in terms of community involvement of nodes, their roles and respective behavioral patterns.

Our model allows for exploratory, descriptive and predictive tasks, including the analysis of communities, roles and their behavioral patterns, the interpretation of communities and roles as well as the prediction of missing links. Such tasks are enabled by posterior inference, for which we design a variational algorithm.

An experimental assessment against state-of-the-art competitors reveals a higher accuracy of our approach on real-world benchmark data sets, both in community detection and link prediction. Role interpretation and behavioral role patterns are also demonstrated.

1 Introduction

The seamless modeling of community discovery [10,15,17,20,24] and role analysis [2,18,22,23] in networks is a progressing area of research, aimed to gain a deeper understanding of topology. Basically, both tasks are beneficial to each other. Indeed, on one hand, role analysis reveals the social functions, that are played by nodes within communities, in order to contribute to the pursuit of the corresponding purposes. Symmetrically, on the other hand, community discovery provides a natural contextualization to identify node roles.

Hitherto, the integration of community discovery and role analysis has been investigated from various perspectives [3–7,9]. However, no emphasis has been placed on two properties of primary relevance in network analysis, i.e., node attributes and behavioral role patterns. Node attributes are descriptive features of nodes. Behavioral role patterns supplement the comprehension of roles as abstract social functions, by providing details of the behavior of nodes within their neighborhoods in the fulfillment of such social functions. As a critical review

© Springer International Publishing AG 2017
A. Bouguettaya et al. (Eds.): WISE 2017, Part I, LNCS 10569, pp. 215–230, 2017.
DOI: 10.1007/978-3-319-68783-4_15

of the currently available literature, node attributes are ignored in [4–7,9]. Also, the studies in [3,6] reinterpret the interaction patterns of roles in terms of susceptibility and authority with respect to certain interaction factors. Yet, the efforts in [4,5,7,9] capture roles simply as abstract behavioral classes, without explaining the respective social functions in terms of behavioral patterns.

We argue that node attributes and behavioral role patterns can help inform a tighter and thorough integration of community discovery and role analysis. Indeed, in principle, attributes allow for refining the affiliations of nodes to both communities and roles. In particular, attributes may unveil affiliation of nodes to communities as well as roles, even with few or no interactions with other nodes. In addition, behavioral role patterns are useful to more accurately model, predict and reason on connections between nodes, in terms both of their involvement in communities and the social functions of their roles in the participated communities. This is of great practical relevance in several domains including (but not limited to) the social, technological, information, biological, ecological, (counter-)intelligence and recommendation ones.

Unfortunately, the combined modeling of network communities, node roles with respective behavioral patterns and attributes raises several issues. Firstly, modeling network structure in terms both of communities and roles with respective behavioral patterns is problematic. On one hand, when roles are ignored, connectivity can be explained by community involvement. In real-world networks, nodes are usually involved in multiple communities and, in such cases, *pluralistic homophily* was found in [26] to be a realistic model of tie formation: links are more likely between nodes with a larger number of affiliations to common communities. On the other hand, in conventional role analysis, ties are typically interpreted according to mixed-membership block-modeling [21], i.e., as interactions between the roles of nodes in compliance with underlying behavioral patterns. Secondly, it is not clear how communities, roles, respective behavioral patterns and attributes are actually interrelated. Thirdly, the unveiled affiliations of nodes to communities and roles are required to be explicative and predictive of network structure.

In this paper, we propose an innovative model-based machine-learning approach to the unsupervised analysis of overlapping communities, roles and respective behavioral patterns in networks with node attributes.

Probabilistic graphical modeling [14] is used to formulate conditional (in)dependencies between the visible and hidden aspects of networks. The visible aspects include node attributes and link structure. Both are treated as observed variables. The hidden aspects are treated as latent variables and correspond to behavioral role patterns and affiliations [16,26]. The latter capture the extent of node involvement in communities and roles, the strength of role associations with communities, as well as the degree of attribute relatedness to roles.

The above conditional (in)dependencies are encoded in CARBONARA (*struCture and AttRiButes of nOdes from commuNities and behAvioral Role pAtterns*), a fully-Bayesian probabilistic generative model of networks. Under CARBONARA, Poisson distributions are placed on the observation of links and attributes, to expedite posterior inference on sparse networks [12]. The parameterizations of the Poisson distributions on links combine pluralistic homophily and the patterns of interactions between roles, by means of node affiliations to communities and roles in addition to the associations of roles with communities. Thus, as in [18], two nodes can play the same roles in spite of different neighborhoods. Moreover, conjugate Gamma priors are placed on all latent affiliations, which favours representation sparseness and, hence, improves interpretability.

The latent variables in CARBONARA are estimated through variational inference [1]. The latter enables four fundamental tasks, i.e., the exploratory analysis of communities, roles and respective behavioral patterns, the interpretation of communities and roles, the prediction of unobserved links as well as the posterior explanation of links in terms of the community and role affiliations of their end nodes, that most likely led to their establishment. The mathematical derivation of mean-field variational inference is provided together with a coordinate-ascent algorithm implementing the variational updates.

A comparative evaluation against state-of-the art competitors on several real-world data sets reveals that using attributes and behavioral role patterns to inform the joint modeling of communities and roles under CARBONARA is beneficial to improved accuracy in both community detection and link prediction. Role interpretation and behavioral role patterns are also demonstrated.

This paper proceeds as follows. Section 2 introduces notation and preliminary concepts. Section 3 presents CARBONARA. Section 4 covers mean-field variational inference and the design of a coordinate-ascent algorithm implementing the variational updates. Section 5 elaborates on the tasks enabled for practical applications by posterior variational inference under CARBONARA. Section 6 is devoted to the quantitative and qualitative evaluation of CARBONARA. Lastly, Sect. 7 concludes and roughs out future research.

2 Preliminaries

The notation used in this paper and some key concepts are presented below.

2.1 Networks, Communities, Attributes, Roles

The network-based representation of a complex system is a graph $\mathcal{G} = \{N, E\}$, where $N = \{1, \ldots N\}$ is a set of nodes (numbered 1 through N) and $E \subseteq N \times N$ is a set of links. Nodes correspond to entities, that interact in the network (e.g., individuals, organizations and so forth). Links represent interactions between nodes and are summarized into a binary adjacency matrix L, whose generic entry $L_{n,n'}$ is 1 iff a link between nodes n and n' is observed and 0 otherwise.

We are interested in the joint modeling and analysis of three primary properties of an input network $\mathcal{G}$, i.e., community structure, node roles and attributes.

Let $A \triangleq \{f_1, \ldots, f_A\}$ be a set of A binary attributes. The boolean indicator $F_{n,a}$ (with $1 \leq a \leq A$) is 1 iff node n exhibits attribute f_a and 0 otherwise.

Notation $\boldsymbol{F} \triangleq \{F_{n,a} | n \in \boldsymbol{N}, f_a \in \boldsymbol{A}\}$ gathers the characterizations of all nodes in terms of their attributes. Noticeably, in practical applications, $\boldsymbol{L}$ and $\boldsymbol{F}$ are, usually, (very) sparse.

Communities are structures of $\mathcal{G}$, which can be formalized as a set $\boldsymbol{C} \triangleq \{C_1, \ldots, C_K\}$ of K overlapping groups of nodes (i.e., $C_k \subseteq \boldsymbol{N}$ with $k = 1, \ldots, K$).

Roles are a set $\boldsymbol{R} \triangleq \{R_1, \ldots, R_H\}$ of H abstract behavioral classes.

2.2 Latent Affiliations

The generic node n can participate in all communities, although with a different involvement. In particular, $\vartheta_{n,k}$ is the degree to which n is involved in the arbitrary community C_k. Likewise, n can play all roles, although with a different attitude. Specifically, $\sigma_{n,h}$ is the extent to which n is suitable for playing the arbitrary role R_h. In turn, roles are specific to communities. The specificity of role R_h to community C_k is $\varphi_{h,k}$. Overall, node affiliations to communities and roles are respectively denoted as $\boldsymbol{\Theta} \triangleq \{\vartheta_{n,k} | n \in \boldsymbol{N}, C_k \in \boldsymbol{C}\}$ and $\boldsymbol{\Sigma} \triangleq \{\sigma_{n,h} | n \in \boldsymbol{N}, R_h \in \boldsymbol{R}\}$. Role affiliations to communities are compactly indicated as $\boldsymbol{\Phi} \triangleq \{\varphi_{h,k} | R_h \in \boldsymbol{R}, C_k \in \boldsymbol{C}\}$.

Attributes are related to roles to a varying degree. The relatedness of attribute f_a to role R_h is $\pi_{a,h}$. Notation $\boldsymbol{\Pi} \triangleq \{\pi_{a,h} | 1 \le a \le A, R_h \in \boldsymbol{R}\}$ means the relevance of all attributes to individual roles.

2.3 Behavioral Role Patterns

The establishment of links between nodes is assumed to be influenced (among others) by their common community affiliations, the strength of such affiliations and the respective behavioral role patterns. The combination of such components to rule tie formation is detailed in Sect. 3. Here, we supplement roles with their corresponding behavioral patterns.

Let R_h be a generic role from $\boldsymbol{R}$. The behavioral role pattern associated with R_h is $\epsilon_h \triangleq \{\epsilon_{h,h'} | R_{h'} \in \boldsymbol{R}\}$, where $\epsilon_{h,h'}$ is the unknown strength of interaction between R_h and $R_{h'}$ (with $1 \le h' \le H$). Intuitively, ϵ_h (with $1 \le h \le H$) captures the behavior involved in the fulfillment of the social function ascribed to role R_h. From this perspective, roles can be more exhaustively set as abstract classes representing social functions [21], which are in turn defined by the corresponding behavioral patterns. Due to the definition of such patterns, all nodes playing the same role behave identically with the roles of their neighbors, in compliance with mixed-membership block-modeling. Notation $\boldsymbol{\Upsilon} \triangleq \{\epsilon_h | R_h \in \boldsymbol{R}\}$ succinctly represents the behavioral patterns of all roles.

2.4 Problem Statement

Given an input network $\mathcal{G}$, a number K of communities and a number H of roles, we aim to perform

- the unsupervised exploratory analysis of $\mathcal{G}$, namely inferring its unobserved aspects Θ, Σ, Υ, Φ and Π;
- the prediction of missing links between nodes of $\mathcal{G}$.

The above tasks are accomplished by inferring a posterior distribution over Θ, Σ, Φ, Υ, Φ and Π given $\mathcal{G}$ (or, equivalently, the respective adjacency matrix L and node attributes F). Actually, the true posterior is approximated by performing variational inference in a generative network model, that explains the formation of $\mathcal{G}$ from a Bayesian probabilistic perspective. Hereinafter, all elements of Θ, Σ, Φ, Υ, Φ and Π are treated as random variables, being unknown and not directly measurable. The devised generative model is proposed in Sect. 3. Variational inference is covered in Sect. 4.

3 The CARBONARA Model

In this section, we present CARBONARA (*struCture and AttRiButes of nOdes from commuNities and behAvioral Role pAtterns*), a Bayesian probabilistic generative model of overlapping communities, roles and respective behavioral patterns in networks with node attributes. CARBONARA explains the observed aspects of a network $\mathcal{G}$ (i.e., topology L and node attributes F), in terms of its latent aspects, namely node affiliations to communities and roles (i.e., Θ and Σ), behavioral role patterns (i.e., Υ), role relevance to communities (i.e., Φ) as well as attribute relatedness to roles (i.e., Π).

Figure 1 illustrates the directed graphical representation in plate notation of the conditional (in)dependencies under CARBONARA between observed and latent aspects of $\mathcal{G}$, respectively, in the form of shaded and unshaded random variables.

The generative process of CARBONARA is assumed to originate $\mathcal{G}$, through a sequence of interactions among the random variables of Fig. 1. As detailed

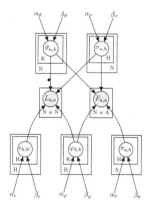

Fig. 1. Graphical representation of CARBONARA

I. For each node $n \in N$
 • For each community $C_k \in C$
   - draw the degree $\vartheta_{n,k}$ to which n is involved in community C_k, i.e., $\vartheta_{n,k} \sim \mathsf{Gamma}(\vartheta_{n,k} | \alpha_\vartheta, \beta_\vartheta)$.
 • For each role $R_h \in R$
   - draw the extent $\sigma_{n,h}$ to which n is suitable to play role R_h, i.e., $\sigma_{n,h} \sim \mathsf{Gamma}(\sigma_{n,h} | \alpha_\sigma, \beta_\sigma)$.

II. For each role $R_h \in R$
 • For each community $C_k \in C$
   - draw the relevance of role R_h to community C_k, i.e., $\varphi_{h,k} \sim \mathsf{Gamma}(\varphi_{h,k} | \alpha_\varphi, \beta_\varphi)$.

III. For each attribute $f_a \in A$
 • For each role $R_h \in R$
   - draw the relatedness of attribute f_a to role R_h, i.e., $\pi_{a,h} \sim \mathsf{Gamma}(\pi_{a,h} | \alpha_\pi, \beta_\pi)$.

IV. For each role $R_h \in R$
 • For each role $R_{h'} \in R$
   - draw the strength of interaction between role R_h and role $R_{h'}$, i.e., $\epsilon_{h,h'} \sim \mathsf{Gamma}(\epsilon_{h,h'} | \alpha_\epsilon, \beta_\epsilon)$.

V. For each node $n \in N$
 • For each node $n' \in N$
   - draw the presence/absence of a link $L_{n,n'}$ between n and n', i.e., $L_{n,n'} \sim \mathsf{Poisson}(L_{n,n'} | \delta_{n,n'})$,
   where $\delta_{n,n'}$ is the rate defined by Eq. 1.

VI. For each node $n \in N$
 • For each attribute $f_a \in A$
   - draw whether or not node n exhibits attribute f_a, i.e., $F_{n,a} \sim \mathsf{Poisson}(F_{n,a} | \delta_{n,a})$, where $\delta_{n,a}$ is the rate defined by Eq. 2.

Fig. 2. The generative process under CARBONARA (at steps I., II., III. and IV., α_ψ and β_ψ with $\psi \in \{\vartheta, \sigma, \phi, \pi, \epsilon\}$ are, respectively, shape and rate hyperparameters)

in Fig. 2, such a process consists of six steps, that preliminarily deal with the realization of the latent affiliations and behavioral role patterns and, eventually, generate G through the realization of the observed links and node attributes.

Essentially, at steps I., II. and III., the generative process samples the strength of the latent affiliations covered in Sect. 2.2 from the respective Gamma priors. At step IV., a Gamma prior is again sampled to draw the behavioral role patterns of Sect. 2.3. Interestingly, Gamma priors at steps I., II., III. and IV. enforce nonnegativity both on affiliation strengths and behavioral role patterns. This improves the interpretability of CARBONARA, by favoring sparseness in its representation. Moreover, it is also beneficial to ensure that the overall strength of affiliation for each node, role and attribute does not amount to 1, which avoids the typical inconvenient of mixed-membership modeling. Consequently, a very strong affiliation of any node, role or attribute does not imply a corresponding drop in the overall strength of all other affiliations of that node, role or attribute.

At step V., the presence/absence of a link between any two nodes is sampled from a Poisson distribution, that is placed over L as the link data likelihood. In particular, the establishment of a link between any two nodes n and n' is governed through a rate $\delta_{n,n'}$, that captures the interaction between n and n' through their affiliations to common communities and respective roles. Precisely,

$$\delta_{n,n'} \triangleq \sum_{k=1}^{K} \sum_{h,h'=1}^{H} \vartheta_{n,k} \sigma_{n,h} \varphi_{h,k} \epsilon_{h,h'} \vartheta_{n',k} \sigma_{n',h'} \varphi_{h',k} \tag{1}$$

Remarkably, Eq. (1) seamlessly reconciles pluralistic homophily with mixed-membership block-modeling for role analysis. Accordingly, the formation of ties between nodes is influenced by their affiliations to common communities, the strength of affiliation to such communities and respective roles, the specificity of roles to communities as well as the patterns of interactions between roles.

Lastly, at step VI., the presence/absence of attribute f_a in the characterization of node n is sampled from a Poisson distribution, that is placed over F as the attribute data likelihood. Specifically, the inclusion of f_a among the attributes of n is governed through the below rate

$$\delta_{n,a} \triangleq \sum_{k=1}^{K} \sum_{h=1}^{H} \vartheta_{n,k} \varphi_{h,k} \sigma_{n,h} \pi_{a,h} \tag{2}$$

$\delta_{n,a}$ captures the adequateness of f_a to n across all roles and communities.

Notably, the choice of Poisson distributions at steps V. and VI. is suited to deal with binary data and beneficial for faster inference on sparse networks [12].

4 Posterior Inference

Given a network $\mathcal{G}$, it is now interesting to determine the latent aspects Θ, Σ, Φ, Υ and Π, that explain the observation of topology L and node attributes F under CARBONARA.

Let $o \triangleq \{\alpha_\vartheta, \beta_\vartheta, \alpha_\sigma, \beta_\sigma, \alpha_\phi, \beta_\phi, \alpha_\pi, \beta_\pi, \alpha_\epsilon, \beta_\epsilon\}$ be the set of all hyperparameters. Inference allows for inverting the generative process of CARBONARA, in order to compute the posterior distribution $\Pr(\Theta, \Sigma, \Phi, \Upsilon, \Pi | L, F, o)$ over the latent aspects Θ, Σ, Φ, Υ and Π, given the observations L and F as well as the hyperparameters o. Unfortunately, as it generally happens with Bayesian models of practical relevance, exact posterior inference is intractable under CARBONARA, due to the complexity of the posterior distribution. Consequently, we resort to variational inference, in order to analytically approximate the true posterior distribution. Variational inference tends to be faster and more easily scalable on large networks than MCMC sampling [1]. In particular, we focus on mean-field variational inference.

We start with the introduction of auxiliary latent variables into CARBONARA. By exploiting the additivity of Poisson random variables, each $L_{n,n'}$ can be rewritten as $L_{n,n'} = \sum_{k=1}^{K} \sum_{h,h'=1}^{H} z_{n,n'}^{(k,h,h')}$, where $z_{n,n'}^{(k,h,h')} \sim Poisson(\vartheta_{n,k} \sigma_{n,h} \varphi_{h,k} \epsilon_{h,h'} \vartheta_{n',k} \sigma_{n',h'} \varphi_{h',k})$. Here, the auxiliary random variable $z_{n,n'}^{(k,h,h')}$ captures the contribution to $L_{n,n'}$ from the interaction between n and n' inside community C_k, when R_h and $R_{h'}$ are their respective roles. Analogously, each $F_{n,a}$ can be rewritten as $F_{n,a} = \sum_{k=1}^{K} \sum_{a=1}^{A} w_{n,a}^{(k,h)}$, with $w_{n,a}^{(k,h)} \sim Poisson(\vartheta_{n,k} \varphi_{h,k} \sigma_{n,h} \pi_{a,h})$ being the contribution to $F_{n,a}$ from the appropriateness of attribute f_a to node n, when the latter plays role R_h within community C_k. Notably, the auxiliary random vectors $z_{n,n'} = \{z_{n,n'}^{(k,h,h')} | C_k \in C, R_h, R'_h \in R\}$ and $w_{n,a} = \{w_{n,a}^{(k,h)} | C_k \in$

C, $R_h \in \mathbf{R}$} preserve the marginal distributions of, respectively, $L_{n,n'}$ and $F_{n,a}$. Moreover, when $L_{n,n'} = 0$ or $F_{n,a} = 0$, the corresponding auxiliary variables are zero. This narrows the focus of approximate posterior inference to only those auxiliary random variables, that are associated with nonzero observations [12].

Let $\mathbf{Z} = \{z_{n,n'} | n, n' \in \mathbf{V}\}$ and $\mathbf{W} = \{w_{n,a} | n \in \mathbf{V}, f_a \in \mathbf{A}\}$ be the sets of auxiliary variables added to CARBONARA. The mean-field family of approximate distributions over $\boldsymbol{\Theta}$, $\boldsymbol{\Sigma}$, $\boldsymbol{\Phi}$, $\boldsymbol{\Upsilon}$, $\boldsymbol{\Pi}$, $\mathbf{Z}$ and $\mathbf{W}$ has the below factorized form

$$q(\boldsymbol{\Theta}, \boldsymbol{\Sigma}, \boldsymbol{\Phi}, \boldsymbol{\Upsilon}, \boldsymbol{\Pi}, \mathbf{Z}, \mathbf{W} | \boldsymbol{\nu}) = \prod_{n \in N, C_k \in C} q(\vartheta_{n,k} | \rho_{n,k}) \prod_{n \in N, R_h \in R} q(\sigma_{n,h} | \lambda_{n,h}) \cdot \prod_{R_h \in R, C_k \in C} q(\varphi_{h,k} | \eta_{h,k}) \cdot$$

$$\prod_{R_h \in R, R_{h'} \in R_{h'}} q(\epsilon_{h,h'} | \gamma_{h,h'}) \cdot \prod_{f_a \in A, R_h \in R} q(\pi_{a,h} | \xi_{a,h})$$

$$\prod_{n \in N, n' \in N} q(z_{n,n'} | \tau_{n,n'}) \cdot \prod_{n \in N, f_a \in A} q(w_{n,a} | \mu_{n,a}) \tag{3}$$

with $\boldsymbol{\nu} \triangleq \{\rho_{n,k}, \lambda_{n,h}, \eta_{h,k}, \gamma_{h,h'}, \xi_{a,h}, \tau_{n,n'}, \mu_{n,a} | n, n' \in \mathbf{N}, C_k \in \mathbf{C}, R_h, R_h' \in \mathbf{R}, f_a \in \mathbf{A}\}$ being the set of all variational parameters, that individually condition the corresponding factors on the right hand side of Eq. 3.

Due to the auxiliary random variables, CARBONARA is a *conditionally conjugate* model. For this class of models, variational parameters can be fitted to the observed network $\mathcal{G}$ through a simple coordinate-ascent variational algorithm [1].

Algorithm 1 sketches the pseudo-code of the coordinate-ascent algorithm, designed to fit the variational parameters $\boldsymbol{\nu}$ of Eq. 3. This is pursued differently, on the basis of the nature of such variational parameters. To elaborate, factors on the right hand side of Eq. 3 have two distinct forms. More precisely, factors $q(z_{n,n'} | \tau_{n,n'})$ and $q(w_{n,a} | \mu_{n,a})$ are multinomial distributions, while all other factors are Gamma distributions.[1] Hence, the generic $\tau_{n,n'}$ and $\mu_{n,a}$ are variational multinomial parameters. In Algorithm 1, these are respectively updated through Eqs. 14 and 15 (reported in Fig. 4), where $\boldsymbol{\Psi}[\cdot]$ is the digamma function (i.e., the first derivative of *log* Γ function). Instead, $\rho_{n,k}$, $\lambda_{n,h}$, $\eta_{h,k}$, $\gamma_{h,h'}$ and $\xi_{a,h}$ are variational Gamma parameters, with each such a Gamma parameter being actually a pair consisting of a shape (*shp*) and a rate (*rate*). Shape parameters are updated via Eq. 4 through Eq. 8 (reported in Fig. 3), whereas rate parameters are updated by means of Eq. 9 through Eq. 13 (reported in Fig. 4).

Essentially, the coordinate-ascent algorithm of Algorithm 1 preliminarily performs an initialization stage (step 1). Subsequently, it enters a loop (steps 2–36), in which the individual variational parameters are optimized one at a time (while all others are kept fixed), as discussed above. Such a loop is reiterated until convergence, which is detected (at step 36), when the difference in the average predictive log likelihood of a validation set falls below 10^{-6}. Interestingly, the coordinate-ascent algorithm is expedited on sparse networks, because of the useful properties of the Poisson distribution. In particular, the computation of variational multinomial parameters (at steps 6 and 13) as well as the sum over

---

[1] The mathematical derivation of the functional forms of factors on the right hand side of Eq. 3 is omitted for brevity.

Algorithm 1. The coordinate-ascent variational algorithm

COORDINATE-ASCENT(L, F, K, H)
  Require: the adjacency matrix L of an observed network $\mathcal{G}$;
      a characterization F of the nodes within $\mathcal{G}$ in terms of their attributes;
      the number K of latent communities;
      the number H of underlying behavioral roles;
  Ensure: the latent aspects Θ, Σ, Υ, Φ and Π.
  1: set all variational parameters $\rho_{n,k}$, $\lambda_{n,h}$, $\eta_{h,k}$, $\gamma_{h,h'}$, $\xi_{a,h}$ equal (except for a random offset) to the prior
      on the corresponding latent variables [12];
  2: **repeat**
  3:    **for all** $n, n' \in N$ such that $L_{n,n'} = 1$ **do**
  4:       **for all** $k = 1, \ldots, K$ **do**
  5:          **for all** $h, h' = 1, \ldots, H$ **do**
  6:             compute $\tau_{n,n'}^{(k,h,h')}$ through Eq. 14;
  7:          **end for**
  8:       **end for**
  9:    **end for**
 10:    **for all** $n \in N$ and $f_a \in A$ such that $F_{n,a} = 1$ **do**
 11:       **for all** $k = 1, \ldots, K$ **do**
 12:          **for all** $h = 1, \ldots, H$ **do**
 13:             compute $\mu_{n,a}^{(k,h)}$ through Eq. 15;
 14:          **end for**
 15:       **end for**
 16:    **end for**
 17:    **for all** $n \in N$ **do**
 18:       **for all** $k = 1, \ldots, K$ **do**
 19:          compute $\rho_{n,k}^{(shp)}$ through Eq. 4 and $\rho_{n,k}^{(rate)}$ through Eq. 9;
 20:       **end for**
 21:       **for all** $h = 1, \ldots, H$ **do**
 22:          compute $\lambda_{n,h}^{(shp)}$ through Eq. 5 and $\lambda_{n,h}^{(rate)}$ through Eq. 10;
 23:       **end for**
 24:    **end for**
 25:    **for all** $h = 1, \ldots, H$ **do**
 26:       **for all** $k = 1, \ldots, K$ **do**
 27:          compute $\eta_{h,k}^{(shp)}$ through Eq. 6 and $\eta_{h,k}^{(rate)}$ through Eq. 11;
 28:       **end for**
 29:       **for all** $h' = 1, \ldots, H$ **do**
 30:          compute $\gamma_{h,h'}^{(shp)}$ through Eq. 7 and $\gamma_{h,h'}^{(rate)}$ through Eq. 12;
 31:       **end for**
 32:       **for all** $f_a \in A$ **do**
 33:          compute $\xi_{a,h}^{(shp)}$ through Eq. 8 and $\xi_{a,h}^{(rate)}$ through Eq. 13;
 34:       **end for**
 35:    **end for**
 36: **until** convergence

$$\rho_{n,k}^{(shp)} = \alpha_\vartheta + \sum_{n' \in N, R_h, R_{h'} \in R} L_{n,n'} \tau_{n,n'}^{(k,h,h')} + \sum_{f_a \in A, R_h \in R} F_{n,a} \mu_{n,a}^{(k,h)} \tag{4}$$

$$\lambda_{n,h}^{(shp)} = \alpha_\sigma + \sum_{n' \in N, C_k \in C, R_{h'} \in R} L_{n,n'} \tau_{n,n'}^{(k,h,h')} + \sum_{f_a \in A, C_k \in C} F_{n,a} \mu_{n,a}^{(k,h)} \tag{5}$$

$$\eta_{h,k}^{(shp)} = \alpha_\varphi + \sum_{n,n' \in N, R_h, R_{h'} \in R: R_{h'} \neq R_h} L_{n,n'} \tau_{n,n'}^{(k,h,h')} + \sum_{n \in N, f_a \in A} F_{n,a} \mu_{n,a}^{(k,h)} \tag{6}$$

$$\gamma_{h,h'}^{(shp)} = \alpha_\epsilon + \sum_{n,n' \in N, C_k \in C} L_{n,n'} \tau_{n,n'}^{(k,h,h')} \tag{7}$$

$$\xi_{a,h}^{(shp)} = \alpha_\pi + \sum_{n \in N, C_k \in C} F_{n,a} \mu_{n,a}^{(k,h)} \tag{8}$$

Fig. 3. Variational updates for Gamma-shape parameters

users and node attributes in the formulas for the calculation of the variational
Gamma parameters (at steps 19, 22, 27, 30 and 33) involve only observed links
and node attributes.

$$\rho_{n,k}^{(rate)} = \left[\left(\sum_{n'\in N, R_h, R_{h'}\in R} \frac{\lambda_{n,h}^{(shp)}}{\lambda_{n,h}^{(rate)}} \frac{\eta_{h,k}^{(shp)}}{\eta_{h,k}^{(rate)}} \frac{\gamma_{h,h'}^{(shp)}}{\gamma_{h,h'}^{(rate)}} \frac{\rho_{n',k}^{(shp)}}{\rho_{n',k}^{(rate)}} \frac{\lambda_{n',h'}^{(shp)}}{\lambda_{n',h'}^{(rate)}} \frac{\eta_{h',k}^{(shp)}}{\eta_{h',k}^{(rate)}}\right) + \left(\sum_{f_a\in A, R_h\in R} \frac{\eta_{h,k}^{(shp)}}{\eta_{h,k}^{(rate)}} \frac{\lambda_{n,h}^{(shp)}}{\lambda_{n,h}^{(rate)}} \frac{\xi_{a,h}^{(shp)}}{\xi_{a,h}^{(rate)}}\right) + \frac{\alpha_\vartheta}{\beta_\vartheta}\right] \quad (9)$$

$$\lambda_{n,h}^{(rate)} = \left[\left(\sum_{n'\in N, C_k\in C, R_{h'}\in R} \frac{\rho_{n,k}^{(shp)}}{\rho_{n,k}^{(rate)}} \frac{\eta_{h,k}^{(shp)}}{\eta_{h,k}^{(rate)}} \frac{\gamma_{h,h'}^{(shp)}}{\gamma_{h,h'}^{(rate)}} \frac{\rho_{n',k}^{(shp)}}{\rho_{n',k}^{(rate)}} \frac{\lambda_{n',h'}^{(shp)}}{\lambda_{n',h'}^{(rate)}} \frac{\eta_{h',k}^{(shp)}}{\eta_{h',k}^{(rate)}}\right) + \left(\sum_{f_a\in A, C_k\in C} \frac{\rho_{n,k}^{(shp)}}{\rho_{n,k}^{(rate)}} \frac{\eta_{h,k}^{(shp)}}{\eta_{h,k}^{(rate)}} \frac{\xi_{a,h}^{(shp)}}{\xi_{a,h}^{(rate)}}\right) + \frac{\alpha_\sigma}{\beta_\sigma}\right] \quad (10)$$

$$\eta_{h,k}^{(rate)} = \left[\left(\sum_{n,n'\in N, R_h, R_{h'}\in R: R_{h'}\neq R_h} \frac{\rho_{n,k}^{(shp)}}{\rho_{n,k}^{(rate)}} \frac{\lambda_{n,h}^{(shp)}}{\lambda_{n,h}^{(rate)}} \frac{\gamma_{h,h'}^{(shp)}}{\gamma_{h,h'}^{(rate)}} \frac{\rho_{n',k}^{(shp)}}{\rho_{n',k}^{(rate)}} \frac{\lambda_{n',h'}^{(shp)}}{\lambda_{n',h'}^{(rate)}} \frac{\eta_{h',k}^{(shp)}}{\eta_{h',k}^{(rate)}}\right) + \left(\sum_{n\in N, f_a\in A} \frac{\rho_{n,k}^{(shp)}}{\rho_{n,k}^{(rate)}} \frac{\lambda_{n,h}^{(shp)}}{\lambda_{n,h}^{(rate)}} \frac{\xi_{a,h}^{(shp)}}{\xi_{a,h}^{(rate)}}\right) + \frac{\alpha_\varphi}{\beta_\varphi}\right] \quad (11)$$

$$\gamma_{h,h'}^{(rate)} = \left[\left(\sum_{n,n'\in N, C_k\in C} \frac{\rho_{n,k}^{(shp)}}{\rho_{n,k}^{(rate)}} \frac{\lambda_{n,h}^{(shp)}}{\lambda_{n,h}^{(rate)}} \frac{\eta_{h,k}^{(shp)}}{\eta_{h,k}^{(rate)}} \frac{\rho_{n',k}^{(shp)}}{\rho_{n',k}^{(rate)}} \frac{\lambda_{n',h'}^{(shp)}}{\lambda_{n',h'}^{(rate)}} \frac{\eta_{h',k}^{(shp)}}{\eta_{h',k}^{(rate)}}\right) + \frac{\alpha_\epsilon}{\beta_\epsilon}\right] \quad (12)$$

$$\xi_{a,h}^{(rate)} = \left[\left(\sum_{n\in N, C_k\in C} \frac{\rho_{n,k}^{(shp)}}{\rho_{n,k}^{(rate)}} \frac{\eta_{h,k}^{(shp)}}{\eta_{h,k}^{(rate)}} \frac{\lambda_{n,h}^{(shp)}}{\lambda_{n,h}^{(rate)}}\right) + \frac{\alpha_\pi}{\beta_\pi}\right] \quad (13)$$

$$\tau_{n,n'}^{(k,h,h')} \propto e^{\Psi[\rho_{n,k}^{(shp)}] - \log\rho_{n,k}^{(rate)} + \Psi[\lambda_{n,h}^{(shp)}] - \log\lambda_{n,h}^{(rate)} + \Psi[\eta_{h,k}^{(shp)}] - \log\eta_{h,k}^{(rate)} + \Psi[\gamma_{h,h'}^{(shp)}] - \log\gamma_{h,h'}^{(rate)} + \Psi[\rho_{n',k}^{(shp)}] - \log\rho_{n',k}^{(rate)}}$$
$$e^{\Psi[\lambda_{n',h'}^{(shp)}] - \log\lambda_{n',h'}^{(rate)} + \Psi[\eta_{h',k}^{(shp)}] - \log\eta_{h',k}^{(rate)}} \quad (14)$$

$$\mu_{n,a}^{(k,h)} \propto e^{\Psi[\rho_{n,k}^{(shp)}] - \log\rho_{n,k}^{(rate)} + \Psi[\lambda_{n,h}^{(shp)}] - \log\lambda_{n,h}^{(rate)} + \Psi[\eta_{h,k}^{(shp)}] - \log\eta_{h,k}^{(rate)} + \Psi[\xi_{a,h}^{(shp)}] - \log\xi_{a,h}^{(rate)}} \quad (15)$$

Fig. 4. Variational updates for Gamma-rate and multinomial parameters

5   Tasks

Fitting CARBONARA to an input network $\mathcal{G}$ through Algorithm 1 results in the approximate posterior distribution of Eq. 3. This enables a variety of exploratory, predictive and descriptive tasks, including those enumerated at Sect. 2.4. Such tasks involve posterior expectations of corresponding random variables (i.e., expectations of the random variables with respect to the approximate posterior).

5.1   Exploratory Network Analysis

Node affiliations reveal the underlying organization of $\mathcal{G}$ into overlapping communities and roles. The degree to which node n participates in community C_k is $\vartheta_{n,k}^* \triangleq E[\vartheta_{n,k}]$. Besides, the extent to which node n is suitable for playing role R_h is $\sigma_{n,h}^* \triangleq E[\sigma_{n,h}]$.

Community structure and roles allow for the posterior explanation of links. Let n and n' be two connected nodes of $\mathcal{G}$, i.e., such that $\langle n, n'\rangle \in \boldsymbol{E}$ and, consequently, $L_{n,n'} = 1$. The probability $P(z_{n,n'}^{(k,h,h')} = 1 | L_{n,n'}, \boldsymbol{\Theta}, \boldsymbol{\Sigma}, \boldsymbol{\Upsilon}, \boldsymbol{\Phi})$ that $\langle n, n'\rangle$ was established with n and n' playing, respectively, roles R_h and $R_{h'}$ within community k is $\tau_{n,n'}^{(k,h,h')}$, according to the mean-field approximation. Therefore, $\langle n, n'\rangle$ can be reasonably explained by the triple (k^*, h^*, h'_*), such that $(k^*, h^*, h'_*) = argmax_{(k,h,h')}\tau^{(k,h,h')}$.

Behavioral patterns supplement roles, by unveiling their functioning in the fulfillment of the respective social functions. Here, the strength of interaction between roles R_h and $R_{h'}$ is $\epsilon_{h,h'}^* \triangleq E[\epsilon_{h,h'}]$.

5.2   Predictive Analysis

The emergence of missing links is forecast through the association of a respective score, that is used for ranking their prospective observation. In particular, the score $s_{n,n'}$ associated with unconnected pairs of nodes n and n' is the posterior expectation of the corresponding Poisson-distribution rate. More precisely, $s_{n,n'} \triangleq E[\delta_{n,n'}]$, where $\delta_{n,n'}$ is defined by Eq. 1.

5.3   Descriptive Analysis

CARBONARA supports the interpretation of roles and communities by discriminatory characterizations.

Roles can be characterized by related node attributes. The relatedness of attribute f_a to role R_h is $\pi^*_{a,h} \triangleq E[\pi_{a,h}]$. A discriminatory characterization of R_h is, hence, any suitable subset of strongly related attributes.

Communities can be characterized by relevant role affiliations (directly) and pertinent node attributes (indirectly). The relevance of role R_h to community C_k is $\varphi^*_{h,k} \triangleq E[\varphi_{h,k}]$. In addition, the pertinence of attribute f_a to community C_k is $\omega^*_{a,k} \triangleq E[\sum_{h=1}^{H} \pi_{a,h}\varphi_{h,k}]$. Accordingly, C_k can be discriminatorily characterized in terms of highly relevant roles and/or especially pertinent attributes.

6   Experimental Evaluation

An empirical assessment of CARBONARA was carried out on three real-world benchmark data sets. *Facebook* (available at https://snap.stanford.edu/data/egonets-Facebook.html) contains 10 ego-networks with 193 ground-truth circles, $4,039$ nodes, $39,367$ attributes and $88,234$ edges. *Twitter* (available at http://snap.stanford.edu/data) consists of 973 ego-networks, $4,869$ ground-truth circles, $81,306$ nodes, $2,090,240$ attributes and $1,768,149$ edges.

Google+ (available at http://snap.stanford.edu/data/egonets-Gplus.html) comprises 133 ego-networks, 479 ground-truth circles, $107,614$ nodes, $1,180,706$ attributes and $13,673,453$ edges.

CARBONARA was tested on the above data sets, in order to empirically verify whether accounting for attributes and behavioral role patterns in the seamless analysis of network communities and roles is actually beneficial to improved accuracy. For this purpose, we compared the performance of CARBONARA against the performance of state-of-the-art competitors both in community detection and link prediction. We considered three classes of state-of-the-art competitors. The first class contains approaches to the simultaneous analysis of communities and roles. We are not aware of approaches belonging to such a class, that also account for both attributes and behavioral role patterns. Therefore, we chose TORTILLA [9] and TOMATOES [7], i.e., two recent Bayesian generative models of networks with overlapping communities and roles. Both enable community discovery and link prediction, though disregarding node attributes as well as behavioral role patterns. The second class includes community-detection

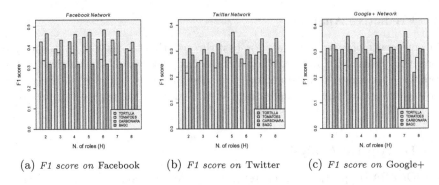

(a) *F1 score on* Facebook   (b) *F1 score on* Twitter   (c) *F1 score on* Google+

Fig. 5. Competitor accuracy in community discovery (F1 score)

methods. For this class we selected BAGC [25], i.e., a model-based approach
to community detection, that deals with the attributes of nodes, while disregarding their roles and respective behavioral patterns. The third class involves
methods for link prediction in networks with node attributes. We focused on
two such competitors, i.e., the approach in [19], which is hereinafter referred to
as AUC-MF, and LRA-SAN [11], in which link prediction is pursued via matrix
factorization.

In all tests, the input parameter K (i.e., the number of communities to discover) was set for the chosen competitors to the number of ground-truth circles
within each network. Instead, the input parameter H (i.e., the number of roles
and respective behavioral patterns) was ranged in CARBONARA, TORTILLA and
TOMATOES within the interval $[2, 8]$.

The performance of all competitors in community discovery was quantified by
the correspondence between discovered and ground-truth communities, according to the standard F1 score. Figure 5 shows the average agreement of the communities detected by all involved competitors inside the individual networks
with the respective ground-truth communities. Notice that the performance of
BAGC is independent of parameter H, since it does not deal with roles (and
respective behavioral patterns). Remarkably, CARBONARA delivers an overcoming community-detection performance for all tested values of H on each chosen
data set.

Link prediction is an insightful test, that allows for comparing the predictive
accuracy of the chosen competitors, by the extent to which these reliably predict
the presence or absence of links between nodes.

The link-prediction performance of all competitors was evaluated through
tests inspired by [13]. Overall, 5 experiments of the predictive performance of the
competing models were carried out over the benchmark networks. Each experiment consisted of two steps. Firstly, the generic input network was separated
into a training set and a held-out test set. In particular, the latter was obtained
by randomly sampling the whole input network to select an equal number of
present and absent links, whose sum amounts to 15% of the overall number of
links in the whole input network.

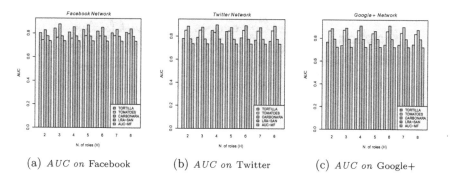

(a) *AUC on* Facebook   (b) *AUC on* Twitter   (c) *AUC on* Google+

Fig. 6. Competitor accuracy in link prediction (AUC)

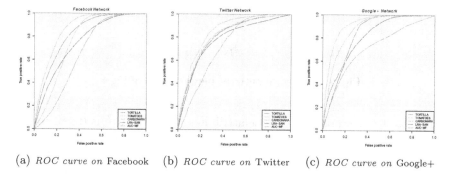

(a) *ROC curve on* Facebook   (b) *ROC curve on* Twitter   (c) *ROC curve on* Google+

Fig. 7. ROC analysis on individual networks from *Facebook*, *Twitter* and *Google+*

Secondly, the links in the held-out test set are predicted by the distinct models inferred from the training set. The details on link prediction in CARBONARA are provided in Sect. 5.2, whereas those about TORTILLA and TOMATOES appear, respectively, in [9] and [7].

Figure 6 summarizes the average AUC values attained by all involved competitors. Notably, CARBONARA overcomes the selected competitors in link prediction for all tested values of H on each chosen data set. An insight into the actual performance of the individual competitors is provided by Fig. 7, which illustrates that the ROC curve of CARBONARA on individual networks from the tested data sets is mostly dominant.

Overall, the empirical findings of Figs. 5, 6 and 7 reveal a consistently higher accuracy of CARBONARA both in community discovery and link prediction on the chosen data sets. This substantiates the rationality of the statistical dependencies under CARBONARA, which accommodate node attributes and behavioral role patterns in the joint modeling of network communities and roles.

Lastly, we inspect the results unveiled in one network from the *Facebook* data set, in order to qualitatively demonstrate the interpretation of roles and respective behavioral patterns.

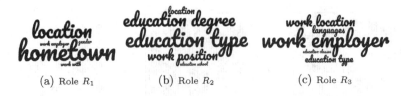

(a) Role R_1 (b) Role R_2 (c) Role R_3

Fig. 8. Role interpretation in terms of related attributes

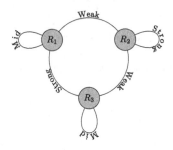

Fig. 9. Behavioral patterns between roles of Fig. 8

Due to the incorporation of node attributes, CARBONARA is (to the best of our knowledge) the first network model, that provides an intelligible interpretation of roles. Figure 8 shows a characterization of three of the detected roles, in terms of mostly related attributes. Notice that, the size of attributes in the context of each role is proportional to their degree of relatedness to that particular role (see Sect. 5.3 for details). Hence, larger sizes denote more strongly related attributes. Accordingly, role R_1 is predominantly characterized by *hometown* and *location*. Role R_2 is primarily marked by *education type* and *education degree*. Role R_3 is mainly discriminated by *work employer* and *work location*.

It is worth to emphasize that, as discussed in Sect. 5.3, a characterization as the one in Fig. 8 is also provided by CARBONARA for communities in terms of pertinent attributes and/or specific roles (skipped due to space requirements).

The behavior exhibited by nodes in the act of playing the roles of Fig. 8 consists in the fulfillment of corresponding social functions. These involve interactions with the roles of other neighboring nodes. The discretized behavioral patterns of Fig. 9 reveal the behaviors implied by the three social functions, in terms of interactions of different strength between node roles.

7   Conclusions

We presented CARBONARA, a new approach to the unsupervised analysis of communities and roles in networks, that takes advantage of node attributes and behavioral role patterns. A comparative experimentation revealed an overcoming accuracy of CARBONARA on real-world benchmark data sets, both in community discovery and link prediction. Role interpretation and behavioral role patterns were demonstrated too.

Further research is required to deal with various types of link and node attributes. Also, it is interesting to study possible adaptations of CARBONARA, that may be beneficially used for higher accuracy in social recommendation [8].

References

1. Blei, D., Kucukelbir, A., McAuliffe, J.: Variational inference: a review for statisticians. arXiv:1601.00670 (2016)
2. Chou, B.-H., Suzuki, E.: Discovering community-oriented roles of nodes in a social network. In: Proceedings of International Conference on Data Warehousing and Knowledge Discovery, pp. 52–64 (2010)
3. Costa, G., Ortale, R.: Mining overlapping communities and inner role assignments through Bayesian mixed-membership models of networks with context-dependent interactions. ACM Trans. Knowl. Discov. Data (to appear)
4. Costa, G., Ortale, R.: A Bayesian hierarchical approach for exploratory analysis of communities and roles in social networks. In: Proceedings of IEEE/ACM International Conference on Advances in Social Networks Analysis and Mining, pp. 194–201 (2012)
5. Costa, G., Ortale, R.: Probabilistic analysis of communities and inner roles in networks: Bayesian generative models and approximate inference. Soc. Netw. Anal. Min. **3**(4), 1015–1038 (2013)
6. Costa, G., Ortale, R.: A unified generative Bayesian model for community discovery and role assignment based upon latent interaction factors. In: Proceedings of IEEE/ACM International Conference on Advances in Social Networks Analysis and Mining, pp. 93–100 (2014)
7. Costa, G., Ortale, R.: A mean-field variational Bayesian approach to detecting overlapping communities with inner roles using poisson link generation. In: Proceedings of International Symposium on Intelligent Data Analysis, pp. 110–122 (2016)
8. Costa, G., Ortale, R.: Model-based collaborative personalized recommendation on signed social rating networks. ACM Trans. Internet Technol. **16**(3), 20:1–20:21 (2016)
9. Costa, G., Ortale, R.: Scalable detection of overlapping communities and role assignments in networks via Bayesian probabilistic generative affiliation modeling. In: Debruyne, C., Panetto, H., Meersman, R., Dillon, T., Kühn, E., O'Sullivan, D., Ardagna, C.A. (eds.) OTM 2016. LNCS, vol. 10033, pp. 99–117. Springer, Cham (2016). doi:10.1007/978-3-319-48472-3_6
10. Fortunato, S.: Community detection in graphs. Phys. Rep. **486**(3–5), 75–174 (2010)
11. Gong, N.Z., Talwalkar, A., Mackey, L., Huang, L., Shin, E.C.R., Stefanov, E., Shi, E., Song, D.: Joint link prediction and attribute inference using a social-attribute network. ACM Trans. Intell. Syst. Technol. **5**(2), 27:1–27:20 (2014)
12. Gopalan, P., Hofman, J., Blei, D.: Scalable recommendation with hierarchical Poisson factorization. In: Proceedings of Conference on Uncertainty in Artificial Intelligence, pp. 326–335 (2015)
13. Henderson, K., Eliassi-Rad, T., Papadimitriou, S., Faloutsos, C.: HCDF: a hybrid community discovery framework. In: Proceedings of SIAM International Conference on Data Mining, pp. 754–765 (2010)
14. Koller, D., Friedman, N.: Probabilistic Graphical Models. Principles and Techniques. The MIT Press, Cambridge (2009)

15. Lancichinetti, A., Fortunato, S.: Community detection algorithms: a comparative analysis. Phys. Re. E **80**, 056117 (2009)
16. Lattanzi, S., Sivakumar, D.: Affiliation networks. In: ACM Symposium on the Theory of Computing, pp. 427–434 (2009)
17. Leskovec, J., Lang, K.J., Mahoney, M.: Empirical comparison of algorithms for network community detection. In: Proceedings of International Conference on World Wide Web, pp. 631–640 (2010)
18. McCallum, A., Wang, X., Corrada-Emmanuel, A.: Topic and role discovery in social networks with experiments on enron and academic email. J. Artif. Intell. Res. **30**(1), 249–272 (2007)
19. Menon, A.K., Elkan, C.: Link prediction via matrix factorization. In: Proceedings of European Conference on Machine Learning and Knowledge Discovery in Databases, pp. 437–452 (2011)
20. Porter, M.A., Onnela, J.-P., Mucha, P.J.: Communities in networks. Not. Am. Math. Soc. **56**(9), 1082–1166 (2009)
21. Ross, R.A., Ahmed, N.K.: Role discovery in networks. IEEE Trans. Knowl. Data Eng. **27**(04), 1112–1131 (2015)
22. Scripps, J., Tan, P.-N., Esfahanian, A.-H.: Exploration of link structure and community-based node roles in network analysis. In: Proceedings of International Conference on Data Mining, pp. 649–654 (2007)
23. Scripps, J., Tan, P.-N., Esfahanian, A.-H.: Node roles and community structure in networks. In: Proceedings of Workshop on Web Mining and Social Network Analysis (WebKDD and SNA-KDD), pp. 26–35 (2007)
24. Xie, J., Kelley, S., Szymanski, B.K.: Overlapping community detection in networks: the state of the art and comparative study. ACM Comput. Surv. **45**(4), 43:1–43:35 (2013)
25. Xu, Z., Ke, Y., Wang, Y., Cheng, H., Cheng, J.: A model-based approach to attributed graph clustering. In: Proceedings of ACM SIGMOD International Conference on Management of Data, pp. 505–516 (2012)
26. Yang, J., Leskovec, J.: Overlapping community detection at scale: a nonnegative matrix factorization approach. In: Proceedings of ACM International Conference on Web Search and Data Mining, pp. 587–596 (2013)

Efficient Approximate Entity Matching Using Jaro-Winkler Distance

Yaoshu Wang[(✉)], Jianbin Qin, and Wei Wang

School of Computer Science and Engineering,
Univeristy of New South Wales, Sydney, Australia
{yaoshuw,jqin,weiw}@cse.unsw.edu.au

Abstract. Jaro-Winkler distance is a measurement to measure the similarity between two strings. Since Jaro-Winkler distance performs well in matching personal and entity names, it is widely used in the areas of record linkage, entity linking, information extraction. Given a query string q, Jaro-Winkler distance similarity search finds all strings in a dataset D whose Jaro-Winkler distance similarity with q is no more than a given threshold τ. With the growth of the dataset size, to efficiently perform Jaro-Winkler distance similarity search becomes challenge problem. In this paper, we propose an index-based method that relies on a filter-and-verify framework to support efficient Jaro-Winkler distance similarity search on a large dataset. We leverage e-variants methods to build the index structure and pigeonhole principle to perform the search. The experiment results clearly demonstrate the efficiency of our methods.

Keywords: Pigeonhole principle · Jaro-Winkler distance · e-variants

1 Introduction

Entity linking (EL) aims to link all mentions to potential entities in a knowledge base (KB) after named entity recognition (NER). However, large amounts of ment ions and entities become the obstacle of the efficient issue. At present, EL methods find mentions and entity candidates using a standard lexicon. Levenshtein distance is a popular distance function to filter strings, but it sometimes does not work on short strings, especially named entities. According to [4], Jaro-Winkler distance (d_{JW}) is an efficient distance function to measure name-matching task.

Mentions and entities are represented as (short) strings in lexicons. Our problem can be modeled as an approximate entity matching problem: Given a lexicon of entity strings and a list of mention strings, we aim to find all potential entity strings in the lexicon for each mention string. In our knowledge, [5] proposed the only approximate entity matching methods by using Jaro-Winkler distance. The idea is to use a trie structure to filter dissimilar strings and prune nodes in trie. The limitation in the work [5] is that it only considers the common characters of strings, and should traverse large amounts of nodes in the trie structure to reach the lower bound of Jaro-Winkler distance.

© Springer International Publishing AG 2017
A. Bouguettaya et al. (Eds.): WISE 2017, Part I, LNCS 10569, pp. 231–239, 2017.
DOI: 10.1007/978-3-319-68783-4_16

In this paper, we solve the approximate entity matching in a new perspective. Our contribution can be summarized as follows.

- We proposed a new lower bound of Jaro-Winkler distance. Instead of indexing single characters, we combined characters into signatures and check the number of common signatures between strings.
- We designed a new index structure and efficient query processing algorithm to support Jaro-Winkler distance.
- We have conducted comprehensive experiments using several named entity datasets. The proposed method has been shown to achieve the best performance among all other ones.

2   Related Work

Jaro-Winkler distance is a general similarity metric used in entity linking [8,11], record linkage [2–4,6], and data cleaning [12]. This paper focuses on how to use Jaro-Winkler distance to performance efficient string similarity search.

For similarity metrics, such as Levenshtein distance, Jaccard similarity, cosine similarity, index-based methods (incl. inverted index based methods and trie based methods) are the most efficient methods of string similarity search so far.

Inverted Index Based Methods. Many state-of-the-art algorithms [13,16,17] adopted inverted index structure to performance query processing. PassJoin [10] partitioned strings according to pigeonhole principle into the set of substrings and indexed these substrings. PPJoin+ [17] utilized tokens as signatures to construct inverted index and introduced positional filtering and suffix filtering to prune false positives. EdJoin [16] and qGramChunk [13] extracted q-grams or q-chunks as signatures to build index, and they designed prefix filterings with shorter prefix size to prune strings. NGPP [15] combined pigeonhole principle and e-variants to generate variant strings, and indexed these strings in the inverted index to improve query processing. Inverted index based methods are very efficient in long strings.

Trie Based Methods. Trie based methods [5,7] adopted the trie structure to perform string similarity search and join. TrieJoin [7] utilized the trie structure to deal with edit distance metric. By adding some pruning strategies, TrieJoin could terminate its method early. LIMES [5] worked in Jaro-Winkler distance metric and counted the common characters between two strings by traversing the trie structure. Trie based methods adapt to short strings, such as short titles, person names and so on.

3   Problem Definition

Definition 1. *Given a list of query mentions M and a lexicon D of entities, the task of approximate entity matching with the Jaro-Winkler distance threshold τ_{JW} is to find all mention-and-entity pairs $< q_m, s_e >$ ($q_m \in M$, $s_e \in D$) that $d_{JW}(q_m, s_e) \geq \tau_{JW}$.*

Notations. We denote the lexicon of entities as $\mathcal{D}$ and collection of mentions as $\mathcal{M}$. We use l_{min} and l_{max} to represent the minimal and maximal lengths of strings in $\mathcal{D}$. $\mathcal{D}[i]$ is represented as the i-th entity string in $\mathcal{D}$. τ_J is the threshold of Jaro distance, and τ_{JW} is the threshold of Jaro-Winkler distance.

4   Lower Bound of Jaro-Winkler Distance

In this section, we first explore the lower bound of Jaro distance and then add the constraint of Winkler distance.

Lower Bound of Jaro Distance. According to Jaro Distance d_J(Eq. 1), given strings s_1 and s_2, we know $0 \le t \le \frac{m}{2}$ according to [4]. Now we set $t = 0$ and design a lower bound of common characters of Jaro distance.

$$d_J = \frac{1}{3}\left(\frac{m}{|s_1|} + \frac{m}{|s_2|} + \frac{m-t}{m}\right) \le \frac{1}{3}\left(\frac{m}{|s_1|} + \frac{m}{|s_2|} + 1\right) = d_J^{lb} \tag{1}$$

If $d_J^{lb}(s_1, s_2) \ge \tau_J$, we have the number of common characters $m \ge T = \frac{(3\tau_j - 1)|s_1||s_2|}{|s_1|+|s_2|}$, which is the lower bound of strings s_1 and s_2. After calculating T, we utilize pigeonhole principle and e-variants [15] to filter strings. Due to no order of matching characters of Jaro distance, we sort all characters according to a specific order L (e.g. the alphabetical order). The construction of L is ignored due to limited space. We calculate the lower bound between s_1 and s_2 as follow.

1. We calculate T described above.
2. We utilize pigeonhole principle to separately partition s_1 and s_2 into k_1 and k_2 parts $P_1 = \{s_1^1, s_1^2, \ldots, s_1^{k_1}\}$ and $P_2 = \{s_2^1, s_2^2, \ldots, s_2^{k_2}\}$. In detail, we design special splitting characters SC to generate partitions. SC is a set of special characters used to partition strings according to L. s is sorted according to L, and then partitioned according to SC as several parts.
3. We separately allocate errors for P_1 and P_2 as $E_1 = \{e_1^1, e_1^2, \ldots, e_1^{k_1}\}$ and $E_2 = \{e_2^1, e_2^2, \ldots, e_2^{k_2}\}$, where $\sum_{i=1}^{k_1} e_1^i = |s_1| - T$ and $\sum_{i=1}^{k_2} e_2^i = |s_2| - T$.
4. We generate all e_1^i-variants for s_1^i of s_1 and e_2^j-variants for s_2^j of s_2 ($1 \le i \le k_1$, $1 \le j \le k_2$). $Var(s_1, T)$ and $Var(s_2, T)$ are denoted as sets of variants of all partitions of s_1 and s_2. For example, $Var(s_1, T) = \bigcup_{i=1}^{k_1} e_1^i\_var(s_1^i)$.

Lemma 1. *New lower bound of Jaro distance.* *For strings s_1 and s_2, and the minimal common characters T, $Var(s_1, T) \bigcap Var(s_2, T) \ne \emptyset$.*

Lower Bound of Jaro-Winkler Distance. Given $d_{JW} \ge \tau_{JW}$ and the definition of d_{JW}, we can deduce the threshold τ_J of Jaro distance $d_J \ge \tau_J = \frac{\tau_{JW} - P' \cdot l}{1 - P' \cdot l}$. Given s_1 and s_2, we calculate P', and then calculate τ_J. Then we use the above lower bound to check common variants of s_1 and s_2.

5   Index Construction for Jaro-Winkler Distance

Given a collection of entities $\mathcal{D}$, the character sequence L, the set of splitting characters SC, and the minimum Jaro threshold τ_J^{min}, we first construct the index I_J of Jaro distance. L is partitioned into $L_1, L_2, \ldots L_K$ parts according to SC.

First, we sort each entity $s_i \in \mathcal{D}$ ($1 \leq i \leq |\mathcal{D}|$) in the special order of L. If same characters exist in one string, we consider them as different characters. Then according to SC, we partition s_i into k_i non-overlapping substrings $P_i = \{s_i^1, s_i^2, \ldots, s_i^{k_i}\}$. Next, given the minimal threshold τ_J^{min}, we calculate the least common number of characters T_i^{min} for s_i (i.e., maximal error). After partition s_i into k_i partitions, we evenly allocate errors for k_i partitions as $E_i^{max} = \{e_i^1, e_i^2, \ldots, e_i^{k_i}\}$. Finally, we generate $Var(s_i, T_i^{min})$ for s_i where $Var(s_i, T_i^{min}) = \bigcup_{j=1}^{k_i} e_i^j\_var(s_i^j)$, and insert all elements $w \in Var(s_i, T_i^{min})$ as keys and tuples (i, s_i, k_i, e) as values into the inverted index, where i is the entity ID of s_i, and e is the number of deleted characters. After inserting all entities in I_J, we sort elements (entities) in lists according to the decreasing order of $|s_i| - e \cdot k_i$. The reason is to apply early termination in query processing step (See Sect. 6).

Winkler distance only considers the first four prefix characters of strings. We construct a 4-level trie structure to index four prefix characters of strings in $\mathcal{D}$. We enumerate all four prefix characters in lexicon $\mathcal{D}$, and insert them into I_{JW}. According to the definition of Jaro-Winkler distance, if $d_{JW}(s_1, s_2) \geq \tau_{JW}$ we can calculate the threshold of Jaro distance at i-th level as $\tau_J^i = \frac{\tau_{JW} - i \cdot l}{1 - i \cdot l}$. For example, the 2nd level of I_{JW} means two strings have 2 common prefix characters, and $\tau_J^2 = \frac{\tau_{JW} - 2l}{1 - 2l}$. Thus i-level node ($0 \leq i \leq 4$) in the index I_{JW} denotes the index of Jaro distance I_J with threshold τ_J^i.

6   Online Query Processing

6.1   Query Processing of Jaro Distance

In this section, we solve the problem of online query processing of Jaro distance. Given a query $q \in \mathcal{M}$, and the threshold τ_J of Jaro distance, we retrieve all strings $s \in \mathcal{D}$ that $d_J(q, s) \geq \tau_J$.

For a query q, we sort and partition q into k_q partitions $P_q = \{q^1, q^2, \ldots, q^{k_q}\}$ according to L and SC. Considering l_{min}, we can get the minimal common characters between q and entities in $\mathcal{D}$ as $T_{min} = \frac{(3\tau_J - 1)|q| \cdot l_{min}}{|q| + l_{min}}$. Then we generate error allocation strategy as $E_{min} = \{e_{min}^1, e_{min}^2, \ldots, e_{min}^{k_q}\}$. Errors are evenly allocated to each partition. Finally, we generate the variant set $Var(q, T_{min})$, probe the index structure and extract corresponding inverted lists from I_J.

T_{min} is the minimal common characters, and the bound is not tight enough. In general, for any entity $s \in \mathcal{D}$, we can deduce the common character rate $R_T = \frac{T}{T_{min}} = \frac{|s| \cdot (|q| + l_{min})}{(|q| + |s|) \cdot l_{min}}$. Thus, the minimal number of common characters between

s and q are $\lceil R_T \cdot T_{min} \rceil$. Meanwhile, according to E_{min}, we have $\sum_{i=1}^{k_q} e^i_{min} = |q| - T_{min}$. For any entity $s \in \mathcal{D}$, we can allocate error as $E = \{e^1, e^2, \ldots, e^{k_q}\}$, where $\sum_{i=1}^{k_q} e^i = |q| - \lceil R_T \cdot T_{min} \rceil$. Due to even allocation strategy of errors, we assign e^i $(1 \le i \le k_q)$ as $e^i = \left\lfloor \frac{|q| - T_{min}}{|q| - \lceil R_T \cdot T_{min} \rceil} \cdot e^i_{min} \right\rfloor$.

We have the following strategies to accelerate query processing, and Algorithm 1 is the pseudo-code of query processing.

1. **Early Termination.** In any list H of I_J, we can get tuples $(eid, s, k, e) \in H$. Assume e is the smallest allocated error of k partitions of string s. Then we have $|s| - e \cdot k \ge T_{eid} \ge T_{min}$, where T_{eid} is the least number of common characters between s and q. According to Sect. 5, elements in list H of index I_J are all sorted according to the decreasing order of value $|s| - e \cdot k$. Thus, we sequentially scan elements in list H until $|s| - e \cdot k \le T_{min}$.
2. **Skipping Elements.** We take the length of string in $\mathcal{D}$ into account. For an entry (eid, s, k, e) in list H of I_J, if $|s| - e \cdot k \ge C = \frac{(3\tau_J - 1)|s||q|}{|s| + |q|}$, we consider s or eid as potential candidate, otherwise, we discard and skip it.

Algorithm 1. $JS(\mathcal{D}, q, \tau_J, L, SC)$

---

1 Sort and partition query q into partition $P = \{q^1, q^2, \ldots, q^k\}$ according to SC;
2 Calculate the minimal common characters T_{min} of q;
3 Allocate minimal errors $E_{min} = \{e^1_{min}, e^2_{min}, \ldots, e^k_{min}\}$, $cand = \{\}$;
4 **for** $q^i \in P$ **do**
5 　　Generate all variant set $e\_var(q^i)$ of q^i;
6 　　**for** $w \in e\_var(q^i)$ **do**
7 　　　　**if** $I[w] \ne \emptyset$ **then**
8 　　　　　　$H \leftarrow I[w]$, $T_{lb} = \frac{(3\tau_J - 1)|q||s_{min}|}{|q| + |s_{min}|}$, where s_{min} is the minimal length of string in H;
9 　　　　　　**for** $(eid, s, k, e) \in H$ **do**
10 　　　　　　　　**if** $|s| - \lceil v \cdot k \rceil \le T_{lb}$ **then**
11 　　　　　　　　　　break;
12 　　　　　　　　**if** $|s| - \lceil v \cdot k \rceil \ge C$ **and then**
13 　　　　　　　　　　$cand = cand \cup \{s\}$;

14 **return** $cand$;

6.2   Query Processing of Jaro-Winkler Distance

In the query processing of Jaro-Winkler distance, query q first traverses the trie structure to find sets of strings that share from 0 to 4 prefix characters with q. For each level, we calculate their corresponding thresholds of Jaro distance. The following step is to use Algorithm 1 to find results.

7   Experiments

7.1   Experimental Setup

The following algorithms are compared in the experiment.

- **JWS, JWS**$_N$ are our methods that apply Pigeonhole principle and e-variants. Here JWS$_N$ means we turn off filtering strategies of our methods.
- **LIMES** [1] [5] is the trie-based method based on Jaro-Winkler distance.
- **Scan** is the baseline method that perform sequential scan on the existing lexicon.

All experiments were conducted on a server with QuadCore AMD Opteron 8378@2.4GHz Processor and 96GB RAM whose system operation is Ubuntu 12.04. In our experiments, we select four publicly available datasets as follow.

- **AMiner-Author** [14] is a collection that contains 1.7 million author names.
- **AIDA-Lexicon** [9] is the lexicon of entities that contain approximate 9 million entities from Wikipedia. We randomly sample 1 million entities as our entity collection.
- **UKB-Lexicon** [1] is the lexicon of entities that contain nearly 4 million entities. We also randomly sample 1 million as one of our datasets
- **IMDB** is the collection that contains approximate 350 thousands actor names.

We measure the overall query response time and candidate size of each method in the following.

7.2   Query Performance

We show the total query running time of four datasets. Figure 1(a), (b), (c) and (d) are query time of four methods with Scan on four datasets. Because LIMES deals with string similarity join problem, we add index construction time into the total running time. JWS$_I$ and JWS$_{NI}$ are denoted as methods that involve the index construction time with and without filtering strategies.

   Among all methods, Scan almost achieves the worst performance, which is at least 10 times slower than JWS$_I$. LIMES has the second worst query performance. For example, in Fig. 1(a), when $\tau = 0.84$, LIMES is 20 times slower than JWS$_I$. Meanwhile, LIMES does not work well with a small threshold. When $\tau = 0.8$, LIMES is very slow and cannot beat Scan. JWS$_I$ and JWS$_{NI}$ are our methods that involve the index construction time. For different τ, we always construct the index with $\tau_{min} = 0.8$. JWS$_{NI}$ is the method with index construction time but without any filtering strategies. When τ is small (e.g., 0.8, 0.82), JWS$_I$ runs almost 2 times faster than JWS$_{NI}$. This is because filtering strategies can largely prune many false positives. However, as τ increases, their gap is becoming smaller, because pigeonhole principle and e-variants have strong pruning power with high values of τ. JWS$_I$ and JWS$_{NI}$ are faster than LIMES,

---
[1] https://github.com/AKSW/LIMES-dev.

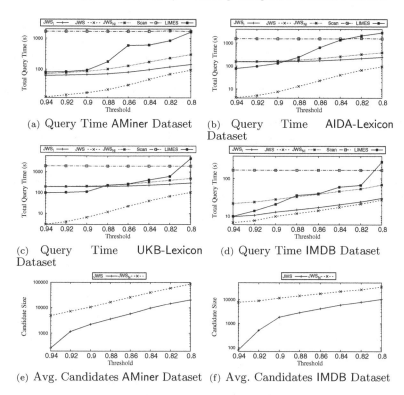

(a) Query Time AMiner Dataset

(b) Query Time AIDA-Lexicon Dataset

(c) Query Time UKB-Lexicon Dataset

(d) Query Time IMDB Dataset

(e) Avg. Candidates AMiner Dataset

(f) Avg. Candidates IMDB Dataset

Fig. 1. Query running time and average candidate number

especially when τ is small. For example, in Fig. 1(b), JWS_I is nearly 3 times faster when $\tau = 0.84$. This is because LIMES needs to search the trie structure to the deeper levels to make sure all results are retrieved if τ is small. To better show the efficiency of Jaro-Winkler similarity search, JWS is the method that excludes index construction time. JWS is much faster than any other methods. For example, in Fig. 1(c), when $\tau = 0.88$, JWS has 12.06 s of total query time, which is at least 20 times faster than the best of others.

7.3   Candidate Number

We show the candidate number of JWS and JWS_N (See Fig. 1(e) and (f)). Without filtering strategies, JWS_N still has pruning power. For example, in Fig. 1(e), JWS_N can still prune 95% strings when $\tau = 0.8$. JWS is the method that involves filtering strategies, and its pruning power is much more powerful. In Fig. 1(e), JWS can prune 98.82% strings when $\tau = 0.8$, and its candidate number is approximately 4 times smaller than JWS_N. As τ increases, the candidate size of JWS decreases rapidly. This is because the filtering strategies work and dominate the whole filtering procedure with the large threshold.

8   Conclusion

In this paper, we proposed a new lower bound of Jaro-Winkler distance to improve query performance. By using signatures that consist of characters in the same partition, we proposed a method of index construction and online query processing. Moreover, we showed the query running time and candidate number. In future work, we consider to improve the verification cost of Jaro-Winkler distance and propose some parallel query processing methods to solve this problem.

Acknowledgement. This work was supported by ARC DP DP130103401 and DP170103710.

References

1. Agirre, E., Barrena, A., Soroa. A.: Studying the Wikipedia hyperlink graph for relatedness and disambiguation. CoRR, abs/1503.01655 (2015)
2. Benjelloun, O., Garcia-Molina, H., Menestrina, D., Su, Q., Whang, S.E., Widom, J.: Swoosh: a generic approach to entity resolution. VLDB J. **18**(1), 255–276 (2009)
3. Christen, P.: Febrl -: an open source data cleaning, deduplication and record linkage system with a graphical user interface. In: SIGKDD, KDD 2008, New York, NY, USA, pp. 1065–1068. ACM (2008)
4. Cohen, W.W., Ravikumar, P., Fienberg, S.E.: A comparison of string distance metrics for name-matching tasks. In: Proceedings of IJCAI-03 Workshop on Information Integration on the Web (IIWeb-03), 9–10 August 2003, Acapulco, Mexico, pp. 73–78 (2003)
5. Dreßler, K., Ngomo, A.N.: On the efficient execution of bounded jaro-winkler distances. Semant. Web **8**(2), 185–196 (2017)
6. Elmagarmid, A.K., Ipeirotis, P.G., Verykios, V.S.: Duplicate record detection: a survey. IEEE Trans. Knowl. Data Eng. **19**(1), 1–16 (2007)
7. Feng, J., Wang, J., Li, G.: Trie-join: a trie-based method for efficient string similarity joins. VLDB J. **21**(4), 437–461 (2012)
8. Galárraga, L., Heitz, G., Murphy, K., Suchanek, F.M.: Canonicalizing open knowledge bases. In: CIKM '2014, New York, NY, USA, pp. 1679–1688. ACM (2014)
9. Hoffart, J., Yosef, M.A., Bordino, I., Fürstenau, H., Pinkal, M., Spaniol, M., Taneva, B., Thater, S., Weikum, G.: Robust disambiguation of named entities in text. In: EMNLP 2011, Stroudsburg, PA, USA, pp. 782–792. Association for Computational Linguistics (2011)
10. Li, G., Deng, D., Wang, J., Feng, J.: Pass-join: a partition-based method for similarity joins. Proc. VLDB Endow. **5**(3), 253–264 (2011)
11. Liu, Y., Shen, W., Yuan, X.: Deola: a system for linking author entities in web document with DBLP. In: CIKM (2016)
12. Prokoshyna, N., Szlichta, J., Chiang, F., Miller, R.J., Srivastava, D.: Combining quantitative and logical data cleaning. Proc. VLDB Endow. **9**(4), 300–311 (2015)
13. Qin, J., Wang, W., Lu, Y., Xiao, C., Lin, X.: Efficient exact edit similarity query processing with the asymmetric signature scheme. In: SIGMOD 2011, New York, NY, USA, pp. 1033–1044. ACM (2011)
14. Tang, J., Zhang, J., Yao, L., Li, J., Zhang, L., Su, Z.: Arnetminer: extraction and mining of academic social networks. In: KDD 2008, pp. 990–998 (2008)

15. Wang, W., Xiao, C., Lin, X., Zhang, C.: Efficient approximate entity extraction with edit distance constraints. In: SIGMOD 2009, New York, NY, USA, pp. 759–770. ACM (2009)
16. Xiao, C., Wang, W., Lin, X.: Ed-join: an efficient algorithm for similarity joins with edit distance constraints. Proc. VLDB Endow. 1(1), 933–944 (2008)
17. Xiao, C., Wang, W., Lin, X., Yu, J.X., Wang, G.: Efficient similarity joins for near-duplicate detection. ACM Trans. Database Syst. 36(3), 15:1–15:41 (2011)

Cloud Computing

Long-Term Multi-objective Task Scheduling with Diff-Serv in Hybrid Clouds

Puheng Zhang[✉], Chuang Lin, Wenzhuo Li, and Xiao Ma

Tsinghua National Laboratory for Information Science and Technology,
Department of Computer Science and Technology, Tsinghua University,
Beijing 100084, China
zhangph14@mails.tsinghua.edu.cn, chlin@tsinghua.edu.cn

Abstract. With the speedy development of E-commerce, requests over the internet from intensive users are soaring, especially in global online shopping festivals. In order to meet the increasing demands of temporary capacity and reduce daily expenses, hybrid clouds are often used, and the task scheduling problem with multi-objectives is further investigated. In this paper, we firstly build a differentiated-service (Diff-Serv) task scheduling model, and formulate a dynamic programming problem, where the state space is too large to be solved by exhaustive iterations. Therefore, we carefully design the value approximation function, and with reference to the reinforcement learning theory, we put forward an approximate dynamic programming (ADP) algorithm so as to conduct the long-term optimization for performance benefit, energy and rental costs. Furthermore, both scheduling quality and scheduling speed are taken into consideration in this algorithm. Experiments with both random synthetic workloads and Google cloud trace-logs are conducted to evaluate the proposed algorithm, and results demonstrate that our algorithm is effective and efficient, especially under bursty requests.

Keywords: Multi-objective optimization · Hybrid cloud · Task scheduling · Approximate dynamic programming (ADP)

1 Introduction

With the rapid development of E-commerce techniques, requests over the internet from intensive users are becoming bursty and unpredictable. In global online shopping festivals, there are always surges in both web visits and deal volumes. For example, the Gross Merchandise Volume (GMV) of Tmall on Singles Day (Double 11) in 2016 amazingly reached 17.79 billion US dollars with an increase of 32% compared with last year, covering 235 countries [4]. However, on-line transactions on normal days are much less [6]. As a result, businesses big and small are turning to hybrid clouds, a composition of private and public clouds that are bound together [14]. Thus, these businesses can "own the base and rent the spike", in order to cut down expenses on normal days and meet sudden increases in temporary capacity needs. In addition, as network applications

© Springer International Publishing AG 2017
A. Bouguettaya et al. (Eds.): WISE 2017, Part I, LNCS 10569, pp. 243–258, 2017.
DOI: 10.1007/978-3-319-68783-4_17

are getting more and more widely used in our daily life, user requests can be classified into several types, such as user management, email, shared printing, and system administration [16]. In Google cloud trace-logs, workloads are distinguished by two indicators: priority and scheduling class [3]. Different types of tasks have different service requirements, and it is more reasonable to treat them with differentiated-service (Diff-Serv) methods in practice. For example, if you want to submit an order on a shopping website, you need to be answered immediately. Meanwhile, the response time for sending an email is not as important. Thus, research on Diff-Serv task scheduling with the objective of finding the maximum benefit in the long run, is urgently needed and of practical value.

In theory, the multi-objective task scheduling problem in a hybrid cloud is complicated. On the one hand, we should arbitrate the tradeoff of multiple metrics, such as performance gain, energy consumption from the private cloud and rental fees from the public cloud. On the other hand, we need to tackle the problem of state explosions. Consider the real-time queue of each instance, and the queue length can take continuous values. Therefore, the number of states for one instance is uncountable and infinite, let alone the combination states of a huge number of instances in both public and private clouds. Classical Markov Decision Processes (MDP) do not work in this scenario due to the curse of dimensionality [10]. What is more, a task scheduling algorithm can not be very intricate or tedious, because a good but time-consuming algorithm does not make any sense for some types of requests. Requests from online trading, for example, need to be handled as soon as possible. Whereas sampling methods, such as Sparrow [7], can quickly return scheduling results but can not guarantee the optimality of the results. Consequently, we should pay equal attention to both scheduling quality and scheduling speed at the same time.

There haven't been many studies dedicated to task distributing problems in hybrid clouds, and some great efforts are as follows.

Niu et al. [6] took advantage of Lyapunov optimization techniques to find the optimal task-distribution under a given budget. Inspired by [6], yet we propose new challenges in the following aspects. Firstly, in [6], they evaluated average response time of the public cloud by the $M/M/1$ queuing model. It implies that the request arrivals conform to Markovian (Poisson) distribution, which may be less accurate in bursty workloads. Independent of Poisson hyperthesis, the effect of our optimization can be further improved. Secondly, they considered the auto-scaling of Amazon Web Services (AWS) cloud, but ignored the cooldown time, during which additional instances are not allowed to be launched or terminated [1]. Cooldown time means that there must be a certain interval of time between two adjacent scalings, and it has a great impact on decision making. Thirdly, they gave the optimal distribution proportion of tasks that are assigned to the public cloud at each time epoch, not specified to each task. In other words, that algorithm scheduled for a suitable distribution proportion at each time epoch, whereas we devote to scheduling at the task level, which is more sophisticated.

Zuo et al. [17] formulated the task allocation problem in a hybrid cloud as an integer programming model, and solved it by a self-adaptive learning particle swarm optimization-based scheduling approach. Ruben et al. [12] proposed a heuristic method to cost-efficiently schedule deadline-constrained applications on both public cloud providers and private infrastructures. However, what [12,17] pursued were the best actions at each time epoch. In a stochastic process, the sum of all the highest rewards at each time epoch may not be the optimum from a long-term perspective. Maybe the choices, which can not result in the highest rewards at the beginning, will lead to maximum gross income in the long run. That is similar to the concept of "Opportunity Cost" in the field of microeconomic theory [15].

Calheiros and Buyya [2] put forward an architecture to cost-effectively schedule applications within their deadlines by considering the whole organization workload at individual task level and proposed an accounting mechanism to determine the share of public cloud resources to be assigned to each user. In E-commerce, it is difficult and nearly impossible to accurately assess the deadline of each task. For example, if you want to make an online deal, the sooner you get response, the better. But how to set the deadline for each request? 0.1 ms or 0.2 ms? If we set the deadline as 0.15 ms, does it mean that 0.16 ms should be rejected? Therefore, it is better to classify requests into several serving-levels (SLs), like Google trace-logs [3], and define different reward functions according to different SLs. In addition, they did not take into account the multiple metrics in hybrid clouds, such as energy costs. Furthermore, in [2] the number of rental instances in a public cloud were fixed, and auto-scaling was not considered.

In conclusion, all previous studies that we have discussed can not address all the requirements and challenges mentioned above. Motivated by the need of high efficient Diff-Serv task scheduling algorithms in hybrid clouds, we put forward an approximate dynamic programming (ADP) algorithm with the objective of finding the long-term optimization for performance gain, rental cost from the public cloud, and energy cost from the private cloud. Main contributions of this work are as follows:

(1) We build a model that can simulate the practical scenario of a hybrid cloud in a better way. We adopt the Diff-Serv method to cope with diversified online requests. The public cloud can be auto-scaled to make full use of resources. We also take into consideration the scaling cooldown time and its influence.

(2) We resort to reinforcement learning theory to solve traditional stochastic dynamic programming (SDP) problems. Thus, we do not rely on inaccurate predictions or assumptions of incoming requests, but use a stochastic gradient method to train parameters of basis functions by constantly learning from sample paths, which indicate particular sequences of real input flows.

(3) We find an efficient way to the conduct the long-term and multi-objective optimizations for task scheduling problems in hybrid clouds with state explosions, rather than only to find local optimums in previous studies. We propose the value function approximation (VFA) that can approximate the hard-to-calculate expectations in MDP [9] and each state value can be easily evaluated.

The remainder of the paper is organized as follows. Section 2 introduces the task scheduling model of a hybrid cloud and formulates the scheduling problem as an SDP problem. In Sect. 3, we take advantage of ADP to solve the problem. In Sect. 4, we conduct simulations and experiments to make QoS evaluations with both random synthetic data and real trace-logs to demonstrate the applicability and superiority of our approach under bursty requests and high throughput. Section 5 concludes the paper.

2   Problem Formulation

In this section, we illustrate models and notions used in this paper, build a Diff-Serv task scheduling model of a hybrid cloud, and formulate an SDP problem.

2.1   System Model

A Diff-Serv task scheduling model in a hybrid cloud is described in Fig. 1. Suppose that a hybrid cloud can receive and handle requests from R regions. In each region, there is a Front-End (FE) proxy that is responsible for collecting all types of requests in this region and retransmitting them to Classifier. Classifier is in charge of separating all requests into different Priority Queues according to the priorities of tasks. Suppose there are N_{pr} priorities in total, and there are N_{pr} Priority Queues. Scheduler should dispatch each task to a public or private instance according to the order of task priorities from high to low. In this paper, we assume the tasks that have been scheduled or processed can not be interrupted.

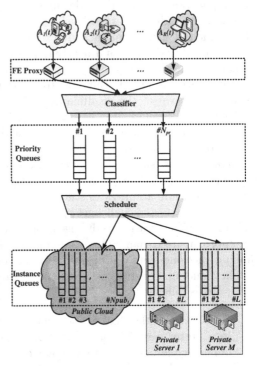

Fig. 1. Diff-Serv task scheduling model in a hybrid cloud

For the pubic cloud, such as AWS cloud [1], the number of rental instances can be adjusted according to specific policies. For example, we can call for enlarging the fleet of Elastic Compute Cloud (EC2) instances whenever the average CPU utilization rate stays above ninety percent for several minutes [1]. We set $Npub_t$ as the number of rental instances that would change with time. Besides, after the previous scaling activity, additional EC2 instances can not be launched or terminated again within a fixed period of time, which is called auto-scaling cooldown time. In AWS cloud, cooldown time is often set to 6 min [1].

For private infrastructures, suppose there are M homogeneous private servers, and each server is virtualized into L instances (virtual machines). Without loss of generality, we assume the processing capacity of each private server is fairly distributed among its hosted instances in this work. Then the processing capacity of each instance is $1/L$ of the total processing capacity of a private server.

2.2   An SDP Problem

The task scheduling problem in the above hybrid cloud model can be formulated as an SDP problem with elements as follows.

(1) **Decision Time Epoch**

Decisions are executed at the beginning of each time slot, i.e., at $t = \tau, 2\tau, 3\tau \cdots$, where τ is the decision time interval.

(2) **System Information and State Variable**

System information can be divided into two categories, endogenous information (En_t) that changes with specific actions at each time epoch, and exogenous information (Ex_t) that is only determined by external factors. (1) Endogenous information En_t refers to tuple $(Npub_t, Qpub_t, Qpri_t, UR_t)^T$ in this work. $Npub_t$ indicates the number of rental instances in the public cloud at time epoch t. $Qpub_t$ and $Qpri_t$ respectively describe the queue length of each instance in the public and private cloud. UR_t represents the average CPU utilization ratio of the public cloud for a period of time till time t. (2) Exogenous information Ex_t means $(R_t, Cpub, Cpri)^T$. $R_t = (R_{1t}, R_{2t}, \cdots, R_{kt}, \cdots)$ refers to the coming requests (tasks) during time $[t-\tau, t)$. Each request R_{kt} is marked as a four tuple, and $R_{kt} = (Arr\_Time, Sch\_C, Priority, Size)^T$. Arr_Time means the arrival time of a task. Sch_C is short for scheduling_class, and roughly represents how latency-sensitive a task is. Sch_C affects machine-local policy for resource access, and can be represented by a single number [3]. $Priority$ denotes the priority of a request. $Size$ indicates instructions of a request measured in Million Instructions (MI). $Cpub$ and $Cpri$ are respectively the processing capacity of a public and private instance. They are measured in Million Instructions Per Second (MIPS), and do not change with time. In this work, state variable S_t is equivalent to endogenous information (En_t).

Suppose that a company owns 20 servers and each server is virtualized into 5 instances, and then there are totally 100 private instances. Suppose at time t, that company rents 200 instances from a public cloud. Each instance possesses a buffer queue. Along with the scheduling process, the value that each queue length can get spans a continuous range, but can be separated into 10^3 discrete values for simplicity according to engineering requirements. Then the total number of states will amount to $10^{3 \cdot (100+200)}$, an absolutely astronomical number! In fact, the queue length of each instance can take continuous values, and we should also consider $Npub_t$ and UR_t. Thus, the number of states is uncountable and infinite.

(3) **Decision Variable**

For each En_t, there is a corresponding action $x_t(En_t)$, which means allocating each request that arrives during $[t - \tau, t)$ to a proper instance from a private or public cloud. Then $x_t(En_t)$ is a vector, comprised of a series of scheduling decisions at the beginning of time t, and X_t represents the value set of x_t.

(4) **Transition Function**

Transition function describes how the system evolves from current state to next state, given the action and exogenous information at time t. In this work, suppose there are N_Ta_t requests to be scheduled at time t, and then endogenous information evolves as described in Eqs. (1), (2), (3) and (4).

$$Qpub_{t+1,n} = max \left\{ \left(Qpub_{t,n} + \sum_{k=1}^{N\_Ta_t} R_{t,k,n} - Cpub \right), 0 \right\} \tag{1}$$

$$Qpri_{t+1,l,m} = max \left\{ \left(Qpri_{t,l,m} + \sum_{k=1}^{N\_Ta_t} R_{t,k,l,m} - Cpri \right), 0 \right\} \tag{2}$$

$$UR_t = Monitor(Qpub_t, Npub_t) \tag{3}$$

$$Npub_{t+1} = ASPolicy(Npub_t, UR_t, T\_CD) \tag{4}$$

In Eq. (1), if the kth task is allocated to instance n of the public cloud at time t, $R_{t,k,n}$ equals the size of that task, otherwise, $R_{t,k,n} = 0$. $Qpub_{t,n}$ represents the queue length of the nth instance in the public cloud.

In Eq. (2), if the kth task is allocated to instance l on server m in the private infrastructure at time t, $R_{t,k,l,m}$ equals the size of that task, otherwise, $R_{t,k,l,m} = 0$. $Qpri_{t,l,m}$ means the queue length of instance l on private server m.

In Eq. (3), $Monitor()$ represents a function to monitor and calculate the average CPU utilization ratio of the public cloud.

In Eq. (4), function $ASPolicy()$ means adopting auto-scaling policy for the public cloud. The policy should comprehensively consider the number of instances ($Npub_t$), average CPU utilization ratio at current time (Ave_UR_t), and the cooldown time (T_CD). These concepts and the auto-scaling mechanism will be elaborated in Sect. 4.1.

(5) **Contribution Function**

Contribution function determines rewards, including both gains received and costs incurred during each time interval. It can be defined in many ways according to specific requirements in business. Whatever the function formulas, they all can be applied into our algorithm. In this work, we focus on performance benefit, rental cost from the public cloud, and energy cost from private infrastructures.

The performance benefit of a request is in proportion to the throughput and inversely in proportion to the response time [8]. At each time epoch, whatever by private or public instances, if task k is handled, the throughput equals the task size, measured in MI. Response time $T_{t,k}$ can be calculated by Eq. (5) if the task is scheduled to the public cloud, and by Eq. (6) if scheduled to the private. T_trans indicates the round-trip network delay of data transmission from Scheduler to the public cloud. Therefore, performance benefit of N_Ta_t tasks during the period of time $[t - \tau, t)$ can be expressed as Eq. (7), where β, η are constant coefficients. According to Eq. (7), the performance gain of a task is related to its scheduling class. As for a task with high Sch_C, shorter response time will lead to higher performance gain. In other words, that task is more time-sensitive, and vice versa.

$$T_{t,k} = \frac{Qpub_{t,n} + R_{t,k}}{Cpub} + T\_trans \tag{5}$$

$$T_{t,k} = \frac{Qpri_{t,l,m} + R_{t,k}}{Cpri} \tag{6}$$

$$PB_t = \sum_{k=1}^{N\_Ta_t} \eta \cdot \left(\frac{R_{t,k}}{T_{t,k}^{1+Sch\_C/\beta}} \right) \tag{7}$$

Besides, power consumption of a private instance can be expressed as Eq. (8). P is the estimated value of the power consumption of one instance. P_{idle} and P_{busy} respectively represent the power consumed when the server is idle and fully utilized. μ is a parameter in proportion to the CPU utilization. The total energy consumption of all private infrastructures during the period of time $[t, t + \tau)$ can be calculated as Eq. (9). It is worth noting that there may be some other formulas that can evaluate the power consumption, and they can be applied to our algorithm only after slight alterations to VFA.

$$P = P_{idle} + \mu \left(P_{busy} - P_{idle} \right) \tag{8}$$

$$EC_t = \sum_{m=1}^{M} \sum_{l=1}^{L} P_{t,l,m} \cdot \tau \tag{9}$$

Finally, the rental cost from the public cloud is directly in proportion to the number of public instances at time t, and can be easily calculated in Eq. (10), where δ is a fixed scale coefficient.

$$RC_t = \delta \cdot Npub_t \cdot \tau \tag{10}$$

In summary, the overall reward during the period of $[t, t + \tau)$ can be calculated in Eq. (11).

$$RE_t = PB_t - EC_t - RC_t \tag{11}$$

(6) **Objective Value Function**

In [11], objective value function of En_t is defined as the supremum over all policies of the expected total rewards from decision epoch t onwards. We define

$x_t^\pi(En_t)$ as the decision made in state En_t under policy π. Then our objective is to find the best policy $\pi^* \in \Pi$ with the largest expected total discounted rewards over the infinite horizon as in Eq. (12). $\gamma \in [0.5, 1)$ is a discount factor, which measures the value at time $t + \tau$ of one unit reward received at time t. That is to say, rewards will be discounted, and one unit of reward only has the present value of γ^t after t periods. Equation (12) can also be expressed by the recursive form in Eq. (13), which is called the Bellman Equation [9].

$$V^{\pi^*} = \max \mathbb{E}^\pi \left\{ \lim_{N \to \infty} \sum_{t=1}^{T} \gamma^{t-1} R_t(En_t, x_t^\pi(En_t)) \right\} \tag{12}$$

$$V_t(En_t) = \max_{x_t \in X_t} \left\{ R_t(En_t, x_t) + \gamma \mathbb{E}\{V_{t+1}(En_{t+1})|En_t\} \right\} \tag{13}$$

To sum up, compared with public rental fee, energy cost from private servers is relatively cheaper. However, if excessive requests are allocated to the private instances, their queue lengths will be pretty long, which will lead to high response delay and low performance profits. Thus, we need to arbitrate between the public and the private in order to seek for the long-term optimal scheduling decision for each task. Consequently, we put forward an ADP algorithm in the next section.

3   An ADP Algorithm

In this section, we elaborate an ADP algorithm to solve the problem mentioned above. The overview of the algorithm is described in Algorithm 1, and the main traits are described as follows.

Algorithm 1. Outline of the ADP algorithm

INPUT: Endogenous and exogenous information (En_t, Ex_t), total iteration number T, discounted factor γ.
OUTPUT: Estimated vector θ_T.
  1: Choose an initial state En_1.
  2: Initialize θ_1, $\overline{V}(En_1)$ and $\phi_f(En_1)$.
  3: **for** $t = 1, 2, \ldots, T$ **do**
  4:     **if** $t > 1$ **then**
  5:         Derive $\phi_f(En)$ from En_t.
  6:         Calculate $\overline{V}_t^x(En_{t-1}^x|\theta)$ by Eq. (15).
  7:     **end if**
  8:     **for** $k = 1, 2, \ldots, N\_Ta_t$ **do**
  9:         Schedule the kth task by solving Eq. (14).
 10:         Update $Qpub$ and $Qpri$.
 11:     **end for**
 12:     Update En_t by Eq. (1), (2), (3) and (4).
 13:     Compute $\hat{v}_t(En_t)$ using Eq. (14).
 14:     Update θ with Eq. (18).
 15:     Set sample path ω_{t+1} as the output requests from priority Classifier.
 16: **end for**
 17: return θ_T.

3.1 Stepping Forward Through Time

In classical dynamic programming, such as MDP, algorithms proceed by stepping backward in time [11]. It means that at each time t, we ought to pre-know essential information at time $t + \tau$. Therefore, most classical algorithms on dynamic programming depend on assumed probability distribution of arriving tasks throughout time. If priori knowledge of any statistical distribution is unpredictable and difficult to obtain in practice, classical algorithms do not work. Accordingly, a new algorithm should make full use of exogenous sample paths, and step forward through time. By continuously mining mass input data information, we can constantly update action-related parameters in vector θ and achieve the globally optimum decisions in the long run.

3.2 Introducing Concept of Post-Decision State Variable (PDSV)

In MDP, Eq. (13) has to be solved for each state En_t. But the states are infinite in this paper, so an exhaustive algorithm does not work. Furthermore, in Eq. (13), it is a kind of complex and tedious work to calculate the expectation $\mathbb{E}\{V_{t+1}(En_{t+1})|En_t\}$. Especially when a state has many possible consequent states is computationally intractable in practical cases to solve out the one-step transition matrix. Thus, the concept of post-decision state variable (PDSV) is introduced in order to avoid the ugly step of approximating the expectation [10]. PDSV En_t^x means the state at time t after we make a decision x and before time $t + 1$. So in this work, PDSV is equivalent to the state value at the next time slot, that is $En_t^x = En_{t+1}$. With PDSV, Eq. (13) can be rewritten as Eq. (14), where $\hat{v}_t(En_t)$ represents the evaluated sample values at time t, and $\overline{V}_{t+1}^x(En_t^x)$ denotes approximate PDSV which will be illustrated in the Sect. 3.3.

$$\hat{v}_t(En_t) = \max_{x_t \in X_t} \{R_t(En_t, x_t) + \gamma \overline{V}_{t+1}^x(En_t^x)\} \qquad (14)$$

3.3 Using Value Function Approximation

To cope with state explosions in dynamic programming, we build an evaluation function to approach the true value of each state by iterations step by step. As in Eq. (15), the evaluation function can be constructed with basis feature function $\phi_f(En)$ and parameter vector θ. $f \in F$, where f is a feature that should be tailored for specific scenarios, and vector $\phi_f(En)$ can be obtained by extracting feature information at each time t from the endogenous state tuple En_t. Vector θ represents parameters that are used for estimating the state value function.

$$\overline{V}_t^x(En_{t-1}^x|\theta) = \sum_{f \in F} \theta_f \cdot \phi_f(En_{t-1}^x) \qquad (15)$$

In this work, rewards are mainly comprised of three parts, performance gain, energy and rental costs. (1) The performance benefit PB_t, whether in the public or the private cloud, should possess the following properties. First,

PB_t is in proportion to the sum of available CPU processing capacities, which are further in proportion to the number of public and private instances at work (N_W_Pub, N_W_Pri). Second, PB_t is in proportion to the number of unhandled tasks in current instance queues (N_Task_Pub, N_Task_Pri). Third, PB_t is in inverse proportion to the sum of queue length of public and private instances ($S\_Q\_L\_Pub, S\_Q\_L\_Pri$). (2) The energy cost EC_t comes from private infrastructures. Similar with the performance gain, the energy consumption is also indirectly in proportion to N_W_Pri, and additionally in proportion to the number of private servers that are at work ($N\_PM\_on\_Pri$). (3) The rental cost RC_t from the public cloud is in proportion to the number of rental instances (N_Pub). Thus, we can totally extract 8 features from a hybrid cloud, and $\phi_f(En)$ can be expressed in Eq. (16).

$$\phi_f(En) = (N\_W\_Pub, N\_W\_Pri, N\_Task\_Pub, N\_Task\_Pri,$$
$$S\_Q\_L\_Pub, S\_Q\_L\_Pri, N\_PM\_on\_Pri, N\_Pub)^T \qquad (16)$$

3.4   Making Decisions Approximately

In line 9 of Algorithm 1, we should find out the most suitable actions that can solve the maximization problem at each time epoch t. In other words, we should allocate a suitable instance for each task that arrives during $[t-\tau, t)$. We find that only one of the following 3 choices may lead to obtain the maximized reward. (1) To choose a public instance with the shortest queue length; (2) To choose a private instance whose hosted server is hibernating, which means that no hosted instances are at work; (3) To choose a private instance with the least queue length. Therefore, there is no need to do comparisons with all possible actions, just three is all right. Besides, what we compare in this algorithm are not just the rewards of three choices, but the sum of $R_t(En_t, x_t)$ and $\gamma \overline{V}_{t+1}^x(En_t^x)$. That is to say, when we are making decisions, both current rewards and future influences are considered at the same time.

3.5   Updating θ_t by Stochastic Gradient

In order to find the most appropriate evaluation function $\overline{V}_t^x(En_{t-1}^x|\theta)$, a stochastic gradient updating strategy, which stems from reinforcement learning theory, is adopted to progressively train vector θ following sample path ω_t. In this paper, ω_t refers to arriving requests at each time epoch. Our aim is to find the most suitable θ^* that produces the minimum expected squared error (MESE) between $\hat{v}_t(En_t)$ and $\overline{V}_t^x(En_{t-1}^x|\theta)$ [9], as Eq. (17). In this way, θ can be updated step by step to approach the optimum θ^*, as is shown in Eq. (18). α_{t-1} represents the step-size, and ∇ denotes the *Nabla* Operator to calculate the gradient.

$$\theta^* = \arg\min_\theta \mathbb{E}\left\{\frac{(\hat{v}_t - \overline{V}_t^x(En_{t-1}^x|\theta))^2}{2}\right\} \qquad (17)$$

$$\theta_t = \theta_{t-1} - \alpha_{t-1}\big(\hat{v}_t(En_t) - \overline{V}_t^x(En_{t-1}^x|\theta_{t-1})\big)\nabla_\theta \overline{V}_t^x(En_{t-1}^x|\theta_{t-1})$$
$$= \theta_{t-1} - \alpha_{t-1}\big(\hat{v}_t(En_t) - \overline{V}_t^x(En_{t-1}^x|\theta_{t-1})\big)\phi(En_{t-1}^x) \qquad (18)$$

4   QoS Evaluation

In this section, we conduct simulations by Matlab 2012a, on a PC with two cores at 3.5 GHz and with memory size of 4GB. After steps of simplifications in Sect. 3, complicated scheduling procedure of each task can be reduced to only comparing three candidates with elementary operations. Obviously, computation complexity of the ADP algorithm can be greatly reduced, and can be solved in polynomial time. Due to space limit, demonstration is omitted here. Thus, scheduling speed is no more a problem with modern computers. We focus on evaluating scheduling quality of our algorithm. We do experiments with both random synthetic workloads and Google trace-logs, and finally compare our algorithm with "Heuristic" (Myopic), "Public_First" and "Local_First" algorithms.

4.1   Simulations on Stochastic Workloads

In order to testify the generality and wide adaptability of the ADP algorithm, we conduct simulations under various conditions.

(1) **Generating Input Variables**

   (a) *Private Instances*

   We assume that a company owns 20 servers ($M = 20$) and each server can be virtualized into 5 instances ($L = 5$). Then there are totally 100 instances for private infrastructures.

   (b) *Public Auto-Scaling Strategy*

   Suppose that in the public cloud, the minimum number of rental instances is set to 20, and the maximum is 500. The CPU utilization rate is real-time monitored, and Ave_UR_t, the average value during the last 300 time epochs $[t-300\tau, t)$, acts as an indicator to scaling actions. The value of Ave_UR_t and the corresponding auto-scaling ratio of the number of public instances are described in Table 1. Except in case 4, where $Npub_t$ stays unchanged, two scaling activities can not happen one after another, and between them there must be an interval time, which is the so-called cooldown time, and set to 360τ in this work.

Table 1. Auto-scaling policy

Case	Average CPU utilization rate	Scaling ratio
1	$Ave\_UR_t = 1$	135%
2	$0.9 \leq Ave\_UR_t < 1$	125%
3	$0.8 \leq Ave\_UR_t < 0.9$	110%
4	$0.7 \leq Ave\_UR_t < 0.8$	100%
5	$0.6 \leq Ave\_UR_t < 0.7$	90%
6	$0.5 \leq Ave\_UR_t < 0.6$	80%
7	$0 \leq Ave\_UR_t < 0.5$	70%

(c) *Priority*

In most cases, the number of priorities (N_{pr}) is an integer not more than 10.

(d) *Task*

Assume that in normal circumstances, the number of tasks during one time epoch and the size of each task both conform to Poisson distribution, and their expectations respectively equal λ_{nor} and σ. Besides, in order to imitate the real traces of online e-commerce transactions, we further suppose that additional flash crowds of tasks would rush in at intervals.

(e) *Step-Size*

A step-size that is too large may lead to unconvergence of the algorithm, whereas if it is small, the procedure of regression will be very slow. In this work, we choose two ways to set the step-size. One is in a constant way, where the step-size equals a constant value throughout time, $\alpha_t = C_{step}$. The other is in a harmonic way, where $\alpha_t = a/(b+t-1)$. The order of magnitude of a, b determines the declining speed of step-size, and a/b represents the original value.

(2) **A Typical Simulation and Related Analysis**

In the first experiment, we set $Cpub = 2.0 \times 10^3$MIPS, $Cpri = 2.6 \times 10^3$MIPS, $\lambda_{nor} = 20$, $\lambda_{bur} = 800$, $\sigma = 2 \times 10^4$MI, $N_{pr} = 5$, $C_{step} = 0.05$, $a = 2500$, and $b = 5000$. We follow the fact that the average rental expense from a public instance is much higher than that from a private one. We further assume that every 8000τ, there are bursty crowds of arriving requests with $\lambda_{bur} = 800$ lasting for 25τ. This combined stochastical workload that changes with time is depicted in Fig. 2. The X-axis is in unit of second, and the Y-axis at t represents the sum length of tasks that arrive during $[t - \tau, t)$. We can see obvious flash crowds periodically. The step-size changing with corresponding t is described in Fig. 3.

Utilizing the ADP algorithm, each of the 8 elements in vector θ can separately converge into an asymptotic value step by step. Examples of $\theta(N\_W\_Pri)$, $\theta(N\_Task\_Pub)$ and $\theta(S\_Q\_L\_Pub)$ are plotted in Figs. 4, 5 and 6. In addition, under the constant step-size rule, the curves may have more intensive oscillations in the end, whereas under the harmonic step-size, the curves may be steep at the beginning. Thus, due to the slightly different convergent values under two step-size rules, we can set $\bar{\theta}$ as the average value in practice.

With the final vector $\bar{\theta}$, we then conduct our ADP algorithm. Meanwhile, we compare our algorithm with three algorithms: (1) "Heuristic" algorithm, which directly uses measures described in Sect. 3.4, and seeks for the highest reward at each time epoch; (2) "Public_First" algorithm, in which requests are firstly allocated to the public cloud, unless its average queue length exceeds a threshold; (3) "Local_First" algorithm, in which requests are firstly scheduled to private infrastructures. Under the same arriving workloads, we simultaneously run the four algorithms, and their reward-changing curves are plotted in Fig. 7.

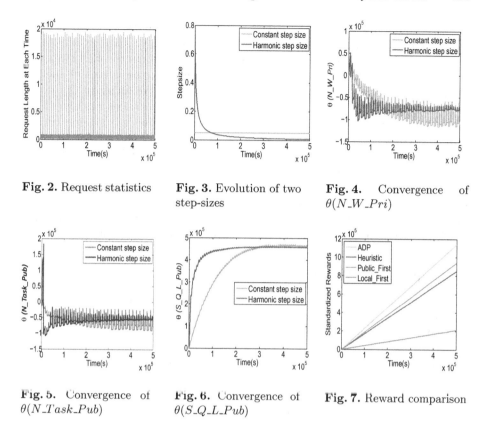

Fig. 2. Request statistics

Fig. 3. Evolution of two step-sizes

Fig. 4. Convergence of $\theta(N\_W\_Pri)$

Fig. 5. Convergence of $\theta(N\_Task\_Pub)$

Fig. 6. Convergence of $\theta(S\_Q\_L\_Pub)$

Fig. 7. Reward comparison

4.2   Comparative Experiments and Analysis

Next, we do simulations in cases of various environments, as listed in Table 2. We consider situations in which CPU processing capacity of a public instance

Table 2. Different input variables

Case	$Cpub$	$Cpri$	λ_{nor}	λ_{bur}	N_{pr}	σ
1	2.0×10^3	2.6×10^3	20	800	5	2×10^4
2	2.5×10^3	2.5×10^3	20	800	5	2×10^4
3	2.6×10^3	2.0×10^3	20	800	5	2×10^4
4	2.0×10^3	2.6×10^3	18	750	4	1.8×10^4
5	2.5×10^3	2.5×10^3	15	950	6	1.1×10^4
6	2.0×10^3	2.6×10^3	12	700	7	1.5×10^4
7	2.0×10^3	2.6×10^3	23	850	3	2.2×10^4
8	3.0×10^3	3.6×10^3	25	700	9	5×10^4
9	2.0×10^3	2.6×10^3	28	920	8	1.5×10^4

is better, equal and worse than that of a private instance. Besides, input data flows stand out distinctly in different cases. In each case, we conduct our ADP algorithm under two step-sizes, and we can get convergent curves of all elements in θ, similar with Figs. 4, 5 and 6. As space is limited, the curves of each θ and reward changing with iterations are omitted here. We focus on the overall rewards after 500 thousand iterations, and the comparisons are shown in Fig. 8. A group of experiments demonstrate that our approach is superior in all circumstances.

In conclusion, our algorithm is particularly suitable for sudden bursty flows, and the reason is simple. Due to cooldown time, the number of public instances can not be auto-scaled again immediately when Ave_UR_t exceeds a threshold, but after a period of time. Whereas our algorithm can simultaneously take into account both current and future influences. Therefore, before crowds come, our algorithm can learn from input data and do preparations ahead of time.

4.3 Experiments Based on Google Trace-Logs

The above experiments manifest the QoS improvement and high adaptability of our approach in stochastic synthetic workloads. In order to testify the feasibility of this ADP algorithm in practical applications, we further carry out experiments with Google cloud trace-logs [3], which describe workload information including 25 million tasks that span 29 days. It is especially difficult and not necessary to exploit all the log data, and therefore we randomly select data from the 5th hour of the 10th day to the 3th hour of the 13th day as our test sample path, which recorded a 70-h period of traces containing 1.5 million tasks. Each task is marked with a four tuple as R_{kt}, mentioned in Sect. 2.2. $Size$ of each task can be calculated by its execution time and resource request for CPU cores (R_R_CPU) as in Eq. (19). T_{finish} and $T_{schedule}$ respectively denote the timestamps of *FIN-ISH* and *SCHEDULE* events, which can be directly derived from the trace-logs [5]. When *EVICT* or *KILL* event of a task happens, we assume it is reset back to the initial state [13].

$$Size = (T_{finish} - T_{schedule}) \times R\_R\_CPU \times Cpub \qquad (19)$$

At each time epoch, the sum length of requests that arrive during each $[t - \tau, t)$ varies dramatically, as depicted in Fig. 9. We can clearly see that they are mixed up with highly bursty requests, and it is pretty difficult to discover any distribution law or to predict the trace-logs. We conduct 3 experiments and set $Cpub$ and $Cpri$ respectively the same as simulation Cases 1–3 in Table 2. The example of a convergence curve, $\theta(N\_W\_Pri)$ in real case2, is plotted in Fig. 10. Reward comparisons of the three cases are separately described in Figs. 11, 12 and 13.

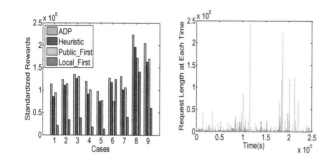

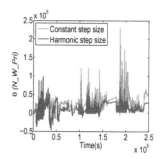

Fig. 8. Overall reward comparisons in all cases

Fig. 9. Request statistics for Google trace-logs

Fig. 10. Convergence of $\theta(N\_W\_Pri)$ in real case 2

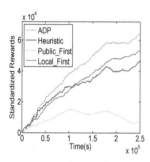

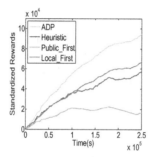

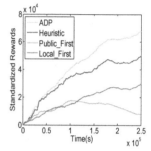

Fig. 11. Real-case1 reward comparison

Fig. 12. Real-case2 reward comparison

Fig. 13. Real-case3 reward comparison

5 Conclusion

In this paper, we investigate the long-term task scheduling problem with Diff-Serv in hybrid clouds. We firstly build a Diff-Serv task scheduling model, construct an SDP problem and then analyze its ingredients. With reference to reinforcement learning theory, we put forward an ADP algorithm, do comprehensive experiments and comparisons to evaluate the algorithm. Experiment results demonstrate that this work provides an effective and efficient way to handle comprehensive task scheduling problems in hybrid clouds, and the heuristic, public and local first algorithms may only be effective under specific situations.

Acknowledgments. This work is supported by the National Natural Science Foundation of China (No. 61472199 and No. 61370132).

References

1. Amazon_Web_Services: Aws auto scaling user guide. http://docs.aws.amazon.com/autoscaling/latest/userguide/as-dg.pdf

2. Calheiros, R.N., Buyya, R.: Cost-effective provisioning and scheduling of deadline-constrained applications in hybrid clouds (2012)
3. Google: Cloud trace-logs. http://code.google.com/p/googleclusterdata/wiki
4. Internetwatch: online-shopping. https://www.chinainternetwatch.com/19280/singles-day-top-categories-2016/
5. Moreno, I.S., Garraghan, P., Townend, P., Xu, J.: An approach for characterizing workloads in Google cloud to derive realistic resource utilization models. In: IEEE Seventh International Symposium on Service-Oriented System Engineering, pp. 49–60 (2013)
6. Niu, Y., Luo, B., Liu, F., Liu, J.: When hybrid cloud meets flash crowd: towards cost-effective service provisioning. In: IEEE INFOCOM 2015 - IEEE Conference on Computer Communications, pp. 1044–1052 (2015)
7. Ousterhout, K., Wendell, P., Zaharia, M., Stoica, I.: Sparrow: distributed, low latency scheduling. In: Twenty-Fourth ACM Symposium on Operating Systems Principles, pp. 69–84 (2013)
8. Peterson, L.L., Davie, B.S.: Computer Networks: A Systems Approach. Elsevier, Amsterdam (2007)
9. Powell, W.B.: Approximate Dynamic Programming: Solving the Curses of Dimensionality, vol. 703. Wiley, Hoboken (2007)
10. Powell, W.B.: What you should know about approximate dynamic programming. Nav. Res. Logistics **56**(3), 239–249 (2009)
11. Puterman, M.L.: Markov Decision Processes: Discrete Stochastic Dynamic Programming. Wiley, Hoboken (2014)
12. Ruben, V.D.B., Vanmechelen, K., Broeckhove, J.: Cost-efficient scheduling heuristics for deadline constrained workloads on hybrid clouds. In: IEEE Third International Conference on Cloud Computing Technology and Science, pp. 320–327 (2011)
13. Wang, J., Bao, W., Zhu, X., Yang, L.T., Xiang, Y.: FESTAL: fault-tolerant elastic scheduling algorithm for real-time tasks in virtualized clouds. IEEE Trans. Comput. **64**(9), 2545–2558 (2015)
14. Wikipedia: Cloud computing. https://en.wikipedia.org/wiki/Cloud_computing#hybrid_cloud
15. Wikipedia: Opportunity_cost. https://en.wikipedia.org/wiki/Opportunity_cost
16. WiseGEEK: What are the different types of network services? http://www.wisegeek.com/what-are-the-different-types-of-network-services.htm
17. Zuo, X., Zhang, G., Tan, W.: Self-adaptive learning PSO-based deadline constrained task scheduling for hybrid IAAS cloud. IEEE Trans. Autom. Sci. Eng. **11**(2), 564–573 (2014)

Online Cost-Aware Service Requests Scheduling in Hybrid Clouds for Cloud Bursting

Yanhua Cao[✉], Li Lu, Jiadi Yu, Shiyou Qian, Yanmin Zhu, Minglu Li,
Jian Cao, Zhong Wang, Juan Li, and Guangtao Xue

School of Electronic, Information and Electrical Engineering,
Shanghai Jiaotong University, Shanghai, China
{buttons,luli_jtu,jiadiyu,qshiyou,yzhu,mlli,cao-jian,
zhongwang,miletu,Xue-gt}@sjtu.edu.cn

Abstract. The hybrid cloud computing model has been attracting considerable attention in the past years. Due to security and controllability of private cloud, some special requests ask to be scheduled on private cloud, when requests are "bursting", the requests may be rejected because of the limited resources of private cloud. In this paper, we propose the online cost-aware service requests scheduling strategy in hybrid clouds (**OCS**) which could make suitable requests placement decisions real-time and minimize the cost of renting public cloud resources with a low rate of rejected requests. All service requests are divided into two categories, the special requests ask to be accepted on private cloud, and the normal requests are insensitive on private or public cloud. In addition, all requests arrive in random, without any prior knowledge of future arrivals. We transform the online model into a one-shot optimization problem by taking advantage of Lyapunov optimization techniques, then employ the optimal decay algorithm to solve the one-shot problem. The simulation results demonstrate that **OCS** is trade-off between cost and rejection rate, meanwhile it can let the resource utilization arbitrarily close to the optimum.

Keywords: Hybrid cloud · Service request scheduling · Cost-aware · Lyapunov optimization

1 Introduction

In recent years, the hybrid cloud model-private cloud plus public cloud [1] is getting more and more attention. It is reported by Gartner Inc, that it becomes an imperative trend of cloud computing, and more than half of large enterprises around the world will deploy hybrid clouds by the end of 2017 [2]. Generally, an enterprise has a private cloud and could elastically lease public clouds' resources when the private cloud resources are not enough. The service requests that users ask for Virtual Machines (VMs) to be hosted on the hybrid clouds, Among service requests, some are definitely specified to be processed on the private cloud with the consideration of data security. For brevity of discussion, these

© Springer International Publishing AG 2017
A. Bouguettaya et al. (Eds.): WISE 2017, Part I, LNCS 10569, pp. 259–274, 2017.
DOI: 10.1007/978-3-319-68783-4_18

service requests are termed as the special requests. The other requests can be scheduled on either private or public clouds, which are called the normal requests. When service requests burst, the private cloud is incapable of responding to all requests due to the limitation of resources. In this case, the special requests will be possibly rejected, while the normal requests can be handled by renting the resources of public cloud. Generally, the higher rejection rate of the special requests, the worse service performance of the hybrid cloud, so we propose the online cost-aware service requests scheduling strategy in hybrid clouds (**OCS**) which could make appropriate requests scheduling decisions with a low rejection rate while minimize the cost of renting public cloud resources.

About service requests scheduling, Quarati et al. [3] present a set of broker strategies for hybrid clouds aimed at the execution of various instances of the weather prediction WRF model subject to different user requirements and computational conditions. It focuses on scheduling four types requests based on the reservation of a static quota of resources of private cloud and assumes that the arrivals of requests are known. This assumption is not realistic and the reservation of a static quota may result in a waste of private cloud resources. Also, Hao et al. [4] design a generalized resources placement methodology that can work across different cloud architectures, with real-time requests arrivals and departures. Maguluri and Srikant [5] propose a load balancing and scheduling algorithm in clouds that is revenue optimal, unknown the duration of service request in advance. However, they are both not for requests scheduling in hybrid clouds, so they could not consider resource limitation of private cloud and the special requests. Our goal is to minimize the cost of hybrid clouds with an acceptable rejection rate.

Fully exploring the concept of "owning the base and renting the insufficient" [6] in hybrid cloud model remains challenging in practice. First, due to security and controllability of private cloud, some special requests ask to be scheduled on private cloud, when requests are "bursting", the requests may be rejected because of the limited resources of private cloud, it is a critical challenge that guarantees low rejection rate and minimizes the cost of renting public cloud resources at the same time. Second, requests arrivals and departures, the number of requests arrivals and type of virtual machine (VM) are all random, it is difficult to predict the requests arrival rate at next time-slot. Third, the original problem model is a time average problem during the whole time-slots, it is also critical challenge to transform the original problem into a one-shot optimization problem.

In this paper, we present the **OCS**. First, our objective is to minimizes the average cost of renting public cloud resources with an acceptable rejection rate. Then we take advantage of Lyapunov optimization techniques to transform the online model into a one-shot optimization problem. Finally, we employ an optimal decay algorithm for solving the one-shot optimization problem. The simulation results demonstrate that our scheduling strategy can let the resource utilization of the private cloud arbitrarily close to the optimum on hybrid cloud scenario and minimize the cost of renting public cloud resources as well as guarantee an acceptable rejection rate when meeting "bursting".

Our main contributions are:

- We formulate the service requests scheduling problem in hybrid clouds as an optimization problem, and consider the rejection rate as the constraint condition.
- We use the Lyapunov optimization techniques to transform the original optimization problem into a one-shot optimization problem and employ an optimal decay algorithm to solve the one-shot optimization problem.
- We conduct simulations and demonstrate our scheduling strategy, and the results show that can make the resource utilization of the private cloud arbitrarily close to the optimum and be trade-off between cost and rejection rate.

The remainder of this paper is organized as follows. We present the system definition and problem model (Sect. 2). We formulate the online requests scheduling problem (Sect. 3). We discuss the evaluation results (Sect. 4). Finally, we describe the related works and conclude the paper (Sects. 5 and 6).

2   System Model

In this section, we define the formulation of the service requests scheduling problem in hybrid clouds. And then, we build an online service requests model for the problem.

2.1   Problem Definition

We now formally define the problem of the online cost-aware service requests scheduling in hybrid clouds (**OCS**). All service requests in the strategy model are divided into two categories, the special requests ask to be processed on private cloud, and the normal requests are insensitive on private or public clouds. We assume that there are multiple requests arrivals in every time-slot, and number of requests, duration of each request, and number of VM in any time-slot are all random. t indexes the time-slot. Suppose that $Rv_i = (a_{i1t}, a_{i2t}, \cdots, a_{iht}; t_i)$ is a vector that denotes the special request only accepted by the private cloud, and $i \in \{1, 2, \cdots, n_t\}$, n_t is the number of special requests in the t-th time-slot. t_i is the duration of the request Rv_i. $Ra_j = (b_{j1t}, b_{j2t}, \cdots, b_{jht}; t_j)$ is a vector that denotes the normal request accepted by the private or public cloud, and $j \in \{1, 2, \cdots, m_t\}$, m_t is the number of normal requests in the t-th time-slot. t_j is the duration of the request Ra_j. We use T_{kt} to denote the available capacity of the k resource on private cloud in the t-th time-slot. Typical resources include CPU, memory, storage disk and I/O bandwidth, when $k = 1$ denotes the resource is CPU, in the same way, $k = 2$, $k = 3$ or $k = 4$ denotes the resource is memory, storage disk or I/O bandwidth respectively. We assume that there are h classes of VMs, the amount of resource k required to host a VM of class v is given by M_{vk}, and $v \in \{1, 2, \cdots, h\}$, a_{ivt} or b_{jvt} denotes the request Rv_i or Ra_j asks for the number of VM_v in the t-th time-slot respectively. We let the c_v denote the unit expenses of VM_v on public clouds in the t-th time-slot. Because the private cloud belongs to the inherent property in the enterprise, we assume that the expenses of requests accepted on the private cloud is zero, so we consider the expenses of renting public cloud resources as the cost for the problem. Y_{it} and Z_{jt} are zero-one decision variables. If $Y_{it} = 1$, the special request Rv_i is accepted in the t-th time-slot by the private cloud, and if $Y_{it} = 0$, the request Rv_i is rejected, and we assume that the threshold of rejection rate is α. When the $Z_{jt} = 1$, the request Ra_j is accepted in the t-th time-slot by the private cloud, and if the $Z_{jt} = 0$, the request Ra_j is accepted by public cloud. The objective of the hybrid clouds service provider is to distribute these service requests so as to minimize its total cost. Specifically, we want the service requests scheduling decisions to be made in a dynamic manner. The problem parameters and decision variables are defined in Table 1.

2.2   Problem Modeling

The problem can be formulated as the following IP model. The cost of the hybrid clouds R_t in the t-th time-slot.

$$R_t = \sum_{j=1}^{m_t} \sum_{v=1}^{h} (1 - Z_{jt}) b_{jvt} c_v t_j, \tag{1}$$

From the perspective of cost, as the cost metric of an enterprise, we minimize the average cost as objection function. So the optimization problem can be formulated as follows:

Minimize:

$$\lim_{T \to +\infty} \frac{1}{T} \sum_{t=0}^{T-1} R_t \tag{2}$$

subject to:

$$Y_{it}, Z_{jt} \in \{0, 1\} \tag{3}$$

$$\sum_{i=1}^{n_t} \sum_{v=1}^{h} Y_{it} a_{ivt} v_{vk} + \sum_{j=1}^{m_t} \sum_{v=1}^{h} Z_{jt} b_{jvt} v_{vk} \leq T_{kt}, \tag{4}$$

$$\lim_{T \to +\infty} \frac{\sum_{t=0}^{T-1} \sum_{i=1}^{n_t} Y_{it}}{\sum_{t=0}^{T-1} n_t} \geq 1 - \alpha \tag{5}$$

The objective function (2) represents the average cost in total T time-slots, and we get the service requests scheduling decisions in each time-slot by optimizing the problem (2). Constraint (3) gives the definitions of the decision variables. Constraint (4) ensures resources capacities of accepted requests by the private cloud are not beyond its spare capacities in each time-slot, $k \in \{1, 2, 3, 4\}$. Constraint (5) restricts average rejection rate of the special requests below the threshold value α. From the formulation, we can see that the problem is an online

Table 1. Key parameters

Notation	Definition
Rv_i	The vector denotes special request
Ra_j	The vector denotes normal request
Y_{it}	$Y_{it} \in \{0,1\}$, when $Y_{it} = 1$, the special request is accepted by the private cloud; And if $Y_{it} = 0$ the request is rejected
Z_{jt}	$Z_{jt} \in \{0,1\}$ when $Z_{jt} = 1$, the normal request is accepted by the private cloud; And if $Z_{jt} = 0$ the request is accepted by public cloud
h	The number of the type of VM
V	Weighting factor
α	The threshold of rejection rate
a_{ivt}	The special request asks for the number of VM_v in the t-th time-slot. $a_{jvt} \in [0, N]$
b_{jvt}	The normal request asks for the number of VM_v in the t-th time-slot. $b_{jvt} \in [0, N]$
t_i	Duration for service request
c_v	The unit expense of VM_v in public cloud
M_{vk}	The k resource capacity of VM_v
T_{kt}	The k resource total spare volume of private cloud in the t-th time-slot, $k \in \{1,2,3,4\}$
n_t	The number of the special requests in t-th time-slot
m_t	The number of the normal requests in t-th time-slot
Ns	The noted solution
d	The queue of decay sequences

service requests scheduling, and it is very difficult to solve the problem (2). We transform the original online model into a one-shot optimization problem by taking advantage of Lyapunov optimization techniques.

3   Online Cost-Aware Service Requests Scheduling Strategy

We consider that there are multiple requests in each time-slot, and the requests arrivals are random. The objective function is to minimize the average cost of renting the public cloud resources. The constraint (4) guarantees that the resource amount of accepted requests below the total capacity of spare k resource in the t-th time-slot, and the constraint (5) however implies that the average rejection rate of the special requests must not exceed the given threshold value α. During the $t + 1$-th time-slot, the below Eq. (6) records the number of the rejected special requests. As a result, a closer examination on the constraint suggests that the inequality (5) can be modeled as a queue $H(t)$. When the rejected number $H(t)$, that is the length of the queue, is stable. We can take advantage of Lyapunov optimization techniques [7] to transform the problem (2) into a one-shot optimization problem.

3.1   Problem Transformation by Lyapunov Optimization

Based on the above analysis, in order to capture the number of rejected special requests which is above the threshold value, we introduce the virtual queue $H(t)$, which is used to accumulate the part of the rejected number that exceeds the maximum permitted number in the t-th time-slot. Initially, we define $H(0) = 0$ and then update the queue according to Eq. (6), inequality (5) guarantees the virtual queue $H(t)$ is bounded. Whereas, if the virtual queue $H(t)$ is unbounded, more and more special requests are rejected, worse and worse service performance of the hybrid clouds is, under the circumstance, it is necessary to upgrade the infrastructure of private cloud for improving performance.

$$H(t+1) = max\{H(t) + n_t(1 - \alpha) - \sum_{i=1}^{n_t} Y_{it}, 0\} \tag{6}$$

Lemma 1. The queue $H(t)$ is stable. $\lim_{T\to+\infty} \frac{H(T)}{T} = 0$.

In order to save space, the proof process is omitted here. Taking advantage of Lyapunov optimization techniques, we consider the Lyapunov function $L(H(t))$ as follows:

$$L(H(t)) = \frac{1}{2}H^2(t). \tag{7}$$

The larger of the $H(t)$, the more number of rejected special requests, the constraint (5) is hard to meet, specifically, the average rejection rate exceeds the threshold α. Therefore, it is essential to keep the average rejection rate being below the threshold value. First we define the one-slot Lyapunov drift of $L(H(t))$ as follows:

$$\Delta(L(H(t))) = E\{L(H(t+1)) - L(H(t))|H(t)\}. \tag{8}$$

Lemma 2. For any $t \in [0, T - 1]$, given any possible control decision, the Lyapunov drift $\Delta(L(H(t)))$ can be deterministically bounded as follows:

$$\Delta(L(H(t))) \leq \frac{1}{2}n_t^2\alpha^2 + H(t)E[n_t(1 - \alpha) - \sum_{i=1}^{n_t} Y_{it}|H(t)]. \tag{9}$$

In order to save space, the proof process is omitted here. According to the Lemma 2, we get the upper bound of $\Delta(L(H(t)))$, after bounding the one-slot Lyapunov drift, we can minimize the upper bound to meet the constraint (5). If we add a penalty term VR_t to both sides of the inequality (9).

$$VR_t + \Delta(L(H(t))) \le \frac{1}{2}n_t^2\alpha^2 + VR_t$$

$$+ H(t)E[n_t(1-\alpha) - \sum_{i=1}^{n_t} Y_{it}|H(t)]. \tag{10}$$

So, the underlying objective of our optimal scheduling decisions is to minimize the bound on the following expression in one-shot optimization problem, that is the each time-slot problem. And the problem (2) can be transformed as follows: Minimize:

$$VR_t + H(t)[n_t(1-\alpha) - \sum_{i=1}^{n_t} Y_{it}] \tag{11}$$

s.t. constraints (3) and (4).

Until now, we have transformed the long-term optimization problem to the above optimization problem in each time-slot. During each time-slot $t \in \{0, 1, \cdots, T-1\}$, we minimize the amount of rejected special requests and the cost of renting public cloud resources. And V is a weighting factor to adjust our emphasis on the rejected amount and the cost of renting public cloud resources, by tuning it, we can choose which objective we need to underline.

3.2 The Analysis of the Relationship Between Cost of Renting and Rejected Requests Number

We now analyze the relationship between cost of renting and number of rejected requests, with any parameter V such that $V > 0$, we assume that R_{op} is the optimal cost of renting in theory during any time-slot $t \in \{0, 1, \cdots, T-1\}$.

$$\frac{1}{T}\sum_{t=0}^{T-1} R_t = R_{op}. \tag{12}$$

Based on the constraint (5), we have

$$\frac{\sum_{t=0}^{T-1}\sum_{i=1}^{n_t}(1-Y_{it})}{\sum_{t=0}^{T-1} n_t} \le \alpha. \tag{13}$$

Due to α is an upper bound of inequality (13), there must exist a β, and $\beta > 0$, which can make the inequality (13) to the following form

$$\frac{\sum_{t=0}^{T-1}\sum_{i=1}^{n_t}(1-Y_{it})}{\sum_{t=0}^{T-1} n_t} \le 1 - \alpha - \beta. \tag{14}$$

The problem (11) implies that we should make decisions to minimize right side of inequality (10). By applying R_{op} to inequality (10), and summing up the inequality (10) over time-slot $t \in \{0, 1, \cdots, T-1\}$, and then dividing both sides by T, we have

$$\frac{V}{T} \sum_{t=0}^{T-1} R_t + \frac{L(H(T)) - L(H(0))}{T}$$

$$\le \frac{1}{2} n_t^2 \alpha^2 + V R_{op} - \frac{\beta}{T} \sum_{t=0}^{T-1} H(t).$$

(15)

Due to the $H(t)$ stability, we get $\lim_{T \to \infty} \frac{H(t)}{T} = 0$,
$H(t)$ is bounded, so $\lim_{T \to \infty} \frac{L(H(t)) - L(H(0))}{T} = 0$.
Due to the $-\frac{\beta}{T} \sum_{t=0}^{T-1} H(t) < 0$, we have

$$\lim_{T \to \infty} \frac{1}{T} \sum_{t=0}^{T-1} R_t \le \frac{n_t^2 \alpha^2}{2V} + R_{op}.$$

(16)

Meanwhile, note that $\frac{V}{T} \sum_{t=0}^{T-1} R_t > 0$, we have

$$\lim_{T \to \infty} \frac{1}{T} \sum_{t=0}^{T-1} H_t \le \frac{\frac{1}{2} n_t^2 \alpha^2 + V R_{op}}{\beta}.$$

(17)

The inequalities (16) and (17) prove that the relationship cost of renting and number of rejected requests is tradeoff. The inequality (16) shows that the average cost of renting under the online service requests scheduling can be pushed closer to the optimum value R_{op} when the V is very large. However, if the V is too large, the $H(t)$ will be unstable, which means that the rejection rate of special requests exceeds the threshold value α. And if the V is too small, the cost of renting public clouds become very large. So within the above online service requests scheduling strategy, tuning the parameter V such that $V > 0$ for all $t \in \{0, 1, \cdots, T - 1\}$, we can minimize both the cost of renting and the number of rejected requests.

3.3   Optimal Decay Algorithm for the Zero-One Integer Linear Programming

Toward the problem (11), it is a representative zero-one integer linear programming problem. Generally, solving the problem by the enumeration method or branch and bound method [8], there are some special methods for the problem, such as for the assignment problem with the Hungarian method. However, there are unknown number of requests in each time-slot for our problem, the performance of solving the problem lies on the amount of requests in each time-slot. When the amount is very huge, the performance of these traditional methods of solving the problem is not good and the algorithm complexity is $O(2^n)$ (n is the number of decision variables). So it is crucial to improve the speed of solving the problem (11). Zhao et al. [9] proposes cards-flipping algorithm for Zero-one integer linear programming, which can examine the former S_i possible solutions in turn and without traversing the objection function values of all possible solutions, and if the S_i-th is the first feasible solution, the solution is optimal, without any calculation for the rest of $2^n - S_i$ possible solutions, the total amount of calculations is S_i. Based on [9], we propose the optimal decay algorithm (**ODA**) for the problem (11). The processes of solving the problem as follows: Firstly, getting the optimal solution of objective function without considering any constraint conditions, this is the loosest solution, we can call it as the noted solution Ns. And secondly, validating the noted solution whether meets the constraints (3) and (4). If not, we get a new noted solution Ns in terms of decay sequence d_0 with minimum decay value in the decay queue d instead of the pre-noted solution. The last, validating it whether satisfies the constraints (3) and (4) or not, if so, the new Ns is the optimal solution, if not, continuing to validate the next new noted solution according to the new d_0 until we find the optimal solution. Refer to Algorithm 1 for details.

d is the queue of decay sequences, and decay sequence is a sequence of 0 and 1, the initial d is $\{(0, \cdots, 0, 1)\}$. Such as there are three decision variables x_1, x_2, x_3 and their coefficients are a_1, a_2, a_3 respective in a minimization problem, if $|a_2| \geq |a_1| \geq |a_3|$, the Cs is the indexes sequence of the sorted decision variables, $Cs = (2, 1, 3)$. If a decay sequence $d_i = (0, 1, 1)$, its decay value $dv_i = |a_1| + |a_3|$. The d_i can be transformed into $d_{i+1} = (1, 0, 1)$ and $d_{i+2} = (1, 1, 1)$ [9].

Lemma 3. The noted solution is the optimal solution.

Proof. **The objective function is minimized**
Assuming the noted solution is Ns, and the i-th element is Ns_i, and a_i is its coefficient, R_n is the objection function value of the problem (11).

$$Ns_i = \begin{cases} 0 & a_i > 0 \\ 1 & a_i < 0. \end{cases} \tag{18}$$

Ns' is the any solution (not noted solution), and the i-th element is Ns'_i, R is the objection function value of any solution. We assume the $Ns_i \neq Ns'_i$. (otherwise, the any solution is noted solution)

① if $a_i > 0$:
In term of the equation (18), the $Ns_i = 0$;
Due to the $Ns_i \neq Ns'_i$, the $Ns'_i = 1$;
So $R_n - R = a_i Ns_i - a_i Ns'_i = 0 - a_i = -|a_i|$.

② if $a_i < 0$:
In term of the equation (18), the $Ns_i = 1$;
Due to the $Ns_i \neq Ns'_i$, the $Ns'_i = 0$;
So $R_n - R = a_i Ns_i - a_i Ns'_i = a_i - 0 = -|a_i|$.
In summary, in term of ① and ②, the $R_n < R$.

For the any $Ns_i \neq Ns'_i$ the objection function value of any solution is always greater than the noted solution's, so the objection function value of noted solution is the minimum value, and the noted solution is the optimal solution.

Algorithm 1. Optimal Decay Algorithm for the 0-1 Integer Linear Programming

---

Input: T_{kt}, Rv_i, Ra_j, $1 \le i \le n_t$, $1 \le j \le m_t$;

Output: $Ns = (Y_{1t}, \cdots, Y_{n_t}, Z_{1t}, \cdots, Z_{m_t})$;

1: get the initial d and the noted solution
    $Ns = [1, 1, \cdots, 1]$
2: **if** Ns meets Constraint (3) and (4) **then**
3:   Ns is the optimal solution
4: **end if**
5: **if** Ns does not meet constraint (3) and (4) **then**
6:   **repeat**
7:      **for** $i = 0$ to $n_t + m_t$ **do**
8:        **if** $(d_0[i] == 1)$ // d_0 is the minimization decay sequence in the queue d
9:        $ind = Cs[i]$ // Cs is the index sequence of the sorted decision variable
10:      $Ns[ind] = abs(Ns[ind] - 1)$ // ind is the index of decision variable
11:     **end for**
12:    **if** Ns does not meet Constraint (3) and (4) **then**
13:      transform d_0 into d_1 and d_2, and then delete d_0 from the queue d
14:      calculate decay values dv_1 and dv_2 and insert them according to their decay
        values ascending into d
15:    **end if**
16:   **until** Ns meets Constraint (3) and (4) or d is empty
17:   **if** Ns meets Constraint (3) and (4) **then**
18:     Ns is the optimal solution
19:   **end if**
20:   **if** d is empty **then**
21:     the problem (14) no solution
22:   **end if**
23: **end if**

Lemma 4: The decay value of d_0 is minimum during no verified decay sequences every time.

Proof. Assuming any decay sequence
$d_i = (q_1, \cdots, q_{i-1}, 1, 1, 0, \cdots, q_n)$,
$q_i = 1$, $q_{i+1} = 1$, the others are 0, $w = (b_1, \cdots, b_n)$,
$b_1 \ge \cdots \ge b_n \ge 0$, $i \in \{1, \cdots, n\}$, $\odot$ is vector inner product.
the d_i is transformed into the below two decay sequences d_{i+1} and d_{i+2},
$d_{i+1} = (0, \cdots, 1, 0, 1, 0, \cdots, 0)$,
$d_{i+2} = (0, \cdots, 1, 1, 1, 0, \cdots, 0)$,
$dv_{i+1} - dv_i = w \odot d_{i+1} - w \odot d_i = b_{i+1} - b_i$,
so $dv_{i+1} \ge dv_i$. $dv_{i+2} - dv_i = w \odot d_{i+2} - w \odot d_i = b_{i-1}$,
so $dv_{i+2} \ge dv_i$.

Due to the decay value of d_0 is minimum in the queue d and the decay values of other decay sequences are not in the queue d and are greater than any decay values in d, so the decay value of d_0 is minimum during no verified decay sequences every time.

In term of Algorithm 1, the total number of decay sequences is 2^{n_t} in the t-th time-slot, so this algorithm can traverse all possible solutions in the worst case. Based on the Lemmas 3 and 4, we know that the algorithm do not need to traverse all possible solutions, and it only calculate and validate the former S_i possible solutions, and without any calculations for the rest of $2^{n_t} - S_i$ possible solutions, so the computational complexity is $O(S_i)$ and reliability is equivalent to enumeration method.

4   Performance Evaluation

In this section, we conduct the simulations to evaluate the performance of the online cost-aware service requests scheduling in hybrid clouds. due to without any priori knowledge of future request arrival pattern, amount of resources and types, and service time, so we use synthetic resources requests generated by statistical function.

4.1   Simulation Setup

We set the capacity of private cloud: 100 vCPUs, 300 Gb RAM, 2000 Gb disk storage, and the prices of VMs are based on Aliyun's pricing model.

Requests Arrival Pattern: We assume the resource requests arrival pattern follows the poisson arrival; The amount of resources and types, and service time are all generated randomly by statistical function. In addition, we also design two other request arrival patterns, one is that the request arrival pattern follows an uniform distribution [10] and the amount of VM and service time of every request also follow an uniform distribution, the other is that the number of requests arrivals is constant and the amount of VMs and service time of every request are also constant [3].

Two Requests Scheduling Strategies for Comparison: In order to explore the performance of the online service requests scheduling in hybrid clouds. We set two simple strategies to be compared with the online cost-aware service requests scheduling in hybrid clouds (**OCS**). One is the "public cloud first", which tends to response all the normal requests to public cloud. The other strategy is the "private cloud first" strategy, it allows to respond the all resource requests to the private cloud as many as possible, which is a totally contrary strategy to "public cloud first". By comparing **OCS** with such two extreme strategies, we can analyze the performance of **OCS** more clearly.

4.2   Performance Evaluation

Cost of Renting. We apply three different optimizing strategies to three request arrival patterns. And the results are displayed in Fig. 1. The request arrival pattern is introduced in Sect. 4.1. In Fig. 1, we can observe that the average cost of the "public cloud first" strategy is the maximum among the three strategies, because all requests except the special requests are all scheduled to the public cloud, the cost of all normal requests is maximum. While the "private cloud first" strategy can enable the cost is minimum, because of all requests scheduled to the private cloud first if the resources of private cloud are enough for these requests, and the cost of resources of the private cloud is free due to the private cloud with enterprise's own property, so it is the optimal strategy of cost minimization. However, in our problem, the "private cloud first" strategy is not good because it makes more number of rejected requests in Fig. 2. Based on Figs. 1 and 2, although it seems that **OCS** is not outstanding in either cost of renting or number of rejected requests, we can discover that **OCS** is trying to make a tradeoff between the cost of renting and number of rejected requests.

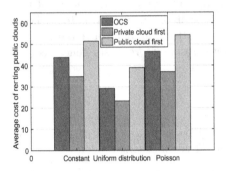
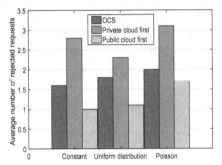

Fig. 1. Average cost of renting public cloud resources

Fig. 2. Average number of rejected requests

Number of Rejected Requests. As illustrated in Fig. 2, the average number of "public cloud first" strategy is minimum, because the "public cloud first" strategy makes the private cloud only to accept the special requests but the normal requests are all scheduled to the public cloud. On the contrary, the average number of "private cloud first" strategy is maximum among the three strategies, due to all requests first scheduled to the private cloud, the resources of private cloud are quickly used up, leading to more special requests rejected. If the objective of problem is to minimize rejection rate, the "public cloud first" is optimal strategy. However, our objective is to minimize the cost of renting public cloud resources, and the average number and cost of renting of **OCS** strategy are both between the above two strategies. Based on Figs. 1 and 2, the average cost of "private cloud first" strategy is minimum but the average number of rejected requests is maximum, and the number of "public cloud first" strategy is minimum yet the average cost of renting public cloud resources is maximum. However, the cost and number of **OCS** are both between the above two strategies, it seems that to cost a little bit more for less number of rejected requests, the **OCS** is searching the balancing point of service quality between the "public cloud first" and "private cloud first".

Resource Utilization of Private Cloud. Figures 3, 4 and 5 show the resource utilization of "public cloud first" strategy is almost always lower than the other

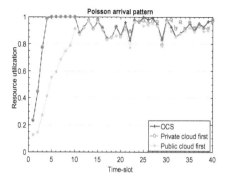

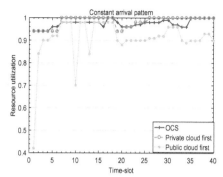

Fig. 3. Resource utilization with poisson arrival pattern

Fig. 4. Resource utilization with uniform distribution arrival pattern

two strategies over time-slots, although the strategy guarantees the lowest rejection rate, it will cause the waste of resources of private cloud. On the contrary, the resource utilization of "private cloud first" strategy is almost highest among the three strategies over time-slots, it is very clearly that any requests are scheduled to the private cloud first if the resources of private cloud are enough. However, the resource utilization of **OCS** strategy is nearly the same with "private cloud first" strategy over time-slots, the **OCS** strategy can minimize the cost of renting public cloud resources under guaranteeing the low rejection rate, meanwhile it can make full use of the private cloud resources. **The cost of renting and number of rejected requests tradeoff.** To analysis the trade-off between the cost of renting public cloud resources and number of rejected requests, we vary the parameter V to choose the metric we want to emphasize, average cost of renting and average number. In Fig. 6, by tuning the value of V to a small one, we can observe that the average number of rejected requests is small while the cost of renting is large. Meanwhile, by setting a large value of V, it brings markedly increase of the average number and decrease of the average cost of renting. Furthermore, when the average cost drops, the average number grows remarkably. And choosing a value of V represents how much we emphasize the average cost compared to the average number of rejected requests and finally get most suitable V to trade-off the average cost of renting public cloud resources and number of rejected requests.

5 Related Work

For scheduling in hybrid clouds, there are main two aspects, scheduling service requests and application tasks. And some works consider the problem of service requests scheduling in clouds [3–5]. We discuss the service requests scheduling problem in this paper, and we recognize that only the work [3] that is similar to our work. [3] takes four types of user resources requests into consideration in hybrid clouds, our study differs in at least three important aspects. First,

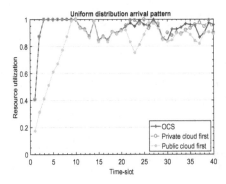

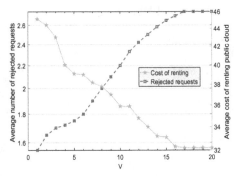

Fig. 5. Resource utilization with constant arrival pattern

Fig. 6. The cost of renting and number of rejected requests trade-off

we model the user service requests scheduling in the hybrid clouds and employ Lyapunov optimization techniques to transform the objection function. Second the model is suitable for all kinds of users service requests. Third, our work proposes an online service requests to schedule the user requests in hybrid clouds. However, [3] only analyzes the historic data and the average number of rejected requests, and the rejection rate can not be guaranteed. Currently, there are a lot of works [11–13] about the application tasks scheduling in clouds, and [13] takes straggler tasks into consideration the tasks scheduling in cloud, our next work will study the straggler tasks in hybrid clouds.

A majority of the existing literatures leverage Lyapunov optimization techniques. For example, with respect to cloud bursting, [6] takes advantage of Lyapunov optimization techniques to design an online decision algorithm for requests scheduling. [14] proposes dynamic service migration and workload scheduling in edge-clouds, and models a sequential decision making markov decision problem, then use the technique of Lyapunov optimization over renewals. In addition, some works [15,16] apply the auction scheduling method in cloud. In this paper, we use the techniques of Lyapunov optimization to transform the average time problem to optimization problem in each time-slot.

6   Conclusion

With increasing the hybrid cloud model in enterprises, due to the limited resources of private cloud, it is necessary to consider the service requests scheduling between private cloud and public cloud. In this paper, we design an online algorithm to schedule users' service requests, meanwhile, the method minimizes the cost of renting public cloud resources with an acceptable rejection rate. By applying Lyapunov optimization techniques, we can make scheduling decisions for all requests in each time-slot. Furthermore, we employ the optimal decay algorithm for the zero-one integer linear programming to solve the optimization problem. The simulation results demonstrate that our scheduling strategy can make the resource utilization of the private cloud arbitrarily close to the optimum and be trade-off between cost of renting public cloud resources and rejection rate of the special requests.

Acknowledgments. This work has been funded by High-Tech Research and Development Program of China under grant No. 2015AA01A202 and partly supported by NSFC (Nos. 61170238, 61420106010).

References

1. Hoseinyfarahabady, M.R., Samani, H.R., Leslie, L.M., Lee, Y.C., Zomaya, A.Y.: Handling uncertainty: pareto-efficient BoT scheduling on hybrid clouds. In: 2013 42nd International Conference on Parallel Processing, pp. 419–428. IEEE (2013)
2. Sill, A.: Socioeconomics of cloud standards. IEEE Cloud Comput. **2**(3), 8–11 (2015)
3. Quarati, A., Danovaro, E., Galizia, A., Clematis, A., D'Agostino, D., Parodi, A.: Scheduling strategies for enabling meteorological simulation on hybrid clouds. J. Comput. Appl. Math. **273**, 438–451 (2015)
4. Hao, F., Kodialam, M., Lakshman, T., Mukherjee, S.: Online allocation of virtual machines in a distributed cloud. In: IEEE INFOCOM 2014-IEEE Conference on Computer Communications, pp. 10–18. IEEE (2014)
5. Maguluri, S.T., Srikant, R.: Scheduling jobs with unknown duration in clouds. In: INFOCOM, 2013 Proceedings IEEE, pp. 1887–1895 (2014)
6. Niu, Y., Luo, B., Liu, F., Liu, J., Li, B.: When hybrid cloud meets flash crowd: towards cost-effective service provisioning. In: 2015 IEEE Conference on Computer Communications (INFOCOM), pp. 1044–1052. IEEE (2015)
7. Neely, M.J.: Stochastic network optimization with application to communication and queueing systems. Synth. Lect. Commun. Netw. **3**(1), 1–211 (2010)
8. Mavrotas, G., Diakoulaki, D.: A branch and bound algorithm for mixed zero-one multiple objective linear programming. Eur. J. Oper. Res. **107**(3), 530–541 (1998)
9. Zhao, N., Wci-Jian, M.I., Wang, D.S.: A new algorithm for zero-one integer linear programming. Oper. Res. Manag. Sci. **21**(5), 111–118 (2012)
10. Zheng, L., Joe-Wong, C., Tan, C.W., Chiang, M., Wang, X.: How to bid the cloud. ACM SIGCOMM Comput. Commun. Rev. **45**(4), 71–84 (2015)
11. Dabbagh, M., Hamdaoui, B., Guizani, M., Rayes, A.: Efficient datacenter resource utilization through cloud resource overcommitment. In: 2015 IEEE Conference on Computer Communications Workshops (INFOCOM WKSHPS), pp. 330–335. IEEE (2015)
12. Huang, Z., Balasubramanian, B., Wang, M., Lan, T., Chiang, M., Tsang, D.H.: Need for speed: cora scheduler for optimizing completion-times in the cloud. In: 2015 IEEE Conference on Computer Communications (INFOCOM), pp. 891–899. IEEE (2015)
13. Ren, X., Ananthanarayanan, G., Wierman, A., Yu, M.: Hopper: decentralized speculation-aware cluster scheduling at scale. In: ACM Conference on Special Interest Group on Data Communication, pp. 379–392 (2015)
14. Urgaonkar, R., Wang, S., He, T., Zafer, M., Chan, K., Leung, K.K.: Dynamic service migration and workload scheduling in edge-clouds. Perform. Eval. **91**, 205–228 (2015)

15. Zhou, Z., Liu, F., Li, Z., Jin, H.: When smart grid meets geo-distributed cloud: an auction approach to datacenter demand response. In: 2015 IEEE Conference on Computer Communications (INFOCOM), pp. 2650–2658. IEEE (2015)
16. Zhang, X., Wu, C., Li, Z., Lau, F.C.M.: A truthful $(1-\varepsilon)$-optimal mechanism for on-demand cloud resource provisioning. In: 2015 IEEE Conference on Computer Communications (INFOCOM), pp. 1053–1061. IEEE (2015)

Adaptive Deployment of Service-Based Processes into Cloud Federations

Chahrazed Labba[1]([✉]), Nour Assy[2], Narjès Bellamine Ben Saoud[1], and Walid Gaaloul[3]

[1] ENSI-RIADI Laboratory, University of Manouba, Manouba, Tunisia
chahrazed.labba@ensi-uma.tn
[2] Eindhoven University of Technology, Eindhoven, The Netherlands
[3] Computer Science Department, Telecom SudParis, Paris, France

Abstract. Service-based processes represent compositions of software services that need to be properly executed by the resources offered by an IT infrastructure within a company. Due to the dynamic changes in their QoS requirements, service-based processes are constantly evolving and demanding new resources. To ensure agility and support more flexibility, it is common today for enterprises to outsource their service-based processes to cloud environments and recently to cloud federations. The main challenge in this regard is to ensure an optimal allocation of cloud resources to process services overtime. In fact, given the diversity of the resources within a federation and the continuous changes of the process QoS needs, the reallocation of cloud resources to process services may result in high computing costs and an increase in the communication overheads. In this paper, we propose a novel adaptive resource allocation approach which can estimate and optimize the final deployment costs. We use agent-based systems to simulate processes' enactment. To cope with the services' QoS changes and dynamically adapt the initial deployment, we propose an extended version of the Pairwise-Movement Fiduccia-Mattheyses (E-PMFM) partitioning algorithm. Our experimental results highlight the efficiency of E-PMFM algorithm and show that deployment costs are sensitive to the initial deployment and the used partitioning algorithm.

Keywords: Cloud federation · Service-based process · Dynamic allocation · Estimation · Optimization

1 Introduction

Service-based processes (SP) are increasingly used for creating new added value to enterprises' business processes by composing services and invoking them dynamically. In composed services, it is crucial to cope with the dynamic QoS changes that can arise at run-time. For example, during peak hours, service component requirements may spike, and as a consequence, additional computational resources need to be allocated. In order to ensure agility, support more flexibility

© Springer International Publishing AG 2017
A. Bouguettaya et al. (Eds.): WISE 2017, Part I, LNCS 10569, pp. 275–289, 2017.
DOI: 10.1007/978-3-319-68783-4_19

and enable greater scalability, it is common today for enterprises to outsource their SPs to cloud environments, and recently, to cloud federations. A cloud federation consists of two or more interconnected cloud providers. It provides a higher scalable infrastructure that allows to optimize resources' delivery and meet the business requirements within a given organization.

To make an efficient use of cloud federations, enterprises need to find ways to efficiently allocate the cloud resources to their SPs services that result in reduced deployment costs. Since services' requirements may be subject to continuous changes during run-time, the initial allocation may need to be efficiently adapted by allocating additional resources and/or de-allocating existing ones. However, given the diversity of the cloud resources within a cloud federation, process services' reallocation may result in high computing costs and an increase in the communication overheads. Consequently, selecting the right resources to achieve an efficient deployment is a challenging task. Efficiency is expressed in terms of reduced overall deployment costs while fulfilling the process QoS requirements.

In the frame of this work, we propose a novel adaptive resource allocation approach to deploy efficiently SPs to cloud federations. Our approach takes as input (i) an SP with dynamic QoS needs, (ii) a cloud federation with its diverse resources, and (iii) a set of external events that impacts the services' QoS requirements overtime and induces demands for new resource allocation. It then estimates and optimizes the overall deployment costs by allocating the right low-cost resources while maintaining a reduced communication overhead. The estimation and optimization are dynamically performed whenever changes in the QoS requirements occur during run-time. To achieve our goal, we use Agent-Based Systems (ABS) to simulate process enactment. We map the task of finding an efficient deployment to a graph partitioning problem. We propose an extension of the Pairwise-Movement Fiduccia-Mattheyses (E-PMFM) partitioning algorithm to cope with the problem of finding a deployment with minimal costs. E-PMFM allows to allocate low-cost cloud resources that fulfill the process QoS needs and reduce the communication overheads between the pairs of interacting services. As communication is a key challenge in cloud environments, E-PMFM focuses mainly on minimizing the amount of the transferred data between the allocated resources within the cloud federation. The implemented approach highlights the efficiency of the proposed E-PMFM algorithm and shows that the final deployment costs of an SP with dynamic QoS needs are sensitive to multiple factors such as the initial deployment and the used partitioning algorithm. Thus far, resource allocation is addressed for SPs from a control flow perspective within either the same cloud provider [6] or the hybrid cloud [5], to improve mainly resource utilization. To the best of our knowledge, this is the first approach that focuses on SPs with dynamic QoS needs and takes into account the features of using an inter-cloud environment.

The rest of the paper is organized as follows: Sect. 2 presents a motivation example. Section 3 provides an overview of the basic concepts. Section 4 presents the proposed deployment approach as well as a detailed description of the extended PMFM. Section 5 describes the experimental results. Related work and Conclusions are presented respectively in Sects. 6 and 7.

2   Motivating Example

In this section we present a running example that will be used to illustrate our approach. We consider the SP shown in Fig. 1 that needs to be deployed in the cloud federation shown in Fig. 2. The SP is modeled using the graphical standard language BPMN. The initial QoS requirements (at time t = t0) of some services are attached to them. The changed QoS requirements during the execution of the process (at time t = t1) are also attached in blue. For example, initially (at t = t0), the service S1 needs a cloud resource with minimum levels of Security(S) $S = 50\%$ and Availability (A) $A = 50\%$. It also needs to be assigned to a Virtual Machine (VM) with a minimal bandwidth $b = 100$ $Mbit/s$, a minimal memory $ram = 2\,GB$ and a minimal CPU $cpu = 5\,Mb$. The cloud federation shown in Fig. 2 interconnects two cloud providers $cloud_1$ and $cloud_2$. Each provider is characterized by a set of hosted virtual machines as well as minimum of security and availability levels. Further, each VM is characterized by its compute price, maximum number of instances and a set of capabilities including cpu, ram and bandwidth. For example, $cloud_2$ has (1) three VMs $\{vm_1, vm_2, vm_3\}$; (2) S level of 60% and; (3) A level of 50%. The VM vm_3 of $cloud_2$ has a compute price of 0,07\$/hour and can host only 2 instances. Moreover, it has a ram of 7 Gb, a 50 Mb of cpu and a bandwidth of 150Mb/S. Further the cloud federation has a maximal bandwidth of 60 Gb/S with 5\$/hour as a price.

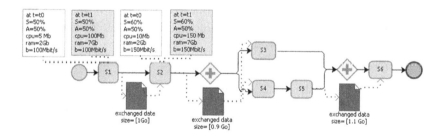

Fig. 1. An SP with initial QoS requirements (at t = t0) and changed QoS requirements during execution (at t = t1) (Color figure online)

A proper SP deployment to a cloud federation consists of (i) starting from an efficient initial allocation of cloud resources to the process services that minimizes the deployment costs and (ii) reacting to the process dynamics overtime, by leasing the right cloud resources and deploying services over them while minimizing the overall costs. Assume that, as initial efficient deployment, both the services S1 and S2 are assigned to vm_3 within $cloud_2$. At (t = t1), an external event occurs; both services go through changes in their QoS requirements (shown

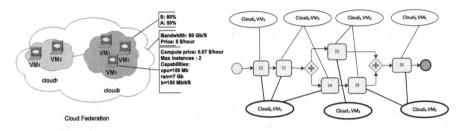

Fig. 2. Cloud federation (Color figure online)

Fig. 3. Possible resources allocation of the SP at time $(t = t1)$ within the cloud federation

in the blue annotations attached to the services in Fig. 1) which exceed the capabilities of the current allocated resource. Consequently new resource allocations are required. At this stage, multiple feasible deployments with different costs are possible. An example of two possible process deployments is illustrated in Fig. 3. The thick and thin ellipsis correspond to the first and second possible deployments respectively. However, the first solution costs (10 $), whereas the second is charged for (7 $). Thus, among all the possible deployments, we are interested in finding the cost-efficient solution that minimizes the compute and data transfer costs.

The next section provides an overview of the basic concepts. Afterwards, the proposed approach to solve the aforementioned issue is presented.

3   Preliminaries and Background

In this section, we provide an overview of the basic concepts introduced in the frame of this work including (i) cloud federation, (ii) dynamic business process and (iii) adaptive deployment of SP.

3.1   Cloud Federation

A federation refers to the union of two or more interconnected cloud providers via a given bandwidth. Each cloud provider offers minimum of security and availability levels for the services. Further, it provides a set of virtual machines characterized by compute and data transfer costs as well as a set of capabilities including cpu, memory and bandwidth. Similarly to [10], we define cloud provider and cloud federation respectively, in Definitions 1 and 2.

Definition 1: *(Cloud Provider)* A cloud provider C is defined as a pair $C = (Cap_C, V_M)$, where:

- $Cap_C = (S, Av)$ represents the competences of C in terms of Security $S \in \mathbb{R}_{>0}$ and Availability $Av \in \mathbb{R}_{>0}$.

- $V_M = \{VM_i \mid i \geq 1\}$ represents the sets of virtual machines provided within C. Each VM is defined as a tuple $VM = (pc, pd, mr, mc, mb)$ where the terms correspond respectively to the compute price (\$/hour), data transfer price (\$/GB), maximum memory capacity (Gb), maximum compute capacity (Gb), and the maximum bandwidth capacity (Mb/s).

Definition 2: *(Cloud Federation)* A cloud federation is defined as $F = (Cap_F, C_F)$ where:

- $Cap_F = (B, pdf)$ is the characteristics of F in terms of bandwidth (Mb/s) and price (\$/hour).
- $C_F = \{C_i \mid i \geq 1\}$ represents the set of the interconnected cloud providers within F.

An example of a cloud federation $F = (Cap_F, C_F)$ is depicted in Fig. 2, where $Cap_F = (60Gb/s, 5\$/hour)$ and $C_F = \{C_1, C_2\}$; $C_2 = (Cap_C, V_M)$ where $Cap_C = (60\%, 50\%)$ and $V_M = \{VM_1, VM_2, VM_3\}$; $VM_3 = (0.07 \$/hour, 0.02 \$/hour, 2, 150\,Mb, 7\,Gb, 150\,Mbit/s)$.

3.2   Service-Based Process

In our work, we define an SP as a process with varying demands for computational cloud resources overtime caused by the variations in the services' QoS requirements. We use ABS [3] to model and simulate the deployment of SPs to cloud federations. An ABS represents a set of agents[1] in interactions according to the sequencing logic defined within the process. An agent is an active entity (real/virtual), having a set of abilities and skills to achieve its goals. It has a set of QoS requirements that need to be fulfilled during the deployment and that changes dynamically due to the emergence of a set of external events. An external event occurs at a specific time t during the execution of an SP and affects the QoS requirements of one or multiple agents. For example, during peak hours, an agent may require additional memory and computing resources.

There exist many design methodologies [9,12] to represent an ABS as well as techniques to map from different process modeling notations (e.g. BPMN) to ABS models [2,7]. The mapping to ABS models is out of scope of this paper. In this work, we assume that an ABS is constructed given a mapping function. To make the approach as general as possible, we give an abstracted representation of an ABS, referred to as *agent-based graph*, that will be used by our algorithm to derive an efficient deployment.

Let $\mathscr{A}$ be a universe of agents. The formal definitions of agent QoS, external events' changes and agent-based graph are given in Definitions 3, 4 and 5 respectively.

Definition 3: *(Agent QoS)* Let $A \in \mathscr{A}$ be an agent. The QoS needs of A are defined as a tuple $QoS = (s, a, rr, rc, rd)$ where s represents the minimal required

---

[1] An agent is equivalent to a service within the process. In the remainder, we will use the term agent instead of service.

security level (%), a depicts the minimal required availability (%), rr, rc and rd are respectively the minimal required memory (Gb), computing capacity (Gb) and bandwidth (Mb/s). We use the shorthand QoS_A to refer to the QoS of the agent A.

For example, in the SP illustrated in Fig. 1, at time (t = t0) QoS_{S1}=(50%, 50%, 5 Mb, 2 Gb, 100 Mbit/s).

Definition 4: (External event change) A change in the QoS requirements of an agent $A \in \mathscr{A}$ caused by an external event is defined as a tuple $\delta = (t, QoS')$ where:

- t represents the time unit at which the change should be applied;
- $QoS' = (s', a', rr', rc', rd')$ are the new QoS requirements.

We use the shorthand $\delta_{(A,t)}$ to refer to the external event change that affects the agent A at time t. We denote by Δ_A the set of external events' changes that occur for the service A during the execution of an SP.

For example, in Fig. 1, $\delta_{(A_1,t1)} = (t1, QoS'_A)$ where QoS'_A=(50%,50%,100 Mb,7 Gb,100 Mbit/s) is the external event change applied to A at time $t1$.

Definition 5: Agent-Based Graph (ABG) An ABG is defined as a tuple $ABG = (Ag, I, Q_{Ag}, \Delta_{Ag}, W_I)$ where:

- $Ag \subseteq \mathscr{A}$ represents the non-empty set of agents composing the system;
- $I \subseteq Ag \times Ag$ represents the set of direct interactions between the agents. An interaction denotes the exchange of messages between the agents;
- $Q_{Ag} = \{QoS_A \mid A \in Ag\}$ is the set of agents' initial QoS requirements (at t=t0);
- $\Delta_{Ag} = \{\Delta_A \mid A \in Ag\}$ is the set of external events changes of the agents in the system;
- $W_i : I \to \mathbb{R}_{\geq 0}$ is a function that assigns for an interaction $i \in I$, a weight $w \in \mathbb{R}_{\geq 0}$ which denotes the amount of the exchanged data between two agents.

3.3   Adaptive Deployment of SP

The adaptive deployment of an SP represented as an ABS into a cloud federation consists of allocating dynamically for each agent in the ABS the appropriate VM within the federation, once an external event occurs.

Let $ABG = (Ag, I, Q_{Ag}, \Delta_{Ag}, W_I)$ be the agent-based graph of an ABS, $F = (Cap_F, C_F)$ be the cloud federation and $\xi = \{t_0, t_1, \ldots, t_k\}$ be the time units at which external events occur. The deployment of ABG into F is formally given in Definition 6.

Definition 6: Adaptive Process Deployment The adaptive process deployment of an ABG into F is a function $D : Ag \times \xi \to C_F \times V_M$ which is invoked at different time units corresponding to the occurrence of external events. It assigns for each agent $A \in Ag$ a virtual machine $VM \in V_M$ in the cloud $C \in C_F$. We denote by $\mathscr{D} = \{(A_1, D(A_1, t)), (A_2, D(A_2, t)), \ldots, (A_n, D(A_n, t))\}$ the set of agents and their allocated resources at time t.

For example, in the process deployment illustrated in Fig. 3, an external event occurs at $t = t1$. The deployment of S1 at t1 is defined as $D(S1, t1) = (C_2, vm2)$.

The deployment of an ABG into F needs to be cost-efficient, i.e. the distribution of the agents during runtime over the leased cloud resources should result in minimal compute and data transfer costs. In the following section, we detail our approach, that, given an ABG and a cloud federation, dynamically finds cost-efficient deployments.

4   Proposed Approach

In this section we present our approach for estimating and optimizing the adaptive deployment of an SP to a cloud federation.

4.1   Overview of the Approach

The main steps of the approach are presented in Algorithm 1. It takes as input an ABG, the cloud federation F, the number of steps required to complete the execution (t_{end} expressed in terms of a time unit) and the time units at which the external event changes occur (the set ξ)[2]. Once the simulation starts, the algorithm checks at each time unit if a new deployment is required (Lines 3–10). If this is the case, the simulation is suspended and our extended E-PMFM is invoked (Line 6 detailed in Sect. 4.2). The E-PMFM computes a cost-efficient deployment according to the changes in the agents' QoS requirements. Then, the resources are reallocated according to the newly computed deployment and the simulation is resumed (Lines 7–8). Since $t0 \in \xi$, the algorithm will also compute the initial deployment.

4.2   E-PMFM Partitioning Algorithm

Our objective is to assign each agent in the ABG to one of the VMs belonging to the federation that fulfills the QoS requirements. At the same time, we aim at finding a deployment in which the total cost of the exchanged data and the compute price between the set of allocated VMs is minimized. We see the problem as a graph partitioning problem in which the VMs of a federation can be represented as a weighted graph. The weight is a function of the amount of the transferred data. Our goal is to find a k partition of the graph where each partition represents the set of VMs within the same cloud that are allocated to some agents in the ABG. The communication costs between the VMs of one partition as well as between partitions should be minimal.

---

[2] Note that we also include t0 in ξ in order to be able to compute the initial deployment.

Algorithm 1. Pseudo code for estimating and optimizing SP cloud deployment

---

Input: ABG,F, t_{end}, $\xi = (t_1, t_2, .., t_k)$
Output: $\mathscr{D}$
1: Start the Simulation
2: $step_{sim} \leftarrow 0$ ▶ Simulation Step
3: **while** $step_{sim} \neq t_{end}$ **do**
4:    **if** $step_{sim} \in \xi$ **then**
5:       Suspend the Simulation
6:       $\mathscr{D} = E\text{-}PMFM(ABG, step_{sim})$
7:       Perform re-allocation according to $\mathscr{D}$
8:       Resume Simulation
9:    **end if**
10:   $step_{sim} \leftarrow step_{sim} + 1$
11: **end while**

To do so, we propose to extend the PMFM algorithm which was firstly introduced in [1]. PMFM focuses mainly on the use of the graph-based bi-partitioning algorithm Fiduccia-Mattheyses (FM) [4]. It takes as input a graph and provides as output a graph partitioned into k subsets such that the number of intra- and inter-edges interconnecting the partitions is minimal. Our extension E-PMFM is different from the original variant in the sense that:

- Our partitioning requirements are different, i.e. we aim at minimizing the cost of communications instead of their number. In this context, it can be the case that the number of communications is high between the partitions but the costs are low and vice versa.
- We are not only looking for partitions with minimal costs, but also we should take into consideration the problem of finding VMs that can fulfill the QoS requirements. In this context, it can be also the case that there exists a k partition with minimal costs, however, the VMs within the partitions do not fulfill the QoS requirements of some agents.

E-PMFM (Algorithm 2) takes as input the *ABG*, the Cloud federation F and the time unit t at which some external event changes should occur. It provides as output a deployment $\mathscr{D}$ (see Definition 6) which assigns each agent in the ABG to a pair of a cloud provider and a VM in the cloud federation. The algorithm is performed in two steps. The first step generates a graph on which the partitioning will be performed (Line 1 detailed in Sect. 4.2). The second step performs our extended E-FM (Lines 2–4 detailed in Sect. 4.2).

First Fit Algorithm. The first step consists of generating a graph on which the partitioning will be performed. Such graph should depict the communication overhead between some initially allocated VMs as well as between their corresponding clouds. Roughly speaking, the graph is nested and consists of three layers. The first layer corresponds to the ABG and shows the communication between the agents. The second layer shows the VMs in the cloud federation and the communication between the allocated ones. The third layer shows the

Algorithm 2. E-PMFM Algorithm

Input: ABG, t
Output: $\mathscr{D}$
1: Generate a deployment graph by applying a first-fit algorithm
2: Generate an initial partitioning of cloud blocks and VM blocks
3: Find a cloud block partitioning
4: Find a VM block partitioning

---

communication between the cloud providers to which the VMs belong. We refer to this graph as *deployment graph*. An example of a deployment graph is shown in Fig. 4. The first layer (graph with smallest nodes' size) in the graph corresponds to the SP in Fig. 3. The second layer shows the VMs in the cloud federation in Fig. 2 as nodes. The VMs are connected if they are allocated to interacting agents. The third layer shows the cloud providers to which the VMs belong as nodes. The edges are annotated with weights that are defined as the data exchange between the agents in the ABG multiplied by the computing and bandwidth prices in the federation.

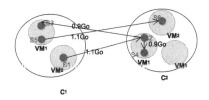

Fig. 4. An example of a deployment graph

In order to generate the deployment graph, we need to find an initial allocation. To do so, we apply a first-fit algorithm (Line 1 in Algorithm 2) which assigns the set of agents to low-cost VMs. In our work, the used algorithm consists of allocating for each agent the low-cost available VM instance within the appropriate cloud provider. The algorithm takes as input, an ordered list of the agents to be deployed as well as an ordered list of the clouds and VMs to be leased. As output, it provides a deployment solution that assigns each agent to the first VM instance found in the ordered list that can fulfill the agent's QoS requirements defined by δ at time t (see Definition 4).

E-FM Extension. The E-FM step performs the partitioning on the generated deployment graph. The partitions, referred to as *blocks*, are defined at two levels: (i) at the level of cloud providers, denoted as *cloud blocks* (third layer in the deployment graph) and (ii) at the level of VMs, denoted as *VM blocks* (second layer in the deployment graph). The cloud blocks show the distribution of the agents between the cloud providers while the VM blocks show the distribution of the agents between the VMs.

The algorithm starts by assuming an initial partitioning where the cloud and VM blocks are defined according to the initial allocation in the deployment graph. For example, in Fig. 4, there exist (i) two cloud blocks that contain $\{A_1, A_3, A_5\}$ and $\{A_2, A_4, A_6\}$ respectively and (ii) four VM blocks that contain $\{A_3, A_5\}$, $\{A_1\}$, $\{A_6\}$ and $\{A_2, A_4\}$ respectively. The algorithm then tries to partition the graph into new blocks that have lower communication overheads by migrating the agents in the deployment graph to different VMs in different cloud providers. This is performed in two phases. In the first phase, the algorithm finds a new partitioning of cloud blocks that minimizes the inter-cloud communication overheads; whereas, in the second phase, a partitioning of VM blocks is created that reduces the communications between the VMs instances within the same cloud provider.

The decision to migrate agents or not (from a cloud provider to another or from a VM to another) is based on a computed *gain*. In our extension, a gain of an agent A is the change in the amount of the communicated data when an agent is moved from a block b_1 to a block b_2. The gain associated to A is defined as:

$$gain(A, b_1, b_2) = WS(A, b_2) - WE(A, b_1) \tag{1}$$

where $WS(A, b_2)$ is the amount of the total communicated data between A and the agents in the block b_2. It corresponds to the sum of the edges' weights connecting A to nodes in the block b_2. $WE(A, b_1)$ is the amount of the total communicated data between A and the agents within the same block b_1. It corresponds to the sum of edges' weights connecting A to nodes in the same block b_1. The gain of A associated with the migration from b_1 to b_2 is only valid if b_2 contains a VM that fulfills the QoS requirements of A.

5   Experiments and Results

The proposed approach is implemented and a set of experiments are performed in order to: (i) analyze the performance results of applying the E-PMFM and its integral version (detailed in Sect. 5.1); and (ii) compare our proposed approach to a naive one consisting of allocating cloud resources without considering the partitioning phase (detailed in Sect. 5.2).

As input we considered three configurations and four different initial allocations (P1, P2, P3, P4) generated using the first fit algorithm for the same ABG. As shown in Table 1, the configurations vary from each other in terms of the total number of agents (services) within each ABG and they share the same infrastructure. The used cloud federation interconnects seven cloud providers having respectively 34, 32, 34, 33, 34, 34 and 30 Vms types. Further, we define in Table 2 the set of external events, which impact the process QoS needs overtime. The experiments were conducted on a laptop with a 64-bit Intel Core 2.10 GHz CPU, 4 GB RAM and Windows 7 as operating system. Due to space limitations, we show the results corresponding to Config. 2 using a maximal number of VM instances equal to 1 Table 1). The conclusions obtained for Config. 1 and Config. 3 are similar to those found with Config. 2.

Table 1. Experimental setting: configurations and cloud federation properties

Configurations	Agents number	Cloud federation		
		Cloud providers number	Number of Vms types	Maximum instances
Config. 1	10	7	34, 32,34, 33,34,34,30	1
Config. 2	50			2,3 4,4 3,5, 3
Config. 3	100			

Table 2. External events

Events	Time (Sim step)	Affected agents (%)	Affected QoS
e1	2	100%	CPU, RAM
e2	4	50%	
e3	6	20%	
e4	8	5%	

5.1 Experiment 1: E-PMFM Vs PMFM

In this section we compare E-PMFM and PMFM using the following efficiency criteria: (i) total link number (NL) inter/intra cloud; (ii) total exchanged data size (DS) inter/intra cloud; and (iii) the number of used VM instances. Both algorithms are executed using as input the four initial allocations generated using the first fit algorithm. Further, both algorithms are performed within two types of infrastructure: (i) Rigid (R) where the partitioning is carried out on a deployment graph that shows only the cloud resources initially leased in the first-fit allocation phase and; (ii) Flexible (F) where the partitioning is performed on the deployment graph where non-initially allocated VMs are included and can be allocated.

E-PMFM vs PMFM (Cloud partitioning): Figures 5 and 6 present respectively the total number of the inter-cloud links and the amount of the inter-cloud exchanged data, before and after performing both the PMFM and E-PMFM partitioning algorithms using flexible and rigid infrastructures. The results show that, regardless the infrastructure type (F or R), E-PMFM provides better results in terms of reduced NL and DS. However, the execution of E-PMFM on a flexible infrastructure gives the best results. In fact, the use of a flexible deployment environment in the partitioning phase, provides more opportunities for the agents with the largest gains to migrate and consequently reduces the inter-cloud communication overheads. Whereas within a rigid infrastructure, there are two cases:

- If the number of the leased VM instances is large enough, the E-PMFM algorithm has more opportunities to move the agents in-between the clouds. For example, the number of the allocated VM instances within P1 and P2 is almost the half of the total agents' number; thus the E-PMFM obtains more reduced inter-cloud communication overheads compared to the initial allocation.

– If the number of the allocated VM instance is small compared to the total agents' number, the partitioning algorithm converged to a local solution, where multiple agents can be moved between the cloud but their QoS requirements cannot be fulfilled. For example, the number of the leased cloud resources within P3 and P4 is almost the quarter of the total agents' number, as shown in Figs. 5 and 6. Therefore, the E-PMFM does not improve the quality of the deployment in such situation.

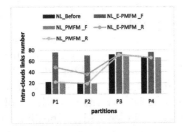

Fig. 5. Inter-cloud link number

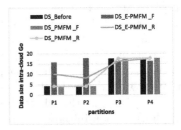

Fig. 6. Inter-cloud data Size

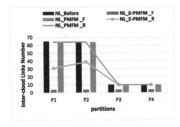

Fig. 7. Intra-cloud link number

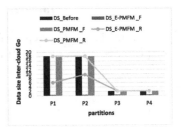

Fig. 8. Intra-cloud data Size

E-PMFM Vs PMFM (VMs partitioning): Figures 7 and 8 present respectively the total number of the intra-cloud links and the amount of the intra-cloud exchanged data, before and after performing both the PMFM and E-PMFM partitioning algorithms. With all the initial partitions, the execution of E-PMFM for cloud partitioning using a flexible infrastructure, resulted in the highest decrease in the inter-cloud communication overheads, which became after the first partitioning intra-cloud exchanges Consequently, as shown in Figs. 5 and 6, E-PMFM with a flexible infrastructure presents highest links number within all the partitions. While, when it comes to the exchanged data, the partitions P3 and P4 presents lowest exchanged amount of data; although, they have the highest links number. Thereby, the minimization of the links number using PMFM does not ensure the reduction in the amount of the exchanged data. Further, as it is shown in the Fig. 9, our modified algorithm uses less Vms instances with all the partitions. In fact, reducing both inter and intra clouds data exchanges allows to cluster the agents and minimize their dispersion within the federation.

5.2   Experiment 2: E-PMFM Vs Naive Approach

In this section, we analyze the performance, in terms of total deployment costs, of our proposed approach against a naive approach that does not consider the partitioning phases during reallocation. A set of external events are defined in Table 2. Further, we selected two different initial allocations (P1 and P2), in order to determine its impact on the overall costs. Due to space limitations, the results corresponds only to Config. 2 in Table 1 on a flexible infrastructure.

As it is shown in Fig. 10 the conducted experiments give the following results:

- The E-PMFM algorithm used with P1 provides the best results in terms of reduced deployment costs overtime.
- From the beginning of the simulation till the (step = 4), both of the performed approaches using P2 as input, gave worse results (increased costs) than the case where P1 is used as initial partition.
- From (step > 4), our approach based on using the E-PMFM algorithm (Algorithm 2) provides better results than the naive approach using both P1 and P2.
- The initial partition generated using the first fit algorithm has an impact on the overall deployment costs. In fact, a good partition from the beginning of the simulation provides better results in terms of reduced deployment costs. Further, considering the use of a graph-based partitioning algorithm to maintain a low communication overhead within the ABG is of great importance to reduce the data transfer costs.

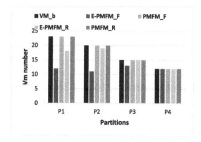

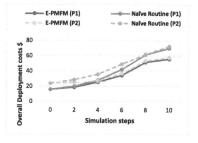

Fig. 9. Number of the used VM instances

Fig. 10. The proposed approach vs a naive routine

6   Related Work

In this section, we discuss the related work of deploying service-based processes as well as business processes into cloud environments. In [13], the authors propose a Petri net decomposition approach to provision service-based processes with platform resources to achieve its appropriate deployment into clouds. Whereas in our work, we focus on the dynamic provisioning of the SP with the suitable

IaaS resources to meet the continuous changes of the process QoS needs. In [8], the authors tackle the deployment challenges of business processes into a cloud infrastructure from the perspective of the process owner. The proposed approach consists of three main phases including process execution prediction, resource allocation and the cost estimation. In comparison to our work, Only execution time and resources reuse are taken into consideration, whereas no strategies are convoked to minimize the communication overhead. Further, they consider the deployment into one cloud provider and they do not address the varying demands for computational cloud resources. In [6], the authors present a self adaptive resource allocation strategy for elastic process execution in a cloud environment. In fact based on a previous knowledge about the current and future process landscape, the cloud resources are automatically allocated or released. The authors focus mainly on avoiding over and under provisioning of the VMs, however, nothing is mentioned about reducing the communication overheads within the process. In [5], a Mixed Integer Linear Programming (MILP) technique is used to ensure a cost-efficient scheduling of an elastic process into a hybrid cloud. The proposed model takes into consideration the data transfer in order to reduce the overall deployment costs. Also in [10], the authors propose a Linear program (LP) in order to determine for a configurable business process the optimal process variant deployment into a cloud federation. The proposed model takes into consideration the compute and the data transfer costs. However in [11], the authors claim that LP are not suitable to resolve the deployment of large-scale applications (large number of components) across computational infrastructures. In fact, the amount of time and resources required to resolve the problem grow exponentially with the size of a given application. Thus, heuristics are required to provide a faster good solution.

7   Conclusion

In this paper, we presented an adaptive cost-efficient approach to deploy SP to cloud federation. We used agent-based systems to simulate SP during runtime. Also we focused on using an extended version of PMFM partitioning algorithm to allocate low cost resources and reduce the communication overhead (data flow) between the agents (services). We compared the E-PMFM to its integral version PMFM and we found that the first algorithm provides better results in terms of both reduced data transfers and number of used VM instances within the federation. Furthermore, we compared our proposed approach to a naive routine, which does not consider the minimization of the communication overheads. Our approach provided better results in terms of reduced costs. Also, the obtained results proved that the initial used allocations have impacts on the final costs of the deployment solution.

As a future work, we plan to improve our proposed algorithm in order to provide effective deployment solutions efficient for services based on configurable processes.

Acknowledgment. This work was financially supported by the PHC Utique program of the French Ministry of Foreign Affairs and Ministry of higher education and research and the Tunisian Ministry of higher education and scientific research in the CMCU project number 15G1413.

References

1. Cong, J., Lim, S.K.: Multiway partitioning with pairwise movement. In: Proceedings of the 1998 IEEE/ACM International Conference on Computer-aided Design, ICCAD 1998, pp. 512–516, ACM, New York, NY, USA (1998)
2. Endert, H., Küster, T., Hirsch, B., Albayrak, S.: Mapping BPMN to agents: an analysis. Agents, Web-Services, and Ontologies Integrated Methodologies, pp. 43–58 (2007)
3. Ferber, J.: Multi-agent Systems: An Introduction to Distributed Artificial Intelligence, vol. 1. Addison-Wesley, Reading (1999)
4. Fiduccia, C.M., Mattheyses, R.M.: A linear-time heuristic for improving network partitions. In: 19th Design Automation Conference, pp. 175–181, June 1982
5. Hoenisch, P., Hochreiner, C., Schuller, D., Schulte, S., Mendling, J., Dustdar, S.: Cost-efficient scheduling of elastic processes in hybrid clouds. In: 2015 IEEE 8th International Conference on Cloud Computing, pp. 17–24, June 2015
6. Hoenisch, P., Schulte, S., Dustdar, S., Venugopal, S.: Self-adaptive resource allocation for elastic process execution. In: 2013 IEEE Sixth International Conference on Cloud Computing, pp. 220–227, June 2013
7. Küster, T., Heßler, A., Albayrak, S.: Towards process-oriented modelling and creation of multi-agent systems. In: Dalpiaz, F., Dix, J., van Riemsdijk, M.B. (eds.) EMAS 2014. LNCS, vol. 8758, pp. 163–180. Springer, Cham (2014). doi:10.1007/978-3-319-14484-9_9
8. Mastelic, T., Fdhila, W., Brandic, I., Rinderle-Ma, S.: Predicting resource allocation and costs for business processes in the cloud. In: 2015 IEEE World Congress on Services, pp. 47–54, June 2015
9. Odell, J., Nodine, M., Levy, R.: A metamodel for agents, roles, and groups. In: Odell, J., Giorgini, P., Müller, J.P. (eds.) AOSE 2004. LNCS, vol. 3382, pp. 78–92. Springer, Heidelberg (2005). doi:10.1007/978-3-540-30578-1_6
10. Rekik, M., Boukadi, K., Assy, N., Gaaloul, W., Ben-Abdallah, H.: A linear program for optimal configurable business processes deployment into cloud federation. In: 2016 IEEE International Conference on Services Computing (SCC), pp. 34–41, June 2016
11. Verbelen, T., Stevens, T., Turck, F.D., Dhoedt, B.: Graph partitioning algorithms for optimizing software deployment in mobile cloud computing. Future Gener. Comput. Syst. **29**(2), 451–459 (2013). special section: Recent advances in e-Science
12. Wooldridge, M., Jennings, N.R., Kinny, D.: The Gaia methodology for agent-oriented analysis and design. Auton. Agent. Multi-Agent Syst. **3**(3), 285–312 (2000)
13. Yangui, S., Klai, K., Tata, S.: Deployment of service-based processes in the cloud using petri net decomposition. In: Meersman, R., Panetto, H., Dillon, T., Missikoff, M., Liu, L., Pastor, O., Cuzzocrea, A., Sellis, T. (eds.) OTM 2014. LNCS, vol. 8841, pp. 57–74. Springer, Heidelberg (2014). doi:10.1007/978-3-662-45563-0_4

Towards a Public Cloud Services Registry

Ahmed Mohammed Ghamry, Asma Musabah Alkalbani, Vu Tran,
Yi-Chan Tsai, My Ly Hoang, and Farookh Khadeer Hussain(✉)

Decision Support and e-Service Intelligence Lab Centre for Artificial Intelligence
School of Software, University of Technology, Sydney, NSW 2007, Australia
{Asma.M.Alkalbani,Vu.Tran-1,Yi-Chan.Tsai}@student.uts.edu.au,
Farookh.Hussain@uts.edu.au

Abstract. Cloud services registry is a cloud services datadase which
contains thousands of records of cloud consumers' reviews and cloud services, such as Platform as a Service (PaaS) and Infrastructure as a Service
(IaaS). The data set is harvested from a web portal called www.serchen.
com. Each record holds detail information about the service such as service name, service description, categories, key features, service provider
link and review list. Each review contains reviewer name, review date
and review content. This work is an extension of our previous work Blue
Pages data set [6]. The data set is valuable for future research in cloud
service identification, discovery, comparison and selection.

Keywords: Cloud services data set · Service discovery · Web harvesting

1 Introduction

With the modernization of IT technologies, cloud computing a critical component for organizations digital business by delivering computing resources as a
service, such as Software as a Service (SaaS), Platform as a Service (PaaS) and
Infrastructure as a Service (IaaS). Cloud computing with on-demand services
brings many benefits to businesses by ensuring service availability and scalability, saving cost, improving efficiency and productivity [11]. In recent years,
there has been an ongoing demand for businesses to adopt public cloud services. A forecast conducted by Cisco predicts that both public and private cloud
markets will continue to grow between 2015 and 2020 [12]. However, the public cloud is expected to grow stronger than the private cloud. Service discovery
and selection are challenging tasks for businesses as a result of the rapid growth
in cloud services market. Using search engines, such as Bing and Google, is a
traditional way to discover cloud services on the World Wide Web (WWW).
This method enables consumers to find cloud services-related information. However, the consumers will need to peruse the content of each website content and
compare service offerings manually to decide the most suitable cloud service for
their needs, such as Database tools. Their search results may return irrelevant
information, which makes the search effort more difficult.

© Springer International Publishing AG 2017
A. Bouguettaya et al. (Eds.): WISE 2017, Part I, LNCS 10569, pp. 290–295, 2017.
DOI: 10.1007/978-3-319-68783-4_20

Another way to locate cloud services information is via web portals such as GetApp (https://www.getapp.com/). web portals offer a significant number of records of cloud service, but it is difficult to obtain up-to-date information when data changes. In recent years, therefore, cloud service discovery has attracted significant attention in the research community. Ontology-based cloud discovery approaches have been proposed in most of researches in cloud service discovery area. An ontology in semantic web technologies is often used to illustrate information and relationships. Cloud services ontology is usually based on the National Institution Standards and Technology (NIST) cloud computing standard and specifications [11]. For instance, a crawler engine is built to harvest web portals using cloud ontology concepts [13].

In [3,4] the authors carried out a categorization for cloud services based on cloud service ontology concepts. Other studies by [2,15] also categorized cloud services using business ontology concepts to help businesses find the most suitable cloud business service.

To identify the most appropriate solutions for cloud service discovery and selection, Alkalbani and Hussain [7] conducted a thorough review of current cloud discovery approaches and proposed eight criteria for comparing the current cloud discovery approaches. In general, a challenge for cloud service discovery approaches is to provide a consistent public registry with the ability to automatically identify and update service information in the registry [7,14].

A cloud services registry can offer a complete list of cloud services from multiple resources to give an insight into service information and facilitate the process of discovering and selecting cloud services with up-to-date information in the market. To address this need, the aim of this work is to provide a comprehensive data set which contains real world SaaS, PaaS and IaaS cloud service offerings. The data set is our future research extension to the Blue Pages data set [6].

To achieve our objectives, we used the latest web scraping technologies to harvest useful cloud service data from a cloud service web portal. web scraping is a well-known technique for harvesting data from the WWW by gathering web page data. Serchen Marketplace, www.serchen.com [1] is an excellent source of cloud services. This website contains thousands of records on cloud services in SaaS, PaaS and IaaS categories. It also has a significant number of reviews from customers about these services. In the developed harvesting tool, sample metadata are collected by tool users to provide a thorough analysis of the selected website. The collected sample metadata is analyzed by an algorithm to learn its structure via HTML page format, which serves as an the indicator to locate exactly where and how to harvest the information that the users expect. The metadata structure learned is the core for the harvesting process, in which thousands of web pages are extracted from the structure using scripting programming techniques. The data set holds all the useful services data obtained from the Internet portal. Each record stores service information such as service name, service description, categories, key features, service provider link and review list. Each review has information about reviewer name, review date and review content. At this stage, we only harvest one portal with a view to developing a tool with the capacity to harvest future web portals. The data set contains more than 6000 cloud services and 5000 customer reviews. We plan to provide a mechanism for updating data to obtain up-to-date information from web portals. The rest of this paper is structured as follows: Sect. 2 discusses some related works on crawling cloud service data sets from the web. The results and analysis presented in Sect. 3. The conclusion and further improvements will be presented in Sect. 4.

2   Related Work

There is very little research in the literature that focuses on providing cloud service data sets. In [9] Han and Sim and [10] Kang and Sim, the authors designed and developed a crawler engine to discover cloud services via the web and provided statistics and details on available cloud services in their local repository. In practice, the authors introduced a cloud service ontology that supported the discovery and collection of cloud services web links via general search engines such as Google and Yahoo. The cloud ontology was also used to filter the collected data set to remove invalid service web links. The final result showed that there are around 5883 valid cloud services available on the web including 1552 cloud services implemented using Service Oriented Computing (SOC). However, the study does not provide a complete view of cloud services published on the web. The reason is that the authors used only the most general and abstract ontology concepts to search via general search engines for cloud services as IaaS, PaaS, SaaS, storage, communication, etc., which might provide only a cross-section of what is available on the web and therefore lead to inaccurate conclusions. Furthermore, the collected data sets lack main service information such as service provider name and service URL; also they include inadequate cloud services information that does not have semantic meaning.

Gong and Sim [8] presented a method to deal with the lack of cloud services information based on crawlers that are designed to collect information from several web resources. The authors targeted the most popular services providers, such as AWS, Rackspace, and Gogrid. A crawler is designed for each cloud service provider which gathers only service specification and service price for each service from the cloud provider's website. The data is stored locally, and a k-mean clustering algorithm performs is performed on them to detect the similarities and differences between cloud services. One of the limitations of this study is that it does relies only on popular service providers where it crawls the world wide web for cloud services is another potential concern because it provides a complete view of what cloud services are available through the web. Another problem with this approach is that it does not take into account the need to provide a publicly available clud market data set.

Another open source crawler engine providing a central repository for cloud services in SaaS was proposed by Alkalbani et al. [5]. This approach used the Nutch-Hadoop crawler to harvest cloud web portals and store data in SQL database. However, the drawback of this data set is the lack of QoS information and

Service Name	Service URL	Service Key Features	Service Categories	Service Description
Xero	www.xero.com	reconcilication	Accounting Software	offices in the United States,
Billy	www.billyapp.com	Accounts Receivable	Accounting Software	cloud-based accounting
Cantorix	www.cantorix.com	Recurring Invoicing	Accounting Software	new online invoicing
Bean Cruncher	beancruncher.com	Accounts Receivable	Accounting Software	Software is your best choice
AccountMate Software	www.accountmate.com/	Multiple Deployment	Accounting Software	AccountMate develops and
CosmoLex	www.cosmolex.com	Billing & Invoicing	Accounting Software	practice management
ZarMoney	www.zarmoney.com/	Accounts Payable	Accounting Software	software with invoicing,
Flare Cloud Accounting	www.flareapps.com	dashboard	Accounting Software	small business's and
docSTAR	www.docstar.com	Software	Accounts Payable	division of Astria Solutions
Nvoicepay	www.nvoicepay.com	International Payments	Accounts Payable	payments solutions to
expex	www.expexinc.com	Expense Management	Accounts Payable	managing your bills in the
Clear Pier	www.clearpier.com		Ad Networks	Premium Performance
Qadabra	www.qadabra.com		Ad Serving	got the magic.Earning more
Epom Ad Server	www.epom.com	Cross-channel ad	Ad Serving	multiplatform ad serving
Admixer	admixer.net	management	Ad Serving	Ltd. Group is a technology
Yext	www.yext.com		Advertising Software	with the award-winning
Onters	onters.com/	video	Affiliate Marketing	agency working in animation
TrackingDesk	www.trackingdesk.com	Analytics	Affiliate Marketing	media management, results
Appsee	www.appsee.com	Recordings	Analytics	analytics platform provides in-
Natero	www.natero.com	Lifetime Value	Analytics	success platform to merge
Altocloud	www.altocloud.com	Messaging	Analytics	and Digital Messaging: Real-
Logi Analytics	www.logianalytics.com		Analytics	self-service analytics,
Morningfame	www.morningfa.me	Analytics	Analytics	the web - like a blog post or a
Stytch	www.stytch.com		Analytics	platform that provides
Teamgraph	teamgraph.io	Analytics	Analytics	solution for CXOs to track
Attribution	attributionapp.com	attribution	Analytics	for all your marketing
Twitter Counter	twittercounter.com	Historical Data	Analytics	in 2008 and today tracks over
Snappii	www.snappii.com	Development	App Building Software	on the principles of
Bubble	bubble.is	mobile apps	App Building Software	more generally the need for
Shoutem	www.shoutem.com	Special Deals	App Building Software	comprehensive mobile app
Mancktech	www.mancktech.com/	software development	App Building Software	software & web development
Contus	www.contus.com	Software	App Building Software	on demand delivery software
FuGenX Technologies	fugenx.com	firms florida	App Building Software	Apps Development Company
Knack	www.knack.com	you need	App Building Software	easily build beautiful, data-
TechValens	www.techvalens.com		App Building Software	LLC is a leading provider in
CATS	www.catsone.com	Portal Branding	Applicant Tracking	based applicant tracking
HireGround	startdate.ca	Search candidate	Applicant Tracking	technology solutions
Bullhorn	www.bullhorn.com/		Applicant Tracking	CRM solutions for
RecruitBPM	www.recruitbpm.com	system	Applicant Tracking	applicant tracking system that
Jobularhty	jobularity.com	job search	Applicant Tracking	Recruitment Platform.
Cloudwalk Hosting, LLC	www.cloudwalks.com	24x7 Free Chat, Email	Application Hosting	leading QuickBooks Hosting
Elucentra Cloud Services	www.elucentra.com	and Phone Support	Application Hosting	Elucentra provides an easy-to-
Cloudvara	www.cloudvara.com	99.99% Server Uptime	Application Hosting	way businesses work by
Solution	www.hitech-cloud.com	Voice Support	Application Hosting	in United States.We are the
Appointment-Plus	plus.com		Scheduling	worldwide leader in online
BookSteam	www.booksteam.com	Online Payments	Scheduling	leading Online Appointment
Timely	www.gettimely.com/	Scheduling	Scheduling	appointment scheduling for
BookedIN	bookedin.com	scheduling	Scheduling	phone? Tired of endless
OnShift	www.onshift.com	Software	Scheduling	software and proactive
SnapAppointments	snapappointments.com	Scheduling	Scheduling	winning online appointment
TimeTap	www.Timetap.com	Multiple Locations	Scheduling	scheduling platform built for
TrekkSoft	www.trekksoft.com	website	Scheduling	with an integrated booking
tuOtempO	www.tuotempo.com	Booking	Scheduling	PATIENT CRM that
Zirtual	www.zirtual.com	executive assistants	Scheduling	U.S. based, virtual assistants,

Fig. 1. List of some cloud service reviews

the fact that it only offers service name and URL [7]. To provide a more meaningful data set, another research in this area focuses on developing a web scraping tool to harvest SaaS offers. However, the data sets mentioned above contains only service name and service URL.

To address the need identified above, we propose to develop a new web scraping tool in this paper that can efficiently harvest cloud services from the cloud service marketplace to create a central cloud services registry. The new approach seeks to take advantage of the flexibility of the data structure displayed on web pages, enabling tool users to control the meaning, content and multi-level structure of the harvested data set. The tool also has the capacity for collecting data from future cloud service web portals using the same methodology, providing the registry with the potential to contain all the cloud services available on the web.

3   Results and Analysis

The primary objective of our work is to provide a cloud service data set to cloud consumers, comprised of IaaS, PaaS and SaaS. The significant level of information will enable users to use the dataset for multiple purposes; For instance, based on the reviews of consumers of cloud services providers, an organization can compare and decide the most suitable services for their functional needs. This study has collected information about 6000 services and 5000 service providers reviews from the Serchen Marketplace. The result shows that there are many cloud services providers offering a huge amount of services for specific areas such as dedicated hosting and web hosting. In contrast, there are fewer cloud services offered in some areas. For example, the web hosting category has the highest number of cloud services at 1500, where as there are zero cloud services offering Anti-Money Laundering Software. The review data set (consumer's comments) in this study is a valuable resource for potential future consumers to choose the service that meets their needs; it is also beneficial for service providers, to evaluate their customer reviews compared with those of their competitors.

4   Conclusion and Future Work

In this paper, we propose a central registry which includes a data set of real cloud services offerings from the web portal. To achieve our objective, we developed a harvesting tool which harvests cloud services information from the web. The data set holds thousands of cloud service records for SaaS, PaaS and IaaS categories, including 6000 cloud services and 5000 customer reviews. The data set is of benefit to cloud customers, service providers and the research community in the areas of cloud service discovery and selection. In future work, we plan to harvest more web portals and develop a comprehensive registry of cloud service listings.

References

1. serchen.com: Compare the best business software cloud services (2017). http:// www.serchen.com/
2. Afify, Y.M., Moawad, I.F., Badr, N., Tolba, M.F.: A semantic-based software-as-a-service (saas) discovery and selection system. In: 2013 8th International Conference on Computer Engineering and Systems (ICCES), pp. 57–63. IEEE (2013)
3. Alfazi, A., Sheng, Q.Z., Qin, Y., Noor, T.H.: Ontology-based automatic cloud service categorization for enhancing cloud service discovery. In: 2015 IEEE 19th International on Enterprise Distributed Object Computing Conference (EDOC), pp. 151–158. IEEE (2015)
4. Alfazi, A., Sheng, Q.Z., Zhang, W.E., Yao, L., Noor, T.H.: Identification as a service: large-scale cloud service discovery over the world wide web. In: 2016 IEEE International Congress on Big Data (BigData Congress), pp. 485–492. IEEE (2016)
5. Alkalbani, A.M., Ghamry, A.M., Hussain, F.K., Hussain, O.K.: Sentiment analysis and classification for software as a service reviews. In: 2016 IEEE 30th International Conference on Advanced Information Networking and Applications (AINA), pp. 53–58, March 2016
6. Alkalbani, A.M., Ghamry, A.M., Hussain, F.K., Hussain, O.K.: Blue pages: software as a service data set. In: 2015 10th International Conference on Broadband and Wireless Computing, Communication and Applications (BWCCA), pp. 269–274. IEEE (2015)
7. Alkalbani, A.M., Hussain, F.K.: A comparative study and future research directions in cloud service discovery. In: 2016 IEEE 11th Conference on Industrial Electronics and Applications (ICIEA), pp. 1049–1056. IEEE (2016)
8. Gong, S., Sim, K.M.: Cb-cloudle and cloud crawlers. In: 2014 5th IEEE International Conference on Software Engineering and Service Science (ICSESS), pp. 9–12. IEEE (2014)
9. Han, T., Sim, K.M.: An ontology-enhanced cloud service discovery system. Proc. Int. MultiConf. Eng. Comput. Sci. 1, 17–19 (2010)
10. Kang, J., Sim, K.M.: Towards agents and ontology for cloud service discovery. In: 2011 International Conference on Cyber-Enabled Distributed Computing and Knowledge Discovery (CyberC), pp. 483–490. IEEE (2011)
11. Mell, P., Grance, T., et al.: The nist definition of cloud computing (2011)
12. Cisco Visual Networking. Cisco Global Cloud Index: forecast and methodology, 2014–2019. (white paper) (2013)
13. Noor, T.H., Sheng, Q.Z., Alfazi, A., Ngu, A.H., Law, J.: Csce: a crawler engine for cloud services discovery on the world wide web. In: 2013 IEEE 20th International Conference on Web Services (ICWS), pp. 443–450. IEEE (2013)
14. Sun, L., Dong, H., Hussain, F.K., Hussain, O.K., Chang, E.: Cloud service selection: state-of-the-art and future research directions. J. Netw. Comput. Appl. 45, 134–150 (2014)
15. Tahamtan, A., Beheshti, S.A., Anjomshoaa, A., Tjoa, A.M.: A cloud repository and discovery framework based on a unified business and cloud service ontology. In: 2012 IEEE Eighth World Congress on Services (SERVICES), pp. 203–210. IEEE (2012)

Query Processing

Location-Based Top-k Term Querying over Sliding Window

Ying Xu[1], Lisi Chen[2], Bin Yao[1(✉)], Shuo Shang[3(✉)], Shunzhi Zhu[4], Kai Zheng[5], and Fang Li[1]

[1] Shanghai Jiao Tong University, Shanghai, China
yaobin@cs.sjtu.edu.cn
[2] Hong Kong Baptist University, Hong Kong, China
[3] King Abdullah University of Science and Technology, Thuwal, Saudi Arabia
jedi.shang@gmail.com
[4] Xiamen University of Technology, Xiamen, China
[5] Soochow University, Soochow, China

Abstract. In part due to the proliferation of GPS-equipped mobile devices, massive svolumes of geo-tagged streaming text messages are becoming available on social media. It is of great interest to discover most frequent nearby terms from such tremendous stream data. In this paper, we present novel indexing, updating, and query processing techniques that are capable of discovering top-k locally popular nearby terms over a sliding window. Specifically, given a query location and a set of geo-tagged messages within a sliding window, we study the problem of searching for the top-k terms by considering both the term frequency and the proximities between the messages containing the term and the query location. We develop a novel and efficient mechanism to solve the problem, including a quad-tree based indexing structure, indexing update technique, and a best-first based searching algorithm. An empirical study is conducted to show that our proposed techniques are efficient and fit for users' requirements through varying a number of parameters.

Keywords: Top-k · Term · Location

1 Introduction

With the proliferation of social media, cloud storage, and location-based services, many researches [11,26–28,30,35,37] in spatial fields are studied. And the amount of messages containing both text and geographical information (e.g., geo-tagged tweets) are skyrocketing. Such messages, which can be modeled as geo-textual data streams, often offer first-hand information for a variety of local events of different types and scale, including breaking news stories in an area, urban disasters, local business promotions, and trending opinions of public concerns in a city.

Data streams from location-based social media bear the following natures: (1) *bursty nature* - messages regarding a particular topic can be quickly buried deep

A. Bouguettaya et al. (Eds.): WISE 2017, Part I, LNCS 10569, pp. 299–314, 2017.
DOI: 10.1007/978-3-319-68783-4_21

in the stream if the user is not fast enough to discover it [17]; (2) *local-intended nature* - users from different locations may post messages related to diverging topics [46]. With thousands of messages being generated from location-based social media each second, it is of great importance to maintain a summary of what occupies minds of users.

To address the problem, existing proposal [31] aims at finding the top-k locally popular terms in content within a user-specified spatio-temporal region. However, in most cases it is difficult for a user to specify a rectangular region on the spatial domain. Instead, a user may prefer a rank-ordered list of terms by taking both term frequency and location proximity into consideration.

Based on the user requirements, we consider a new kind of top-k term query, Location-based Top-k Term Query (LkTQ), that returns top-k locally popular terms by taking into account both location proximity and term frequency for geo-textual data over a sliding window.

Figure 1 provides a toy example of LkTQ. Let us consider 10 geo-tagged tweets located on the map of China. The point with square label indicates the query location. The points with circle labels are geo-textual messages. For each geo-textual message, we present its textual information and corresponding distances to the query point. The results of the LkTQ are the k most locally popular terms based on a location-aware frequency score, which are shown in Fig. 1b. The score of a term is computed by a linear combination of the term frequency and location proximities between the query and the messages containing the term.

(a) Messages and distances   (b) Tag Cloud

Fig. 1. Example of Querying in China

A straightforward approach for answering an LkTQ is to evaluate all terms of messages within the current sliding window. Specifically, for each of such terms we compute the location-aware frequency score between the term and the query. This approach, however, will be very expensive for a large number of geo-textual messages. For efficiently processing an LkTQ, we need to address the following challenges. First, it is computationally expensive to return the exact result of LkTQ. Hence, we need seek approximate solutions with high accuracy. Second, the location-aware frequency score measures both term frequency and term location proximity in a continuous fashion. Therefore, it is non-trivial to propose a hybrid indexing structure and its corresponding algorithm that could effectively prune the search space based on term frequency and location proximity simultaneously. Thrid, because of the sliding-window scenario of LkTQ, the indexing mechanism must be able to handle geo-textual data streams with high arrival rate.

Our contributions are summarized as follows:

1. We define a new problem of processing LkTQ that searches for the top-k locally popular terms by taking into account both term frequencies and location proximities from geo-textual dataset.
2. A hybrid quad-tree based indexing structure that has low storage and update cost and a searching algorithm with effective pruning strategies are proposed to enable the fast and accurate top-k term search. Specifically, since it is impossible to store every messages in such a big streaming data, we augment each quad-tree node with a summary file for summarizing the term frequencies. The non-leaf node maintains an upper bound error by storing the merging summaries of its child nodes. Misra-Gries summary (MG summary) [21] and Space-Saving summary (SS summary) [19,20] are two simple and popular summaries for frequency estimation and heavy hitters problems. Due to the merge processing [1] of MG summaries is lightweight and has a guarantee on the accuracy of frequency [31], and there are a lot of merging manipulations in quad tree nodes, we adopt the MG summary instead of the SS summary.

The rest of this paper is organized as follows: In Sect. 2, preliminaries and some related works are introduced. In Sect. 3, we provide our proposed solution on the problem. An experimental analysis is presented in Sect. 4. A conclusion is presented in Sect. 5.

2 Preliminaries and Related Work

2.1 Problem Definition

Let D be a 2D Euclidean space, W be a sliding window, S be a set of geo-textual messages located within D and W. Each geo-textual message is denoted by $o = (pos, text)$, where pos is a point location in D, and $text$ is text information. An LkTQ q is represented by a tuple (loc, k) where loc indicates the query location and k denotes the number of result terms. It returns k terms with the highest *location-aware frequency score* of messages within W.

The location-aware frequency score of a term t in the sliding window W is defined as a linear combination of the distance and the frequency of the term in W:

$$FS(t) = \alpha \times \frac{freq(t)}{|W|} + (1 - \alpha) \times (1 - \frac{d(q, W_t)}{d_{diag} \times |W_t|}) \tag{1}$$

where $freq(t)$ is the number of messages containing term t, $|W|$ is the number of messages in the sliding window W, $d(q, W_t)$ is the sum of distance between the query and the messages that contain t in window W, d_{diag} is the diagonal length of the rectangular region R, $|W_t|$ denotes the number of messages in W that contain t, and α $(0 \leq \alpha \leq 1)$ is a parameter which balances the weight between the term frequency and the location proximity.

2.2 Related Work on Top-k Spatial-Keyword Query

Location-based queries [10,33,34,39,42], similarity search [11–13,38,41] or optimal path search [25,29,36] are hot issues in recent years. Top-k spatial-keyword query (e.g., [4,8,14,22,40,43–45]) is a more comprehensive problem of previous works. It retrieves k most relevant geo-textual objects by considering both location proximity (to query location) and textual similarity (to query keywords). Hybrid indices are developed to store the location and text information of objects, which use both location information and text information to prune search space during the query time. Most of such indices combine spatial index (e.g., R-tree, quad-tree) and the inverted file for storing location and text information, respectively. However, these studies aim at retrieving top-k geo-textual objects, which is different from the problem of retrieving top-k terms.

2.3 Frequent Item Counting

In stream data processing, aggregation is a widely studied problem. Existing aggregation techniques are commonly categorized into counter-based techniques and sketch-based techniques.

Counter-based techniques monitor all the items with a fixed number of counters, each message for an individual counter in a subset of S. When an item in the monitored set comes, its counter is updated. If the item is not in the monitored set and the counters are full, then some other actions will be taken in different algorithms. For instance, Space-Saving algorithm can find any item with the minimum counter value, replace the new item with it, and then increase the counter by 1.

Another popular algorithm - MG summary is very simple to implement. Given a parameter k, since an MG summary stores $k - 1$ (item, count) pairs, there are three cases when dealing with a new coming item i in the stream.

1. if i has already maintained in the current counters, increase its counter value by 1;
2. if i is not in the monitoring list and the number of counters does not reach k, insert i into the summary and set its counter value to 1;
3. if i is not in the monitoring list and the summary has maintained k counters, we decrement all the counter value of messages in the monitored set by 1 and remove all the messages whose counter value is equal to 0.

Other notable counter-based algorithms include LossyCounting [18] and Frequent [7,16].

Sketch-based techniques monitor all the messages rather than a subset of S using hashing techniques. Messages are hashed into the space of counters, and the hashed-to counters will be updated for every hit of the corresponding item. The CountSketch algorithm [3] solves the problem of finding approximate top keywords with success probability $(1-\delta)$. The GroupTest algorithm [6] aims at searching queries about hot items and achieves a constant probability of failure, δ. And it is generally accurate. Count-Min Sketch [5] is also a representative Sketch-based technique.

Sketch-based techniques have less accuracy and less guarantees on frequency estimation than counter-based techniques due to hashing collision. Moreover, they do not provide guarantee about relative order in the continuous stream. Therefore, we adopt counter-based techniques in our work.

2.4 Related Systems

There are several recent systems using related techniques. Skovsgaard [31] designs a framework supporting indexing, updating and query processing which are capable of return the top-k terms in posts in a user-specified spatio-temporal range. The called adaptive frequent item aggregator (AFIA) system is implemented through multiple layers of grids to partition space into multiple granularities. In each grid cell, a precomputed summary is maintained. The system also performs a checkpoint to prevent the situation where a counter enters the top-k counters along with its possible error as a standalone system employing spatial-temporal indexing.

BlogScope [2] is a system which collects news, mailing list, blogs, and so on. It supports finding and tracking the objects, events or stories in real world, monitoring most of the hot keywords as well as the temporal/spatial bursts. The biggest drawback of BlogScope is that it cannot aggregate keywords according to user-specified spatio-temporal region. Moreover, it has weak timeliness which only support the search in a few minutes.

NewsStand [32] and TwitterStand [24] are two similar systems. NewsStand is a news aggregator of spatio-textual data, collecting geographical contents from RSS feeds into story clusters. Users are expected to retrieve and search some stories related to the query keywords within the geographical region. The difference between NewsStand and TwitterStand is that TwitterStand uses Tweets as data source instead of RSS feeds. They both adopt a spatio-textual search engine, which supports spatio-temporal searching not long, on a small ProMED dataset. However, both of the systems have a not high rate of updating.

3 Proposed Solution

The details of our algorithm are presented in this section for handling LkTQ. We first introduce the data indexing model, which is used to store all the data messages in a sliding windows (Sect. 3.1). Then, we show the process of the query searching in Sect. 3.2.

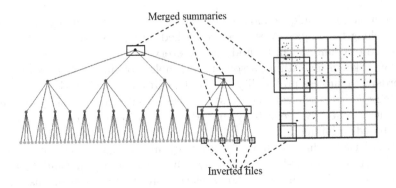

Fig. 2. The basic structure of a quad-tree.

3.1 Data Indexing Model

We use a quad-tree based indexing structure to store all the geo-textual messages in a stream for faster indexing. The basic idea of the quad-tree is to divide the underlying space into different levels of cells. Specifically, it recursively divides the space into four congruent subspaces, until the tree reaches a certain depth or stops on a certain condition. Quad-tree is widely used in image processing, spatial data indexing, fast collision detection in 2D, sparse data, etc. The basic structure in our algorithm is shown in Fig. 2. One thing to mention, the different color of the nodes corresponds to the certain quadrant cells in the rectangle of right side. And the information stored in leaf nodes are inverted files, while non-leaf nodes stores merged summaries.

Quad-tree has a very simple structure, and when the geo-textual messages distribution is relatively uniform, it has relatively high insertion and query efficiency. The black points in the figure is the messages which locate in their expected region. In our algorithm, we set M as the largest number of messages in a leaf node. In other words, if the number of messages stored in a node is more than M, this node will become a non-leaf node and be split into four leaf nodes with equal sizes.

3.2 Query Processing

Overview. According to our problem definition in Sect. 2.1, we proceed to describe the framework we use to get top-k terms with the highest scores rapidly and accurately in a specified situation adapting the social geo-tagged stream data over a sliding window.

Index Updating over Sliding Window. Different from the region-based term query [31], the location of LkTQ is a point instead of a specified spatial region. We aim to find the locally most popular k terms in a comprehensive consideration with location proximity and frequency. If the sliding window is not full, when a new message comes and is inserted to a leaf node of the quad-tree, the summaries in this node will be updated. Then, its parent node will update its merged summaries. The process will be done upward recursively until the root node of quad-tree gets the newest merged summaries. If the sliding window is full, when a message in the stream comes and is inserted, a message with the earliest time stamp should be deleted. Then, the index updating process is the same as the condition when the sliding window is not full.

Summary Merging. Each leaf node of quad-tree stores the summaries of all the textual information of contained messages. The details of MG algorithm [21] is shown in Algorithm 1. [1] has proved that MG summary and SS summary are isomorphic and SS summary can be transferred by MG summary. Recall that, for the reason that the merge processing of MG summaries is easy and efficient, while there are a lot merging manipulations in quad-tree, we adopt the MG summary instead of the SS summary. The process of merging MG summaries is pretty simple. First, we combine two summaries by adding up the corresponding counters. This step results in up to $2k$ counters. Then a prune operation is manipulated: take the $(k + 1)$-th largest counter, and subtract its counter value from all the counters. Finally, we remove all the non-positive counters. This is a process with constant number of sorts and scans of summaries of size $O(k)$.

Algorithm 1. Misra-Gries(counters k, stream T)

---

1   $n \leftarrow 0$;

2   $T \leftarrow \emptyset$;

3   **foreach** i **do**

4      $n \leftarrow n + 1$;

5   **if** $i \in T$ **then**

6        $c_i \leftarrow c_i + 1$;

7   **else if** $|T| < k - 1$ **then**

8        $|T| \leftarrow |T| \cup \{i\}$;

9        $c_i \leftarrow 1$;

10   **else**

11   **for** *all* $j \in T$ **do**

12          $c_j \leftarrow c_j - 1$;

13   **if** $c_j = 0$ **then**

14            $|T| \leftarrow |T| \backslash \{j\}$

In this algorithm, both leaf nodes and non-leaf nodes store summaries of the messages in it. In leaf nodes, summaries are computed using the process stated in Algorithm 1, while in parent nodes, summaries come from the merging processing of all its child node using the method we describe above.

Computing the Term Score. Given a term, we have two steps to obtain its score:

1. First, we need to compute the score in each node employing the summaries stored in each node. Eq. (1) defines the formula to calculate the score. For convenience, we divide the score calculation formula as the "Frequency part" ($\frac{freq}{|W|}$) and the "Distance part" ($1 - \frac{d(q, W_t)}{d_{diag} \times |W_t|}$). Essentially, the score is a linear integration of the two parts. As the MG summaries estimate the frequency of any item with error at most $n/(k+1)$ (n is the number of all the messages), we add the maximum error to $freq$ to calculate the "Frequency part". $d(q, W_t)$ is the sum of distance between the query and the messages that contains the term t, here, we use the minimum distance between the query and the four edges of the node which contains the term as an upper bound value of a message.

   Since a term may occur more than one time in a node, we need to consider the redundant calculation of the same term in distance calculation. Then, the "Distance part" involves the division part by the number of the messages that contain the same term in the same node. Finally, we calculate the sum of the two parts with a linear weight parameter α and normalize it into $[0, 1]$. Obviously, in this way, we get an upper bound score for each term in each node.

2. After we get all the scores in each node for a term, the score of the term can be integrated. It is computed by adding the score of several nodes to make the score value as big as possible. The rule must be kept that the nodes involved should cover the whole area of the given region (the quad-tree).

Best First Querying Algorithm. Pseudo code of the detailed algorithm implementation is provided in Algorithm 2.

Algorithm 2. GetTopKTerms(QuadTree tree, Query query)

---

1 $C \leftarrow$ *a priority queue storing candidate words;*
2 *Result $\leftarrow$ the list storing the top k results;*
3 **foreach** $i = C.poll()$ *AND Result.size()* $<$ *query.getK()* **do**
4 $\quad$ *tree.traverse(root, node) :*
5 $\quad$ *i.score'* $\leftarrow$ *score of i in one of node's children;*
6 $\quad$ *i.score'* $\leftarrow$ *getWordScore(tree, query, i, α);*
7 $\quad$ **if** *i.score'* $<$ *i.score* **then**
8 $\quad\quad$ *replace(i.score, i.score');*
9 $\quad$ *until node is leaf and get the exact score : i.score_exact;*
10 $\quad$ *C.add(i);*
11 $\quad$ **if** *C.poll().score* $==$ *i.score_exact* **then**
12 $\quad\quad$ *Result.add(i);*
13 *Return Result;*

α is a preference parameter to balance the location proximity and term frequency. In Line 2, C is a priority queue that stores all the candidate terms. To get the candidate terms, we extract the summaries in root node of the quad-tree. However, if the candidates are stored in many nodes while the user specified k is only a very small number, the redundant computing of a large number of term scores will incur much additional time cost on calculating useless results. So we come up with a pruning method to avoid unnecessary computations while ensuring that we do not miss any candidate terms.

The pruning process is as follows: after we get the exact k from user, we recompute the score of the k-th term, making the score of "Distance part" to 0 as a lower bound. Then, from the $(k+1)$-th term in the root summary (since the summaries are all sorted), we recompute their scores of "Distance part" as full values as their upper bound. When i-th $(i > k)$ term has an upper bound score that is still smaller than the lower bound score of the k-th term, we believe that all the terms after the i-th term have no possibilities to get onto the top of the priority queue in the near future of k times of manipulation of Lines 4–13 in Algorithm 2.

Lines 4–13 show the process to find the exact score of a term. For each candidate which is popped from the top of the priority queue, we traverse the whole tree from the root to leaf nodes. If we find a smaller score in a child node than in the parent node, we replace the current score with the new smaller score and insert the new score into the queue until we get a small enough score which is equal to the top element in the priority queue. Then, this term with an exact score will be added in our result list - see Lines 12–13.

4   Experiments and Analysis

We conduct experiments to evaluate the solution and compare with other feasible methods. All the experiments are conducted on a workstation with Intel(R) Xeon(R) CPU E5-2643 0 @3.30 GHz and 64 GB main memory on a 64-bit Windows operating system. And the whole framework is implemented in Java.

The dataset consists of tweets collected in the United States. It has 20,000,000 messages, each of which contains a timestamp, a list of terms, and the longitude and latitude of the tweet (i.e., the geographical tag set by user). Notice that the result of each set of experiments is averaged over 10 independent trails with different query inputs.

4.1   Baselines

We use the following exact algorithm as the baseline for making comparison and validation of our approach. The indexing structure of the baseline is also based on the quad-tree. Specifically, in each leaf node of the quad-tree, we store the exact frequency of each term. When a message arrives, we update the frequency in corresponding node. To get the frequency information of a non-leaf node, we traverse the quad-tree recursively until we reach the leaf node. This approach can return the exact result of an LkTQ. Therefore, it can be used as a measure of querying accuracy in subsequent experiments.

4.2   Index Updating of Quad-Tree

First, we conduct an experiment that evaluates the performance when insert and remove a message in the sliding window. Since we aim to find the top-k term over a sliding window, when the sliding window is full, each time a new message is generated, an old message should be deleted.

We find that the two manipulations in baseline and our approach scarcely cost time as this is based on a well-constructed quad-tree. Therefore, we conduct another experiment to find out the time cost of constructing a quad-tree with all the term frequencies computed and index updating. Experiment results are shown in Fig. 3.

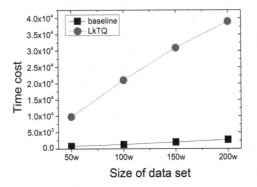

Fig. 3. Time cost of updating index on varied size of data

Specifically, for baseline, constructing the quad-tree includes counting and merging all the term frequencies while for our approach, the construction stage includes computing all the MG summaries of all the nodes in the quad-tree. As we can see, the time cost is much higher in our approach than in baseline. However, we conduct more experiments to prove that, even in this situation, our approach is much more efficient than the baseline.

4.3   Varying Message Capacity in Quad-Tree Leaf Node

Recall that when we construct a quad-tree to index all the messages, we have a condition to determine when we split the node and generate new child nodes. The condition is that when the number of messages in a node reaches M, then the node becomes a parent node and should split. We conduct an experiment to vary the maximum number of messages stored in a leaf node, so that we can find out what is the best message capacity of leaf node with better performance.

Other parameter settings are: the targeted k is 20, $\alpha = 0.7$ and the number of counters in MG summary is 500. Specially, the number of counters is set to 500 mainly for large data sets to reduce summary error.

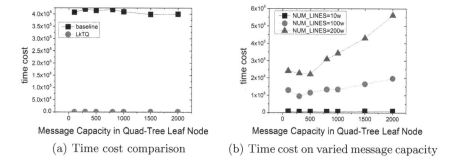

(a) Time cost comparison   (b) Time cost on varied message capacity

Fig. 4. Varying message capacity in quad-tree leaf node

Figure 4 shows the results. Figure 4a is the comparison results when the data set size is 10,000. M is ranged from 100 to 2000. Our approach is significantly faster than the baseline. It has a little fluctuation in varying M. The message capacity of quad-tree leaf node has no big influence on the performance in baseline. Once it fix M, the tree is fixed and the score is stable to be computed. However, M actually influences the performance of our algorithm. In theory, the bigger the M is, the smaller the depth of the quad-tree is. Because when computing score in each node, we use the nearest edge to the query in "Distance part", if the tree is deeper, then the distance will be smaller and the number of leaf node is larger. As Fig. 4b shows, as M increases, time cost is higher. When M is getting larger, the cost of splitting is larger. There is a little turning down when M is around 300 and 500. In this range, it has almost the best performance.

4.4   Varying Targeted k

In this experiment, we vary the targeted k. The targeted k is actually specified by users, and other fixed parameters are set as follows: $\alpha = 0.7$, the maximum number of messages in each leaf node M is 1,000, and the number of counters in MG summary is 100. Although M around 300 to 500 has the most excellent results, 1,000 is chose for controlling the quad-tree depth and for more accurate results. Because, experiments are conducted to prove that, when M is close to 1,000, the results will be consistent when other parameters varied.

Figure 5 shows the results. The range of targeted k is set according to the normal requirements of users. The performance of our algorithm is remarkably better than the baseline, which counts one by one (Fig. 5a). The size of data set in Fig. 5a is 10,000, however, baseline need approximately seven minutes to return the results. The time cost of baseline is on a stable and inefficient level which is

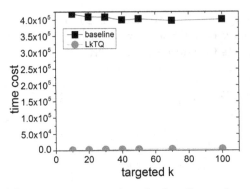

(a) Time cost comparison for baseline and our algorithm

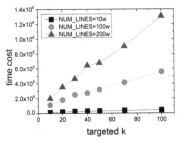

(b) Time cost on varied sizes of data

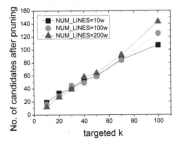

(c) No. candidates after pruning

Fig. 5. Varying targeted k

around 400,000 ms. For more larger data set, baseline has an extremely slow running speed, for instance, dealing with 5,000 messages, it needs about 12 million milliseconds and it costs nearly 60 million milliseconds to handle 100,000 messages, which is very inefficient. So we do not show the non-competitive results.

Actually, as expected, the time cost of our approach increases as the target-k increases. It is not obvious to see for the great disparity of the time cost on tick labels. Therefore, another further experiment has proved this shown in Fig. 5b. Moreover, as the size of data set is getting larger, the tendency is more conspicuous. Specifically, to find out the origin of the fastness, we conduct another experiment to validate the number of candidates after our pruning algorithm according to k is truly close to k. The result is shown in Fig. 5c as proof.

4.5   Accuracy Versus Baseline

Accuracy is a vital factor which users concern. The accuracy experiment results of our algorithm versus baseline are shown in Fig. 6. We measure the fraction of the correct top-k returning from our algorithm for different sizes of data sets. Since the baseline has such inefficient running speed, we choose relatively small data sets, which however, does not influence the high performance of our algorithm. When targeted-k is set to a low value, our approach produces pretty accurate results and can guarantee 80% correctness. As targeted-k becomes large, the accuracy is a little decreasing. However, the lowest accuracy is above 0.39 even when targeted-k is 100 and enable satisfy most of users' requirements.

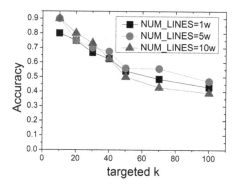

Fig. 6. Accuracy on varied sizes of data

4.6 Varying Parameter α

α is a parameter which balances the weight of the score computing formula. Varying parameter α is to adjust the influence rate of distance and term frequency. It depends on users to determine their preferences. Through experiments, it is proved that the results of our algorithm are sensitive in a range of $(0.9, 1.0)$. Certainly, when α is set to 0 or 1, then the results represent the unilateral influence of distance or frequency. Specifically, the sensitive range of α is influenced by the distribution of data sets. However, the experiments we conduct prove that our algorithm can be sensitive to the results by varing α so that it can satisfy the preferences of users.

5 Conclusion

We propose a new approach for supporting querying the top-k locally popular and valuable terms in social stream data with a huge amount of geo-tagged tweets. A comprehensive definition of term score considering both distance with queries and the term frequencies is presented. Quad-tree is used for indexing and extended to employ MG summaries to count term frequencies rapidly. Query processing adopts a best-first algorithm to pick up candidate terms and obtain exact term score for results. An empirical experiment is conducted to validate our algorithm and offers performance and accuracy of top-k term querying in geo-textual social data streams over a sliding window and the framework is capable of returning results accurately and rapidly.

Acknowledgement. This work was supported by the NSFC (U1636210, 61373156, 91438121 and 61672351), the National Basic Research Program (973 Program, No. 2015CB352403), the National Key Research and Development Program of China (2016YFB0700502), the Scientific Innovation Act of STCSM (15JC1402400) and the Microsoft Research Asia.

References

1. Agarwal, P.K., Cormode, G., Huang, Z., Phillips, J., Wei, Z., Yi, K.: Mergeable summaries. In: PODS (2012)
2. Bansal, N., Koudas, N.: BlogScope: a system for online analysis of high volume text streams. In: VLDB (2007)
3. Charikar, M., Chen, K., Farach-Colton, M.: Finding frequent items in data streams. In: Widmayer, P., Eidenbenz, S., Triguero, F., Morales, R., Conejo, R., Hennessy, M. (eds.) ICALP 2002. LNCS, vol. 2380, pp. 693–703. Springer, Heidelberg (2002). doi:10.1007/3-540-45465-9_59
4. Cong, G., Jensen, C.S., Wu, D.: Efficient retrieval of the top-k most relevant spatial web objects. PVLDB **2**(1), 337–348 (2009)
5. Cormode, G., Muthukrishnan, S.: An improved data stream summary: the count-min sketch and its applications. J. Algorithms **55**(1), 58–75 (2005)
6. Cormode, G., Muthukrishnan, S.: What's hot and what's not: tracking most frequent items dynamically. TODS **30**(1), 249–278 (2005)
7. Demaine, E.D., López-Ortiz, A., Munro, J.I.: Frequency estimation of internet packet streams with limited space. In: Möhring, R., Raman, R. (eds.) ESA 2002. LNCS, vol. 2461, pp. 348–360. Springer, Heidelberg (2002). doi:10.1007/3-540-45749-6_33
8. Felipe, I.D., Hristidis, V., Rishe, N.: Keyword search on spatial databases. In: ICDE (2008)
9. Finkel, R.A., Bentley, J.L.: Quad trees a data structure for retrieval on composite keys. Acta Inform. **4**(1), 1–9 (1974)
10. Li, F., Yao, B., Kumar, P.: Group enclosing queries. TKDE **23**(10), 1526–1540 (2011)
11. Li, F., Yao, B., Tang, M., Hadjieleftheriou, M.: Spatial approximate string search. TKDE **25**(6), 1394–1409 (2013)
12. Li, F., Yi, K., Tao, Y., Yao, B., Li, Y., Xie, D., Wang, M.: Exact and approximate flexible aggregate similarity search. VLDBJ **25**(3), 317–338 (2016)
13. Li, Y., Li, F., Yi, K., Yao, B., Wang, M.: Flexible aggregate similarity search. In: SIGMOD (2011)
14. Li, Z., Lee, K.C.K., Zheng, B., Lee, W., Lee, D.L., Wang, X.: IR-Tree: an efficient index for geographic document search. TKDE **23**(4), 585–599 (2011)
15. Lian, X., Chen, L.: Shooting top-k stars in uncertain databases. VLDBJ **20**(6), 819–840 (2011)
16. Karp, R.M., Shenker, S., Papadimitriou, C.H.: A simple algorithm for finding frequent elements in streams and bags. TODS **28**(1), 51–55 (2003)
17. Ozsoy, M.G., Onal, K.D., Altingovde, I.S.: Result diversification for tweet search. In: Benatallah, B., Bestavros, A., Manolopoulos, Y., Vakali, A., Zhang, Y. (eds.) WISE 2014. LNCS, vol. 8787, pp. 78–89. Springer, Cham (2014). doi:10.1007/978-3-319-11746-1_6
18. Manku, G.S., Motwani, R.: Approximate frequency counts over data streams. In VLDB (2002)

19. Metwally, A., Agrawal, D., El Abbadi, A.: Efficient computation of frequent and top-k elements in data streams. In: Eiter, T., Libkin, L. (eds.) ICDT 2005. LNCS, vol. 3363, pp. 398–412. Springer, Heidelberg (2004). doi:10.1007/978-3-540-30570-5_27

20. Metwally, A., Agrawal, D., El Abbadi, A.: An integrated efficient solution for computing frequent and top-k elements in data streams. TODS 31(3), 1095–1133 (2006)

21. Misra, J., Gries, D.: Finding repeated elements. Sci. Comput. Program. 2(2), 143–152 (1982)

22. Rocha-Junior, J.B., Gkorgkas, O., Jonassen, S., Nørvåg, K.: Efficient processing of top-k spatial keyword queries. In: Pfoser, D., Tao, Y., Mouratidis, K., Nascimento, M.A., Mokbel, M., Shekhar, S., Huang, Y. (eds.) SSTD 2011. LNCS, vol. 6849, pp. 205–222. Springer, Heidelberg (2011). doi:10.1007/978-3-642-22922-0_13

23. Nutanong, S., Tanin, E., Zhang, R.: Incremental evaluation of visible nearest neighbor queries. TKDE 22(5), 665–681 (2010)

24. Sankaranarayanan, J., Samet, H., Teitler, B.E., Lieberman, M.D., Sperling, J.: Twitterstand: news in tweets. In: GIS (2009)

25. Shang, S., Ding, R., Yuan, B., et al.: User oriented trajectory search for trip recommendation. In: EDBT (2012)

26. Shang, S., Ding, R., Zheng, K., et al.: Personalized trajectory matching in spatial networks. VLDBJ 23(3), 449–468 (2014)

27. Shang, S., Zheng, K., Jensen, C.S., et al.: Discovery of path nearby clusters in spatial networks. TKDE 27(6), 1505–1518 (2015)

28. Shang, S., Chen, L., Wei, Z., et al.: Collective travel planning in spatial networks. TKDE 28(5), 1132–1146 (2016)

29. Shang, S., Chen, L., Jensen, C.S., et al.: Searching trajectories by regions of interest. TKDE 29(7), 1549–1562 (2017)

30. Shang, S., Chen, L., Wei, Z., et al.: Trajectory similarity join in spatial networks. PVLDB 10(11), 1178–1189 (2017)

31. Skovsgaard, A., Sidlauskas, D., Jensen, C.S.: Scalable top-k spatio-temporal term querying. In: ICDE (2014)

32. Teitler, B.E., Lieberman, M.D., Panozzo, D., Sankaranarayanan, J., Samet, H., Sperling, J.: Newsstand: a new view on news. In: GIS (2008)

33. Wang, Z., Wang, D., Yao, B., Guo, M.: Probabilistic range query over uncertain moving objects in constrained two-dimensional space. TKDE 27(3), 866–879 (2015)

34. Xiao, X., Yao, B., Li, F.: Optimal location queries in road network databases. In: ICDE (2011)

35. Xie, D., Li, F., Yao, B., Li, G., Zhou, L., Guo, M.: Simba: efficient in-memory spatial analytics. In: SIGMOD (2016)

36. Xie, D., Li, G., Yao, B., Wei, X., Xiao, X., Gao, Y., Guo, M.: Practical private shortest path computation based on oblivious storage. In: ICDE (2016)

37. Yao, B., Li, F., Kumar, P.: Reverse furthest neighbors in spatial databases. In: ICDE (2009)

38. Yao, B., Li, F., Hadjieleftheriou, M., Hou, K.: Approximate string search in spatial databases. In: ICDE (2010)

39. Yao, B., Li, F., Kumar, P.: K nearest neighbor queries and KNN-joins in large relational databases (almost) for free. In: ICDE (2010)

40. Yao, B., Tang, M., Li, F.: Multi-approximate-keyword routing in GIS data. In: GIS (2011)

41. Yao, B., Li, F., Xiao, X.: Secure nearest neighbor revisited. In: ICDE (2013)

42. Yao, B., Xiao, X., Li, F., Wu, Y.: Dynamic monitoring of optimal locations in road network databases. VLDBJ **23**(5), 697–720 (2014)

43. Zhang, C., Zhang, Y., Zhang, W., Lin, X.: Inverted linear quadtree: Efficient top k spatial keyword search. In: ICDE (2013)

44. Zhang, D., Chan, C., Tan, K.: Processing spatial keyword query as a top-k aggregation query. In: SIGIR (2014)

45. Zhang, D., Tan, K., Tung, A.K.H.: Scalable top-k spatial keyword search. In: EDBT, pp. 359–370 (2013)

46. Zhao, K., Chen, L., Cong, G.: Topic exploration in spatio-temporal document collections. In: SIGMOD (2016)

A Kernel-Based Approach to Developing Adaptable and Reusable Sensor Retrieval Systems for the Web of Things

Nguyen Khoi Tran[1]([✉]), Quan Z. Sheng[2], M. Ali Babar[1], and Lina Yao[3]

[1] The University of Adelaide, Adelaide, SA 5000, Australia
`nguyen.tran@adelaide.edu.au`
[2] Macquarie University, Sydney, NSW 2109, Australia
[3] The University of New South Wales, Sydney, NSW 2052, Australia

Abstract. In the era of the Web of Things, a vast number of sensors and data streams are accessible to client applications as Web resources. Web Sensor Retrieval systems (WSR) help client applications to access Web-enabled sensors needed for their operation dynamically in an ad-hoc manner. Due to the diversity of sensors and query types, a functional WSR instance must be adaptable to different usage and deployment scenarios to ensure its utility. In this paper, we focus on the systematic reuse of components to enable adaptable WSR. In particular, we propose a modular architecture for WSR and develop a kernel to support the development and composition of WSR modules. We demonstrate our solution with a reference WSR instance deployed on a Raspberry Pi 3. This instance provides five types of queries on eight types of sensors deployed across two sensor platforms. We provide our kernel and reference WSR instance as open-source under MIT license.

Keywords: Web of Things · Discovery · Search · Sensor · Architecture

1 Introduction

The *Internet of Things (IoT)* is the next exciting technology revolution since the Internet, in which software systems can network with sensors and actuators on global scale via the Internet to carry out their programming directly in the real world [8]. The emerging *Web of Things (WoT)* [14] is an implementation of IoT in which things in the real world are abstracted as Web resources and accessed with HTTP protocol according to REST architecture style (Fig. 1). For instance, in a WoT-based building management application, networked sensors and actuators can be presented as JSON documents which are addressed with URLs and accessed with HTTP GET requests. With WoT, the sensing and actuating infrastructure deployed in a building can support many applications that were not planned in the design time. Vice versa, a building management application can be adapted to many buildings. While Web-enabling sensors and

© Springer International Publishing AG 2017
A. Bouguettaya et al. (Eds.): WISE 2017, Part I, LNCS 10569, pp. 315–329, 2017.
DOI: 10.1007/978-3-319-68783-4_22

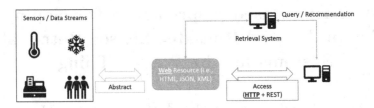

Fig. 1. Sensors and data streams are accessible to client applications as Web resources in the Web of Things. Web Sensor Retrieval systems discover and resolve queries on these resources.

actuators incurs overheads in processing time and resources, simplicity of application development due to the familiarity of the Web makes WoT a favorable solution in most scenarios [5].

To adapt themselves to different sensing and actuating infrastructure with minimal users' intervention, software systems require the ability to retrieve needed sensors and actuators from the Web. For instance, to list all empty meeting rooms near a user, the building management system requires access to passive infrared sensors that detect motion in meeting rooms to deduce their availability. While sensors can be pre-configured, the building management system is more flexible with the ability to detect and retrieve sensors in run-time. We define *Web Sensor Retrieval system (WSR)* as a system providing the ability to discover and resolve queries on Web-enabled sensors (i.e., *Web Sensors*). In a smart building, instances of WSR can be deployed on each floor to manage its sensors, or on a remote cloud service to manage all Web sensors in the building. WSR instances can also be integrated into the building management systems to interact with Web sensors in the vicinity of users.

Beside scalability in both direction, *diversity of sensors* and *queries imposing on them* are major problems in the development of WSR. In a smart building, different types of sensors such as temperature, humidity, luminance, power consumption, noise and image are deployed. They have different forms of descriptions and different quality metrics, such as refresh rate and power consumption. They produce observations with different data types and formats. Even if they produce observations with the same syntax, the real world feature that they observe might be different. For instance, the temperature in a meeting room has different meaning and behavior than the operational temperature of an HVAC system. Queries on these sensors are also diverse. They vary from simple ID look-up to combinations of textual description, spatial location and numeric value of sensors [3] (e.g., finding all available parking spots that are less than 500 meters away from the campus). Queries can also be imposed on features and characteristics of observation streams produced by sensors, such as finding power consumption sensors that show abnormalities in the last 24 h. Each type of sensor and sensor query requires different techniques to detect, index and query. Despite this diversity, a functional WSR instance must be able to retrieve any Web sensors needed by client applications.

While the diversity challenge can be addressed by equipping WSR instances with every available mechanisms and algorithms to work with any sensor and query, this solution is not appropriate for WSR instances deployed in resource-constrained scenarios (e.g., [18,19]). An alternative solution is enabling *systematic reuse* of existing research and development efforts to compose WSR instances that are *adaptable* to different usage and deployment scenarios. For instance, assume that WSR is organized into self-contained modules that encapsulate its activities (e.g., detecting, indexing, and scoring sensors), and assume that related techniques and mechanisms in the literature have already been encapsulated as modules. Then, a WSR instance that queries temperature-humidity sensors (e.g., DHT22) on seasonal feature of their observations can be composed from modules that detect, parse, index DHT22 sensors and resolve queries on seasonal feature of time series. Extending the WSR instance to support other types of sensors and queries can be done by changing its modules. This approach also increases the usage and external evaluation of research efforts in WSR.

In this paper, we propose a kernel-based approach to facilitating the systematic reuse of research results in developing adaptable WSR. We have designed and developed a *WSR Kernel* that defines modules, which encapsulate activities of WSR, and valid ways to compose them into functional WSR instances. These definitions, in form of abstract classes, provide the foundation to develop modules that can be integrated into other WSR instances that share the kernel. Bootstrapping utilities for composing modules into WSR instances, according to the configuration of developers, are also provided by the kernel. Our approach is inspired by modern operating systems, in which all functionality and operations are developed as plug-able components around a micro kernel. Our approach allows WSR to be extended and adapted to different scenarios with modules acquired via package management systems similarly to Ubuntu OS with APT and Python with PIP.

Our major contributions are as the following:

- *Defining the architecture of an adaptable and reusable Web Sensor Retrieval system*. This architecture presents modules comprising a functional WSR instance. Each module encapsulates a specific activity in the discovery and search processes of WSR.
- *Designing and developing a kernel for WSR to facilitate systematic reuse of modules and adaptation*. This kernel provides abstract classes on which modules are developed, and utilities to compose them into a WSR instance without modifying their source code.
- *Developing a reference WSR instance using our kernel-based approach*. This reference WSR instance is capable of resolving five types of queries on eight types of sensors that deployed across multiple hosts. It can be deployed on low-power computing devices such as Raspberry Pi.
- *Providing the kernel and reference WSR instance as open source under MIT license*. These resources can be accessed at our repository[1].

---

[1] https://bitbucket.org/nguyen_tran__/wsr.

2   Related Works

Effective management of large-scale things over the Web relies on the discovery and search capability provided by retrieval systems [14]. Most existing research prototypes and industrial solutions are built specifically for a predefined scenario. These systems can be organized into three groups based on the type of entity being sought. The first group, commonly referred to as Discovery Service, consists of systems that *retrieve data records* of physical objects which are scattered across supply chains. Discovery Service is a component of the EPCglobal architecture [4]. BRIDGE project (funded by European Commission) investigated Discovery Service [1], and a part of its results contribute to the IETF draft on the Extensible Supply-Chain Discovery Service (ESDS) [12]. The second group of retrieval systems *works with actuation services.* These systems provide semantic models for things [2] and their functionality [7]. Based on these models, they retrieve actuation services that are semantically relevant to users.

The third group retrieval systems works with Web sensors. Early research prototypes, including MAX [19], Microsearch [15] and Snoogle [18], focus on searching for sensor nodes based on their embedded textual content. They investigate the distribution of information retrieval techniques over a network of low-power sensor nodes. As WoT and Web sensor emerge, more works focus on querying sensors based on their observations at the query time. The key challenge of these projects is predicting sensors' observations to avoid costly sensor pull operations. It is solved with regression using SVM [20], fuzzy sets [17] and patterns in observation streams [10]. Other works focus on integrating spatial information (e.g., ThinkSeek [13], IoT-SVK [3]) and quality metrics [11] into sensor queries. In general, each research prototype exceeds at a different aspects of WSR. However, reusing and composing them into a more optimal WSR instance are not supported.

Industrial solutions relevant to Web sensor search consist of open-source Information Retrieval (IR) systems and databases. IR systems such as Elasticsearch[2], Apache Solr[3], Xapian[4] and Indri[5] are efficient in processing large corpora. However, they lack efficient processing for streams and spatial data, which are crucial for Web sensors. Time series databases such as InfluxDB[6] and Warp10[7] are more suitable. They provide efficient storage for observation streams and query capability for their metadata. However, most stated systems require hardware with strong computing capability, which prevents them from being deployed at the edge of WoT. Moreover, substantial amount of development effort is required to build WSR instances around these systems. This effort might not be reusable.

---

[2] https://www.elastic.co/.

[3] http://lucene.apache.org/solr/.

[4] https://xapian.org/.

[5] https://www.lemurproject.org/indri/.

[6] https://www.influxdata.com/.

[7] http://www.warp10.io/.

3   Background

In this section, we present an overview of WSR in term of its main tasks: discovery and search.

3.1   Discovery

Discovery is the process of *detecting* sensors and, optionally, *collecting* them into *sensor collections* (Fig. 2a). This process varies by its scope, trigger, and collection activity. The scope of discovery process either extends across the globe via the Internet (i.e., *global scope*) [10, 13] or limits to the wireless communication range of the device hosting the WSR instance (i.e., *local scope*) [18, 19]. Global discovery in WoT can be considered WoT crawling [13]. The discovery process can be executed continuously as a background task, or triggered by a user's request. This feature affects the implementation of sensor collection activity. Discovery processes implemented as background tasks tend to collect and store the information of detected sensors. On the other hand, processes triggered by users tend to return the detected sensor information to users without storing.

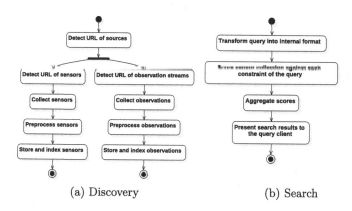

(a) Discovery   (b) Search

Fig. 2. Discovery and search activities of WSR.

For small sensing infrastructure, discovery process alone is sufficient. WoT applications can filter the small list of detected resources by themselves. However, larger sensing infrastructure and more complex queries, such as finding "power consumption sensors on the forth floor of the campus building that recorded irregularities in the last 24 h", require more sophisticated query mechanisms.

3.2   Search

The *Search* process resolves queries on discovered sensors (Fig. 2b). Let s be a sensor, and S be the collection of discovered sensors. For each sensor query q, a set of sensors $r_q \subseteq S$ that are relevant to the query exists. The task of search process is constructing the result set $\widetilde{r_q} \subseteq S$ that approximates the unknown r_q. This task is done by evaluating the relevance of each discovered sensor s against the given query q with a relevance function $f(s, q)$. If the relevance function produces binary result, the result set comprises of sensors scored positive and the process is considered *"sensor selection"* or "lookup" (Eq. 1). In case the relevance function produces a real number, the result set contains sensors whose score higher than a predefined threshold α (Eq. 2). This process is called *"sensor scoring"*.

$$\text{Selection: } \widetilde{r_q} = \{s \in S | f(s, q) = 1\} \text{ where } f(s, q) \in (0, 1) \tag{1}$$

$$\text{Scoring: } \widetilde{r_q} = \{s \in S | f(s, q) \geq \alpha\} \text{ where } f(s, q) \in \Re \tag{2}$$

While the sensor search process shares the formalization with text information retrieval, it is different in three ways. First, sensor search relies on rich metadata and descriptive information that varies according to sensor's context (e.g., spatial location, interacting users). Second, content of sensors (observation streams) is unbounded in size and constantly changing. On the other hand, content of text documents are commonly fixed. Finally, text content is small and infrequent in sensors. Therefore, sensor search process relies on the sensor models in addition to text descriptions.

Sensor Model: Sensor models define constituting components and metadata of sensors, such as real-world features that they observe. This information determines types of query that a WSR instance can resolve. Notable sensor models in the existing literature includes W3C's Web Thing model [16] and OGC's Sensor Thing API model [9]. In our project, WSR is designed based on a subset of the Sensor Thing API model. Each sensor (s) has an unbounded number of datastreams (str), which store time-stamped observations (obs) (Eq. 3). Each datastream observes a property of a real-world feature. Consider a temperature sensor deployed in a meeting room. Its datastream observes the "apparent temperature" property of the "meeting room" feature. Each observation is a time-stamped measurement of the feature's property.

$$s = \{str | str = \{obs | obs \text{ is a time-stamped observation}\}\} \tag{3}$$

It should be noted that this sensor model is not restricted to real-world, physical phenomenons such as temperature and humidity. It can model any measurable properties and features, such as daily performance of a software development team or daily revenue of a business.

4   Architecture of a Web Sensor Retrieval System

In this section, we propose a modular architecture for Web Sensor Retrieval systems. We develop our architecture from overlapping activities between existing

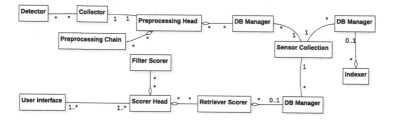

Fig. 3. Architecture of an adaptable and reusable Web Sensor Retrieval system.

WSR prototypes. We model each activity as a module and define a valid combination of these modules into a WSR instance. The existence of common activities between different types of WSR instances is the rationale of our approach. These activities represent building blocks of WSR. By interchanging these blocks, we can form different variations of WSR to adapt to any usage and deployment scenarios. These building blocks facilitate the systematic reuse of research and development efforts in building adaptable WSR.

Figure 3 presents modules comprising WSR. *Detector* and *collector* modules encapsulate detection and collection activities of the discovery task. Each collector possesses a chain of preprocessing modules to perform optional cleaning and transformation on sensor data before inserting it into *sensor collections* via *DB managers*. *Indexer* modules construct and maintain indexes of sensor collections. They can either be invoked one time or continuously in response to specific events. The search process is encapsulated into chains of *scorer* modules, which are linked in parallel or sequence. A scorer can be either a *retriever* which generates search results, or a *filter* which only reduces the set of results from other scorers. The *user interface* module passes queries from users to scorer chains and displays the search results. It utilizes different forms of interaction, from RESTful API to Web Socket to Augmented Reality (AR).

4.1   Reusable Processing Chain

A notable feature of our architecture is reusable processing chains, which model the preprocessing of collected sensors and query scoring. Each processing chain consists of a chain head and unbounded number of chain members. Each member encapsulates an independent and reusable processing such as denoising, accumulating data and scoring. Chain heads arrange the execution of members in either sequence or parallel. They invoke the chain and perform the final processing step. In a sequential chain, the head acts as the last member of the chain. In a parallel chain, the head acts as an aggregator or results from chain members. Figure 4 presents parallel and sequential arrangement of a processing chain with two members. By organizing complex processing into independent modules, the reusable processing chain model facilitates the systematic reuse of processing mechanisms and algorithms across WSR instances.

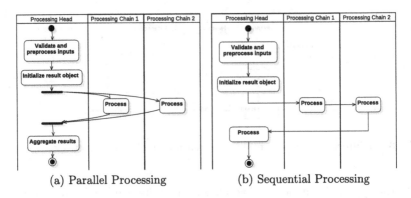

(a) Parallel Processing   (b) Sequential Processing

Fig. 4. Parallel and Sequential arrangement of user-defined processing.

4.2 Adaptability of the Modular Architecture

Majority of existing WSR prototypes can be mapped onto our modular WSR architecture. WSR instances that have explicit discovery tasks, search tasks and sensor collections, such as ThinkSeek [13], Dyser [10] and IoT-SVK [3], have nearly 1:1 mapping onto our architecture. For instance, WoT crawlers in ThinkSeek and Dyser are mapped into detectors and collectors. Three forms of sensor scoring on textual description, spatial location and numerical values of IoT-SVK [3] can be mapped into a parallel scorer chain consisting of three chain members. WSR instances that operate without sensor collections (e.g., Disco-WoT [6]) or on predefined sets of sensors (e.g., CASSARAM [11]), can be mapped into a subset of our architecture that excludes detector and collector. For WSR prototypes that distribute the processing over multiple sensor nodes (e.g., MAX [19] and Snoogle [18]), each node can be mapped onto our architecture as a WSR instance.

5 Kernel-Based Approach to Developing WSR

Our modular WSR architecture facilitates systematic reuse of existing efforts in WSR instances via interchanging modules. However, it does not support the development of reusable modules and sharing them between WSR instances. A common solution to this problem is developing an open-source library of mechanisms used in each module, so that researchers can integrate them into their own WSR instance. Drawbacks of this approach includes the lack of support for composing modules into a WSR instance, and the bottleneck at the integration of new mechanisms into the library.

We propose an alternative *kernel-based approach*, which is inspired by modern operating systems, to develop, share and compose modules. In our approach, each WSR instance is built around a *WSR kernel* that automates the composition of modules into a functional system. The kernel also defines interface, attributes and key tasks of modules, which are implemented by WSR developers. It simplifies

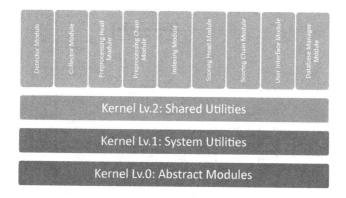

Fig. 5. Kernel of a Web Sensor Retrieval system

the development of modules by separating system management logic from the processing defined by users, such as scoring sensors and recognizing context from observations. By providing the definitions and facilities to develop modules, the kernel also improves their interoperability. An implementation of our kernel in Python is available on our repository (see footnote 1).

The WSR kernel has three levels (Fig. 5). Level 0 contains abstract modules which provide the foundation to develop reusable WSR modules. Level 1 contains system utilities for composing modules into WSR instances. They validate configuration scripts given by users, confirm the existence of required modules, initialize threads to accommodate modules and setup the communication between them. Figure 7 presents the initialization process carried out by the Lv.1 kernel. Level 2 provides basic utilities that modules might need, such as HTTP client and server. This level is more prone to change than the lower ones.

5.1   Abstract Modules

Figure 6 presents the hierarchy of abstract modules defined at the Lv.0 kernel. All concrete modules utilized in WSR instances are developed from the leaves of this hierarchy. The root `AbsModule` defines three sets of parameters that are used in the initialization of all modules. *System parameters* contain shared memory objects, queues and pipes for communicating with other modules. They are specified by the Lv.2 kernel in the bootstrapping process. *Module parameters* and *sub-modules* are specified by WSR instance developers in configuration scripts. These parameters are accessible to the user-defined processing methods.

Each abstract module contains concrete methods for performing system management tasks and invoking *abstract methods* which contain user-defined processing. For instance, in the `AbsRunnableModule` that represents a module running on a separated thread, the `loop()` concrete method repetitively invokes the abstract method `proc()` after every fixed number of milliseconds specified in the configuration script. When developing a module based on

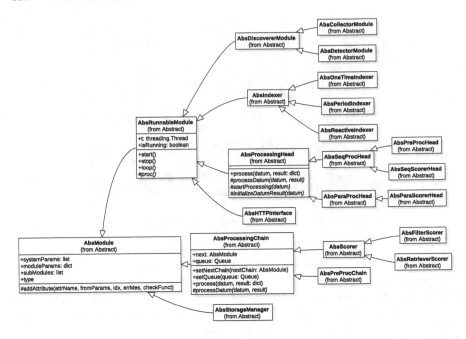

Fig. 6. Class diagram of abstract modules defined in Lv.0 kernel. Only attributes and methods of essential classes are presented.

`AbsRunnableModule`, developers only need to implement `proc()` method. The underlying thread management and looping are managed by the LV.1 kernel and the `AbsRunnableModule`.

5.2   Bootstrapping Process

Every WSR instance is defined by a configuration script. It lists modules comprising the instance, provides locations of their executables and declares their parameters.

The bootstrapping process composes executables of modules into a functional WSR instance. The process starts with validating the given configuration script and the existence of listed modules. Figure 7 presents the bootstrapping process of an ordinary WSR instance. During this process, each module performs its own validation to ensure that all of its attributes are provided in the configuration file. At the end of a successful bootstrap, a WSR instance is composed and started.

6   Evaluation

Due to the lack of quantitative measurements to assess the ability of a architecture in supporting diverse sensor types and queries, we evaluate our solution

Fig. 7. The bootstrapping process carried out by WSR kernel.

with a demonstration. In this demonstration, we assume the role of researchers who are developing a complex WSR instance to support a personal building management system. This WSR instance must interact with eight types of sensors deployed across two sensor platforms. One of which belongs to the building, the other belongs to the city. The instance must resolve five types sensor query: *(i)* searching by ID, *(ii)* by metadata, *(iii)* by observed property, *(iv)* by feature of interest (e.g., spatial location) and *(v)* by current observation. It must be deployed on a Raspberry Pi 3 (RPi3), which also acts as a Web gateway for sensors in the building, and accessible by module devices (Fig. 9).

Two key requirements of our WSR instance are the adaptability to work with new types of sensors and modify scoring mechanisms in the future. Our WSR architecture and kernel address these demands. The effectiveness of our approach is assessed via the degree that it meets our expectation as WSR instance developers. To be specific, we expect to be able to focus on major tasks of WSR (e.g., scoring a sensor) without having to consider the underlying mechanisms of the system, such as how query are passed from the user interface and how scorer threads are synchronized. In the next section, we present major modules that we have developed and the support that the WSR kernel provides in each one.

6.1   Reference WSR Instance

Major software and hardware components are presented in Fig. 9. To emulate Web sensors of the building and the city, we replay sensor datasets from Intel Lab[8] and City Pulse project[9] with a Web Sensor platform (WSP) that we developed. This platform models and presents each sensor as a Web resource according to OGC Sensor Thing API standard. We deploy one instance of WSP on RPi3 to emulate the sensor platform of the building. The other one is deployed on a workstation to emulate the sensor platform of the city. We utilize an off-the-shelf REST client application, deployed on a tablet PC, to emulate a WSR client.

Constituting modules and their composition in the reference WSR instance is presented in Fig. 8. Our WSR instance utilizes Mongo DB for storing, indexing and querying Web sensors. The kernel provides `AbsStorageManager` class to encapsulate sensor collections. We develop the `MongoDBStorageManager` module

---

[8] http://db.csail.mit.edu/labdata/labdata.html.
[9] http://ict-citypulse.eu/.

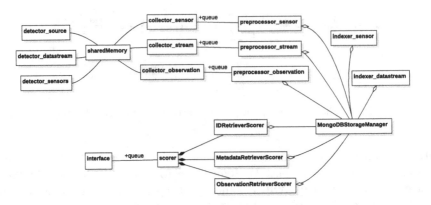

Fig. 8. UML class diagram of the reference WSR instance. "Queue" type associations indicate that instances of involving classes are connected with thread-safe queues. Normal associations indicate connection via function-calls.

from this abstract class. Abstract methods for connecting to the database and performing CRUD operations are implemented using `PyMongo` library[10].

The discovery process is performed by three detectors, three collectors and three preprocessor chains. The `detector_source` module extracts URLs pointing to WSP instances from module parameters and puts them into the shared memory object, provided by the kernel. Remaining detectors extract URLs of sensors and datastreams with the assumption that sources are compliant to OGC Sensor Thing API. Collectors are extended from `AbsCollectorModule`. We implement the abstract method `collect()` to download the JSON document at the detected URL with the REST client provided in LV.2 of the kernel. The shared memory object is made available, thread-safe, to the `collect()` method by the kernel without requiring inputs from the developer. Each collector module is connected to a preprocessing chain via a thread-safe queue. We implement a chain member from `AbsPreProcChain` module to remove symbols that violate the syntax of Mongo DB from collected JSON documents. The chain head utilizes `MongoDBStorageManager` module to insert processed data to the sensor collection.

The search process is implemented as a parallel chain of retriever scorers. For scorers, we extend `AbsRetrieverScorer` module from the kernel and implement its `retrieve(query, result)` with the query capability of Mongo DB. For scorer head, we extend `AbsParaScorerHead` and implements its aggregation method. WSR kernel manages the passing of queries to `retrieve()` and returning of result lists to aggregation method in the scorer head. The user interface is implemented using `AbsHTTPInterface`. Figure 10 demonstrates the interaction with our WSR instance from a REST client.

From these modules, we compose a WSR instance by declaring its components in a JSON file and invoking the bootstrapping utility. This utility

---

[10] https://api.mongodb.com/python/current/.

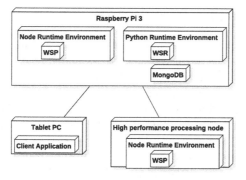

Fig. 9. Deployment of our reference WSR instance.

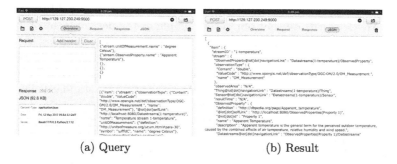

(a) Query   (b) Result

Fig. 10. Screen shots from the REST client application, demonstrating a query for sensors that observe apparent temperature and report in degree Celsius.

connects modules based on their name. For instance, modules `collector_[type]` and `preprocessor_[type]` are connected via a queue if their types are identical. Figure 8 presents the architecture of the reference WSR instance. Its source code in Python is available on our repository (see footnote 1).

6.2   Discussion

The WSR kernel simplifies the development of WSR instances by hiding irrelevant system management tasks from developers to reduce the cognitive loads and the amount of codes to develop. In the reference WSR instance, developers only implement **36 methods (500 lines of codes)** in the total **162 methods (1655 lines)** making up the system. In other word, developers using the WSR kernel only need to work on 22% of their WSR instance. This figure will be *even lower* if most modules are already available from other projects. Because of loose coupling of modules and their independence from the composition process, the resulting WSR instance is reusable and adaptable. It can be extended with additional modules. For instance, WoT crawlers [13] can be integrated as a source detector module. Influx DB can be added to provide more efficient stream storage. Its integration can be done similarly to the existing Mongo DB.

7   Conclusion and Future Works

In the Web of Things, sensors and their observations are accessible as Web resources. Web Sensor Retrieval systems provide software applications the ability to discover and search for relevant Web sensors. They allow applications to adapt to diverse and dynamic sensing infrastructure in WoT. Diversity of sensors, observations and types of query are major challenges in developing WSR. Over-provisioning is not an efficient solution and inapplicable in resource-constrained WSR instances. In this paper, we address the diversity challenge by enabling systematic reuse of research and development efforts to develop adaptable WSR instances. We develop a WSR kernel that defines modules of WSR and provides bootstrapping utilities to compose these modules into specific WSR instances. We evaluate the feasibility of our solution with the reference WSR instance that provides five types of queries on eight types of sensors from two different sensor sources. We demonstrate that our kernel simplifies the development of modules and allows them to be integrated without modifying the source code. The kernel and reference WSR instance are available on our repository (see footnote 1).

Future improvements of our solution include simplifying the ability to share modules online via package managers. We also consider standardizing query interfaces of WSR instances. With standardized interface, WSR instances can be federated to provide the needed computing capability and coverage for the Web of Things. Supporting Web-enabled actuators is also a potential extension to the WSR kernel. Finally, we consider integrating security and privacy protection into modules and inter-module communications to support these critical requirements from the architecture level.

References

1. BRIDGE: WP02 - High Level Design for Discovery Services (2007). http://www.bridge-project.eu/index.php/workpackage2/en/
2. Christophe, B., Verdot, V., Toubiana, V.: Searching the 'Web of Things'. In: Proceedings of the 5th IEEE International Conference on Semantic Computing (ICSC), pp. 308–315. IEEE (2011)
3. Ding, Z., Chen, Z., Yang, Q.: IoTSVKSearch: a realtime multimodal search engine mechanism for the internet of things. Int. J. Commun. Syst. **27**(6), 871–897 (2014)
4. EPCglobal: The EPCglobal Architecture Framework (2005). http://www.gs1.org/sites/default/files/docs/architecture/architecture_1_3-framework-20090319.pdf
5. Guinard, D., Trifa, V., Mattern, F., Wilde, E.: From the internet of things to the web of things: resource-oriented architecture and best practices. In: Uckelmann, D., Harrison, M., Michahelles, F. (eds.) Architecting the Internet of Things, vol. 1, 1st edn, pp. 97–129. Springer, Heidelberg (2011). doi:10.1007/978-3-642-19157-2_5
6. Mayer, S., Guinard, D.: An extensible discovery service for smart things. In: Proceedings of the Second International Workshop on Web of Things, pp. 1–6. ACM (2011)
7. Mrissa, M., Mdini, L., Jamont, J.P.: Semantic discovery and invocation of functionalities for the web of things. In: Proceedings of the IEEE 23rd International Conference on Enabling Technologies: Infrastructure for Collaborative Enterprises (WETICE), pp. 281–286. IEEE (2011)

8. Ngu, A.H.H., Gutierrez, M., Metsis, V., Nepal, S., Sheng, Q.Z.: IoT middleware: a survey on issues and enabling technologies. IEEE Internet Things J. (IoT-J) **4**(1), 1–20 (2017)

9. Open-Geospatial-Consortium: OGC SensorThings API Part I: Sensing - OGC Implementation Standard (2016). http://docs.opengeospatial.org/is/15-078r6/15-078r6.html

10. Ostermaier, B., Romer, K., Mattern, F., Fahrmair, M., Kellerer, W.: A real-time search engine for the web of things. In: Proceedings of the 1st International Conference on the Internet of Things (IOT), pp. 1–8. IEEE (2010)

11. Perera, C., Zaslavsky, A., Christen, P., Compton, M., Georgakopoulos, D.: Context-aware sensor search, selection and ranking model for internet of things middleware. In: Proceedings of the IEEE 14th International Conference on Mobile Data Management (MDM), vol. 1, pp. 314–322. IEEE (2013)

12. Rezafard: Extensible Supply-chain Discovery Service Problem Statement (2008). http://tools.ietf.org/html/draft-rezafard-esds-problem-statement-03

13. Shemshadi, A., Sheng, Q.Z., Qin, Y.: ThingSeek: a crawler and search engine for the internet of things. In: Proceedings of the 39th International ACM SIGIR Conference on Research and Development in Information Retrieval, pp. 1149–1152. ACM (2016)

14. Sheng, Q.Z., Qin, Y., Yao, L., Benatallah, B.: Managing the Web of Things: Linking the Real World to the Web, vol. 1. Morgan Kaufmann, Burlington (2017)

15. Tan, C.C., Sheng, B., Wang, H., Li, Q.: Microsearch: when search engines meet small devices. In: Indulska, J., Patterson, D.J., Rodden, T., Ott, M. (eds.) Pervasive 2008. LNCS, vol. 5013, pp. 93–110. Springer, Heidelberg (2008). doi:10.1007/978-3-540-79576-6_6

16. Trlfa, V., Gulnard, D., Carrera, D.: Web thing model (2015). http://www.w3.org/Submission/wot-model/

17. Truong, C., Romer, K., Chen, K.: Fuzzy-based sensor search in the web of things. In: Proceedings of the 3rd International Conference on the Internet of Things (IOT), pp. 127–134. IEEE (2012)

18. Wang, H., Tan, C.C., Li, Q.: Snoogle: a search engine for pervasive environments. IEEE Trans. Parallel Distrib. Syst. **21**(8), 1188–1202 (2010)

19. Yap, K.K., Srinivasan, V., Motani, M.: MAX: human-centric search of the physical world. In: Proceedings of the 3rd International Conference on Embedded Networked Sensor Systems, pp. 166–179. ACM (2005)

20. Zhang, P., Liu, Y., Wu, F., Liu, S., Tang, B.: Low-overhead and high-precision prediction model for content-based sensor search in the internet of things. IEEE Commun. Lett. **20**(4), 720–723 (2016)

Reliable Retrieval of Top-k Tags

Yong Xu, Reynold Cheng[(✉)], and Yudian Zheng

The University of Hong Kong, Pok Fu Lam, Hong Kong
yxu94@connect.hku.hk, {ckcheng,ydzheng2}@cs.hku.hk

Abstract. Collaborative tagging systems, such as Flickr and Del.icio.us, allow users to provide keyword labels, or *tags*, for various Internet resources (e.g., photos, songs, and bookmarks). These tags, which provide a rich source of information, have been used in important applications such as resource searching, webpage clustering, etc. However, tags are provided by casual users, and so their quality cannot be guaranteed. In this paper, we examine a question: given a resource r and a set of user-provided tags associated with r, can r be correctly described by the k most frequent tags? To answer this question, we develop the metric *top-k sliding average similarity (top-k SAS)* which measures the reliability of k most frequent tags. One threshold is then set to estimate whether the reliability is sufficient for retrieving the top-k tags. Our experiments on real datasets show that the threshold-based evaluation on top-k SAS is effective and efficient to determine whether the k most frequent tags can be considered as high-quality top-k tags for r.

Experiments also indicate that setting an appropriate threshold is challenging. The threshold-based strategy is sensitive to a little change of the threshold. To solve this problem, we introduce a parameter-free evaluation strategy that utilizes machine learning models to estimate whether the k most frequent tags are qualified to be the top-k tags. Experiment results demonstrate that the learning-based method achieves comparable performance to the threshold-based method, while overcoming the difficulty of setting a threshold.

1 Introduction

Collaborative tagging systems, such as *Flickr, Delicious.com, Last.fm*, and *Bibsonomy*, enable web resources (e.g., songs, photos, video clips, and bookmarks) to be annotated and shared among Internet users. Users, or *taggers*, can provide text labels or *tags* to describe the resources that they have uploaded or viewed. Typically, a set of selected tags, which provide a succinct summary of the associated resource, are displayed to users. These tags form a rich source of information for various novel applications, including webpage ranking [1], social interest discovery [8] and resource recommendation [14].

In this paper, we investigate the *top-k tag* problem: Given an integer k, a resource r, and a set of tags about r, which are the k most representative tags? If these tags are selected well, they can provide good description about the resource r, which could benefit the performance of the tag-based applications

© Springer International Publishing AG 2017
A. Bouguettaya et al. (Eds.): WISE 2017, Part I, LNCS 10569, pp. 330–346, 2017.
DOI: 10.1007/978-3-319-68783-4_23

mentioned above. Here, k can be freely set according to the requirement of specific application. For example, Last.FM usually only displays 5 tags, while Del.icio.us shows more tags.

Our work study the use of frequency to find the top-k tags. The frequency of a tag reflects users' consensus on describing a resource, such as how common, popular, or descriptive a tag is among taggers [13,18]. Figure 1 shows all the 37 distinct tags of the URL "Google hottrends" (http://www.google.com/trends/hottrends), sorted in descending order of frequencies, on a dataset obtained from Delicious.com (This dataset, provided by the authors of [17], contains the URLs and their tags in the whole year of 2007). Observe that the keywords with the five highest frequencies (i.e., "google", "search", "trends", "statistics", and "internet") constitute appropriate descriptions about this URL. This implies that frequency is a good indicator to select the top-k tags.

Sometimes, however, frequency cannot be used to return top-k tags reliably. In Fig. 1, supposing that k is large (e.g., 30), not all the keywords with the 30 highest frequencies make sense. Notice that only the first 17 tags appear twice or more times; the remaining 20 tags have been provided only once. Any 13 of these 20 tags could qualify to be one of the top-30 labels. However, some of the tags, such as "firefox: toolbar", appear to be a bad choice. Indeed, noisy, misspelled, and ambiguous tags are commonly found in collaborative tagging systems [6,10,16]. As we can see in this example, using frequency inappropriately may allow low-quality tags to be included in the top-k tags.

To address the above paradox of utilizing tag frequency, we propose to evaluate whether these k tags with the highest frequencies can be qualified as the top-k tags of a resource, instead of directly returning the k most frequent tags. We first design a metric, *top-k sliding average similarity (top-k SAS)*, to quantify the reliability of the k-most-frequent tags. A high top-k SAS implies that the k tags with the highest occurrence frequencies form a good description of r. Then a threshold is set to determine whether the reliability is sufficient or not. Only if top-k SAS score passes the threshold, the k-most-frequent tags can be considered as the top-k tags of the resource.

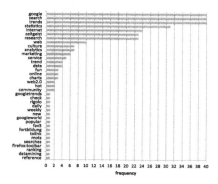

Fig. 1. Tag frequency distribution.

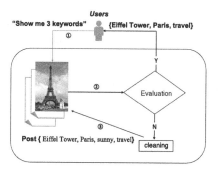

Fig. 2. Reliable retrieval of top-k tags.

Figure 2 interprets the process of top-k tags retrieval. Similar to existing collaborative tagging systems, this application allows web user to retrieve the top-k tags of a resource r he is interested in (step 1). Instead of directly returning the current k-most frequent tags, the system first tests how much the tag frequency distribution of r is reliable to be used. If top-k SAS passes the threshold and returns "yes", then the k-most-frequent tags of r are returned to the user (Step 2). Otherwise, the system will conduct cleaning on existing data (Step 3), then re-test whether the k-most-frequent tags can be returned. The process continues until the evaluation result is satisfied. A possible way of cleaning is to hire a group of users with high quality to annotate the resources. In this paper, we mainly focus on the evaluation part and leave the cleaning strategies in future discussion.

Comprehensive experiments (Sect. 3.3) on real datasets demonstrate that the evaluation is both reliable and efficient in retrieving the top-k tags of a resource. However, we find the performance is quite sensitive to the threshold. A little change of the threshold has huge effect on the result. And for different applications (e.g., Del.icio.us and Last.FM), there is hardly a universal threshold. Thus, setting an appropriate threshold becomes a challenge of the threshold-based evaluation method. This motivates us to explore another strategy that can automatically do the evaluation. Finally, we model the problem as a binary classification problem and apply machine learning methods to solve it.

In summary, we make the following contributions in this paper. (1) We propose the metric, top-k Sliding Average Similarity (top-k SAS), that measures the reliability of retrieving the k most frequent tags of a resource as top-k tags. (2) We introduce a threshold-based approach to estimate whether the top-k SAS is satisfactory or not. Experiment results on real datasets show that, under appropriate threshold, threshold-based top-k SAS is both effective and efficient. It achieves satisfactory accuracy while saving considerable tagging cost. It is also faster (5 times faster when $k = 5$) to compute than existing metric. (3) To address the challenge of setting threshold, we propose a learning-based evaluation strategy to evaluate whether the k-most-frequent tags are reliable or not. Experiments show that learning-based method can get comparable performance as threshold-based method while overcoming the difficulty of setting threshold.

The rest of the paper is as follows. Section 2 discusses the related work. We then present the top-k SAS measure and threshold-based evaluation method in Sect. 3. The details of learning-based method is introduced in Sect. 4. Section 5 concludes.

2   Related Works

Let us now review the related work in collaborative tagging and existing measures for evaluating the quality of tags.

Collaborative Tagging. Collaborative tagging has emerged as a novel way to organize web resources for searching, filtering, navigation and sharing. User-generated tags arouse a lot of research interest and various applications as well.

In [1], Bao et al. optimize page ranking using collaborative tags. They observe that collaborative annotations provide good description of corresponding web-pages, then propose algorithms that incorporate tagging information into page ranking. Ramage et al. use tag data to improve web page clustering [12]. Li et al. [8] study social network in collaborative tagging system (i.e. del.icio.us) and utilize the patterns of frequent co-occurrences of tags to capture topics of users' interest. Eirini Giannakidou et al. [3] combine tag data, social knowledge and content-based information to improve resource clustering.

However, since users can freely assign tags to resources, the quality of social tags is hard to control. In [16], Robert Wetzker. et al. investigate the tagging dynamics of *Del.icio.us* system, and point out that social bookmarking systems are vulnerable to various forms of spam. There are problems related to synonym, polysemy and misspelling errors as well [10], which affect the utility of social tags. Lipczak et al. [9] propose to recommend appropriate tags based on resource content and user profile. Another concern of collaborative tagging systems comes from the large amount of "under-tagged" resources that receive very few posts. Yang et al. [17] observe that there are in *Del.icou.us* dataset of 2007, over 10 million URLs were tagged just once, while only a few URLs were tagged more than 10, 000 times. Those resources cannot be reliably used, thus could affect the quality of many tag-based applications mentioned above. In this paper, we also observe the problem that when a resource receives not enough posts, its top frequent tags may not be reliable to be returned as the top-k tags. And in order to check the reliability of frequent tags, in this paper, we propose two evaluation methods, one threshold-based approach and the other learning-based approach, that are both effective and efficient.

Stability of Collaborative Tagging. Our work is inspired by previous research on the stabilization of tag frequency. In [4,5], the authors observe that if a resource receives more posts, the frequencies of its tags become more stable, and the tag distribution appears as a long-tail curve. There are several works that study the tag stabilization phenomenon in collaborative tagging systems. In [2,7,11], the authors try to generate the tag distribution and use power law model to fit it. Yang et al. [17] proposed the concept of tag stability to measure the quality of a resource's tags. They studied the relative frequency distributions (*rfd*) of tags per resource, and further develop a metric, Moving Average (MA) score, to measure the tagging stability by calculating the average cosine similarity of adjacent rfds. Wagner et al. [13] studied the limitations and potentials of three other state-of-the-art methods for measuring semantic stability of tag distributions: *Stable Tag Proportions* [4], *Stable Tag Distributions* [5] and *Power Law Fits* [2], then proposed a *Rank-based Semantic Stability Method* based on Rank Biased Overlap [15] similarity measure.

After studying all existing works, we find most of them can be summarized in an uniform formula which we calls "SAS" (sliding average similarity). Instead of directly adopting SAS, in our work, we propose a top-k SAS to measure the stability of the frequencies of k most frequent tags. The details of SAS and top-k SAS are discussed in Sect. 3.1.

3   Top-k SAS and Evaluation

In this section, we present the details of top-k SAS and how it is utilized to reliably return the top-k tags of a resource through threshold-based evaluation approach. Comprehensive experiments are conducted to evaluate the performance of top-k SAS together with threshold-based evaluation strategy.

3.1   Design of Top-k SAS

Let us first explain the intuition behind top-k SAS. We use the term "post" to denote a tagger's action of uploading a set of text labels to the collaborative system. Figure 3 shows the (relative) frequency values of five most frequent tags, over the number of *posts* collected for resource "Google hottrends" during a period of one year. We can see that the more posts are made, the less is the fluctuation of frequency values. In this experiment, after about 61 posts, the frequencies of these five tags become "stable". Intuitively, after taggers give posts for a sufficient number of times, the statistics (in this case, frequency) reflect the popularity and importance of tags among taggers. Furthermore, if the frequencies of these five tags are observed to be consistently higher than those of other tags, then these tags can be regarded as "high-quality" top-k tags.

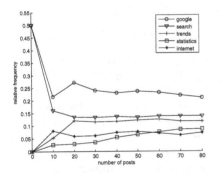
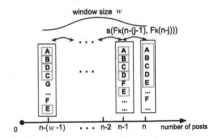

Fig. 3. Frequencies of five tags in "Google hottrends"

Fig. 4. Interpretation of top-k SAS

Based on above observation, we develop the *top-k SAS* which measures how "stable" the k-most-frequent tags are. The formal definition of top-k SAS is inspired by state-of-the-art research on tagging stability. For example, Halpin et at. [5] adopted KL-divergence to measure the stability of tag distributions. Yang et al. [17] proposed a metric, Moving Average (MA) score, to measure the tagging stability by calculating the average cosine similarity of adjacent relative frequency distributions. Most of the existing methods can be presented in an uniform formula which we call *Sliding Average Similarity (SAS)*, denoted by $g(n, \omega, d, s)$. Given a resource r, the formulation of **SAS** is summarized as follows.

$$g(r, n, \omega, d) = \frac{1}{\omega - d} \sum_{j=0}^{\omega-d-1} s(F(n - j - d), F(n - j)), \qquad (1)$$

where $F(n)$ is the tag frequency distribution when a resource receives n posts. $g(n, \omega, t, s)$ is assessing tagging stability by calculating the average similarity of $(\omega - d)$ pairs of consecutive tag distributions $(F(n - j - d), F(n - j))$. d is the step size between two consecutive tag distributions. s is the similarity metric used in stability assessment. ω is the window size for taking the average value of similarity between pairs of consecutive tag distributions.

In this paper, we develop top-k SAS instead of directly adopting SAS because our aim is to check the stable status of the k most frequent tags rather than the whole bunch of tags. Obviously, SAS is a more strict measurement (i.e. when k frequent tags become stable, the whole set of tags are not guaranteed to become stable), thus requiring more tagging effort to stabilize the whole set of tags. The advantage of top-k SAS over SAS can be found in experiment results in Sect. 3.3.

Let us now present top-k SAS. Let $F_k(m)$ be a k-size vector that describes the frequencies of top-k tags after the m-th post is received. The **top-k SAS** is defined as

$$g(r, k, n, \omega, d) = \frac{1}{\omega - d} \sum_{j=0}^{\omega-d-1} s(F_k(n - j - d), F_k(n - j)), \qquad (2)$$

where

$$F_k(m) = (f(v_1, n), \ldots, f(v_k, n)), m = n - (\omega - 1), \ldots, n \qquad (3)$$
$$v_i = V_k(n)[i], i = 1, \ldots, k \qquad (4)$$

Figure 4 illustrates the idea of top-k SAS, where $k = 5$ and $d = 1$. We compare the similarity between the pair of $F_k(m - 1)$ and $F_k(m)$ values (for $m \in [n - w + 2, n]$), over a window ω. Notice that if a resource receives less than ω posts, g will be set to 0, which means the resource with few posts is really not stable. To determine $F_k(m)$, we first set the k-most-frequent tags after receiving n posts, $V_k(n)$, as the domain. In this example, $V_k(n)$ are $\{A, B, C, D, E\}$. Then each $F_k(m)$ is composed of the relative frequencies of $V_k(n)$ on receiving m posts. If one tag v_i doesn't appear at m-th post, its relative frequency will be set to 0.

3.2 Threshold-Based Evaluation on Top-k SAS

The above section presents the formal definition of top-k SAS to measure the stability of the k-most-frequent tags of a resource. Given a top-k SAS score, another important problem is to determine whether the score is satisfactory so that k-most-frequent tags can be returned as the top-k tags. Or else, more posts are still required. Most of the existing works only analyze tagging stabilization qualitatively [2,7,11], few present how to compute the stability in a quantitative way [13,17], but seldom study how to check whether a resource becomes stable based on the stability score.

To address the problem, we propose a threshold-based strategy in which a threshold t is set as stability standard. This is inspired by Yang. et al. [17]. In their work, when users keep on assigning posts to a resource, a point is set as stable point once SAS score passes a threshold 0.9999. The idea is quite straightforward and easy to implement. Top-k SAS increases with more and more posts received. Once Top-k SAS achieves the threshold, the k most frequent tags are assessed to be stable so that they can be returned as the top-k tags of the resource.

3.3   Experiments

Dataset. We select two datasets to perform experiment. One is *Del.icio.us*, provided by the authors of [17]. The dataset recorded the tag sequences of 5,000 selected URLs throughout the year 2007, with an average of 112 posts each URL. The 5,000 URLs are selected from all the URLs of Del.icio.us bookmarks in 2007. The selection criteria is that the frequencies of all their associated tags achieved stability at the end of 2007, where achieving stability means SAS passes a threshold 0.9999 [17] with ω set as 20. The other dataset *Last.FM* is publicly available from HetRec 2011 workshop (http://ir.ii.uam.es/hetrec2011/datasets. html), which contains 186,480 bookmarks from August 1, 2005 to May 9, 2011. This dataset initially contains many under-tagged resources, thus, we select a set of stable resources as done for *Del.icio.us* dataset. The only difference is the threshold of criteria. We use 0.999 as selection threshold instead because 0.9999 is extremely high for *Last.FM* dataset. Among the whole 4024 resources, only 12 resources whose SAS passes 0.9999 and the sample is too small to be used. Finally, we use 0.999 for *Last.FM* dataset and 159 resources are selected.

Hardware. We implemented our experiments in Python, and tested them in OS X 10.9.4 with a 2.6 GHz Intel Core i5 processor and 8 GB of RAM.

Evaluation Metric. To evaluate the effectiveness of top-k SAS, we compare two stability scoring measures: (1) top-k SAS, proposed in this paper, and (2) SAS, as introduced in Eq. 1. Note that SAS is essentially a special case of top-k SAS where frequency information is assessed on *all* tags instead of k highest frequent ones. When conducting comparison between top-k SAS and SAS, we use threshold-based method to determine the stable status of a resource.

We will compare top-k SAS and SAS from two main aspects. During the tagging process of a resource r, there will be one point at which top-k SAS tells it stable, and the other stable point determined by SAS. First, we will compare the quality of the k highest frequent tags at these two points. Let $G_k(r)$ be the true top-k tags of resource r. Suppose that it takes r to call for m posts to achieve the top-k stability threshold t, then we use $S_k(r,m)$ to denote its set of the k most frequent tags from these m posts. Then the accuracy of $S_k(r,m)$, is:

$$accuracy@k = \frac{|S_k(r,m) \bigcap G_k(r)|}{k} \tag{5}$$

We use the above equation to measure the quality of the k most frequent tags extracted from resource r. In our experiments, since the frequencies of tags for each resource r achieve stability by the end of 2007, we treat the k most frequent tags of r obtained at the final post of 2007 as G_k. We also compare the number of posts, **#posts**, needed to achieve the stability threshold by top-k SAS and SAS respectively.

In the default setting, for both top-k SAS and SAS, we set the window size ω to 10, step size d to 1, and similarity metric s as cosine similarity. In *Del.icio.us* dataset, we set the default *stability threshold t* to 0.9999, and 0.999 for *Last.FM* dataset. We will show the effect of w, t, d, s on these measures in the following Section.

Results

Effect of Posts. Let us first examine the impact of the number of posts on the "stable portion" of resources (i.e., the fraction of the resources whose top-k SAS scores has exceed t.) Here we set $k = 10$, and the conclusion is similar for other k. Figure 5 shows the result of *Del.icio.us* dataset ($t = 0.9999$). As we can see in Fig. 5(a), when the average number of posts for a resource is 90 or more, more than 70% of resources are highly stable to 0.9999. Figure 5(b) shows a comparison between accuracy and top-10 SAS over different number of posts. We can see that after 90 posts, the accuracy is also very high. As discussed before, in many collaborative tagging systems, posts can be given by taggers in a voluntary manner. This means that for these 70% of resources, no further action is necessary, and their 10-most-frequent tags can be returned to users as top-10 tags right away. For the remaining 30% of resources, they can be further processed by workers (e.g., in the crowdsourcing-enabled system illustrated in Fig. 2). Thus, top-k SAS can be used to help to prune away a large number of resources from being crowdsourced to workers, potentially saving tremendous crowdsourcing costs.

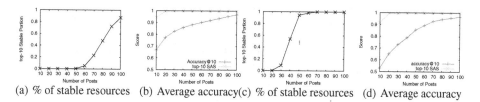

(a) % of stable resources   (b) Average accuracy(c) % of stable resources   (d) Average accuracy

Fig. 5. Effect of posts, Del.ici.ous   **Fig. 6.** Effect of posts, Last.FM

The similar result can be found in *Last.FM* dataset, as shown in Fig. 6(c) and (d). When the average number of posts for a resource is 40, nearly 60% of resources can achieve the stability threshold 0.999.

Effect of k. Now we examine the effect of k on *accuracy@k* (Eq. 5) and the number of posts needed to reach threshold t for both datasets. Figure 7 shows the result of *Del.icio.us* dataset. Figure 7(a) shows that with k increasing, *accuracy@k* slightly decreases, which is reasonable because it might be more difficult for two larger sets to have similar proportion of common tags than smaller sets. Over different values of k, top-k SAS yields an accuracy of 0.9 or more which is a sufficiently high quality. In Fig. 7(b), we see that the number of posts required to reach t increases with k.

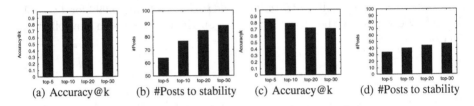

(a) Accuracy@k   (b) #Posts to stability   (c) Accuracy@k   (d) #Posts to stability

Fig. 7. Effect of k, Del.ici.ous   **Fig. 8.** Effect of k, Last.FM

Table 1 presents the comparison between top-k SAS and SAS on *accuracy@k* and #posts. Accuracy with top-k SAS is just slightly lower than SAS. However, as shown in Table 2, the number of posts required for smaller values of k is much less than if SAS is used. For example, top-5 SAS needs only 63 posts, which is 37% less than that of SAS. The reason is that SAS requires all of tags to achieve stable, which calls for more posts. In return, a larger accuracy can be obtained. This indicates that top-k SAS is able to return accurate top-k tags with a lower tagging effort, compared with SAS.

Figure 8 shows the effect of k on *Last.FM* dataset. We can observe similar results as *Del.icio.us* dataset. The *accuracy@k* decrease when k is larger, and it takes more posts in average for a resource to achieve stability when k increases. Tables 3 and 4 show the comparison between top-k SAS and SAS. The result is consistent to the observation of *Del.icio.us* dataset.

Table 1. *Accuracy@k*, top-k SAS vs. SAS, Del.icio.us (t = 0.9999)

	Accuracy@5	Accuracy@10	Accuracy@20	Accuracy@30
Top-k SAS	0.94	0.93	0.90	0.89
SAS	0.98	0.96	0.94	0.93

Table 2. #posts, top-k SAS vs. SAS, Del.icio.us (t = 0.9999)

	Top-5 SAS	Top-10 SAS	Top-20 SAS	Top-30 SAS	SAS
#posts	63	76	84	88	99

Table 3. *Accuracy@k*, top-k SAS vs. SAS, Last.FM (t = 0.999)

	Accuracy@5	Accuracy@10	Accuracy@20	Accuracy@30
Top-k SAS	0.86	0.79	0.72	0.71
SAS	0.92	0.84	0.75	0.73

Table 4. #posts, top-k SAS vs. SAS, Last.FM, (t = 0.999)

	Top-5 SAS	Top-10 SAS	Top-20 SAS	Top-30 SAS	SAS
#posts	34	40	44	47	50

Effect of Threshold t. We also study the impact of threshold t on top-k stability on *Del.icio.us* and *Last.FM* dataset. As what have explained in Sect. 3.3, the threshold standard 0.9999 is so high that almost all of the resources in the original *Last.FM* dataset can't be used for experiments. Thus, we use 0.999, a strict standard for *Last.FM*, to select the data for experiment usage. To study the effect of threshold, for each dataset, we set 3 different thresholds for each dataset. For example, thresholds for *Del.icio.us* are 0.9999, 0.999, and 0.99. And for *Last.FM* dataset, the thresholds are set to 0.999, 0.99 and 0.9.

Figure 9 shows the result of *Del.icio.us*. A lower t will require less posts to achieve stability, but as a result, the accuracy decrease. For example, when t is set to 0.999, an average number of 40 posts can make most of the 5000 resources get stable. Well, the accuray will sacrifice and decrease to 0.8. We see that when t is lowered to 0.99, the number of posts is significantly reduced, but accuracy is lowered to 0.7 which is barely satisfactory. Thus, the top-k stability is quite sensitive to the use of t; a slight decrease of t causes the accuracy to degrade. Figure 10 represents the similar result on *Last.FM* dataset. The result also implies that setting threshold should be very careful especially for different applications because the data characteristics may vary.

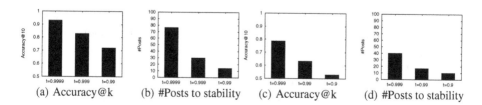

(a) Accuracy@k   (b) #Posts to stability   (c) Accuracy@k   (d) #Posts to stability

Fig. 9. Effect of k, Del.ici.ous   **Fig. 10.** Effect of k, Last.FM

The above result invokes an interesting question: how to set a proper threshold that can make good balance between accuracy and #posts? It seems that there is hardly an universal instruction to threshold setting. For example, $t = 0.99$ works fine for *Del.icio.us* dataset. The accuracy passes 0.7 with lower than 20 posts in average. While, for *Last.FM* dataset, $t = 0.99$ is not that satisfactory. This inspires us to come up with a general practice. In next section, we will introduce the details of a parameter-free method, learning-based evaluation.

Effect of Window Size ω. We next examine the effect of window size ω (Eq. 2) on top-k stability. Due to page limitation, we only show the result on *Del.icio.us* dataset. We use $k = 10$ here, but the observations are similar for other values of k. Figure 11 shows that in general, when ω increases, *accuracy* grows slightly, as the average similarity computation for top-k SAS becomes more accurate. This comes at a higher price with a larger number of posts. The value of ω of 6 is a good balance between accuracy and number of posts. Observe that top-10 SAS comes close to SAS in terms of accuracy, but saves a significant portion (about 25%) of posts needed to detect top-k stability. This is consistent with our results in Fig. 7.

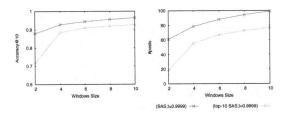
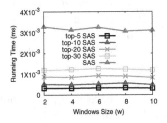

Fig. 11. Effect of ω on *Del.ici.ous*   **Fig. 12.** Efficiency comparison

Effect of s and d. We have done experiments to evaluate the effect of similarity metric and step size. However, due to the page limitation, we could not display the details of effect of s and d, but summarize the conclusion here.

For s, we implement Kendall's Tau which is usually used to assess similarity between two ranked lists. Comparing with cosine similarity, the absolute value of Kendall's Tau is smaller. The core difference between Kendall's Tau and cosine similarity is that the former considers the rank of each item while the latter considers the specific value. Over all 5000 resources, the average top-k SAS by cosine similarity is 0.99995, while for Kendall's tau 0.9893. When top-k SAS with cosine similarity passes the threshold and tells that the top-k tags become stable, top-k SAS by Kendall's Tau judges it not stable and requires more posts to achieve stability. This implies that the stability standard for different similarity metrics should be different.

We also vary the step size d to see its effect. We find that when step size increases, top-k SAS decreases. This makes sense because two tag distributions far apart have more difference than two adjacent ones. A larger step size improves accuracy in the cost of more posts, as the effect of window size ω.

Efficiency. Finally, we examine the efficiency of top-k SAS measures. Figure 12 shows that top-k SAS is quick to compute. For example, top-5 SAS can be computed in 3×10^{-4} ms on average. The window size ω has little effect on the performance. When k increases, more time is needed, because similarity computation for k-dimensional vector pairs is more costly. SAS is the slowest; it takes 10 times longer than top-5 SAS and 3 times than top-30 SAS to finish.

Summary. The experiment results show that top-k SAS outperforms SAS. First, top-k SAS is more reliable than SAS. It requires fewer posts to determine that the k-most-frequent tags are good representatives of top-k tags. This is important to a "crowdsourcing-enabled" tag retrieval system (Fig. 2), since fewer workers need to be invoked, thereby saving the cost. Moreover, top-k SAS is much faster than SAS to compute.

4   Learning-Based Stability Evaluation

The experiment results in Sect. 3 demonstrate that threshold-based top-k SAS achieves good results. However, a big challenge is that the performance is so sensitive to threshold that it's hard to set a proper threshold. There is not a one-for-all threshold for different domains of applications.

   To solve the problem, in this section, we introduce a learning-based method that can automatically evaluate the stable status instead of manually setting threshold for every application. The goal is to evaluate whether the k most frequent tags become stable or not, which can be modeled as a binary classification problem. Figure 13 interprets the idea of learning-based method. First, we pre-process the original resources and their post sequences to get the labeled dataset. Then we use the labeled dataset to train a classifier using common classification models (i.e. Logistic Regression, Support Vector Machine). Now when a new sample comes, the classifier can directly evaluate whether the k most frequent tags of the resource are stable or not. Experiments are conducted to demonstrate its performance and the comparison between threshold-based and learning-based methods.

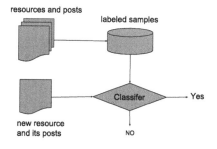

Fig. 13. Learning-based stability evaluation

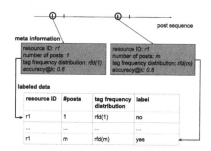

Fig. 14. Labeled data preparation

4.1 Dataset Preparation and Representation

The tagging process of a resource r forms a post sequence. Each time the resource receives a new post, we can capture its meta information such as the current number of posts r receive, the tag frequency distribution, the *accuracy@k* score (defined in Sect. 3.3), etc. The meta information will be utilized to prepare the dataset for learning. For example, given a resource r with totally n posts, we take the meta information after receiving each post as a sample. The sample will be labeled as yes if its *accuracy@k* passes 0.8, and no if *accuracy@k* is lower than 0.8. In the experiment, we use *accuracy@10* because 10 tags is a general setting in collaborative tagging websites to describe a resource. And 0.8 means 8 out of 10 tags are the real top-10 tags, which indicates the k most frequent tags are already of high quality to be returned as the top-k tags.

Figure 14 illustrates how we prepare the labeled data. After processing the 5000 URLs of *Del.icio.us* dataset, we got 562,048 samples with label "stable" or "unstable". For the 159 resources on *Last.FM*, we finished with 15,360 samples.

For both datasets, we split up the data into 80% training and 20% testing. Here stratified sampling (https://en.wikipedia.org/wiki/Stratified_sampling), instead of random sampling, is used because there are more stable samples than unstable ones in the labeled dataset.

4.2 Simple Feature Extraction and Selection

In the above representation of records, resourceID, #posts, relative frequency distribution (rfd) are three original features of each sample, and stable status is what we want to predict. Before exploiting specific models for prediction, we conduct simple feature extraction and selection to remove non-useful features and extract new features from functions of the original features.

Among the original features, resourceID itself has no relation with stable status and will be removed. Then we transform rfd sequence to top-k SAS score using top-k SAS measure. #posts remains as one feature because it has relation with the target. For example, when #posts is larger, the sample is more likely to become stable. But when #posts is small, the sample appears unstable. The following is the representation of processed samples. In our example, we use top-10 SAS as the feature which aligns with *accuracy@10* in labeling the dataset.

#posts	top-k SAS	Stable status
1	0.9534	0
2	0.9728	0
.	.	
m	0.9999	1
1	0.9433	0
.	.	

Table 5. Performance on *Del.icio.us*

	Precision	Recall	F_1	AUC
Logistic regression	0.92	0.98	0.95	0.79
Random forest	0.93	0.95	0.94	0.81

Table 6. Performance on *Last.FM*

	Precision	Recall	F_1	AUC
Logistic regression	0.86	0.90	0.82	0.80
Random forest	0.84	0.86	0.85	0.78

4.3   Learning and Evaluation

Our goal is, given the meta information of each sample, to predict whether its k most frequent tags are stable to be returned as top-k tags. This can be considered as a typical binary classification problem.

There are models that can perform binary classification, such as Logistic Regression, Supporting Vector Machine (SVM), Random Forest, Deep Learning, etc. Each model has its own advantage. Random forest is especially powerful when the number of features is large. Deep learning is a good choice for large dataset, but it takes a long time for training. SVM runs slow when training on large volume of data. In our problem, since the features are simple and dataset is large, we choose Logistic Regression and Random Forest to perform the classification.

Considering there are more stable samples than unstable ones, the overall classification accuracy may not be an appropriate measure of performance. If an unstable sample is classified to be stable (False Positive), the accuracy will sacrifice. If a stable sample is classified to be unstable (False Negative), it will require more tagging cost. Thus, we use various metrics for evaluation from different perspectives, such as precision, recall, F_1 score (https://en.wikipedia.org/wiki/Precision_and_recall) and AUC.

4.4   Results

Here we present the results on the samples from *Del.icio.us* and *Last.FM* datasets. Table 5 shows the performance of Logistic Regression and Random Forest on *Del.icio.us* dataset. We can find that F_1 achieves 0.95 which indicates a good overall classification result. In Logistic Regression method, the recall value is high to 0.98, well, precision is slightly lower. This implies that the model intends to predict a sample to be stable when it is still unstable. Comparing with Logistic Regression, Random forest performs a little bit better on precision but slightly worse on recall. Table 6 shows the result of *Last.FM* dataset. The performance is slightly worse, which is reasonable due to the relatively small dataset for learning.

We also evaluate the average *accuracy*@10 and the number of posts needed to achieve stable status over the whole dataset. For each resource, we use its post sequence to simulate the tagging process. Once it is classified to be stable by the classifier, we record the #posts and *accuracy*@10 for each resource. Tables 7 and 8 show the statistics. For *Del.icio.us* dataset we use Random Forest classifier, with an average 14 posts, resource will be classified to be stable, and the *accuracy*@10 reaches around 0.75. It is slightly lower than *accuracy*@10 of 0.8 which we use to label a resource stable. It is reasonable because some resources get lower accuracy because they are misinterpreted as stable. While, for all the right predictions, we record the minimal number of posts with which the *accuracy*@10 can just achieve the passing-line accuracy 0.8. Thus, the overall *accuracy*@10 is lower than 0.8. The similar result can be found in *Last.FM* dataset.

Table 7 also shows the comparison between threshold-based and learning-based method on *Del.icio.us* dataset. We can see that although the result of learning-based method is not as good as the result when large threshold is set, it is comparable to that when $t = 0.99$. The *accuracy*@10 from learning-based method is higher than that when $t = 0.99$ with the same number of posts. The similar results can be found on *Last.FM* dataset in Table 8. The *accuracy*@10 from learning-based method is slightly lower than $t = 0.999$, but helps save 7 posts. And the accuracy is better than other two threshold settings. The results once again indicates that the performance of threshold-based method is quite sensitive to the threshold setting. Different applications require different threshold setting strategies. While the learning-based method shows its advantage in automatically determining the stable status regardless of various application domains.

Table 7. Comparison between threshold-based method and learning-based method, Del.icio.us

	$t = 0.9999$	$t = 0.999$	$t = 0.99$	Learning-based method
accuracy@10	0.93	0.83	0.72	0.74
#posts	76	30	14	14

Table 8. Comparison between threshold-based method and learning-based method, Last.FM

	$t = 0.999$	$t = 0.99$	$t = 0.9$	Learning-based method
accuracy@10	0.79	0.63	0.53	0.75
#posts	40	17	10	33

4.5 Summary

In this section, we present the results of learning-based method to evaluate the stable status of tag distributions. Results show that the learning-based classification works well in predicting the stable status given the meta tagging information of a resource. Learning-based method can effectively address the challenge of setting threshold for top-k SAS. Currently, we only utilize two simple features for learning the stable status. In the future, we will explore more on the data characteristics of collaborative tagging and make more effort on feature engineering to improve prediction accuracy.

5   Conclusion

In this thesis, we raise a question concerning the quality of keywords (top-k tags) of resources in collaborative tagging systems. We investigate a simple but representative meta information, tag frequency, and propose a metric, top-k Sliding Average Similarity (SAS), to measure the reliability of the k-most-frequent tags being retrieved as the real top-k tags. The empirical study on two real datasets show that top-k SAS provides a more accurate and efficient metric for the top-k tags retrieval problem than existing stability measurement.

Another important problem we investigate into is how to evaluate whether the stability is to a satisfactory extent so that the k most frequent tags can be returned as the top-k tags. Two methods are introduced to solve the problem. One is threshold-based inspired by existing work [17]. The other one is a parameter-free method based on machine learning classification model. Comprehensive experiments are conducted on both methods. The threshold-based method performs well if only proper threshold can be set. The learning-based method delivers comparable result to threshold-based one while overcoming the difficulty of setting threshold.

Acknowledgement. Xu Yong, Reynold Cheng, and Yudian Zheng were supported by the Research Grants Council of Hong Kong (RGC Projects HKU 17229116 and 17205115) and the University of Hong Kong (Projects 102009508, 104004129, and 201611159247). We would like to thank the reviewers for their insightful comments. We would also like to thank Prof. Wang Chien Lee (The Pennsylvania States University) for his valuable advice for the initial solution.

References

1. Bao, S., Xue, G., Wu, X., Yu, Y., Fei, B., Su, Z.: Optimizing web search using social annotations. In: WWW (2007)
2. Cattuto, C., Loreto, V., Pietronero, L.: Semiotic dynamics and collaborative tagging. Proc. Nat. Acad. Sci. (2007)
3. Giannakidou, E., Kompatsiaris, I., Vakali, A.: SEMSOC: SEMantic, social and content-based clustering in multimedia collaborative tagging systems. In: 2008 IEEE International Conference on Semantic Computing, pp. 128–135. IEEE (2008)
4. Golder, S., Huberman, B.: Usage patterns of collaborative tagging systems. J. Inf. Sci. **32**, 198–208 (2006)
5. Halpin, H., Robu, V., Shepherd, H.: The complex dynamics of collaborative tagging. In: WWW (2007)
6. Kennedy, L.S., Chang, S.F., Kozintsev, I.V.: To search or to label?: predicting the performance of search-based automatic image classifiers. In: Proceedings of the 8th ACM International Workshop on Multimedia Information Retrieval (2006)
7. Kipp, M.E., Campbell, D.G.: Patterns and inconsistencies in collaborative tagging systems: an examination of tagging practices. ASIST (2006)
8. Li, X., Guo, L., Zhao, Y.E.: Tag-based social interest discovery. In: WWW (2008)
9. Lipczak, M., Hu, Y., Kollet, Y., Milios, E.: Tag sources for recommendation in collaborative tagging systems. ECML PKDD Discov. Chall. **497**, 157–172 (2009)

10. Marchetti, A., Tesconi, M., Ronzano, F., Rosella, M., Minutoli, S.: Semkey: a semantic collaborative tagging system. In: WWW (2007)
11. Mathes, A.: Cooperative classification and communication through shared metadata. University of Illinois (2005)
12. Ramage, D., Heymann, P., Manning, C.D., Garcia-Molina, H.: Clustering the tagged web. In: WSDM (2009)
13. Wagner, C., Singer, P., Strohmaier, M., Huberman, B.: Semantic stability and implicit consensus in social streams tagging. In: WWW (2014)
14. Wan, C., Kao, B., Cheung, D.W.: Location-sensitive resources recommendation in social tagging systems. In: CIKM (2012)
15. Webber, W., Moffat, A., Zobel, J.: A similarity measure for indefinite rankings. TOIS (2010)
16. Wetzker, R., Zimmermann, C., Bauckhage, C.: Analyzing social bookmarking systems: a del.icio.us cookbook. In: ECAI Mining Social Data Workshop (2008)
17. Yang, X.S., Cheng, R., Mo, L., Kao, B., Cheung, D.W.: On incentive-based tagging. In: ICDE (2013)
18. Yi, K.: Harnessing collective intelligence in social tagging using delicious. ASIST (2012)

Estimating Support Scores of Autism Communities in Large-Scale Web Information Systems

Nguyen Thin(✉), Nguyen Hung, Svetha Venkatesh, and Dinh Phung

Centre for Pattern Recognition and Data Analytics, Deakin University,
Geelong, Australia
{thin.nguyen,hung,svetha.venkatesh,dinh.phung}@deakin.edu.au

Abstract. Individuals with Autism Spectrum Disorder (ASD) have been shown to prefer communication at a socio-spatial distance. So while rarely found in the real world, autism communities are popular in Web-based forums, convenient for people with ASD to seek and share health related information. Reddit is one such avenue for people of common interest to connect, forming communities of specific interest, namely sub-reddits. This work aims to estimate support scores provided by a popular subreddit interested in ASD – www.reddit.com/r/aspergers. The scores were measured in both the quantities and qualities of the conversations in the forum, including conversational involvement, emotional, and informational support. The support scores of the subreddit *Aspergers* was compared with that of an average subreddit derived from entire Reddit, represented by two big corpora of approximately 200 million Reddit posts and 1.66 billion Reddit comments. The ASD subreddit was found to be a supportive community, having far higher support scores than did the average subreddit. Apache Spark, an advanced cluster computing framework, is employed to speed up processing of the large corpora. Scalable machine learning techniques implemented in Spark help discriminate the content made in *Aspergers* versus other subreddits and automatically discover linguistic predictors of ASD within minutes, providing timely reports.

Keywords: Big data · Apache Spark · Large-scale distributed computing · Support scores · Autism communities

1 Introduction

Individuals who are affected by Autism Spectrum Disorder (ASD) often have social interaction and communication difficulties [1]. Web-based forums open a new support avenue where people affected by ASD could find their voice in online environments with the safety of social and spatial distances. In this research, Reddit is used as one such avenue, hosting more than 300,000 subreddits, each of which is home to people of common interest to join and discuss their own stories.

© Springer International Publishing AG 2017
A. Bouguettaya et al. (Eds.): WISE 2017, Part I, LNCS 10569, pp. 347–355, 2017.
DOI: 10.1007/978-3-319-68783-4_24

Then the term "online autism communities" is defined as subreddits that involve users affected by or interested in ASD, related to or providing supports for people with ASD. Online autism communities have been recently studied, such as in examining the topics the users are interested in and the way they express their arguments in terms of language styles; however, to date, little is known about support scores that characterize these discussions. Also, a comparison with other communities on the platform was not conducted, which probably shows how supportive the community is in relation with others in the online forum setting.

This study aims to estimate several aspects of support, including engaging, emotional, and informational support in an online community interested in ASD, namely www.reddit.com/r/aspergers – *"For safe and helpful conversation with people who have Aspergers or live with someone who does. We also welcome people with other autism spectrum conditions."* This work will then investigate into the differences in the support scores between the subreddit with the whole Reddit. Apache Spark [9], an advanced framework in cluster computing, will be employed to process the big corpora.

A key contribution of this work is to introduce coarse-grained measures of support for online communities. Another contribution is in the demonstration on the efficiency of advanced cluster computing in extracting support features and discriminating the content made in *Aspergers* versus those made in other sudreddits. The implication includes the automatic discovery of linguistic predictors of target online communities, supporting timely decision-making.

The current paper is organized as follows. Section 2 outlines the proposed methods, data, and experimental set-up. The difference in support scores between subreddit *Aspergers* and entire Reddit is described in Sect. 3. This section also reports the performance of an Apache Spark cluster in both extracting support features and classifying content made in *Aspergers* versus other subreddits. Section 4 concludes the paper.

2   Method

2.1   Data

Subreddit *Aspergers*. In this paper Reddit data was chosen since it allows people to create or join communities of common interest.

While searching for Reddit communities interested in autism, two popular subreddits were shown: *r/autism*[1] and *r/aspergers* [2]. The focus of r/autism is on news or articles related to ASD. On the other hand, r/aspergers is the venue mainly for discussions on their own stories among people who have aspergers or live with someone having ASD. As the main objective of the work is to estimate the supportiveness the members would perceive through sharing their own stories, r/aspergers is chosen into the study.

---

[1] https://www.reddit.com/r/autism/.

[2] https://www.reddit.com/r/aspergers/.

We used PRAW, a Python Reddit API wrapper to crawl posts and comments from the subreddit.[3] We crawled 19,517 posts and 310,401 comments, made by 18,816 members (redditors) of *Aspergers* from March 2010 to March 2017.

Entire Reddit. *Aspergers* will be analyzed in relation with other subreddits. To represent the entire Reddit, two large corpora were downloaded: 196,531,736 Reddit posts (251 GB, time-stamped from 24 January 2006 to 31 August 2015)[4] and 1,659,361,605 Reddit comments (908 GB, time-stamped from 15 October 2007 to 31 May 2015)[5].

2.2 Support Scores

Response Activities. The activities of commenting reflect the degree of participation of members in the forum, including the percent of posts receiving at least a comment; the average number of comments per post; and how quick a reply by other members firstly occurs after posting. These quantities could be considered as coarse proxies of support. These numbers for *Aspergers* will be compared with that of entire Reddit, derived using the big corpora.

Emotional Support. To roughly estimate emotional support of the forum, the affective information expressed in post and comment content should be measured. For the affective aspect, we used ANEW lexicon [2,5] to extract the sentiment conveyed in the content. Words in this lexicon are rated in terms of valence, arousal, and dominance. The valence of words is on a scale of one, *very unpleasant*, to nine, *very pleasant*. The arousal is measured in the same scale, one for *least active* and nine for *most active*. The dominance also ranges from one, *submissive*, to nine, *dominant*.

LIWC Features. To learn the linguistic features of the content made in Reddit, we examined the proportions of words in 68 psycho-linguistic categories as defined in the LIWC package [7], such as linguistic, social, affective, cognitive, perceptual, biological, relativity, personal concerns and spoken. These outputs of the package could be inferred into support scores. For instance, the usage of *first-person singulars* (e.g., I, me, mine), a type of function word, was used as the marker of *self-disclosure* (informational support) in health support forums.

2.3 Difference Between *Aspergers* and Entire Reddit

To examine how *Aspergers* differs from other subreddits, same support features for both *Aspergers* and entire Reddit will be pulled out. In particular, firstly, support scores a subreddit may provide will be roughly estimated through response activities. Secondly, affective scores and language features for the posts and comments will be extracted. These characteristics of *Aspergers* will be compared with that of an average subreddit, derived from the two big corpora.

---

[3] https://praw.readthedocs.org/en/stable/, retrieved March 2017.
[4] http://bit.ly/1MvQobz, downloaded 1 October 2015.
[5] http://bit.ly/1RmhQdJ, downloaded 1 October 2015.

2.4   Computing Environment

The cluster used in the experiments consists of eight worker nodes. Each node features a dual eight-core (32 vcores) Intel® Xeon® E5-2670@2.60 GHz processors, 128 gigabytes of main memory, and CentOS 7.2 operating system. Apache Spark [9] is chosen as the cluster computing platform for the experiments. We will examine the computing framework in both (1) extracting the support features and (2) discriminating posts and comments made in *Aspergers* versus other subreddits, providing predictors of the community. For the first task, the support features for *Aspergers* and the entire Reddit will be extracted using Apache Spark. The performance of extracting the features by a single machine and by a Spark cluster of eight worker nodes was compared. Running time was used as the measure of performance. For the second task, several machine learning algorithms implemented in Spark will be used to classify the content made in *Aspergers* versus others.

3   Experimental Results

3.1   Support Scores

Response Activities. The raw measures of support via commenting behavior for *Aspergers* and entire Reddit (indeed the corpora of approximately 200 million posts and 1.66 billion Reddit comments) were extracted, shown in Table 1. In all the measures, *Aspergers* is more supportive than an 'average' subreddit.

In particular, firstly, on average each post in *Aspergers* receives 16 replies, which is twice of the number for entire Reddit, showing a much higher level of conversational involvement by members of the subreddit. Secondly, only 2.4 % of *Aspergers* posts receive no comment, whereas more than 40 % of Reddit posts have no response. Finally, it takes only 3.3 h for a post in *Aspergers* to be firstly commented by other Redditors, while, for an average Reddit post, this figure is 12.4 h, somehow reflecting higher support in *Aspergers* community, in comparison with an average subreddit.

Table 1. Response activities within *Aspergers*, in comparison with entire Reddit.

	Aspergers	Reddit
Number of posts	19,517	196,531,736
Number of comments	310,401	1,659,361,605
Average number of comments per post	16	8
Number of posts receiving comments	19,041	114,814,771
Response ratio	97.6%	58.4%
Response time (in hours)	3.3 ± 37.0	12.4 ± 152.9

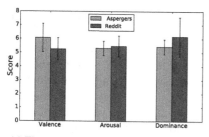

(a) The mean of sentiment scores for posts.

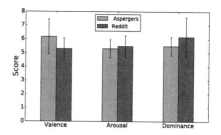

(b) The mean of sentiment scores for comments.

Fig. 1. The mean of sentiment scores for posts made in *Aspergers* and in entire Reddit.

Emotional Support. We examine the difference in sentiment scores and language styles between subreddit *Aspergers* and entire Reddit.

Figure 1 shows the mean of sentiment scores for posts and comments made in subreddit *Aspergers* and in entire Reddit. It is observed that the mean valence of the community was higher than that of an average subreddit. So, on average, more pleasant words were used in *Aspergers* than in an average subreddit. It could be because for *Aspergers*, especially in early stages of the diagnosis, words with quite high valence, such as *hope* (valence: 7.05) or *hopeful* (valence: 7.10), were used more than average. Then when passing years dealing with the disorder, they would feel *proud* (valence: 8.03) or *grateful* (valence: 7.37).

On the other hand, the arousal score of *Aspergers* is lower than that of an average subreddit. This probably implies that members of *Aspergers* tend to use more low active words in their posts and comments than does an average redditor.

Likewise, the dominance score of *Aspergers* is lower. It is likely that most members of the community tend to use more submissive and less dominant words in their posts and comments.

Psycho-Linguistics. Psycho-linguistic features for Reddit posts and comments were extracted using LIWC package [7]. Then the relative difference in the use of a feature between *Aspergers* and an 'average' subreddit is determined as

$$diff = \frac{mean_{Aspergers} - mean_{Reddit}}{(mean_{Aspergers} + mean_{Reddit})/2} * 100 \qquad (1)$$

where $mean_{Aspergers}$ and $mean_{Reddit}$ are the average of the feature for posts or comments made in subreddit *Aspergers* and in entire Reddit, respectively. If *diff* > 0, the feature is used more in *Aspergers* than in an average Reddit, and vice versa.

In Posting. Out of 68 LIWC features, 36 features have the relative differences of more than 30%, computed following Eq. 1.

The most different feature is the number of words (*word count*) used in a post, at 130% difference. Message length was considered as the proxy for the amount of communication [8]. Higher word count was found to be a signal of better communication and better group performance [8].

First-person singulars are also among the preference of members of *Aspergers* subreddit, in comparison with an average subreddit, with 77.8% difference. This shows a high level of *self-disclosure* in the forum, similar to mental health related subreddits [3], providing a high level of *informational support* to *Aspergers*'s members. The higher degree of *self-disclosure* in *Aspergers* is also confirmed by the higher use of *social* (for all of its sub-categories *family*, *friends*, and *humans*) and *cognitive* (all of its sub-categories but *inhibition*) words [3].

It is apparent to observe that *anxiety* words (e.g., worried, nervous, anxious) were used much more by *Aspergers* than by an average subreddit in posting, with 96.2% gap. Indeed the link between autism and anxiety has been found in literature, and thus *anxiety* related topics are expected to be discussed either by the sufferers or by the carers within *Aspergers* subreddit. Accordingly, words on *health* were used more by *Aspergers*, with 60.3%. This could be considered parts of *informational support*.

On the other hand, the lower in the use of *inhibition* words (e.g., block, constrain, stop) would attract comments with higher degree of emotional, instrumental, information and prescriptive advice [3].

In Commenting. About half of the LIWC features have the relative differences of more than 20%, computed following Eq. (1).

Similar to posting behavior, more words were used to reply to a post in *Aspergers*, likely more time was spent on commenting, showing a high level of conversational engagement by members of the community. In particular, the average length of comments (in number of words) made in *Aspergers* is 66.0 ($\pm$89.5), more than twice that of an average subreddit (28.6 ± 49.7).

The degree of *self-disclose*, a type of *informational support*, is also found to be highly used in commenting within the autism community. It is evidenced by higher values, than an average subreddit, of *first-person singulars*, *social*, and *cognitive* words in the comments. Higher use of *health* words also contribute to the level of *informational support* provided by the community.

On the other hand, comments made by *Aspergers* members have lesser *swear* words (e.g., d∗mn, p∗ss, f∗ck), with the gap of 98.5% compared with the averaged subreddit. This is likely to make the community peaceful to join as well as to positively contribute to the *emotional support* the community might provide.

3.2   Performance of Cluster Computing

Feature Extraction. The time needed for extracting the affective scores for the Spark cluster of eight nodes is 18 h and 45 min and for a single node of the cluster is 39 days, 2 h, and 12 min, which is *50 times* slower. Similarly, it took the Spark cluster only 3 h and 21 min to complete extracting LIWC features and took the single node 18 days, 20 h, and 34 min, which is *134 times* slower. Apart from collecting the resources of joining nodes in the cluster, one possible reason for the better performance by the cluster is that Spark supports *parallel* operations [9]. This type of operations is commonly used in independently extracting features for each data point of the large corpora.

Classifications. While the feature extraction job is simple to be scaled up by Spark, as it can be conducted independently for each data point, though not straightforward, the cluster framework is also utilized to implement parallel and scalable machine learning algorithms. We will examine this capacity through learning linguistic predictors of autism communities, probably providing tools to identity online communities of interest. The linguistic features are already extracted from Sect. 3.1 and will be used as the initial predictors in the classifications of posts and comments made in *Aspergers* versus those made in other subreddits.

To create a balanced dataset for the classification, the same number of posts and comments were sampled from the whole Reddit corpora, excluding those made by *Aspergers* subreddit. This results in two balanced datasets: approximately 40,000 posts and 600,000 comments, which are the inputs to *Aspergers* versus others classifications. The classifiers consist of popular built-in ones implemented in Spark MLlib: logistic regressions with stochastic gradient descent (SparkLR-SGD) and limited-memory BFGS (SparkLR-LBFGS); Naive Bayes; SVM; decision tree; and random forest [4]. Outside of the MLlib, we also used Spark OLR algorithm (One-pass Logistic Regression), which is recently proposed and claimed to be state of the art in both predictive performance and the execution time [6]. F-measure is used as the predictive performance and the execution time is also reported.

Figure 2 shows the result of the classifications. For predictive performance, LIWC features could gain more than 80% F-measure in post classifications and more than 70% F-measure in comment classifications. The F-measure is comparable for all the classifiers, except Spark Naive Bayes's, which is far worse than others.

For execution time, OLR is the fastest in comment prediction and the second fastest in post prediction. OLR is *five times* faster than the slowest (SparkLR-

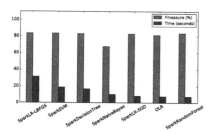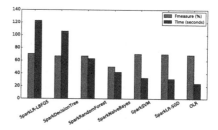

(a) Post classification as *Aspergers* versus others with LIWC as the predictors.

(b) Comment classification as *Aspergers* versus others with LIWC as the predictors.

Fig. 2. Performance of the Spark cluster in the classifications by different classifiers.

LBFGS) in the comment classification (24 s versus 123 s). The key difference helping OLR to run fast is to parallelize the computation of sufficient statistics in Spark [6].

From the coefficients of a logistic regression model learned in the post classification, top positive predictors of *Aspergers* include *word count, first person singular*, and *cogmech* (cognitive processes). This is in line with the difference found in Sect. 2.2, and can be interpreted in terms of the support scores. These features were also found as top positive predictors of *Aspergers* in the comment classification. In this comment prediction, *affective processes* (e.g., happy, grateful) is also a strong positive predictor of *Aspergers*, implying a high *emotional support* in the community. The implication from this experiment is that the predictors of target online communities of our interest are discovered within minutes, likely providing timely responses.

4   Conclusion

This study investigated support scores of an online community interested in autism in comparison with other online communities. An advanced computing framework was used to process the big corpora. The community *Aspergers* was found to be a supportive community, having much better support scores than did an average subreddit. Indeed, the conversational involvement, emotional support, and self-disclose scores of the community are higher than that of an average subreddit, showing an encouraging atmosphere within the group, inspiring the fellows to stick with conversations. The results of this study suggest that data mining of online communities has the potential to detect meaningful patterns for understanding the representations of ASD conditions online. The work also demonstrates the efficiency of an advanced computing framework in dealing with big data, supporting timely decision-making.

Acknowledgment. This work is partially supported by the Telstra-Deakin Centre of Excellence in Big Data and Machine Learning.

References

1. American Psychiatric Association.: Diagnostic and Statistical Manual of Mental Disorders, 5th edn. American Psychiatric Publishing, Arlington (2013)
2. Bradley, M.M., Lang, P.J.: Affective norms for English words (ANEW): instruction manual and affective ratings (1999)
3. De Choudhury, M., De, S.: Mental health discourse on Reddit: self-disclosure, social support, and anonymity. In: Proceedings of the International AAAI Conference on Weblogs and Social Media, pp. 71–80 (2014)
4. Meng, X., Bradley, J., Yavuz, B., Sparks, E., Venkataraman, S., Liu, D., Freeman, J., Tsai, D.B., Amde, M., Owen, S.: MLlib: machine learning in Apache Spark. J. Mach. Learn. Res. **17**(34), 1–7 (2016)
5. Nguyen, T.: Mood patterns and affective lexicon access in weblogs. In: Proceedings of the ACL Student Research Workshop, pp. 43–48 (2010)

6. Nguyen, V., Nguyen, D.T., Le, T., Venkatesh, S., Phung, D.: One-pass logistic regression for label-drift and large-scale classification on distributed systems. In: Proceedings of International Conference on Data Mining (ICDM), pp. 1113–1118 (2016)
7. Pennebaker, J.W., Francis, M.E., Booth, R.J.: Linguistic inquiry and word count (LIWC) [Computer software]. LIWC Inc (2007)
8. Tausczik, Y.R., Pennebaker, J.W.: The psychological meaning of words: LIWC and computerized text analysis methods. J. Lang. Soc. Psychol. **29**(1), 24–54 (2010)
9. Zaharia, M., Chowdhury, M., Franklin, M.J., Shenker, S., Stoica, I.: Spark: cluster computing with working sets. In: Proceedings of the USENIX Conference on Hot Topics in Cloud Computing, p. 10 (2010)

Spatial and Temporal Data

DTRP: A Flexible Deep Framework for Travel Route Planning

Jie Xu[1]([⊠]), Chaozhuo Li[1], Senzhang Wang[2], Feiran Huang[1], Zhoujun Li[1], Yueying He[3], and Zhonghua Zhao[3]

[1] School of Computer Science and Engineering, Beihang University, Beijing, China
{xujie1020,lichaozhuo,huangfr,lizj}@buaa.edu.cn
[2] Nanjing University of Aeronautics and Astronautics, Nanjing, China
szwang@nuaa.edu.cn
[3] China National Computer Network Emergency Response Technical
Team/Coordination Center of China, Beijing, China
{hyy,zhaozh}@cert.org.cn

Abstract. Route planning aims at designing a sightseeing itinerary route for a tourist that includes the popular attractions and fits the tourist's demands. Most existing route planning strategies only focus on a particular travel route planning scenario but cannot be directly applied to other route planning scenarios. For example, previous next-point recommendation models are usually inapplicable to the must-visiting problem, although both problems are common and closely related in travel route planning. In addition, user preferences, POI properties and historical route data are important auxiliary information to help build a more accurate planning model, but such information are largely ignored by previous studies due to the challenge of lacking an effective way to integrate them. In this paper, we propose a flexible deep route planning model DTRP to effectively incorporate the available tourism data and fit different demands of tourists. Specifically, DTRP includes two stages. In the model learning stage, we introduce a novel multi-input and multi-output deep model to integrate the rich information mentioned above for learning the probability distribution of next POIs to visit; and in the route generation stage, we introduce the beam search strategy to flexibly generate different candidate routes for different traveling scenarios and demands. We extensively evaluate our framework through three travel scenarios (next-point prediction, general route planning and must-visiting planning) on four real datasets. Experimental results demonstrate both the flexibility and the superior performance of DTRP in travel route planning.

Keywords: Travel route planning · Location recommendation · Trajectory mining

1 Introduction

Tourists traveling to a new place often face the problem of "travel route planning", which aims to design a sightseeing itinerary route that should cover the

A. Bouguettaya et al. (Eds.): WISE 2017, Part I, LNCS 10569, pp. 359–375, 2017.
DOI: 10.1007/978-3-319-68783-4_25

most interesting visitor attractions and fit the various travel demands of tourists [5]. Travel route planning is an important while time-consuming travel preparation activity before the tourists visit a new city for the first time [24]. A desirable route planning strategy not only can help tourists better enjoy their trips, but also contribute to free them from the time-consuming and trivial preparation works.

Existing route planning works can be roughly categorized into three categories. The first category of works utilize the relative geographical distances among the Point of Interests (POIs) to locate the shortest possible tour paths, which is also known as the "Traveling Salesman Problem" [3,5,17]. The second category treats the route planning as an recommendation problem, which aims to generate the candidate routes based on the historical travel route data of the visitors [1,27–30]. The last type of related literatures are kind of hybrid methods, which usually propose an optimization model to incorporate both POI properties and historical route data [4,11,13,19].

Although a bunch of related works studied the travel route planning problem, there are two major limitations for the existing methods. First, previous works only focus on a specific travel route planning scenario. According to the data analysis, we summarize three common route planning scenarios as shown in Table 1. Existing models are mostly designed for one scenario. For example, [1,5] try to solve the "General Planning" problem while [6,12,15] focus on the "Next-Point Recommendation" problem. It is difficult to directly apply these models to other scenarios. Hence, a more general and flexible route planning model which can be suited to different scenarios is needed. Secondly, there can be various types of data available coming from different sources, including user preferences, POI properties, and historical route data of other users, and these data can be helpful to design an desirable traveling route for a new user. However, previous single data source based models are not effective to incorporate the muti-sourced data. Besides, user preferences, POI properties and historical route data potentially encode different types of information. Traditional shallow methods, such as topic model [9] and hidden markov model [2], cannot effectively capture the highly non-linear complex correlations among them either [26].

In this paper, we aim to propose a general and flexible deep route planning model which can effectively incorporate user preferences, POI properties, and historical route data while fit various demands of tourists. This task is difficult to address due to the following challenges. Firstly, there are different kinds of constraints including travel time, cost, and must-visit POI constraint in the above mentioned scenarios, which ought to influence the generation process of traveling routes. It is non-trivial to propose a general and flexible model which can satisfy these constraints. Secondly, although combining user preferences, POI properties and historical route data into the route planning process is expected to achieve better performance, it is not obvious how best to do this under a unified framework.

To address the above challenges, we propose a general and flexible Deep Travel Route Planning model (DTRP). We formulate this task as a sequence

2. Bonamy, M., Bousquet, N.: Reconfiguring independent sets in cographs. CoRR, abs/1406.1433 (2014)

3. Bonamy, M., Bousquet, N.: Token sliding on chordal graphs. CoRR, abs/1605.00442 (2016)

4. Bonamy, M., Bousquet, N., Feghali, C., Johnson, M.: On a conjecture of Mohar concerning Kempe equivalence of regular graphs. CoRR, abs/1510.06964 (2015)

5. Bonsma, P.: The complexity of rerouting shortest paths. Theoret. Comput. Sci. **510**, 1–12 (2013)

6. Bonsma, P.: Independent set reconfiguration in cographs. In: Kratsch, D., Todinca, I. (eds.) WG 2014. LNCS, vol. 8747, pp. 105–116. Springer, Cham (2014). doi:10.1007/978-3-319-12340-0_9

7. Bonsma, P., Kamiński, M., Wrochna, M.: Reconfiguring independent sets in claw-free graphs. In: Ravi, R., Gørtz, I.L. (eds.) SWAT 2014. LNCS, vol. 8503, pp. 86–97. Springer, Cham (2014). doi:10.1007/978-3-319-08404-6_8

8. Booth, K., Lueker, G.: Testing for the consecutive ones property, interval graphs, and graph planarity using PQ-tree algorithms. J. Comput. Syst. Sci. **13**(3), 335–379 (1976)

9. Demaine, E.D., et al.: Polynomial-time algorithm for sliding tokens on trees. In: Ahn, H.-K., Shin, C.-S. (eds.) ISAAC 2014. LNCS, vol. 8889, pp. 389–400. Springer, Cham (2014). doi:10.1007/978-3-319-13075-0_31

10. Diestel, R.: Graph Theory. Graduate Texts in Mathematics, vol. 173, 3rd edn. Springer, Heidelberg (2005)

11. Feghali, C., Johnson, M., Paulusma, D.: A reconfigurations analogue of Brooks' theorem and its consequences. CoRR, abs/1501.05800 (2015)

12. Feghali, C., Johnson, M., Paulusma, D.: Kempe equivalence of colourings of cubic graphs. CoRR, abs/1503.03430 (2015)

13. Flum, J., Grohe, M.: Parameterized Complexity Theory. Springer, New York Inc., New York (2006). doi:10.1007/3-540-29953-X

14. Gopalan, P., Kolaitis, P.G., Maneva, E., Papadimitriou, C.H.: The connectivity of Boolean satisfiability: computational and structural dichotomies. SIAM J. Comput. **38**(6), 2330–2355 (2009)

15. Hearn, R., Demaine, E.: PSPACE-completeness of sliding-block puzzles and other problems through the nondeterministic constraint logic model of computation. Theor. Comput. Sci. **343**(1–2), 72–96 (2005)

16. Ito, T., Demaine, E., Harvey, N., Papadimitriou, C., Sideri, M., Uehara, R., Uno, Y.: On the complexity of reconfiguration problems. Theoret. Comput. Sci. **412**(12–14), 1054–1065 (2011)

17. Kamiński, M., Medvedev, P., Milanič, M.: Complexity of independent set reconfigurability problems. Theoret. Comput. Sci. **439**, 9–15 (2012)

18. Mouawad, A.E., Nishimura, N., Raman, V., Simjour, N., Suzuki, A.: On the parameterized complexity of reconfiguration problems. In: 8th International Symposium on Parameterized and Exact Computation, IPEC 2013, pp. 281–294 (2013)

19. van den Heuvel, J.: The complexity of change. In: Blackburn, S.R., Gerke, S., Wildon, M. (eds.) Surveys in Combinatorics 2013, pp. 127–160. Cambridge University Press, Cambridge (2013)

20. Yamada, T., Uehara, R.: Shortest reconfiguration of sliding tokens on a caterpillar. In: Kaykobad, M., Petreschi, R. (eds.) WALCOM 2016. LNCS, vol. 9627, pp. 236–248. Springer, Cham (2016). doi:10.1007/978-3-319-30139-6_19

Computing Maximum Cliques in B_2-EPG Graphs

Nicolas Bousquet[1] and Marc Heinrich[2]([⊠])

[1] G-SCOP (CNRS, Univ. Grenoble-Alpes), Grenoble, France
[2] LIRIS (Université Lyon 1, CNRS, UMR 5205), Lyon, France
marc.heinrich@univ-lyon1.fr

Abstract. EPG graphs, introduced by Golumbic et al. in 2009, are edge-intersection graphs of paths on an orthogonal grid. The class B_k-EPG is the subclass of EPG graphs where the path on the grid associated to each vertex has at most k bends. Epstein et al. showed in 2013 that computing a maximum clique in B_1-EPG graphs is polynomial. As remarked in [Heldt et al. 2014], when the number of bends is at least 4, the class contains 2-interval graphs for which computing a maximum clique is an NP-hard problem. The complexity status of the Maximum Clique problem remains open for B_2 and B_3-EPG graphs. In this paper, we show that we can compute a maximum clique in polynomial time in B_2-EPG graphs given a representation of the graph.

Moreover, we show that a simple counting argument provides a $2(k+1)$-approximation for the coloring problem on B_k-EPG graphs without knowing the representation of the graph. It generalizes a result of [Epstein et al. 2013] on B_1-EPG graphs (where the representation was needed).

1 Introduction

An *Edge-intersection graph of Paths on a Grid* (or *EPG graph* for short) is a graph where vertices can be represented as paths on an orthogonal grid, and where there is an edge between two vertices if their respective paths share at least one edge. A turn on a path is called a *bend*. EPG graphs were introduced by Golumbic et al. [11]. They showed that every graph can be represented as an EPG graph. The number of bends on the representation of each vertex was later improved in [13]. EPG graphs have been introduced in the context of circuit layout, which can be modeled as paths on a grid. EPG graphs are related to the knock-knee layout model where two wires may either cross on a grid point or bend at a common point, but are not allowed to share an edge of the grid. In [11], the authors introduced a restriction on the number of bends on the path representing each vertex. The *class B_k-EPG* is the subclass of EPG graphs

N. Bousquet—Supported by ANR Projects STINT (ANR-13-BS02-0007) and LabEx PERSYVAL-Lab (ANR-11-LABX-0025-01).

M. Heinrich—Supported by the ANR-14-CE25-0006 project of the French National Research Agency.

H.L. Bodlaender and G.J. Woeginger (Eds.): WG 2017, LNCS 10520, pp. 140–152, 2017.
https://doi.org/10.1007/978-3-319-68705-6_11

where the path representing each vertex has at most k bends. Interval graphs (intersection graphs of intervals on the line) are B_0-EPG graphs. The class of trees is in B_1-EPG [11], outerplanar graphs are in B_2-EPG [14] and planar graphs are in B_4-EPG [14]. Several papers are devoted to prove structural and algorithmic properties of EPG-graphs with a small number of bends, see for instance [1,2,7,10].

While recognizing and finding a representation of a graph in B_0-EPG (interval graph) can be done in polynomial time, it is NP-complete to decide if a graph belongs to B_1-EPG [6] or to B_2-EPG [15]. The complexity status remains open for more bends. Consequently, in all our results we will mention whether the representation of the graph is needed or not.

Epstein et al. [8] showed that the k-coloring problem and the k-independent set problem are NP-complete restricted to B_1-EPG graphs even if the representation of the graph is provided. Moreover they gave 4-approximation algorithms for both problems when the representation of the graph is given. Bougeret et al. [4] proved that this there is no PTAS for the k-independent set problem on B_1-EPG graphs and that the problem is $W[1]$-hard on B_2-EPG graphs (parameterized by k). Recently, Flavia et al. [3] showed that every B_1-EPG graph admits a 4-clique coloring and provides a linear time algorithm that finds it, given the representation of the graph.

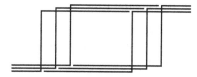

Fig. 1. A complete graph K_6 minus a matching.

Maximum Clique problem on EPG graphs. A *claw* of the grid is a set of three edges of the grid incident to the same point. Golumbic et al. proved in [11] that a maximum clique in a B_1-EPG graph can be computed in polynomial time if the representation of the graph is given. This algorithm is based on the fact that, for every clique X of a B_1-EPG graph, either there exists an edge e of the grid such that all the vertices of X contain e, or there exists a claw T such that all the vertices of X contain at least two of the three edges of T. In particular, it implies that the number of maximal cliques in B_1-EPG graphs is polynomial. Epstein et al. [8] remarked that the representation of the graph is not needed since the neighborhood of every vertex is a weakly chordal graph. When the number of bends is at least 2, such a proof scheme cannot hold since there might be an exponential number of maximal cliques. Indeed, one can construct a complete graph minus a matching in B_2-EPG (see Fig. 1) which has $2^{n/2}$ maximal cliques. So to compute a maximum clique on B_k-EPG graphs for $k \geq 2$, a new proof technique has to be introduced.

EPG graphs are closely related to two well known classes of intersection graphs, namely k-interval graphs, and k-track interval graphs on which the maximum clique problem have been widely studied. A k-*interval* is the union of k

distinct intervals in the real line. A *k-interval graph*, introduced in [16], is the intersection graph of k-intervals. A *k-track interval* is the union of k intervals on k-distinct lines (called tracks). A *k -track interval graph* is an intersection graph of k-track intervals (in other words, it is the edge union of k interval graphs on distinct lines). One can easily check, as first observed in [13], that B_{3k-3}-EPG graphs contain k-track interval graphs and B_{4k-4}-EPG graphs contain k-interval graphs.

Since computing a maximum clique in a 2-interval graph is NP-hard [9], the Maximum Clique Problem is NP-hard on B_4-EPG graphs. So the complexity status of the Maximum Clique problem remains open on B_k-EPG graphs for $k = 2$ and 3. In this paper, we prove that the Maximum Clique problem can be decided in polynomial time on B_2-EPG graphs when the representation of the graph is given. The proof scheme of [11] cannot be extended to B_2-EPG graphs. Indeed, there cannot exist a bijection between local structures, like claws, and maximal cliques since there are examples with an exponential number of different maximum cliques. Our proof is divided into two main lemmas. The first one ensures that we can separate so-called Z-vertices (vertices that use paths on two rows) from U-vertices (vertices that use edges of two columns). The second ensures that if a graph only contains Z-vertices, then all the maximal cliques are included in a polynomial number of subgraphs; subgraphs for which a maximum clique can be computed in polynomial time.

Coloring B_k-EPG graphs. We also provide an upper bound on the number of edges of B_k-EPG graphs for any value of k. This bounds ensures that there is a polynomial time algorithm that colors the graph with $2(k+1) \cdot \chi(G)$ colors in polynomial time without knowing the representation of the graph, where $\chi(G)$ is the chromatic number of G. In particular, it provides a simple coloring algorithm using at most 4 times the optimal number of colors on B_1-EPG graphs without knowing its representation. It improves the algorithm of [8] where the representation was needed.

A class of graphs $\mathcal{C}$ is *χ-bounded* if there exists a function f such that $\chi(G) \leq f(\omega(G))$ for every graph G of $\mathcal{C}$ with $\omega(G)$ the size of a maximum clique in G. Combinatorial bounds on the chromatic number of generalizations of interval graphs received a considerable attention initiated in [12]. The bound on the degeneracy of the graph ensures that the class B_k-EPG is χ-bounded and $\chi(G) \leq 2(k+1) \cdot \omega(G)$. As a by-product, it also ensures that graphs in B_k-EPG contain either a clique or a stable set of size $\sqrt{\frac{n}{2(k+1)}}$ which improves a result of [2] in which they show that every B_1-EPG graph admits a clique or a stable set of size $\mathcal{O}(n^{1/3})$.

2   Preliminaries

Let a, b be two real values with $a \leq b$. The *(closed) interval* $[a, b]$ is the set of points between a and b containing both a and b. The interval that does not contain b is represented by $[a, b)$ and the one that does not contain a by $(a, b]$.

Table 1. Three typical route planning scenarios.

Scenario	Description
General Planning	For a tourist who comes to a new city for tour, he/she needs an appropriate route which includes the hot places and with as little as possible detours. Also the traveling time budget is limited
Next-Point Recommendation	For a tourist who has visited several POIs already, he/she needs to be recommended which place to visit next, considering the tourist preferences, POI popularities and time limitation
Must-Visiting Planning	There are several POIs that a tourist wants to visit very much. Thus a travel route plan covering these must-visiting POIs is needed in this case. Tourists need a route which includes all these POIs and has the least detours

generation task which is suitable to be addressed by deep models. DTRP includes two major stages: model learning stage and route generation stage. In the model learning stage, we try to learn the probability of next POIs given the user preferences and previous POIs in a route. In the route generation stage, based on the learned probability distribution, we introduce the beam search strategy to flexibly integrate different constraints into the route generation process to ensure the generalization of the proposed framework. For each type of the above mentioned route planning problems, we design an appropriate generation strategy considering user preferences and demands.

To summarize, we make the following contributions:

- We introduce a novel multi-input and multi-output deep learning model to learn the probability distribution of next visiting POIs given the previous ones in a travel route. The proposed model can effectively utilize the multi sourced tourism data.
- Based on the learned probability distributions, we further introduce beam search algorithm to generate candidate routes for each proposed scenario, in which the various types of user demands can be flexibly integrated.
- We extensively evaluate our approach on four dataset. Experimental results show the superior performance of DTRP over state-of-the-art methods.

The rest of this paper is organized as follows. Section 2 summarizes the related works. Section 3 formally defines the problem of route planning. Section 4 introduces the proposed framework DTRP in details. Section 5 presents the experimental results. Finally we conclude this work in Sect. 6.

2   Related Work

Route planning has attracted a lot of research interests recently, and existing related works can be roughly categorized into three categories. The first category of related works are the variants of the "Traveling Salesman Problem" (TSP), which tries to locate the shortest path connecting a given set of geographical points. Beirigo and Santos [3] proposes a parallel Iterated Local Search (ILS) heuristic to search for promising candidate routes in a realistic travel network. Brilhante et al. [5] proposes to generate the budgeted trajectory by composing the popular itineraries with a specific instance of the Traveling Salesman Problem, aiming to find the shortest path crossing the popular itineraries. Matai et al. [17] utilizes TSP greedy heuristic method which greedily constructs the solution by always selecting the trajectory with the closest POI.

The second category of related works consider the route planning task as an recommendation problem, which aim to generate candidate routes based on visitor's historical route data. Zheng and Xie [29] recommends the top-m popular travel sequences in the given region utilizing the HITS-based inference model. Zheng et al. [30] investigates the traffic flow among different regions of attractions by exploiting Markov chain model and analyzes tourists' travel pattern of a tour destination for recommendation.

The third category of related works usually propose an optimization problem to incorporate both POI properties and historical route data, which are kind of hybrid works. Rodríguez et al. [19] recommends each user a trajectory best suited to their needs by using an interactive multi-criteria technique and a mathematical model. Lim [13] utilizes a variant of Orienteering Problem to maximize the overall profit of the recommended trajectory with a mandatory POI category based on user interest as additional constraint. Kurashima et al. [11] incorporates user preference and present location information into the probabilistic behavior model by combining topic models and Markov models. Brilhante et al. [4] models the task of planning personalized touristic tours as an instance of the Generalized Maximum Coverage problem, maximizing a measure of interest for the tourist given her preferences and visiting time budget.

The major limitation of related works is that they cannot effectively incorporate historical routes, user preferences and POI properties under an unified framework. In addition, existing works are mostly designed for a specific tour scenario and it is difficult to directly apply these models to a new tour scenario with different travel demands.

3   Problem Definition

In this section, we will first introduce several key terminologies, and then formally define the studied problem of travel route planning.

3.1   Key Terminologies

POI Property Vector: Given the set of POIs $\mathcal{P} = \{p_1, \ldots, p_{|\mathcal{P}|}\}$, where each element p_i represents a specific POI, we define the property vector of POI p_i as: $v_{p_i} \in \mathbb{R}^{1 \times m}$, where m is the dimension of property vector. Vector v_{p_i} preserves property information of POI p_i, such as category, popularity, latitude/longitude coordinates and stay time.

User Preference Vector: Given the POI category set $\mathcal{C}$ and a tourist u_i in the tourist set $\mathcal{U}$, we define the user preference vector $v_{u_i} \in \mathbb{R}^{1 \times |\mathcal{C}|}$ as the normalized interests of u_i on different POI categories. Elements in v_{u_i} represent the interests distribution of u_i calculated by the number of previous visiting times to different POI categories.

Historical Travel Routes: Historical travel route set $\mathcal{T}$ contains travel routes shared by other tourists who have visited the city before. A historical travel route is defined as $T_u = ((p_1), \ldots, (p_k))$, which is an ordered sequence of k POIs visited by tourist u.

3.2   Problem Definition

Based on the above definitions, we formally define the studied problem as follows.

Definition 1. *(Travel Route Planning): Given the POI set $\mathcal{P}$, historic travel route set $\mathcal{T}$ and the corresponding tourist set $\mathcal{U}$, we aim to generate a personalized travel route for each user. Specifically, we firstly try to learn the probabilities of visiting the next POI given the previous POIs and the tourist preferences: $P(p_t|u_i, p_{t-1}, p_{t-2}, \cdots), \forall p_t \in \mathcal{P}$. Then based on the learned probability, we try to generate a desirable candidate route for a new user considering his travel preferences and demands.*

4   DTRP: Deep Traveling Route Planning Framework

In this section, we will present the details of the proposed DTRP model. We will first briefly introduce the framework of DTRP, and then we will elaborate the details of the deep model utilized in the modelc learning stage. Finally in the route generation stage, we will introduce the beam search based route generation strategy, which can be applied in different travel scenarios.

4.1   Framework

Figure 1 shows the framework of DTRP. Given a city for travel, we first crawl the travel routes shared by other tourists, and collect the corresponding POI

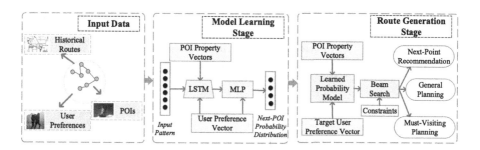

Fig. 1. Framework of DTRP.

properties and user preferences (left part of Fig. 1). In the model learning stage, the historical routes are transformed into a set of sequence patterns as the input of our model. Next we introduce a multi-input and multi-output deep model to learn the probability of visiting the next POI given the input pattern (the middle part of Fig. 1). In this stage, we utilize a LSTM model to encode the POI property vectors, user preference vector and route sequence information into a low-dimensional sequence representation vector, and then a MLP model is utilized to predict the probabilities of visiting the next POIs by incorporating the generated sequence vector and user preference vector. Finally in the route generation stage (the right part of Fig. 1), based on the learned probability model from the previous stage, we utilize the beam search strategy to generate the candidate routes by incorporating the constraints in different scenarios. Based on the preferences of a target user, we take the candidate POIs with the highest probability as the first POI to visit in the future route. Considering user preferences and the generated POIs in the temporal routes, we iteratively insert POIs with high probability for the next place to visit into the temporal routes. The pruning method is flexible as it can be utilized to form the generated routes satisfying different constraints in different scenarios.

4.2   Model Learning Stage

In this subsection we present the deep learning model which is designed to learn the probability of visiting the next POI given previous ones in a travel route. We aim to propose a data-driven model to mine knowledge from the historical routes to help plan future travel routes for new users.

Given a user and his historical routes, firstly we transform each input route into a set of POI patterns and their next POIs. As shown in Table 2, given a travel route, the user and the first several POIs are merged as an input pattern, and the next visited POI is treated as the future POI for prediction.

Based on the input pattern $x = \{u, p_1, p_2, \ldots, p_n\}$, we utilize a deep learning model to learn the probability of next POI to visit: $\{P(p_i|x), \forall p_i \in \mathcal{P}\}$. As shown

Table 2. Generation of patterns.

(a) Example of the input route	
Trajectory	$\{p_1, p_4, p_9, p_3\}$
User	u_{10}

(b) Patterns extracted from above data	
Pattern	Next POI
u_{10}	p_1
u_{10}, p_1	p_4
u_{10}, p_1, p_4	p_9
u_{10}, p_1, p_4, p_9	p_3

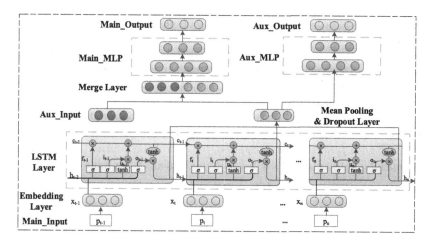

Fig. 2. Architecture of the probability learning model.

in Fig. 2, the proposed model is a multi-input and multi-output model which is convenient for the manipulation of the interwined data streams. The main input is the generated patterns, and the auxiliary input is the user preference vector. We utilize the LSTM (Long Short Term Memory) model [20] to encode the input pattern and the corresponding POI properties into a distributed vector. LSTM can process input sequences of arbitrary length [18]. Comparing to the normal RNN (Recurrent Neural Network) models, LSTM can learn long-term dependencies by introducing a memory cell to preserve the state of long time before [8]. The main output contains the final probability distribution on the POIs for next position considering the user preferences. The auxiliary output predicts the probability of POIs for the next position only based on the sequence representation vector. With the auxiliary output part, label information is utilized earlier in the model, which is a good regularization mechanism for deep models. Besides, the auxiliary output part also ensures the LSTM layer to be trained smoothly. We design a loss function for each output and linearly combine them as the final loss function.

Next we will introduce the details of each layer in the model:

- **Main_Input:** The generated pattern $x = \{u, p_1, p_2, \ldots, p_n\}$ with the length $n + 1$ is the main input. Note that the length of the input pattern can be arbitrary.
- **Embedding Layer:** This layer projects each POI p_i in the input pattern to its property vector $v_{p_i} \in \mathbb{R}^{1 \times m}$. User u in the pattern is embedded into his preference vector $v_u \in \mathbb{R}^{1 \times |\mathcal{C}|}$, in which $\mathcal{C}$ is the set of POI categories. LSTM model requires the embedding vectors of components in the input sequence the same size. Hence we adopt the padding strategy [23] to insert zeros to the shorter embedding vectors, After the padding process, the components in the input pattern are projected into the $maximum(m, |\mathcal{C}|)$-dimensional vectors.

Here we assume $m > |\mathcal{C}|$, hence the shape of embedding vectors outputted by this layer is $\mathbb{R}^{1 \times m}$.

- **LSTM Layer:** The LSTM layer encodes the input pattern into a d-dimensional representation vector through the recursive of a transition function on a recurrent state h_t. For the $t-$th POI p_t, the input of LSTM layer is the representation vector $v_{p_t} \in \mathbb{R}^{1 \times m}$, and the output is the recurrent state $h_t \in \mathbb{R}^{1 \times d}$. The recurrent state h_t is a non-linear transformation of input vector and its previous recurrent state h_{t-1}. LSTM unit utilizes an input gate i_t, an forget gate f_t and an output gate o_t to control the information transferred. The specific parameterization of LSTM is defined by the following equations:

$$i_t = \sigma(W^{(i)} v_{p_t} + U^{(i)} h_{t-1}) + b^{(i)}$$
$$f_t = \sigma(W^{(f)} v_{p_t} + U^{(f)} h_{t-1}) + b^{(f)}$$
$$o_t = \sigma(W^{(o)} v_{p_t} + U^{(o)} h_{t-1}) + b^{(o)}$$
$$u_t = tanh(W^{(u)} v_{p_t} + U^{(u)} h_{t-1}) + b^{(u)}$$
$$c_t = i_t \otimes u_t + f_t \otimes c_{(t-1)}$$
$$h_t = o_t \otimes tanh(c_t)$$

where $\sigma(\cdot)$ is a sigmoid function, $\otimes$ is element-wise multiplication and $tanh(\cdot)$ is a hyperbolic tangent function. $W^{(i)}, W^{(f)}, W^{(o)}, W^{(u)} \in \mathbb{R}^{d \times m}, U^{(i)}, U^{(f)}, U^{(o)}, U^{(u)} \in \mathbb{R}^{d \times d}, b^{(i)}, b^{(f)}, b^{(o)}, b^{(u)} \in \mathbb{R}^{d \times 1}$ are the parameters.

- **Mean Pooling and Dropout Layer:** Following recent works [16], we utilize the mean pooling layer to generate pattern representation vector $h \in \mathbb{R}^{1 \times d}$ by averaging the recurrent state h_t outputted in every time t: $h = \frac{1}{n} \sum_{t=1}^{n} h_{(t)}$. After the mean pooling process, we introduce the dropout layer [25] to avoid overfitting through its regularization effect. While copying the input to the output, the dropout layer sets some entries selected randomly to zero with a probability of 0.5 in our experiment.

- **Merge Layer:** This layer has two input vectors: the sequence representation vector $h \in \mathbb{R}^{1 \times d}$ from the dropout layer and the user preference vector $v_u \in \mathbb{R}^{1 \times |\mathcal{C}|}$. Merge layer concatenates these two input vectors as a single output vector.

- **Main_MLP:** The main MLP part is a two-layer feed-forward neural network model, which takes the representation vector $r_{merge} \in \mathbb{R}^{1 \times (d+|\mathcal{C}|)}$ as input. The output vector $y' \in \mathbb{R}^{1 \times |\mathcal{P}|}$ contains the conditional probability over all POIs given the input pattern and the user preferences. The formulation of this layer is:

$$\mathcal{O} = W_2(tanh(b_1 + W_1 o_{merge})) + b_2$$

where $b_1 \in \mathbb{R}^{l \times 1}, b_2 \in \mathbb{R}^{|\mathcal{P}| \times 1}, W_1 \in \mathbb{R}^{l \times (d+|\mathcal{C}|)}, W_2 \in \mathbb{R}^{|\mathcal{P}| \times l}$ are parameters to be learned, and l is the dimension of hidden layer. Then we introduce the softmax function as the activation function of output layer:

$$y'_i = P(p_i|x) = \frac{e^{\mathcal{O}_i}}{\sum_{k=1}^{|\mathcal{P}|} e^{\mathcal{O}_k}}$$

– **Aux_MLP Layer:** The aux_MLP part is also a two-layer feed-forward neural network taking the pattern representation vector $h \in \mathbb{R}^{1 \times d}$ as input, and the output vector is $y'' \in \mathbb{R}^{1 \times |\mathcal{P}|}$. The formulation of this layer is:

$$\mathcal{O}' = W_2'(tanh(b_1' + W_1'h)) + b_2'$$

$$y_i'' = \frac{e^{\mathcal{O}_i'}}{\sum_{k=1}^{|\mathcal{P}|} e^{\mathcal{O}_k'}}$$

where $b_1' \in \mathbb{R}^{l' \times 1}$, $b_2' \in \mathbb{R}^{|\mathcal{P}| \times 1}$, $W_1' \in \mathbb{R}^{l' \times d}$, $W_2' \in \mathbb{R}^{|\mathcal{P}| \times l'}$ are parameters to be learned, and l' is the dimension of the hidden layer.

To the input pair of pattern x and its next point p_j, we introduce the categorical cross-entropy function as the loss function:

$$L = -\sum_{k=1}^{|\mathcal{P}|} (y_k \cdot log(y_k') + \lambda \cdot y_k \cdot log(y_k''))$$

$$y_k = \begin{cases} 1, & k = j \\ 0, & k \neq j \end{cases}$$

in which λ is the weight of the auxiliary loss. We utilize the SGD (stochastic gradient descent) algorithm to minimize the loss L.

After the model learning process, we can obtain a learned model which is able to predict the probability of POIs for the next position given an input pattern. The prediction results are preserved in the main output vector. This stage forms up a powerful basis to generate desirable candidate routes.

4.3   Route Generation Stage

In this subsection, we present route generation strategies designed for each route planning scenario shown in Table 1. We first describe the POI prediction process for the "Next-Point Recommendation" problem; then we introduce beam search method to generate candidate routes for "General Planning" problem; finally the generation approach for "Must-Visiting Planning" problem is presented.

For the "Next-Point Recommendation" problem, the learned probability of visiting next POI from Sect. 4.2 can be directly utilized to recommend the next POIs. As shown in Algorithm 1, the inputs of the algorithm include a POI pattern and the user preference vector. If the input pattern only contains the target user as $\{u_i\}$, this task is still meaningful because it recommends the first POI purely relying on the user preferences. The output of Algorithm 1 is the main output vector y' which contains the probabilities of visiting the next POI. According to the output vector, top-K POIs with the highest probabilities as well as fitting the visit time limitation, are selected as the final recommendations.

"General Planning" problem is different from "Next-Point Recommendation", as it requires a complete route rather than a single POI as the output. In order to solve the route generation problem, we introduce the greedy beam search procedure with pruning techniques to find top-K candidates [22].

Algorithm 1. Next_-Point Prediction Algorithm

Input:

1      $main\_input$: an input pattern $x : \{u_i, p_1, p_2, \cdots, p_t\}$;

2   aux_input : user preference vector v_{u_i} ;

Output:

3      $main\_output$: POI probability vector $y' \in \mathbb{R}^{1 \times |\mathcal{P}|}$

4 model = Load_model()   // load the learned probability model from model learning stage ;

5 $inputs = [main\_input, aux\_input]$;

6 $y' = $ model.predict($inputs$) ;

7 return y'

Algorithm 2 presents the details of route generation method for the "General Planning" problem.

The inputs of the general planning algorithm include target user preference vector v_{u_i}, travel time budget d, acceptable time range ϵ, and beam size n. The output is a set of generated candidate routes along with their probabilities. Given a generated candidate route T_n, function p_n returns the probability scores calculated by the learned probability model, and function t_{T_n} returns the entire travel time cost for this route.

After the initialization of the candidate route set $\mathcal{A}$ and the temporal route set $\mathcal{T}$, we insert the initial temporal route (u_i) into $\mathcal{T}$. In each iteration from line 9 to 29 in Algorithm 2, for each old temporal route T_o in $\mathcal{T}$, we utilize the learned model from Sect. 4.2 to estimate the probabilities of its next POIs as shown in line 12. For each POI in top-n highest probability set $S_{nextpois}$, we append it into T_o to form a new temporal route T_n. Then the probability score and time cost of the new route T_n are calculated in line 15 and 16. We utilize a pruning technique to incorporate user's constraints into the generation process. If the time cost exceeds the budget, the new route T_n is discarded. If its time cost satisfies the budget, T_n is considered as an appropriate candidate route and is appended into set $\mathcal{A}$. If the time cost is lower than the upper limit, we insert the new route T_n into $\mathcal{T}$ to further process it. After the update procedure, in line 27 we remove the old route T_o from $\mathcal{T}$. This iteration will be repeated until there are no routes in $\mathcal{T}$. Finally $\mathcal{A}$ contains all the candidate routes which satisfy the user's constraints with the top-k highest visiting probabilities. Finally the candidate routes with the highest probabilities in $\mathcal{A}$ are selected as the final results.

The solution of "Must-Visiting Planning" problem is similar to Algorithm 2 with two major modifications. Assume the must-visit POI set $P_{must} = \{p_1, p_2, \cdots\}$ contains all the POIs that must be visited. Firstly for line 12 in Algorithm 2, besides the n POIs with the highest probabilities, we also insert all POIs in P_{must} into the set $S_{nextpois}$ to guarantee the generated candidates contain the entire possible space. Secondly in the pruning step, if T_o has included a POI p_i in P_{must}, p_i will not be appended into T_o to form the new route T_n, which is helpful to avoid repetitive computation.

Algorithm 2. General Planning Algorithm

   Input:
1   User preference vector v_{u_i} ;
2   Travel time budget d;
3   Acceptable time range ϵ;
4   Beam size n;
   Output:
5   Candidate route set $\mathcal{A}$;
6 Set the $\mathcal{A} = \Phi$;
7 Set the temporal route set $\mathcal{T} = \Phi$;
8 Insert initial route $\{u_i\}$ into $\mathcal{T}$;
9 **repeat**
10   **for** $T_o \in \mathcal{T}$ **do**
11        $y' = Next\_Prediction(T_o, v_{u_i})$;
12        $S_{nextpois} \leftarrow$ get the top-n POIs from y' ;
13   **for** poi in $S_{nextpois}$ **do**
14            $T_n = T_o.\text{append}(poi)$;
15            $p(T_n) = p(T_o) \times y'_{poi}$;
16            $t(T_n) = t(T_o) + t(poi) + travel\ time$;
17   **if** $t(T_n) > d + \epsilon$ **then**
18   continue ;
19   **end**
20   **if** $d - \epsilon \leq t(T_n) \leq d + \epsilon$ **then**
21   insert T_n into $\mathcal{A}$;
22   **end**
23   **if** $t(T_n) < d + \epsilon$ **then**
24   insert T_n into $\mathcal{T}$;
25   **end**
26   **end**
27   remove T_o from $\mathcal{T}$
28   **end**
29 **until** $\mathcal{T} = \Phi$;
30 **return** $\mathcal{A}$

---

5 Experiments

In this section we evaluate the performance of the proposed model. We first introduce the datasets and baseline methods used in this paper. Then for the three mentioned travel scenarios, we thoroughly evaluate the proposed model on four datasets. Finally we analyze the quantitative experimental results.

5.1 Experiment Setup

Datasets. The dataset used in this paper is the Flickr User-POI Visits Dataset published in [14][1], which extract geo-tagged photos of eight cities from

---

[1] https://sites.google.com/site/limkwanhui/datacode#ijcai15.

YFCC100M dataset [21]. Based on the geo-tagged photos, we map the photos to the specific POIs location, and thus comprises a set of users and their visits to various POIs. We group the consecutively visited POIs whose visit time differences are less than 8 h as travel routes [21]. In our experiment, we use the tourism data of four cities: Toronto, Vienna, Edinburgh and Glasow. Table 3 shows the details of these datasets.

Table 3. Description of the dataset

City	POI photos	Users	POIs	Travel sequences	Categories
Toronto	39,419	1,395	29	6,057	6
Vienna	34,515	1,155	28	3,193	8
Edinburgh	33,944	1,454	28	5,028	6
Glas	11,434	601	27	2,227	7

In the datasets, each route is correlated with a specific user and contains a set of ordered POIs. Each POI is represented as its property vector including its ID, category (e.g. shopping, culture, religious and park), geographical coordinates, traveling cost time and popularity. Each user is represented as his preference vector, which is estimated from the categories of POIs he has visited.

Baseline Methods. We compare the proposed DTRP model with the following four baseline methods:

- **Multinomial Model:** Multinomial model [13] predicts the next POI based on its popularity score of each POI, which can be calculated by using the multinomial probability distribution over POIs. This model only considers the POI popularity but ignores the user's current location and preferences.
- **Order-1 Markov Model:** Markov model [7] predicts the next POI based on the user's current location. Given the current POI, the most often visited POI is recommended. This model considers user's current location but ignores user's interests.
- **Topic Model:** Topic model [10] predicts the next POI based on the user's preferences, which is calculated by the LDA model. This model considers the user's interest but ignores the user's current location and previous POIs.
- **RNN:** Comparing with proposed multi-input and multi-output model, this model only has single input (main input) and single output (main output) [15].

Parameter Setup. In DTRP model, the output dim of LSTM and MLP are both set to 64, the main MLP and the aux_MLP are both two-layer feed-forward neural network, the dropout rate is set to 0.2, and the weight of the auxiliary loss $\lambda = 0.2$. The size of greedy beam search is set to $n = 3$, and the acceptable time range is set to $\epsilon = 0.5$ h.

5.2   Next-Point Recommendation

For the "Next-Point Recommendation" problem, we try to predict the next POI given the user and previous POIs (pattern) in a route. 80% of the travel routes are randomly selected as the training data, and the rest are considered as the test data. We select the following metrics to evaluate the performance of different methods:

- **Accuracy@k:** Accuracy@k measures whether the next actual POI is included in the top k recommended POIs. Assume $\mathcal{R}(T)$ is the top k recommended POIs of the input pattern T, v is the actual POI, and N is the number of test data, accuracy@k is defined as: $accuracy@k = \frac{1}{N} \sum_{T=1}^{N} |v \cap \mathcal{R}(T)|$.

- **MAP:** Mean average precision (MAP) [15] is the standard evaluation metric for ranking tasks, which can evaluate the quality of the whole ranked lists. Let $I(T)$ be where the actual next POI appears in the ranked list of the input pattern T, MAP is defined as: $MAP = \frac{1}{N} \sum_{T=1}^{N} \frac{1}{I(T)}$.

Table 4. Next-point recommendation performance on four datasets.

Toronto				Vienna			
Algorithm	Accu@1	Accu@3	MAP	Algorithm	Accu@1	Accu@3	MAP
Multinomial model	0.1080	0.2582	0.2615	Multinomial model	0.1518	0.2508	0.2784
Topic model	0.1094	0.2680	0.2623	Topic model	0.1545	0.2895	0.2793
Markov model	0.1436	0.3152	0.2981	Markov model	0.1742	0.3447	0.3325
RNN	0.3139	0.6270	0.5105	RNN	0.2996	0.5838	0.4875
DTRP	**0.3657**	**0.65**	**0.5411**	DTRP	**0.3333**	**0.6075**	**0.5128**
Edinburgh				Glasow			
Algorithm	Accu@1	Accu@3	MAP	Algorithm	Accu@1	Accu@3	MAP
Multinomial model	0.1206	0.2727	0.2715	Multinomial model	0.0927	0.2735	0.2642
Topic model	0.1296	0.2859	0.2889	Topic model	0.1039	0.2935	0.2751
Markov model	0.1647	0.3436	0.3186	Markov model	0.1158	0.3236	0.2933
RNN	0.2916	0.5853	0.4686	RNN	0.3440	0.6366	0.5132
DTRP	**0.3314**	**0.6055**	**0.5090**	DTRP	**0.3705**	**0.6501**	**0.5483**

Table 4 shows the results of different prediction methods. We can clearly see that the deep models (RNN and DTRP) consistently outperforms other shallow models, which proves the strong feature learning capacity of deep models. By introducing the auxiliary learning part, DTRP outperforms RNN by 4% in Edinburgh and 3% in other datasets, which proves the effectiveness of the multi-input and multi-output deep model. Overall, one can see that the proposed DTRP model can effectively incorporate user preferences, POI properties and historical routes.

5.3   General Planning

In the "General Planning" task, we aim to generate a complete route as the final result. Based on the proposed next-point recommendation methods, we utilize the beam search strategy shown in Sect. 4.3 to generate next POI one by one to form the final route. To evaluate the quality of the generated routes, we choose the following metrics:

- **Route Interest:** The route interest of the generated route T to a user u is defined as: $Int_u(T) = \sum_{p \in T} Int_u(p)$, in which $Int_u(p) = \frac{n(u, \{p.cat\})}{n(u)}$ denotes the user's preference on the category of POI p. $n(u)$ represents the total number of the POIs visited by the users.
- **Route Popularity:** The popularity of a single POI is defined as the number of times it has been visited by the tourists, and the route popularity is defined as the summation of the popularities of the contained POIs.
- **Edit Distance (ED):** The edit distance is an evaluation metric to measure the minimum number of edit operations from one sequence to another. We apply ED to measure the difference between generated candidate route and the actual route.

Table 5 shows the evaluation results. From the results, we can see that the proposed DTRP model beats the best shallow baselines by about 20% in the route interest by incorporating the user preference and POI properties. Besides, DTRP outperforms RNN by 4% on the metric of edit distance. The experimental results show that the proposed model can better satisfy the user's preferences and generate the high quality routes with the lowest edit distance.

Table 5. General route planning performance on four datasets.

Toronto				Vienna			
Algorithm	Interest	Popularity	ED	Algorithm	Interest	Popularity	ED
Multinomial model	0.2508	**0.1281**	1.1852	Multinomial model	0.4671	**0.1371**	1.4109
Topic model	0.2557	0.0814	1.0771	Topic model	0.4865	0.0931	1.1951
Markov model	0.2658	0.0829	1.0232	Markov model	0.4922	0.1078	1.1820
RNN	0.4416	0.0874	0.6670	RNN	0.6936	0.1001	1.1176
DTRP	**0.4729**	0.0902	**0.6294**	DTRP	**0.7187**	0.1089	**1.0841**
Edinburgh				Glasow			
Algorithm	Interest	Popularity	ED	Algorithm	Interest	Popularity	ED
Multinomial model	0.3354	**0.1322**	1.4077	Multinomial model	0.2728	**0.1229**	1.3323
Topic model	0.3931	0.1007	1.3827	Topic model	0.2968	0.0931	1.1136
Markov model	0.3914	0.1130	1.2909	Markov model	0.3013	0.1092	0.9864
RNN	0.6008	0.1073	1.1297	RNN	0.4705	0.0896	0.7230
DTRP	**0.6230**	0.1168	**1.0885**	DTRP	**0.4926**	0.0989	**0.6999**

5.4 Must-Visiting Planning

In this task users can specify several POIs that must visit, which add extra constraints to the route generation process. The must visited POIs are randomly picked from the actual routes. Here we utilize tour precision, recall and F1-score as the evaluation metrics:

- **Route Precision:** Route precision denotes the proportion of POIs in the generated route that are also in the actual route, which is defined as $P(T) = \frac{|P_g \cap P_r|}{|P_g|}$.
- **Route Recall:** Route recall denotes the proportion of POIs in the actual P_r that are also in the generated route P_g, which is defined as $R(T) = \frac{|P_g \cap P_r|}{|P_r|}$.
- **Route F1-score:** Route F1-score denotes the harmonic mean of the precision and recall of the generated route T, defined as $F_1 = \frac{2 \times P(T) \times R(T)}{P(T) + R(T)}$.

Table 6. Must-visiting route planning performance on four datasets.

Toronto				Vienna			
Algorithm	Precision	Recall	F1_score	Algorithm	Precision	Recall	F1_score
Multinomial model	0.4496	0.5124	0.4727	Multinomial model	0.4316	0.4522	0.4361
Topic model	0.4898	0.5716	0.5213	Topic model	0.4854	0.5501	0.5095
Markov model	0.5444	0.6229	0.5748	Markov model	0.5183	0.5445	0.5251
RNN	0.5905	0.6947	0.6385	RNN	0.6077	0.6806	0.6460
DTRP	**0.6261**	**0.7114**	**0.6590**	DTRP	**0.6359**	**0.7034**	**0.6621**
Edinburgh				Glasow			
Algorithm	Precision	Recall	F1_score	Algorithm	Precision	Recall	F1_score
Multinomial model	0.4097	0.4737	0.4344	Multinomial model	0.4903	0.5312	0.5063
Topic model	0.4871	0.5537	0.5128	Topic model	0.5026	0.5855	0.5359
Markov model	0.4968	0.5277	0.5062	Markov model	0.5352	0.5785	0.5501
RNN	0.5763	0.6606	0.6156	RNN	0.6743	0.7518	0.7109
DTRP	**0.6000**	**0.6894**	**0.6342**	DTRP	**0.6947**	**0.7793**	**0.7283**

Table 6 shows the results. One can see that our model consistently outperforms other methods on the precision and recall. F1-score is a balanced representation of both precision and recall measurements, and higher precision and recall will result in a higher F1-score. The F1-score of DTRP beats best baseline by 3%, which reveals the effectiveness of DTRP in the "Must-visiting Planning" task.

6 Conclusion

This paper proposes a flexible deep route planning framework DTRP to incorporate available tourism data and fit the tourist's demands. We formulate this task as a sequence prediction problem, which aims to learn the probability of visiting the next POIs given the user preferences and previously visited POIs in a route. In the model learning stage we propose a novel deep learning model to learn the probability distribution of the next POIs for visiting. In the route generation stage we introduce beam search strategy to generate the travel route based on the learned probability model. Experimental results on three popular scenarios over four datasets demonstrate the effectiveness of DTRP.

Acknowledgment. This work was supported by the National Natural Science Foundation of China (Grand Nos. U1636211, 61672081, 61602237, 61370126), National High Technology Research and Development Program of China (No. 2015AA016004), fund of the State Key Laboratory of Software Development Environment (No. SKLSDE-2017ZX-19), and the Director's Project Fund of Key Laboratory of Trustworthy Distributed Computing and Service (BUPT), Ministry of Education (Grant No. 2017KF03).

References

1. Arase, Y., Xie, X., Hara, T., Nishio, S.: Mining people's trips from large scale geo-tagged photos. In: International Conference on Multimedea 2010, pp. 133–142 (2010)
2. Ashbrook, D., Starner, T.: Using GPS to learn significant locations and predict movement across multiple users. Pers. Ubiquit. Comput. **7**(5), 275–286 (2003)
3. Beirigo, B.A., Santos, A.G.D.: A parallel heuristic for the travel planning problem. In: International Conference on ISDA, pp. 283–288 (2016)
4. Brilhante, I., Macedo, J.A., Nardini, F.M., Perego, R., Renso, C.: Where shall we go today?: planning touristic tours with tripbuilder. In: CIKM (2013)
5. Brilhante, I.R., Macedo, J.A., Nardini, F.M., Perego, R., Renso, C.: On planning sightseeing tours with trip builder. Inf. Process. Manag. **51**(2), 1–15 (2015)
6. Chen, M., Yu, X., Liu, Y.: Mining Moving Patterns for Predicting Next Location. Elsevier Science Ltd., Amsterdam (2015)
7. Gao, H., Tang, J., Liu, H.: Exploring social-historical ties on location-based social networks. In: AAAI (2012)
8. Graves, A.: Generating sequences with recurrent neural networks. arXiv preprint arXiv:1308.0850 (2014)
9. Iwata, T., Watanabe, S., Yamada, T., Ueda, N.: Topic tracking model for analyzing consumer purchase behavior. In: IJCAI, pp. 1427–1432 (2009)
10. Kurashima, T., Iwata, T., Irie, G., Fujimura, K.: Travel route recommendation using geotags in photo sharing sites. In: CIKM, pp. 579–588 (2010)
11. Kurashima, T., Iwata, T., Irie, G., Fujimura, K.: Travel route recommendation using geotagged photos. Knowl. Inf. Syst. **37**(1), 37–60 (2013)
12. Lian, D., Xie, X., Zheng, V.W., Yuan, N.J., Zhang, F., Chen, E.: CEPR: a collaborative exploration and periodically returning model for location prediction. ACM Trans. Intell. Syst. Technol. **6**(1), 1–27 (2015)
13. Lim, K.H.: Recommending tours and places-of-interest based on user interests from geo-tagged photos. In: SIGMOD PhD Symposium, pp. 33–38 (2015)
14. Lim, K.H., Chan, J., Leckie, C., Karunasekera, S.: Personalized tour recommendation based on user interests and points of interest visit durations. In: IJCAI (2015)

15. Liu, Q., Wu, S., Wang, L., Tan, T.: Predicting the next location: a recurrent model with spatial and temporal contexts. In: AAAI, pp. 194–200 (2016)
16. Liu, Y., Sun, C., Lin, L., Wang, X.: Learning natural language inference using bidirectional LSTM model and inner-attention. arXiv preprint arXiv:1605.09090 (2016)
17. Matai, R., Singh, S., Mittal, M.L.: Traveling salesman problem: an overview of applications, formulations, and solution approaches. Comput. In: Traveling Salesman Problem, Theory and Applications, vol. 12 (2010)
18. Mikolov, T., Karafit, M., Burget, L., Cernocky, J., Khudanpur, S.: Recurrent neural network based language model. In: Interspeech, pp. 1045–1048 (2010)
19. Rodrłguez, B., Molina, J., Prez, F., Caballero, R.: Interactive design of personalised tourism routes. Tour. Manag. **33**(4), 926–940 (2012)
20. Sundermeyer, M., Schlter, R., Ney, H.: LSTM neural networks for language modeling. In: Interspeech, pp. 601–608 (2012)
21. Thomee, B., Shamma, D.A., Friedland, G., Elizalde, B.: The new data and new challenges in multimedia research. Commun. ACM **59**(2), 64–73 (2015)
22. Tillmann, C., Ney, H.: Word reordering and a dynamic programming beam search algorithm for statistical machine translation. Comput. Linguist. **29**(1), 97–133 (2006)
23. Vanhoucke, V., Mao, M.Z.: Improving the speed of neural networks on CPUs. In: Deep Learning and Unsupervised Feature Learning Workshop NIPS (2011)
24. Vansteenwegen, P., Souffriau, W., Berghe, G.V., Oudheusden, D.V.: The city trip planner: an expert system for tourists. Expert Syst. Appl. **38**(6), 6540–6546 (2011)
25. Wager, S., Fithian, W., Wang, S., Liang, P.: Altitude training: strong bounds for single-layer dropout. Adv. Neural Inf. Process. Syst. **1**, 100–108 (2014)
26. Wang, D., Cui, P., Zhu, W.: Structural deep network embedding. In: KDD, pp. 1225–1234 (2016)
27. Wang, S., He, L., Stenneth, L., Yu, P.S., Li, Z.: Citywide traffic congestion estimation with social media. In: SIGSPATIAL/GIS, pp. 1–10 (2015)
28. Wang, S., He, L., Stenneth, L., Yu, P.S., Li, Z., Huang, Z.: Estimating urban traffic congestions with multi-sourced data. In: IEEE International Conference on Mobile Data Management, pp. 82–91 (2016)
29. Zheng, Y., Xie, X.: Learning travel recommendations from user-generated GPS traces. ACM Trans. Intell. Syst. Technol. **2**(1), 2 (2011)
30. Zheng, Y.T., Zha, Z.J., Chua, T.S.: Mining travel patterns from geotagged photos. ACM Trans. Intell. Syst. Technol. **3**(3), 1–18 (2012)

Taxi Route Recommendation Based on Urban Traffic Coulomb's Law

Zheng Lyu[1]([✉]), Yongxuan Lai[1]([✉]), Kuan-Ching Li[1,2], Fan Yang[3], Minghong Liao[1], and Xing Gao[1]

[1] School of Software, Xiamen University, Xiamen, China
xmdxlz@stu.xmu.edu.cn, {laiyx,liao,gaoxing}@xmu.edu.cn
[2] Department of Computer Science and Information Engineering,
Providence University, Taichung, Taiwan
kuancli@pu.edu.tw
[3] Department of Automation, Xiamen University, Xiamen, China
yang@xmu.edu.cn

Abstract. With the advances and availability of networking and data processing technologies, the number of researches supporting taxi as a mean of transportation and further optimization of their route selection is increasing and broadly discussed. For the taxis, when they are cruising on the street the drivers looking for passengers, most drivers rely on their experience and intuition for the guideline to optimize their cruise routes and increase profit. This approach, however, is not efficient and usually increases the traffic load in urban cities. A solution is highly required to match and recommend appropriate cruising routes to taxis so that aimless cruising would be avoided and the drivers income would be increased. In this paper, we propose a route recommendation algorithm based on the *Urban Traffic Coulomb's Law* to model the relationship between the taxis and passengers in urban traffic scenarios. Different from existing route recommendation methods, the relationship among taxis and passengers are fully taken into account in the proposed algorithm, e.g. the attractiveness between taxis and passengers and the repulsion among taxis. It collects useful information from historical trajectories, and calculates the traffic attraction for cruising taxis, based on which optimal road segments are recommended to drivers to pick up desired passengers. Extensive experiments are conducted on the road network based on massive real-world trajectories to verify the effectiveness, and evaluations demonstrate that the proposed method outperforms among existing methods and can increase the drivers' income by more than 8%.

Keywords: Taxi · Trajectories · Cruising route recommendation · Urban traffic Coulomb's law

This work is supported by the Natural Science Foundation of China (61672441), the National Key Technology Support Program (2015BAH16F01).

A. Bouguettaya et al. (Eds.): WISE 2017, Part I, LNCS 10569, pp. 376–390, 2017.
DOI: 10.1007/978-3-319-68783-4_26

1   Introduction

Taxis are playing an important role in daily lives of urban citizens. Different from other public transportation like buses or subways, taxis do not follow fixed routes periodically. Drivers have to plan their own routes right after passengers are dropped off [19], and their incomes are largely determined by the selection of these routes. Nevertheless, it is not easy for a driver to schedule and select the best route to maximize his/her earning. If they select a route that has no/fewer passengers, they would waste their time and petrol, which also decreases the profit of taxi drivers and increases traffic load in urban cities. Thus, a solution to help taxi drivers select better and more effective cruising routes is highly required. It would mutually benefit both drivers and passengers, as well as improve the efficiency of urban transportation systems.

Fortunately, with the advancement of smart devices and networking technologies, the Location-Based Services (LBS) are widely used in our daily lives [3,9,15]. Most taxis are equipped with GPS localizer, and geographic positions are reported to the operating company periodically, e.g. 1–3 times per minute [17], the occupancy as well as other information of each taxi are also recorded. Having these datasets, it is possible to create new recommendation strategies to help drivers select their cruising routings and pick up more passengers [6,19,21]. In the passenger-pickup scenario, taxis hunt for passengers and passengers wait for taxis, and drivers have to compete with each other to pick up more passengers. With this respect, there are similarities between the traffic network and the electrostatic field in physics. Stated in Coulomb's law [5], there are positive and negative charges in electrostatic field, as like charges repel and unlike charges attract. Similarly, the concept of *Urban Traffic Coulomb's law* is developed, where urban traffic network is viewed as a big electrostatic field and taxis and passengers as the positive and negative charges. A taxi 'repels' another taxi as they are competing to find passengers, and a passenger 'attracts' a taxi as they are looking for each other. Some analogous models have been applied to other areas, such as the application "the electrostatic model" in the trip distribution [10].

As we can note, there are two aspects in the *Urban Traffic Coulomb's law*: the former is the attractiveness between taxis and passengers, and the latter is the repulsion among taxis. Most existing researches matched to the first aspect, road segments with more passengers and hotspots are recommended as cruising routes [4,6,7,12,16,19,21]. Yet they failed to consider the second aspect, i.e., the conflict and competition among vacant taxis, which would lead to ineffectiveness of the route recommendation strategy. In this study, we propose a taxi cruising route recommendation algorithm, which takes into account both aspects of the *Urban Traffic Coulomb's law* for the algorithm design. The proposed algorithm collects useful information from historical trajectories, and calculates the traffic attraction for cruising taxis, based on which optimal road segments are recommended to drivers to pick up desired passengers. Extensive simulations conducted based on a real trajectory dataset verified the effectiveness of the proposed method, and results show that it can improve taxis' income by more than 8% when compared with the ground truth and other methods.

The major contributions of this work are as follows:

- The concept of *Urban Traffic Coulomb's law* is proposed to model the relationship between taxis and passengers in urban cities. Taxis are viewed as positive charges and passengers as negative charges. Formulas that calculate the *traffic charges* and *traffic forces* are also given within this concept.
- A cruising route recommendation method is proposed for taxi drivers based on *Urban Traffic Coulomb's law*. It first collects useful information from trajectories, then calculates the attractiveness from the regions around, based on which taxis are routed to optimal road segments to pick up desired passengers. The proposed algorithm considers both the attractiveness between taxis and passengers and the competitions among both taxis and passengers.
- Extensive experiments were conducted on large number of real-world historical GPS trajectories to verify the effectiveness of the proposed method. Experimental results showed that the proposed scheme can effectively increase the income of taxi drivers compared to other methods.

The remaining of this paper is organized as follows. First, related discussions are presented in Sect. 2, whilst preliminaries and introduction of *Urban Traffic Coulomb's law* are depicted in Sect. 3. Next, Sect. 4 presents the detailed description of the proposed algorithm, followed by experiments studies and analysis delivered in Sect. 5. Finally, Sect. 6 concludes the paper and presents some future directions.

2   Related Work

Recently, the traditional pattern of taxis picking up passengers has changed with the emergence of some taxi service platforms, such as Didi Taxi and Uber. Taxis can pick up passengers with the guide of these platforms when there is a need, while they still need to plan their own cruising routes when there is not. Therefore, it is necessary to make an efficient recommendation for taxis. Researches on recommendation system of taxis using GPS trajectories could be mainly divided into two categories: macroscopic and microscopic recommender systems.

2.1   Macroscopic Recommender Systems

In macroscopic recommender systems, only the driving directions are provided to taxi drivers, rather than the complete driving routes. For example, an improved ARIMA method to forecast the spatial-temporal variation of passengers in a hotspot is proposed in [8] to help taxi drivers find new passengers. Li et al. [7] study the passenger-finding strategies (hunting/waiting) of taxi drivers in Hangzhou. In this work, L1-Norm SVM is used to select features for classifying the passenger-finding strategies in terms of performance. In [6], Hsueh *et al.* recommend the next cruising location through a model combined with waiting time, distance and some other factors. Similarly, Powell et al. [11] propose an approach to suggest most profitable (grid-based) locations for taxi drivers by constructing a Spatio-Temporal Profitability map (STP), on which, the nearby regions of the driver are scored according to the potential profit calculated by the historical data.

2.2 Microscopic Recommender Systems

Compared with macroscopic recommender systems, the microscopic one provide taxi drivers with actual driving routes. A recommendation system for both taxi drivers and passengers is proposed in [18]. For taxi drivers, the system recommends a good parking place with shorter waiting time and longer distance for the next trip. Another work presented in [4] scores each road segment and thus obtains a cruising route with the highest score using a heuristic algorithm called MSCR. In [12], pick-up points from large-scale real-world GPS dataset using a clustering algorithm and a Markov Decision Process (MDP) model is proposed to provide the strategy for taxis.

The proposed approach is categorized in the microscopic one, which is more flexible and reliable on recommending the next driving road segment. Compared to the methods mentioned above, the strengths are mainly embodied in the following aspects. First, existing researches are mainly focused on the pick-up points or the live trips, which neglect the competition among vacant taxis. Thus these methods tend to recommend places with potential passengers but strong competitions, such as airport and train station, which not only may not be able to increase the profit, but also lead to the unbalanced deployment of taxis. In the proposed approach, we take the density of vacant taxis into consideration and reasonably recommend a place with larger chance rather than more passengers. Second, most previous methods are focused solely on historical trajectories, while the proposed approach combines the historical trajectories and real-time trajectories together for the recommendation, which would improve the precision of recommendation results.

3   Preliminary

3.1   Problem Statement

Road network can be built in many ways [2,13,14]. In this research, the road network is characterized by a set of intersections and road segments (e.g. extracted from OpenStreetMap [1]), represented by a graph $G = (I, S)$, where $I = I_1, I_2, ..., I_n$ and $S = S_1, S_2, ..., S_m$, standing for a finite set of n intersections and m road segments. A road segment S_i contains five properties, the identification $S_i.id$, the direction $S_i.dir$, the length $S_i.length$, the starting intersection $S_i.s$, and the ending intersection $S_i.e$. Similarly, a road intersection I_i is associated with three properties, the identification $I_i.id$, the longitude $I_i.lon$ and the latitude $I_i.lat$. A route is a directed sequence of L road segments, $W_i = S_1, S_2, ..., S_L$, where $W_i.s = S_1.s, W_i.e = S_L.e$ and $S_j.e = S_{j+1}.s$ for $1 \leq j < L$ which means that consecutive road sections contained in a route should share an intersection.

In fact, a taxi driver only makes decisions when he/she arrives at an intersection. Given a road network $G = (I, S)$, a position trajectories T, an operation trajectories O, the recommendation approach is to find the optimal route W for taxis. Specifically, providing the best road segment to taxi drivers as soon as he/she reaches an intersection.

3.2   Coulomb's Law

Coulomb's law is one of the most important laws in electrostatics to quantify the force between two electrostatic charges. It states that the magnitude of the electrostatic force between two point charges is directly proportional to the scalar multiplication of the magnitudes of charges and inversely proportional to the square of the distance between them; the direction of the force is along the straight line that join them. In short, *Coulomb's law* could be simply formulated as Eq. (1):

$$F = k \times \frac{q_1 \times q_2}{r^2} \tag{1}$$

where k is the Coulomb's constant, q_1 and q_2 are the signed magnitudes of the charges, the scalar r is the distance between the charges and F is the electric force between q_1 and q_2.

3.3   Urban Traffic Coulomb's Law

Urban traffic network can be considered as a special category of "electric field" by isomorphing passengers and taxis as charges with different signs. Suppose the urban traffic network is divided into several regions with "electric charge"s, the cruise route of each taxi is affected by the "electric charge" from surrounding regions. It is clear that the idea of *Coulomb's Law* could be utilized in this situation, denominating as *Urban Traffic Coulomb's Law* (UTCL), being defined as follows:

Definition 1. *Traffic Charge:* *A traffic charge c of region r is mainly composed of the taxis and passengers in r, corresponding to the "electric charges" in r in "Coulomb's Law". Traffic charge describes the ability on attracting taxis of a region, which is positive in most cases.*

Here we give an abstract concept and the formal definition used in this paper is given in Sect. 4.2.

Definition 2. *Traffic Force:* *Traffic force describes the direction and magnitude of the attractiveness on taxis from a region, corresponding to the "electric force" in "Coulomb's Law". Specifically, the traffic force $\overrightarrow{F}$ from region r_2 to the taxi in region r_1 can be stated as a mathematical expression:*

$$\overrightarrow{F(r_1, r_2, t)} = k \times \frac{C_{r_2 t}}{R_{r_1 r_2}^2} \times \frac{\overrightarrow{r_1 r_2}}{|r_1 r_2|} \tag{2}$$

where t is the current time slot, k is a constant to balance the result, $C_{r_2 t}$ is the traffic charge of r_2 in t, $R_{r_1 r_2}$ is the distance between the r_1 and r_2, $\frac{\overrightarrow{r_1 r_2}}{|r_1 r_2|}$ stands for the direction from r_1 to r_2.

4   Routes Recommendation Using UTCL

Figure 1 gives an overview of the proposed approach, which mainly consists of three components: metadata collection and calculation, real-time update, and online recommendation. The on-board devises on taxies collect both spatial and temporal metadata, send them to server. Then the historical *traffic charges* are calculated based on the historical trajectories. During the process of online route recommendation, the real-time *traffic charges* are first calculated based on the real-time trajectories, and the *traffic forces* are calculated as soon as an empty taxi reaches an intersection, using the historical *traffic charges* along with the real time *traffic charges*. The road segment nearest to the direction of the *traffic force* would be recommended to the driver at that intersection. The process will repeat until the taxi picks up a passenger.

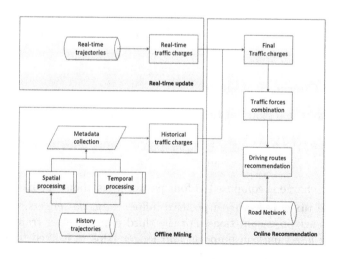

Fig. 1. System overview

4.1 Spatio-Temporal Processing and Metadata Collection

Metadata collection means to collect and extract useful statistic information from trajectories, such as the density of passengers and taxis in a given area. Considering the influence of spatio-temporal context, two types of work need to be done before metadata collection, i.e., spatial processing and temporal processing.

Spatial processing is the first step, which means to divide a map into a few smaller regions with comparable areas, and then a grid-based method processing is conducted [11]. For example, the map is divided by $0.001°$ of longitude and $0.001°$ of latitude in the experiment, where each unit is viewed as a region. Temporal processing is another important step before metadata collection. Intuitively, the distribution of traffic is changing throughout the time, e.g., the number of passengers waiting for taxis in the daytime is much larger than that during midnight. Therefore, metadata collection should based on a pre-defined time division. The influence of these factors will be presented in Sect. 5.3.

Once spatial processing and temporal processing are completed, metadata on each region and each time slot can be collected from trajectories. Several metadata used in the proposed method are listed below (Table 1):

Table 1. Notations.

Notation	Description
P_{tr}	The passenger density in region r at time slot t
P_t	The average passenger density in all non-zero regions at time slot t
M_{tr}	The average money of one trip in region r at time slot t
V_{tr}	The density of vacant taxis in region r at time slot t
A_{tr}	The density of all taxis in region r at time slot t
S_{tr}	The average speed of all cars (taxis) in region r at time slot t

4.2 Traffic Charge Storage

The *traffic charge* in region r at time slot t, i.e. C_{rt}, defined by Eq. (3) as follows:

$$C_{tr} = \frac{P_{tr}}{P_t} \times (2 - \frac{V_{tr}}{A_{tr}}) \times (1 + \frac{S_{tr}}{S_{max}}) \times (1 + \frac{M_{tr}}{M_{max}}) \qquad (3)$$

The traffic charge is composed of four parts. $\frac{P_{tr}}{P_t}$ is about the passengers and $\frac{V_{tr}}{A_{tr}}$ about the taxis, reflecting a positive influence on passengers, as well as a negative influence on vacant taxis. For the third part, a better traffic condition ($\frac{S_{tr}}{S_{max}}$) means less time consumption and longer trips in a fixed time leads to more income. Last, a region with higher revenue per trip ($\frac{M_{tr}}{M_{max}}$) is preferred. For the last three parts, we use '2 minus' to guarantee the value are within the range between 1 and 2, thereby preventing undue influence of the aspects. From

Eq. (3), we can find a region with higher *traffic charge* is considered to have more potential passengers, less rival taxis, better traffic condition and higher revenue per trip.

As stated at Sect. 4.1, the traffic condition changes with time, thus a better taxi route recommendation should be consider both the historical regularity and real-time change of traffic. UTCL calculates the historical *traffic charges* (extracted from the historical trajectories) and real-time *traffic charges* (extracted from the real-time trajectories) respectively and then mix them up to the final *traffic charges*. The definition of final *traffic charges*, represented as C_f, is shown as:

$$C_f = w \times C_h\,(ts) + (1 - w) \times C_r\,(f) \tag{4}$$

where w stands for weight of historical *traffic charge*, C_h stands for the historical *traffic charge*, ts stands for the corresponding time slot division, $1 - w$ stands for weight of real-time *traffic charge*, C_r stands for the real-time *traffic charge* and f stands for the corresponding update frequency. The optimum weights are partly determined by the ratio between ts and f and may change in different circumstance.

4.3   Combination of Traffic Forces

The concept of *traffic force* from one region to another is already given in Definition 2. A taxi is affected by *traffic forces* from all regions in urban traffic network, while some regions far away have little effect. Therefore, there makes little sense to model the entire regions. To solve this problem, we define a sub-region around the taxi's current location, e.g., 1 square kilometers, which contains a few but not all regions [11]. We use the concept of *Attraction* to represent the vectorial sum of all these *traffic forces* in a sub-region. Thus, according to Eq. (2), the definition of *Attraction* is given as follows.

Definition 3. Attraction: *Attraction is the aggregated result of traffic forces, including both directions and forces. Attraction on taxi i in region r_1 from sub-region M during time slot t could be derived by the formula below:*

$$\overrightarrow{A\,(r_1, t)} = \sum_{r_m}^{M} k \times \frac{C_{r_m t}}{R_{r_1 r_m}{}^2} \times \frac{\overrightarrow{r_1 r_m}}{|r_1 r_m|} \tag{5}$$

where r_m stands for the regions in sub-region M and other variables are the same as those in Eq. (2). Attraction can be 0, which just means the choice of different directions makes no difference.

4.4   Route Recommendation

As a vacant taxi cruises on the street, the taxi driver can only take actions at the intersections [20]. Therefore, for a recommendation system, the recommended

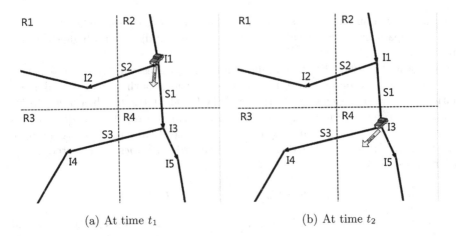

(a) At time t_1 (b) At time t_2

Fig. 2. An illustration of route recommendation. I1, I2, I3, I4 and I5 stand for five intersections. S1, S2, S3 stand for three road segments. R1, R2, R3, R4 stand for four regions.

results should be given at the same time, exactly the way the proposed method works. Additional details about the recommendation algorithm are presented using the examples shown in Fig. 2.

Imagine a vacant taxi driver arrives at I_1 at time slot t_1. At this moment, he can select to drive along S_1 or S_2. The proposed algorithm will judge these two choices and recommend him the better one. First, the proposed algorithm will acquire the *traffic charges* from the database and further calculate the *attraction*. After comparing the direction of the *attraction* with the direction of $\overrightarrow{I_1 I_2}$ and $\overrightarrow{I_1 I_3}$, the proposed algorithm will choose the road segment nearest to the direction as the recommended road segment. Hereby, the force of *attraction*, which stands for the attraction degree, would not be used in the proposed method. Suppose S_1 is the next recommended road segment and the driver arrives at I_3 at time slot t_2. The same procedure presented above will be performed again but with different time slot and region. As shown in Fig. 2, S_3 is the recommended one this time. The above steps will be under processing until the driver picks up the next passenger and a new recommendation process will start as soon as the passenger leaves the taxi. The concrete formula is given in Eq. (6):

$$S_{rec} = \arg \min_{S_j \in S | S.si = I_{id}} \theta_{(S_j.dir, D)} \tag{6}$$

where I_{id} stands for the intersection the taxi locates, S_j stands for the road segment started with I_{id}, D stands for the direction of traffic force and $\theta_{(S_j.dir, D)}$ stands for the angle between the road segment S_j and the traffic force D.

The recommendation procedure is personalized since each driver is located at a different location. From a macro perspective, this mechanism can prevent sending the same information to multiple drivers, which may result in localized competition and a non-equilibrium state. The recommendation procedure is shown in Algorithm 1.

As a taxi drops off his most recent passenger, the algorithm will first acquire his current location, including region and intersection (lines 3–4). Then we use the historical traffic charge and real-time traffic charge to calculate the final traffic charge (line 5). Next, attraction will be calculated given the position and time according to Eq. (5) (line 6). Finally, the recommended road segment will be obtained using the method proposed in Sect. 4.4 and provided to the driver (lines 7–8). This process will stop when he picks up the next passenger (line 2).

Algorithm 1. Online recommendation using UTCL

Input:
n_t, the number of time slots;
n_r, the number of regions;
M, the size of subregion;
$c_h[n_t][n_r]$, the array of historical traffic charge;
$c_r[n_t][n_r]$, the array of real-time traffic charge;
w, the weight of historical traffic charge($1-w$, the weight of real-time traffic charge)
lon_c, current longitude;
lat_c, current latitude;
T_c, current time;
$isOccupied$, the taxi's current status, initialized to 0;

Output:
W, recommended route;
1: $W \leftarrow \emptyset$
2: **while** ($isOccupied = 1$) **do**
3:     $region \leftarrow getRegion(lon_c, lat_c)$;
4:     $intersection \leftarrow getIntersection(lon_c, lat_c)$;
5:     $c_f[n_t][n_r] \leftarrow getCharge(c_h, c_r, w)$;
6:     $attraction \leftarrow getAttraction(region, T_c, c_f, M)$;
7:     $rs \leftarrow getRecommendSegment(attraction, intersection)$;
8:     $W \leftarrow W + \{rs\}$;
9:     **if** (a passenger has been picked up by the taxi) **then**
10:       $isOccupied \leftarrow 1$;
11:     **else**
12:       send W to the driver;
13:     **end if**
14: **end while**

4.5 Complexity Analysis

As shown in Eq. (4), the historical *traffic charges* and the real-time *traffic charges* together make up the final *traffic charges*, which is updated in a certain frequency. The computational complexity can be expressed as $O(T+R)$[1], composed of the metadata collection of the real-time trajectories and the calculation of the final *traffic charge* for each region.

---

[1] T is the amount of real-time trajectories, R is the number of regions.

The computation on *attraction* and routes recommendation are real-time. The computational complexity of this process is $O(M)^2$. The proposed approach is practical and can be also be deployed on smart mobile phones.

5   Experimental Analysis

5.1   Environment Setup and Simulation

The proposed method is evaluated using the taxi trajectory data of Xiamen (China) during July 2014, containing about 5,000 taxis. Besides, an associated taxi operation dataset is also used. Totally, there are over 220 million GPS position records and 8 million taxi live trips. The dataset is split up by two, from July 1st to July 21st being the training set and the remaining is considered as test set. The region is limited to $[118.066E,118.197E] \times [24.424N,24.561N]$ and partitioned into 131×137 grids with equal intervals. The sub-region is set to 0.02° of longitude and 0.02° of latitude. During data cleaning processing, the GPS records outside and those containing contradictory in time are removed from the dataset. To evaluate the proposed method, we use the maps available provided by OpenStreetMap to build a road network. Totally, 52479 road segments and 49773 intersections are stored in PostgreSQL. After building a road network, it is important to locate GPS trajectories into corresponding ways, which is also called map matching. We will leave this issue to the extension of this paper due to limited space. Besides, time slot and the update frequency of *traffic charges* are both set to 1 h. w_h is set to 0.8 and 0.2 for w_r.

To verify the effectiveness of the proposed method, we selected 50 most active taxis from 8 a.m. to 12 p.m in the test set, 5 taxis each day. We simulate the moving paths of virtual taxi drivers from the starting locations of these 50 drivers in the road network and count their income according to the real passengers that appeared in the test set.

5.2   Evaluation

The probability distribution function of the ground truth (the real income from the test set) and the simulation results using the proposed method are shown in Fig. 3. It is obvious that taxis with UTCL have a bigger probability to make higher revenue(60–80 per hour), while taxis with the ground truth are more likely to make lower revenue (20–50 per hour) compared to UTCL. As overall, the taxis following the routes recommended by the proposed method have a better earning ability than those following their own routes, which demonstrate the effectiveness of the proposed method.

The experimental results of some similar methods are also presented in Fig. 4 for comparison. The first one is the modified 'HITS' model [22], which recommends taxi drivers to the regions with highest profit without taking into consideration other factors. The next one is the method proposed in [11], called 'STP'

---

[2] M is the number of regions in the sub-region.

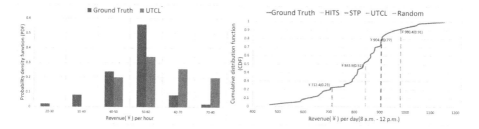

Fig. 3. Comparison of taxis' earning ability between UTCL and the ground.

Fig. 4. Comparison of average earning ability of taxis on UTCL, STP, HITS and randomized method.

for convenience, which utilize a formula to calculate the profitability scores in an STP map and recommends the region with the highest scores. The original 'STP' belongs to the macroscopic recommendation that only recommends the driving direction. We improve the original 'STP' to the road-segment level and make a contrast with UTCL. Besides, we also utilize a randomized method, that randomly selects a road segment when meeting an intersection. As shown in Fig. 4, the average revenue from the randomized method is 712.4, smaller than that with the ground truth(i.e. 822.7), indicating the effect of drivers' experience. The effectiveness of using a passenger-finding strategy is shown when compared with the results without a passenger-finding strategy. The average revenue earned by the taxi with UTCL is within top 10% of all taxis, performs best among these three methods. Specifically, compared to STP and HITS, the revenue of taxis has been increased by more than 8% when using UTCL, which indicate higher competitiveness of UTCL method.

5.3 Impact Factor Analysis

Partition Granularity. Partition granularities determine the spatial accuracy on recommendation. From Fig. 5, we could note the differences on different partition granularities. Simulation results from a smaller granularity like PG = 0.001 (each region is partitioned into 0.001° of longitude and 0.001° of latitude. So does the "PG = 0.005" and "PG = 0.01".) are more likely to reach a high revenue(i.e. 1000–1300), while results with larger granularity tend to get a low revenue (i.e. 0–800), e.g., 45% for PG = 0.01 and 30% for PG = 0.05. This is mainly due to larger granularity may lead to fewer regions and rougher calculations, thus resulting in ambiguous recommendation. In simulation experiments performed, PG = 0.01 signifies to divide Xiamen into 14 × 14 grids with an area of nearly 1 km². It is hard to make a targeted recommendation in such a big area. Nevertheless, a proper partition granularity should not be too small either, which may lead to few trajectory data on each grid and thus affect the recommendation.

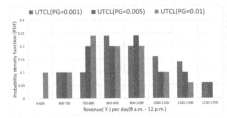
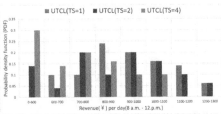

Fig. 5. Simulation results on different partition granularity.

Fig. 6. Simulation results on different time slots.

Time Slots. Time slot determines the temporal accuracy of recommendation. The differences on different time slots are shown in Fig. 6. From Fig. 6, we can note that the ratio of high revenue taxi drivers (i.e. 1000–1300) from TS = 2 (time slot is set for 1 h. So does the "TS = 1" and "TS = 4".) is about the same as that from TS = 1, while the ratio of high revenue taxi drivers from TS = 4 is much lower. This is reasonable since more specific time may capture more accurate traffic change and thus make a better recommendation. The distribution of traffic may not change too much during 2 h, but surely a huge change will happen during 4 h. This illustrates that a better time division can improve the recommendation.

Analysis of Traffic Condition. Figure 7 shows the differences with or without the factor of traffic condition. Compared to UTCL without traffic condition, the distribution from UTCL is more concentrated in higher revenue. As we can observe, there are almost 55% taxi drivers making more than 900 from the simulation with traffic condition, whilst only 45% drivers can make that much from the simulation without traffic condition. The situation is straightforward, since a better traffic condition means a bigger chance to pick up more passengers. Simulation results demonstrate that it is useful to take traffic condition into consideration.

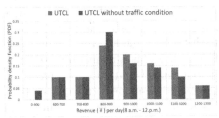
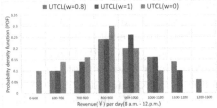

Fig. 7. Simulation results on using or without using traffic condition.

Fig. 8. Simulation results on different ratio between historical trajectories and real-time trajectories.

Analysis of the Ratio Between Historical and Real-Time Trajectories.
As shown in Fig. 8, the recommendation effect would not achieve the best when using historical trajectories ($w = 1$) or real-time trajectories ($w = 0$) alone. In this experiment, given the condition that 1 h for the time slot and the update frequency of *traffic charges*, the best outcome is achieved when w is set to 0.8. Since the final traffic charges are composed of the historical traffic charges and the real-time traffic charges, one appropriate ratio between them can reflect the distribution of traffic more accurately, thus make a better recommendation.

6 Conclusion and Future Works

In this paper, we proposed a new framework of recommending cruising routes based on *Urban Traffic Coulomb's Law* where taxis and passengers are viewed as different types of charges. *Traffic charges* and *attractions* are calculated for each region at different time slots according to *Urban Traffic Coulomb's law*, then the cruising routes for drivers are computed by comparing the differences between *attractions* and the headings of adjacent road segments. Different from other methods, the relationship among taxis and passengers are fully taken into account in the proposed algorithm. Besides, the real-time trajectories are also used for recommendation while previous methods are focused on the historical trajectories alone. Experimental results show that the proposed method can effectively provide taxi drivers with better routes. Compared to other methods, drivers with this proposed method outperforms by more than 8%.

At this work, we divide the urban area into grids with the same size for the ease of traffic charge calculation and recommendation. Yet the *traffic charges* at road-segment levels could be considered for our future work. Also, at this paper we only consider the income of taxis drivers for the recommendation, and the waiting time and preferences of passengers are not fully considered. We would import these factors in our future work. We are optimistic that further study on the *Urban Traffic Coulomb's law* is promising. It would increase the drivers' income and the overall traffic efficiency in urban areas.

References

1. Openstreetmap. http://www.openstreetmap.org/. Accessed 30 Oct 2016
2. Brakatsoulas, S., Pfoser, D., Salas, R., Wenk, C.: On map-matching vehicle tracking data. In: International Conference on Very Large Data Bases, Trondheim, Norway, 30 August - September, pp. 853–864 (2005)
3. Chow, C.Y., Mokbel, M.F.: Trajectory privacy in location-based services and data publication. ACM SIGKDD Explor. Newslett. **13**(1), 19–29 (2011)
4. Dong, H., Zhang, X., Dong, Y., Chen, C.: Recommend a profitable cruising route for taxi drivers. In: IEEE International Conference on Intelligent Transportation Systems, pp. 2003–2008 (2014)
5. Hetherington, D.: Coulomb's law. Phys. Educ. **32**(4), 277 (1997). http://stacks.iop.org/0031-9120/32/i=4/a=025

6. Hsueh, Y.L., Hwang, R.H., Chen, Y.T.: An effective taxi recommender system based on a spatiotemporal factor analysis model. In: International Conference on Computing, Networking and Communications, pp. 28–40 (2014)
7. Li, B., Zhang, D., Sun, L., Chen, C., Li, S., Qi, G., Yang, Q.: Hunting or waiting? Discovering passenger-finding strategies from a large-scale real-world taxi dataset. In: IEEE International Conference on Pervasive Computing and Communications Workshops, pp. 63–68 (2011)
8. Li, X., Pan, G., Wu, Z., Qi, G., Li, S., Zhang, D., Zhang, W., Wang, Z.: Prediction of urban human mobility using large-scale taxi traces and its applications. Front. Comput. Sci. China 6(1), 111–121 (2012)
9. Niu, B., Li, Q., Zhu, X., Cao, G., Li, H.: Achieving k-anonymity in privacy-aware location-based services. In: Proceedings - IEEE INFOCOM, pp. 754–762 (2014)
10. Patricksson, M.: The Traffic Assignment Problem: Models and Methods (3), pp. 271–272 (1994)
11. Powell, J.W., Huang, Y., Bastani, F., Ji, M.: Towards reducing taxicab cruising time using spatio-temporal profitability maps. In: Pfoser, D., Tao, Y., Mouratidis, K., Nascimento, M.A., Mokbel, M., Shekhar, S., Huang, Y. (eds.) SSTD 2011. LNCS, vol. 6849, pp. 242–260. Springer, Heidelberg (2011). doi:10.1007/978-3-642-22922-0_15
12. Qian, S., Zhu, Y., Li, M.: Smart recommendation by mining large-scale GPS traces, pp. 3267–3272 (2012)
13. Quddus, M.A.: High Integrity Map Matching Algorithms for Advanced Transport Telematics Applications. Imperial College London, London (2007)
14. Quddus, M.A., Ochieng, W.Y., Noland, R.B.: Current map-matching algorithms for transport applications: state-of-the art and future research directions. Transp. Res. Part C Emerg. Technol. 15(5), 312–328 (2007)
15. Shao, J., Lu, R., Lin, X.: Fine: A fine-grained privacy-preserving location-based service framework for mobile devices. In: 2014 Proceedings IEEE INFOCOM (2014)
16. Takayama, T., Matsumoto, K., Kumagai, A., Sato, N., Murata, Y.: Waiting/cruising location recommendation based on mining on occupied taxi data. In: WSEAS International Conference on Mathematical and Computational Methods in Science and Engineering, pp. 225–229 (2010)
17. Yuan, J., Zheng, Y., Xie, X., Sun, G.: Driving with knowledge from the physical world. In: Proceedings of the 17th ACM SIGKDD International Conference on Knowledge Discovery and Data Mining, pp. 316–324. ACM (2011)
18. Yuan, J., Zheng, Y., Zhang, L., Xie, X., Sun, G.: Where to find my next passenger. In: International Conference on Ubiquitous Computing, pp. 109–118 (2011)
19. Yuan, N.J., Zheng, Y., Zhang, L., Xie, X.: T-finder: a recommender system for finding passengers and vacant taxis. IEEE Trans. Knowl. Data Eng. 25(10), 2390–2403 (2013)
20. Zhang, D., He, T.: pCruise: reducing cruising miles for taxicab networks. In: 2012 IEEE 33rd Real-Time Systems Symposium (RTSS), pp. 85–94, December 2012
21. Zhang, M., Liu, J., Liu, Y., Hu, Z., Yi, L.: Recommending pick-up points for taxi-drivers based on spatio-temporal clustering. In: International Conference on Cloud & Green Computing, pp. 67–72 (2012)
22. Zheng, Y., Zhang, L., Xie, X., Ma, W.Y.: Mining interesting locations and travel sequences from GPS trajectories. In: International Conference on World Wide Web, WWW 2009, Madrid, Spain, April, pp. 791–800 (2009)

Efficient Order-Sensitive Activity Trajectory Search

Kaiyang Guo[1], Rong-Hua Li[1], Shaojie Qiao[2], Zhenjun Li[1(✉)],
Weipeng Zhang[1], and Minhua Lu[3]

[1] College of Computer Science and Software Engineering, Shenzhen University,
Shenzhen, China
435647651@qq.com, 15323940@qq.com, 493745134@qq.com, rhli@szu.edu.cn
[2] Chengdu University of Information Technology, Chengdu, China
qiaoshaojie@gmail.com
[3] College of Biomedical Engineering, Shenzhen University, Shenzhen, China
luminhua@szu.edu.cn

Abstract. In this paper, we study the problem of order-sensitive activity trajectory search. Given a query containing a set of time-order target locations, the problem is to find the most suitable trajectory from the trajectory database such that the resulting trajectory can achieve the minimum distance from the query. We formulate the problem using two different order-sensitive distance functions: the sum-up objective function, and the maximum objective function. For the sum-up objective function, we propose a dynamic programming (DP) algorithm with time complexity $O(mn^2)$ where m is the length of the trajectory and n is the number of query locations. To improve the efficiency, we also propose an improved DP algorithm. For the maximum objective function, we propose exact and approximation algorithms to tackle it. The approximation algorithm achieves a near-optimal performance ratio, and it improves the time complexity from $O(mn^2)$ to $O(n \log(d/\epsilon))$ in comparison with the DP algorithm. Extensive experimental studies over both synthetic and real-world datasets demonstrate the efficiency and effectiveness of our approaches.

1 Introduction

With the rapid development of wireless communication and positioning technology, a large amount of data recording the location of moving objects, known as *trajectories*, are widely generated and managed in numerous application domains. This inspires tremendous efforts made on analyzing large scale trajectory data from a variety of aspects in the last few years. Representative work includes designing effective trajectory indexing structures [2,12], efficient trajectory matching [14–16,21], and mining patterns from trajectories [7,8], to name a few. There are also a variety of applications based on trajectory data, such as locating potential friends through motion trajectories in everyday life [9], route planning [10], motion recognition through data collected by sensors attached to

© Springer International Publishing AG 2017
A. Bouguettaya et al. (Eds.): WISE 2017, Part I, LNCS 10569, pp. 391–405, 2017.
DOI: 10.1007/978-3-319-68783-4_27

the human body [19], and climate change prediction [13] and so on. The movement of objects has a certain pattern, such as people in the workday is usually along the same path to get off work, and in the holiday may go shopping or watching movies. Therefore, through the people's trajectory, interesting behavior habits can be dug out and provided personalized help for improving their lives.

In the different applications of the trajectory data, a very important and fundamental operation is the trajectory similarity search. For instance, imaging such a scenario, the user wants to start from the A location, passing by the locations B, C, D, and finally reach the E point, how to find a short and reasonable route. Our idea does not to re-plan a totally new route, but selects an existing taxi trajectory in the city to find out the most reasonable path. One reasonable solution is to select such most reasonable route among the database containing the records of existing city taxi trajectory. Compared to re-planning a new trajectory, the choice of an existing path has several advantages. First of all, it satisfies the requirement of a relatively short distance. Second, since most taxi drivers are familiar with the route of the city, the path will avoid traffic route more easily, or take some shortcuts to shorten the travel time. Finally, we do not need to consider the issue of path reachability between locations A and E, because the trajectory database stores the real routes that have been actually passed.

In this paper, we studied the problem of *order-sensitive activity trajectory search*, that is, given a trajectory database $\mathcal{T}$ and a query q consisted of several two-dimensional locations, our goal is to find the shortest trajectories from $\mathcal{T}$ w.r.t. the query q. To measure the distance between two trajectories, we propose two distance functions to calculate the order-sensitive distances. The first one function is based on the sum-up order-sensitive distance, and the second one is based on the maximum order-sensitive distance. We formulate two problems of *order-sensitive activity trajectory search* based on the sum-up distance function and the maximum distance function respectively, denoted by SumOATS-problem and MaxOATS-problem. To these aims, we develop several efficient algorithms to find the exact and approximate answers.

Both the SumOATS-problem and MaxOATS-problem are to find the optimal solution, which captures the characteristics of the optimal sub-structure. Thus, we propose dynamic programming (DP) algorithms to solve these two problems. By searching for a point within the range determined by traversing the candidate trajectory, we can calculate the shortest distance and get the optimal solution of the current state, and then use the current state to calculate the next state until the final result. The time complexity of the DP algorithms are $O(mn^2)$, where m denotes the length of the trajectory, and $n = |q|$ is the number of query locations in q. Clearly, such DP algorithms are very costly.

To speed up the algorithm efficiency, we propose several well-designed strategies to improve the basic DP algorithms. For the SumOATS-problem, we propose an improved DP algorithm called OptDPSum by marking known results and avoiding a lot of computation. For the MaxOATS-problem, we propose an approximation algorithm based on the carefully-designed binary search technique. The time complexity of the approximation algorithm is $O(n\log(d/\epsilon))$ where $\epsilon \geq 0$ is a given parameter to control approximation ratio. Moreover, in order to improve the accuracy achieved by the approximation algorithm, we propose a practically efficient algorithm that can compute the exact answer by iteratively invoking the approximation algorithm.

We conduct extensive experiments over real-world datasets. The results show that our algorithms using the improvement strategies achieve at least three orders of magnitude faster than the basic DP algorithms. The main contributions of this paper are summarized as follows.

- We investigate the problem of *order-sensitive activity trajectory search*, and study it with different problem formulations using two distance functions SumOATS-problem and MaxOATS-problem respectively.
- We propose a basic DP algorithm to solve the SumOATS-problem. To improve the efficiency, we also proposed an improved DP algorithm to significantly avoid unnecessary computations.
- We propose a near-optimal approximate algorithm named ApproxDis that takes $O(n\log(d/\epsilon))$ time complexity to tackle the MaxOATS-problem.
- We also propose an exact algorithm ExactDis to compute the MaxOATS-problem. Unlike the basic DP algorithm, both the ApproxDis and ExactDis algorithms are based on a new parametric search technique.
- We conduct extensive experiments on real-world and synthetic datasets to evaluate the proposed algorithms, and the results confirm our theoretical contributions.

The rest of the paper is organized as follows. Section 2 reviews the related work. We formulate our problems in Sect. 3. In Sect. 4, we propose the exact DP algorithms to solve SumOATS-problem. In Sect. 5, we propose the exact and approximation algorithms to tackle MaxOATS-problem. We report the experimental results in Sect. 6. Finally, we conclude this work in Sect. 7.

2   Related Work

Our work is closely related to the similarity search problem for trajectory data. The key of the similarity search problem is to define the distance function. In the literature, there are a number of distance metric that have been proposed. Notable examples are listed below. The basic way to measure the distance between two trajectories is the Euclidean distance [1] over locations of the trajectory. However, this metric only consider the geometrical distance between the trajectories, and it does not take the particular features of the trajectory into account. Another notable distance metric is the Dynamic Time Warping (DTW) distance [20] which applies dynamic programming for aligning two trajectories in such a way that their overall distance is minimized. Several other notable distance metrics include longest common subsequence [18], edit distance with real penalty (ERP) [3], edit distance on real sequences (EDR) [4], and enhanced techniques for evaluating the similarity of time series are also studied in [11,17]. All the above mentioned distance metrics do not consider the time order constraint. In the literature, there exist many algorithms to compute these distance functions efficiently [5,6].

To capture the time order constraint, Chen et al. [5] proposed a distance metric. In their work, the trajectory similarity query problem is formulated as the k-best connected trajectory (k-BCT) query Q which is represented by a set of locations $Q = \{q_1, q_2, \ldots, q_m\}$. The distance between the query location q_i and a trajectory $R = \{p_1, p_2, \ldots, p_l\}$ is the minimum Euclidean distance between q_i and p_j. They also propose a DP algorithm to compute their distance metric. Note that in their distance metric, two different query locations may match the same trajectory location, thus it may not be suitable for many real-world applications. In our distance metrics, we do not allow the query locations to match a repeated trajectory point.

3   Problem Statement

In this section, we will define the notion of trajectory query used in this paper, followed by two different distance metrics for trajectory search.

A *trajectory* is a sequence of two-dimensional spatial points indexed by the time moments, denoted by $T = s_1 s_2 \ldots s_n$ where s_i ($1 \leq i \leq n$ representing the time index) is a location represented by a two-dimensional spatial point. The length of a trajectory T is n. A trajectory $T' = s_{c_1} \ldots s_{c_k}$ where $1 \leq c_1 < c_2 < \cdots < c_k \leq n$ is called a sub-trajectory of T, i.e., $T' \subseteq T$. We can easily extend the notation of trajectory to multiple-dimensional points.

An *activity trajectory search query* is a sequence of two-dimensional spatial points ordered by its activity priority, denoted by $q = \{t_1, t_2, \cdots, t_m\}$.

Given a trajectory T, the sub-trajectory of T starting from i-th point to j-th point is denoted by $T[i, j] = s_i, \cdots, s_j$ where $1 \leq i \leq j \leq n$. Given an activity trajectory search query $q = \{t_1, t_2, \cdots, t_m\}$ and a point $t_k \in q$, we define the minimum distance between the point t_k and the sub-trajectory $T[i, j]$ of T as

$$d(T[i,j], t_k) = \min_{i \leqslant h \leqslant j} \{d(s_h, t_k)\}, \tag{1}$$

where $d(s_i, t_j)$ denotes the Euclidean distance between the locations s_i and t_j.

Based on the above distance function, we define the *order-sensitive distance* between a query q and a trajectory T based on two different functions of sum-up distance and maximum distance, which are respectively denoted by *sum-up order-sensitive distance* and *maximum order-sensitive distance* as follows.

Definition 1 (Sum-up order-sensitive distance).

$$\mathsf{SumDis}(T, q) = \sum_{j=1}^{m} d(T[l_i, l_{i+1}], t_j), \tag{2}$$

where $1 \leq l_1 \leq \ldots \leq l_m \leq l_{m+1} \leq n$.

Definition 2 *(Maximum order-sensitive distance).*

$$\text{MaxDis}(T, q) = \max_{1 \leqslant j \leqslant m} d(T[l_i, l_{i+1}], t_j), \tag{3}$$

where $1 \leq l_1 \leq \ldots \leq l_m \leq l_{m+1} \leq n$.

Problem Formulation. Given a query q and a trajectory T, the Order-sensitive Activity Trajectory Search (*OATS*) problem is to find a set of points in trajectory T such that the order-sensitive distance between T and q achieves the minimum. There exist two different functions of order-sensitive distance. On one hand, the problem formulation adopts the sum-up order-sensitive distance function, denoted by SumOATS-problem, i.e., computing SumDis(T, q). On the other hand, the problem formulation adopts the maximum order-sensitive distance function can be similarly formulated as solving objective function MaxDis(T, q), called MaxOATS-problem.

In the following sections, we will propose several algorithms to address above two problems. Specifically, we respectively consider the problems using the functions of sum-up order-sensitive distance in Sect. 4 and maximum order-sensitive distance in Sect. 5.

4   Algorithms for the SumOATS-Problem

In this section, we propose DP algorithms for exactly solving SumOATS-problem. We first develop a DP method based on the state transition equation varied by separated segments. To further improve the algorithm efficiency, we revise the state transition equation and design another DP algorithm by avoiding the unnecessary computation.

4.1   The Dynamic Programming (DP) Algorithm

First of all, to analyze the definition of SumOATS-problem, we know that the SumOATS-problem is to divide the candidate trajectory T into m segments (m is the number of the midpoints of the query), then takes the Minimum distance of points in the candidate trajectory segment to the corresponding candidate trajectory points, and finally sums up the minimum distance. Thus, our key problem is how to determine the location of each sub-point, and then calculate the shortest distance within each segment, and then sum up the results.

We can think of it as a multi-stage decision problem, and divide the process into several interrelated phases. At each stage of it, we need to make a decision, that is, to select the segmentation points so that the results of each stage is optimal, and the whole process to achieve optimal results. Of course, the decision-making at all stages of the selection is not arbitrarily determined, it depends on the current state, and it also affects the future development.

The most important step in the DP algorithm is to determine the state transition equation. Assuming that we now compute the m-th point of the query

---

Algorithm 1. DPSum(T, q)

1: **for** $i = 1$ to n **do**
2:     $rec[i][1] \leftarrow d(T[1,i],t_1)$;
3: **for** $k = 1$ to m **do**
4:     **for** $i = k$ to n **do**
5:         $d_{\max} \leftarrow 0$; $d_{\min} \leftarrow \infty$;
6:         **for** $j = k$ to $i - 1$ **do**
7:             $d_{\max} \leftarrow rec[j][k-1] + d(T[j+1,i],t_k)$;
8:             **if** $d_{\max} < d_{\min}$ **then**
9:                 $d_{\min} \leftarrow d_{\max}$;
10:        $rec[i][k] \leftarrow d_{\min}$;
11: **return** $rec[n][m]$;

(the query q is preceded by $m - 1$ points, at least one point is assigned to each point), then the m-point of the query q corresponds to the starting point of the segment on the candidate trajectory T from the $m - 1$th point to the last point. We calculate the shortest distance d from the $m - 1$th point to the last point in the query q, and then add the shortest distance d to the value of the previous state. The state transition equation of the dynamic programming algorithm is as follows:

$$ds(j, k) = \min_{i=k}^{j-1}\{ds(i, k - 1) + d(T[i+1,j],t_k)\}. \tag{4}$$

The detailed description of our algorithm is shown in Algorithm 1. Steps 1–2 initialize the value, $rec[n][m]$ represents the SumDis distance from the current candidate trajectory T of length n to the target trajectory q of length m. From the definition of the SumDis function, if q has only one point t, then the SumDis distance is the shortest distance of all the points in the T. Next, the SumDis distance in the case of $n * m$ is calculated by two-layer cycle, in the last layer of the cycle through the state transition equation to calculate the current state of the SumDis distance. That is, the sum of the previous state and the shortest distance between the current k-th point and the $(j+1)$-th point in the candidate trajectory is selected as the SumDis distance of the current state. It can be shown that the time complexity of Algorithm 1 is $O(mn^2)$.

4.2   The Optimized DP Algorithm

Even though the SumOATS-problem can be exactly solved by Algorithm 1, the time cost is $O(mn^2)$, which is very costly. To further speed-up the algorithm efficiency, we optimize its computation in this subsection.

Note that in the third loop of Algorithm 1, the algorithm aims to find a break point. After a careful analysis, we find that we can record the location of the break point which can be used as a starting point in the third loop of Algorithm 1 to find the next beak point (instead of a starting point at k). By using this strategy, we can significantly reduce the unnecessary computations in the third loop.

Algorithm 2. OptDPSum(T, q)

1: $loc \leftarrow 0$;
2: **for** $i = 1$ **to** n **do**
3:     $rec[i][1] \leftarrow d(T[1,i], t_1)$;
4: **for** $k = 1$ **to** m **do**
5:     **for** $i = k$ **to** n **do**
6:         $d_{max} \leftarrow 0$; $d_{min} \leftarrow \infty$;
7:         **for** $j = loc$ **to** $i - 1$ **do**
8:             $d_{max} \leftarrow rec[j][k-1] + d(T[j+1,i], t_k)$;
9:             **if** $d_{max} < d_{min}$ **then**
10:                $d_{min} \leftarrow d_{max}$;
11:                $loc \leftarrow j$;
12:         $rec[i][k] \leftarrow d_{min}$;
13: **return** $rec[n][m]$;

The detailed description of our algorithm is shown in Algorithm 2. In step 9 of the algorithm, the position of the break point is recorded in loc, and then in step 7, the start position of the third loop is set to the loc.

5 Algorithms for the MaxOATS-Problem

In order to attack the MaxOATS-problem, we develop three exact and approximate algorithms using several carefully-designed strategies in this section. We first propose an exact dynamic programming algorithm based on the state transition equation varied by separate segments. To speed up the computation, we design an approximate algorithm using binary search strategy that avoids the search of all candidate answers. Finally, we proposed a new exact algorithm which makes use of an iteratively-refined strategy to invoke the approximate algorithm for flitting bad answers and quickly obtaining the exact solution.

5.1 The Exact DP Algorithm

Similarly, we first analyze the definition of MaxOATS-problem as follows. First, we divide a trajectory T into m segments (m is the number of query locations). And then we take the shortest distance from each segment to the corresponding query point. Finally, we select the largest distance among all the shortest distances.

Clearly, it is still a multi-stage decision problem. Thus, we can solve the problem via dynamic programming. The optimal sub-structure of this problem is very similar to SumOATS-problem. We can derive the state transition equation as follows.

$$ds(j,k) = \min_{i=k}^{j-1} \max\{\{ds(i, k-1), d(T[i+1,j], t_k)\}\}. \tag{5}$$

The detailed description of our algorithm is shown in Algorithm 3. Clearly, the time complexity of Algorithm 3 is also $O(mn^2)$.

Algorithm 3. DPmax(T, q)

1: $loc \leftarrow 0$;
2: **for** $i = 1$ **to** n **do**
3:     $rec[i][1] \leftarrow d(T[1, i], t_1)$;
4: **for** $k = 1$ **to** m **do**
5:     **for** $i = k$ **to** n **do**
6:         $d_{\max} \leftarrow 0$; $d_{\min} \leftarrow \infty$;
7:         **for** $j = loc$ **to** $i - 1$ **do**
8:             $d_{\max} \leftarrow max\{rec[j][k-1], d(T[j+1, i], t_k)\}$;
9:             **if** $d_{\max} < d_{\min}$ **then**
10:                $d_{\min} \leftarrow d_{\max}$;
11:                $loc \leftarrow j$;
12:        $rec[i][k] \leftarrow d_{\min}$;
13: **return** $rec[n][m]$;

Algorithm 4. DisTesting($\tilde{d}$, m, k_0, L, R)

1: $k \leftarrow k_0$;
2: **for** $i = L$ **to** R **do**
3:     **if** $d(s_i, t_k) \leq \tilde{d}$ **then**
4:         $k \leftarrow k + 1$;
5:     **if** $k > m$ **then**
6:         **return** true;
7: **return** false;

5.2 The Approximation Algorithm

To improve the time complexity of the dynamic programming algorithm, we propose a near-optimal approximate algorithm, called ApproxDis. Unlike the DP algorithm, the ApproxDis algorithm is based on a dramatically different technique. Parameter search is a very common method to solve the optimal solution problem. Its essence is to add parameters to the original problem. We solve the problem with parameters, and then continue to adjust the parameters, and ultimately find the optimal solution. The general idea of our technique is that to solve the minimization problem, we first solve a related decision problem which is described as follows. Given a distance d, can we find a candidate trajectory T with n points such that the distance between a target point and its nearest point on T is no larger than d? If that is the case, we refer to d as a feasible solution for the decision problem. It is easy to see that the optimal distance should be the minimal feasible solution of the decision problem. Once the decision problem can be efficiently solved, we can use a binary search procedure to find the minimal feasible solution. Below, we first present an algorithm to solve the decision problem.

We solve the decision problem by using Algorithm 4. Algorithm 4 aims to determine whether the trajectory can be divided into m segments such that the maximal distance between the query point and its corresponding segment is no larger than a given distance d. If that is the case, we known that the optimal

Algorithm 5. ApproxDis(T, q, ϵ)

1: $d_{\max} \leftarrow \max\limits_{1 \leqslant i \leqslant m} \{d(s_i, t_i)\}$;
2: $d_{\min} \leftarrow 0$;
3: **while** $d_{\max} - d_{\min} \geq \epsilon$ **do**
4:     $\tilde{d} \leftarrow (d_{\min} + d_{\max})/2$;
5:     **if** DisTesting($\tilde{d}$, m, 1, 1, n) **then**
6:         $d_{\max} \leftarrow \tilde{d}$;
7:     **else**
8:         $d_{\min} \leftarrow \tilde{d}$;
9: **return** $d_{\max}$;

distance must be no larger than d, because d is a feasible solution. We omit the detailed the description of Algorithm 4 for brevity. It is easy to derive that the time complexity of Algorithm 4 is linear.

After solving the decision problem, we need to find the minimum distance to meet the requirements. We find that the decision problem is monotonic. That is to say, for any $d_1 > d_2$, if d_2 is a feasible solution, then d_1 is also a feasible solution. Then, based on the monotonic property of the decision problem, we can apply a binary search procedure to find the minimal feasible solution.

The ApproxDis algorithm is detailed in Algorithm 5. Since τ is a real value, we may not obtain the minimal feasible solution exactly by using binary search. Therefore, in Algorithm 5, we use a parameter ϵ to balance the tradeoff between the accuracy and running time of the algorithm. In the experiments, we will show how ϵ affects the accuracy and the running time of the algorithm. Note that in Algorithm 5, we define 0 and the maximum distance between candidate trajectories and target trajectories as the upper and lower bounds of the algorithm. And then we change the parameter d when invoking Algorithm 4. After the error value is less than the parameter ϵ, the algorithm returns the approximate solution. In addition, it is straightforward to show that the ApproxDis algorithm can guarantee a solution that is at most ϵ larger than the optimal solution. The time complexity of the ApproxDis algorithm is $O(n \log(d/\epsilon))$. Here d is the maximal distance between any two trajectory points. Additionally, it is easy to verify that the space complexity of ApproxDis is $O(m)$.

5.3 The New Exact Algorithm

In order to improve the accuracy of the approximation algorithm, we propose an exact algorithm to solve the MaxDis problem. Clearly, for the MaxDis problem, the final result must be the distance from a point in the query q to a point in the candidate trajectory T. Therefore, we can find the exact solution from all the $m \times n$ distances.

The detailed description of the new exact algorithm is shown in Algorithm 6. Specifically, we test all the $m \times n$ distances using Algorithm 4. For a distance d, Algorithm 6 invokes Algorithm 4 to check whether it is a feasible solution. The algorithm maintains the best distance found so far (i.e., $d_{\min}$ in Algorithm 6),

Algorithm 6. ExactDis(T, q)

1: $d_{min} \leftarrow +\infty$;
2: **for** $j = 1$ **to** m **do**
3:     **for** $i \leftarrow 1$ **to** n **do**
4:         $d \leftarrow d(s_i, t_j)$;
5:         **if** $d < d_{min}$ **then**
6:             **if** DisTesting($d, m, 1, 1, n$) **then**
7:                 $d_{min} \leftarrow d$;
8: **return** d_{min};

---

and also applies the current minimum feasible distance to prune the unpromising distances (line 6). Since the time complexity of Algorithm 4 is $O(n)$, the total time cost of Algorithm 6 is bounded by $O(mn^2)$. Compared to the DP algorithm (Algorithm 3), our new exact algorithm can significantly prune the search space as verified in the experiments.

6   Experiments

In this section, we conduct extensive experiments to evaluate the proposed algorithms. All algorithms are implemented in C++ and tested on a computer with 3.30 GHz Intel(R) Core(TM) i5-4590 CPU and 8 GB memory running Windows 8.1 operation system.

Datasets. In the experiment, we use two datasets. The first one is a real-world taxi trajectory dataset. This is a sample of T-Drive trajectory dataset that contains a one-week trajectories of 10,357 taxis. The total number of points in this dataset is about 15 million and the total distance of the trajectories reaches 9 million kilometers. The T-Drive taxi trajectory dataset was generated by over 10,000 taxis in a period of one week in Beijing. We download this dataset from this website[1]. We use $n \in [500, 100000]$ to represent the length of the trajectory in the real taxi trajectory data. The second one is the synthetic trajectory dataset. For the synthetic dataset, we set n to be a value in the range $[500000, 2000000]$.

Comparison Methods and Evaluation Metrics. We test and compare our algorithms on two problems SumOATS-problem and MaxOATS-problem. For the SumOATS-problem, we compare two methods in this paper, namely, DPSum and OptDPSum. DPSum is the dynamic programming algorithm in Algorithm 1 as the baseline. OptDPSum is Algorithm 2 improving from DPSum. For the MaxOATS-problem, we compare three exact and approximation algorithms, namely, DPmax, ApproxDis, and ExactDis. DPmax is a dynamic programming algorithm (Algorithm 3) as the baseline. ApproxDis is an approximation algorithm in Algorithm 5. We set the parameter $\epsilon = 0.1$ for the ApproxDis algorithm. ExactDis is an exact algorithm (i.e., Algorithm 6). We use the running time as an evaluation metric to compare the efficiency.

---
[1] https://www.microsoft.com/en-us/research/publication/t-drive-trajectory-data-sample/.

6.1 Efficiency Evaluation

Here, we report the running time of the DPSum, OptDPSum, DPmax, ApproxDis, and ExactDis algorithms. The results varied by the number of trajectory points in T as m and the number of query points as n on all datasets are shown in Fig. 1. For the SumOATS-problem, OptDPSum is significantly faster than DPSum. OptDPSum achieves 3–10 times of efficiency performance improvement than DPSum on average. With the increasing m, the efficiency performance improvement of OptDPSum reduces a little bit. Finally, when n is the same, the running time of DPSum and OptDPSum algorithm increase with the increasing m. For the MaxOATS-problem, ApproxDis and ExactDis methods are 231,009 and 12,022 times faster than the DPmax algorithm respectively. In other words, ApproxDis, and ExactDis methods are at least four orders of magnitude faster than DPmax approach in real taxi trajectory datasets. As desired, the running time of the ApproxDis algorithm increases with the increasing n. When the number of query points as n keeps unchanged, the running time does not vary much regardless of how m changes.

6.2 Scalability Testing

In this experiment, we test the scalability of our 5 methods, namely, DPSum, DPSum, DPmax, ApproxDis, and ExactDis algorithms. To this end, we vary both the number of trajectory points in T as m from 500 to 2,000,000 and the number of query points as n from 5 to 30 to evaluate the running time of our algorithms in taxi trajectory data dataset and synthetic trajectory dataset. For the SumOATS-problem, we report the results in Fig. 1. When the datasets is less than 10000, both the DPSum and DPSum algorithms are scalable well. While the size of datasets are large than 10000, DPSum algorithm cannot obtain the solution in one day. For the MaxOATS-problem, the results are reported in Fig. 2, when the size of datasets is large than 1000, DPmax cannot get the result in one day in our datasets, while ApproxDis and ExactDis algorithms can compute the results even when $n = 2000000$. The reason is that the time complexity of our algorithms are sub-linear. The results further confirm that our algorithm can be scalable on massive data very well.

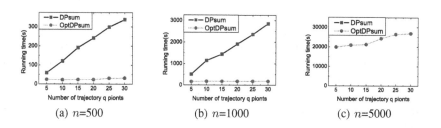

(a) n=500   (b) n=1000   (c) n=5000

Fig. 1. Efficiency testing for the SumOATS-problem (vary n).

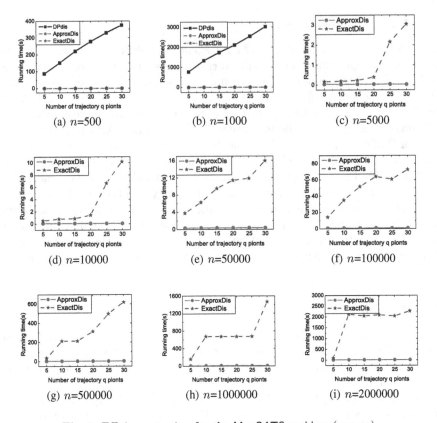

Fig. 2. Efficiency testing for the MaxOATS-problem (vary n).

6.3 Parameter Sensitive Evaluation

In this experiment, we study how the parameter ϵ affects the efficiency and effectiveness of the ApproxDis algorithm. For this purpose, we vary ϵ from 0.01 to 1. We evaluate the effectiveness using the absolute value of the difference between result of ApproxDis and ExactDis as an error. The smaller error is, the better performance is. To evaluate the efficiency of ApproxDis algorithm, we also use running time as the metric. We set $m = 10$ and $n = 50000$. The results of running time are shown in Fig. 1. As can be seen, the running time decreases with the increasing ϵ. When $\epsilon \leq 0.01$, the running time of ApproxDis decreases very fast. However, when $\epsilon > 0.01$, the curve performs relatively smooth. We show the error results of ApproxDis with different ϵ. In Fig. 1, we can see that the accuracy is getting lower with the increase of ϵ. When $\epsilon = 0.01$, the results of ApproxDis are relatively accurate. As a result, we choose $\epsilon = 0.1$ as the parameter of ApproxDis algorithm, which make a good balance between the running time and accuracy (Fig. 3).

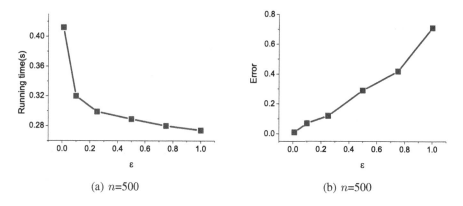

(a) n=500   (b) n=500

Fig. 3. Effect of the parameter ϵ (Vary n)

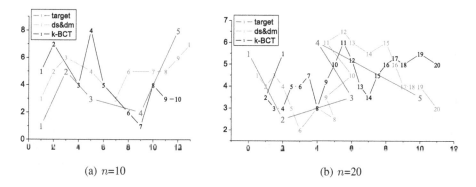

(a) n=10   (b) n=20

Fig. 4. Results in synthetic data

6.4   Case Study

In this experiment, we evaluate the effectiveness of our methods by the case study on real-world dataset. We compare the method proposed in [5] called the k Best-Connected Trajectory (k-BCT), with our two distance functions SumDis (ds) and MaxDis (dm). Specifically, we choose a short query and then use the nearest trajectories from the database measured by k-BCT, SumDis, and MaxDis.

The results are reported in the real-world and synthetic datasets are reported in Figs. 4 and 5 respectively. Note that due to the long length of the trajectory (which is difficult to visualize), we visualize a part of the resulting trajectories. We can see that compared with the trajectory using k-BCT distance, the trajectories using SumDis (ds) and MaxDis (dm) distance are closer to the query. Note that the trajectories of SumDis and MaxDis distances are very similar in the shape, structure and direction, thus we visualize them in one trajectory to avoid confusion. These results indicate the effectiveness of our distance metrics.

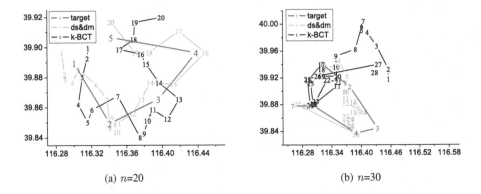

(a) n=20   (b) n=30

Fig. 5. Results in real-world trajectory data

7   Conclusion

We study the problem of *order-sensitive activity trajectory search*, and formulate it into two objective functions using two order-sensitive distance functions respectively as SumOATS-problem and MaxOATS-problem. For the SumOATS-problem, we propose two dynamic programming algorithms to solve it. For the MaxOATS-problem, we proposed three exact and approximation algorithms to tackle it. The near-optimal approximate algorithm ApproxDis takes $O(n \log(d/\epsilon))$ time complexity, which significantly improves the basic dynamic programming algorithm taking $O(mn^2)$ time. Extensive experiments on real-world and synthetic datasets are conducted to evaluate the proposed algorithms and confirm our theoretical contributions.

Acknowledgement. The work was supported in part by (i) NSFC Grants (61402292, U1301252, 61033009), NSF-Shenzhen Grants (JCYJ20150324140036826, JCYJ20140418095735561), and Startup Grant of Shenzhen Kongque Program (827/000065). Dr. Zhenjun Li and Prof. Minhua Lu are the corresponding authors of this paper.

References

1. Agrawal, R., Faloutsos, C., Swami, A.: Efficient similarity search in sequence databases. In: Lomet, D.B. (ed.) FODO 1993. LNCS, vol. 730, pp. 69–84. Springer, Heidelberg (1993). doi:10.1007/3-540-57301-1_5
2. Cai, Y., Ng, R.T.: Indexing spatio-temporal trajectories with Chebyshev polynomials. In: SIGMOD (2004)
3. Chen, L., Ng, R.T.: On the marriage of lp-norms and edit distance. In: VLDB, pp. 792–803 (2004)
4. Chen, L., Özsu, M.T., Oria, V.: Robust and fast similarity search for moving object trajectories. In: SIGMOD, pp. 491–502 (2005)
5. Chen, Z., Shen, H.T., Zhou, X., Zheng, Y., Xie, X.: Searching trajectories by locations: an efficiency study. In: SIGMOD (2010)

6. Faloutsos, C., Ranganathan, M., Manolopoulos, Y.: Fast subsequence matching in time-series databases. In: SIGMOD (1994)
7. Jeung, H., Yiu, M.L., Zhou, X., Jensen, C.S., Shen, H.T.: Discovery of convoys in trajectory databases. PVLDB 1(1), 1068–1080 (2008)
8. Lee, J., Han, J., Whang, K.: Trajectory clustering: a partition-and-group framework. In: SIGMOD (2007)
9. Li, Q., Zheng, Y., Xie, X., Chen, Y., Liu, W., Ma, W.: Mining user similarity based on location history. In: GIS, p. 34 (2008)
10. Li, R., Qin, L., Yu, J.X., Mao, R.: Optimal multi-meeting-point route search. IEEE Trans. Knowl. Data Eng. 28(3), 770–784 (2016)
11. Morse, M.D., Patel, J.M.: An efficient and accurate method for evaluating time series similarity. In: SIGMOD, pp. 569–580 (2007)
12. Ni, J., Ravishankar, C.V.: Indexing spatio-temporal trajectories with efficient polynomial approximations. IEEE Trans. Knowl. Data Eng. 19(5), 663–678 (2007)
13. Sefidmazgi, M.G., Sayemuzzaman, M., Homaifar, A.: Non-stationary time series clustering with application to climate systems. In: Jamshidi, M., Kreinovich, V., Kacprzyk, J. (eds.) Advance Trends in Soft Computing. SFSC, vol. 312, pp. 55–63. Springer, Cham (2014). doi:10.1007/978-3-319-03674-8_6
14. Shang, S., Chen, L., Jensen, C.S., Wen, J., Kalnis, P.: Searching trajectories by regions of interest. IEEE Trans. Knowl. Data Eng. 29(7), 1549–1562 (2017)
15. Shang, S., Ding, R., Yuan, B., Xie, K., Zheng, K., Kalnis, P.: User oriented trajectory search for trip recommendation. In: EDBT (2012)
16. Shang, S., Ding, R., Zheng, K., Jensen, C.S., Kalnis, P., Zhou, X.: Personalized trajectory matching in spatial networks. VLDB J. 23(3), 449–468 (2014)
17. Sherkat, R., Rafiei, D.: On efficiently searching trajectories and archival data for historical similarities. PVLDB 1(1), 896–908 (2008)
18. Vlachos, M., Gunopulos, D., Kollios, G.: Discovering similar multidimensional trajectories. In: ICDE, pp. 673–684 (2002)
19. Yang, A.Y., Jafari, R., Sastry, S., Bajcsy, R.: Distributed recognition of human actions using wearable motion sensor networks. JAISE 1(2), 103–115 (2009)
20. Yi, B., Jagadish, H.V., Faloutsos, C.: Efficient retrieval of similar time sequences under time warping. In: ICDE (1998)
21. Zheng, K., Shang, S., Yuan, N.J., Yang, Y.: Towards efficient search for activity trajectories. In: ICDE (2013)

Time Series Classification by Modeling the Principal Shapes

Zhenguo Zhang[1,2], Yanlong Wen[1], Ying Zhang[1], and Xiaojie Yuan[1(✉)]

[1] College of Computer and Control Engineering, Nankai University,
38 Tongyan Road, Tianjin 300350, People's Republic of China
yuanxj@nankai.edu.cn
[2] Department of Computer Science and Technology, Yanbian University,
977 Gongyuan Road, Yanji 133002, People's Republic of China

Abstract. Time series classification has been attracting significant interests with many challenging applications in the research community. In this work, we present a novel time series classification method based on the statistical information of each time series class, called Principal Shape Model (PSM), which can quickly and effectively classify the time series even if they are very long and the dataset is very large. In PSM, the time series with the same class label in the training set are gathered to extract the principal shapes which will be used to generate the classification model. For each test sample, by comparing the minimum distance between this sample and each generated model, we can predict its label. Meanwhile, through the principal shapes, we can get the intrinsic shape variation of time series of the same class. Extensive experimental results show that PSM is orders of magnitudes faster than the state-of-art time series classification methods while achieving comparable or even better classification accuracy over common used and large datasets.

Keywords: Principal shapes · Time series · Fitting · Classification

1 Introduction

Time series research has attracted significant interests in the data mining community, due to the fact that series data are presented in a wide range of our daily life. As a fundamental research, time series classification has been studied extensively and many algorithms have been proposed during the last decade. Recent empirical evidence has strongly suggested that the simple nearest neighbor classifier is very difficult to beat [5]. Thus, the vast majority of time series classification research has focused on alternative distance measures for 1-NN classifiers based on raw data, compression data or smoothing data [20]. In these kinds of methods, the similarity of time series is a key point for the classification task. *Ding et al.* [5] present a good survey of similarity calculation methods for time series. While the nearest neighbor classifier has the advantages of simplicity and not requiring extensive parameter tuning, it does have several disadvantages.

© Springer International Publishing AG 2017
A. Bouguettaya et al. (Eds.): WISE 2017, Part I, LNCS 10569, pp. 406–421, 2017.
DOI: 10.1007/978-3-319-68783-4_28

One is its time and space consumption for large datasets, the other is it does not give us a reasonable explanation about the relationship between a time series and a class. Unlike the NN-based methods taking all series points into consideration, another kinds of methods treat the subsequences of time series as a basis for classification. Recent years, the shapelets-based methods [6,14,17,21] have been studied most, which attempt to find the subsequences with high discriminative power as features of one class. The idea is that the time series in different classes can be distinguished by their local subsequences instead of the whole ones. These kinds of methods provide interpretable results, which can help researchers understand the data.

Due to the influences of many factors, the time series of a class are slightly different at some time point. It is common for data acquisition. The above two kinds of methods, i.e., 1-NN classifier and shapelets-based methods, describe these differences by distance between time series or between subsequences. It is a reasonable assumption that two time series in the same class have a small distance. In 1-NN classifier, the distances between the test sample and all training time series should be calculated first and the one with the minimum value should have the same label with the test sample. For shapelets-based methods, all subsequences with any length are shapelet candidates and the most important step to find out shapelets is to calculating the distances between a candidate and all training time series [21]. Even though there are some pruning strategy to accelerate this process [17], the process of finding the most discriminative subsequences is computationally expensive.

Unlike existing methods, we utilize an entirely different method and build a model to describe these differences of the time series of a class. The label of a test sample can be obtained by comparing the similarity of the instance and the models. By this way, the large amount of distance calculation is unnecessary. Moreover, the variation characteristics of time series of a class can also be represented by the model. In this paper, the variation characteristics of time series is described by a linear model. Firstly, We extract the intrinsic variation characteristics of each class to generate the principal shapes based on its statistical information. Then each class corresponds to a model that constructed by its principal shapes. All time series of a class are considered as a linear combination of principal shapes. The similarity of a test time series and a class models is represented by an objective function. By optimizing these objective functions and comparing the similarity values, we can predict the class label of this test time series. The experimental results on a large number of time series datasets demonstrate that our method is orders of magnitudes faster than other methods, while achieving comparable or even better classification accuracies.

In summary, our contributions include the following:

- We utilize the principal shapes to describe the variation characteristics of time series in the same class, which gives a good understand for time series.
- The distance between the model constructed by principle shapes and a test sample is employed to predict the test sample's label. Compared with 1-NN based method, this technique can reduce the amount of distance calculation.

- We empirically validate the performance of our proposed method on commonly used time series datasets and some large datasets. The results show promising results compared to other methods.

The remainder of this paper is organized as follows: We provide some background into time series classification and review the common used algorithms in Sect. 2. Section 3 describes the principal shape model in detail and illustrates the principal shapes by a toy example. The effectiveness and efficiency of the proposed method are also discussed. Section 4 demonstrates the PSM is effective by adequate experiments. Conclusions and future work are drawn in Sect. 5.

2   Background and Related Work

2.1   Time Series Classification

A time series is an ordered set of real-valued variables. Usually, the time series are recorded in temporal order at fixed intervals of time. Time series classification is defined as the problem of building a classifier from a labelled training time series set to predict the label of new time series. A time series x_i with m values is represented as $x_i = (x_{i,1}, x_{i,2}, \ldots, x_{i,m})^T$ and an associated class label l_i. The training set is a set of n labelled pairs (x_i, l_i) with C classes: $X = \{(x_1, l_1), (x_2, l_2), \ldots, (x_n, l_n)\}$, where $l_i \in C$. For classification algorithm, the normalization of time series is necessary and the commonly used approach is z-normalization: $norm(x) = \frac{x - mean(X)}{std(X)}$. For ease of explanation, we still use x to denote $norm(x)$. Given an unlabelled time series dataset, the objective is to classify each sample of this dataset to one of the predefined classes.

The main difference between time series classification problems and the general classification task is the order of observations is very important. Therefore, the algorithm for time series classification requires specific techniques to meet this characteristic. Two kinds of methods are studied in recent years. One is global-based method by taking the whole time series into account, while the other considers the usage of time series subsequences, called local-based method.

2.2   Global-Based Methods

In global-based methods, the classifier takes the whole time series as features and predicts the label of a new time series based on its similarity metrics where distance is measured directly between time series points. Two kinds of distances, Euclidean distance (ED) and Dynamic Time Warping (DTW) distance, have been widely and successfully used as the similarity metrics [10,11,15,18]. ED is usually time and space efficient, but it often gets a poor classification accuracy [18]. DTW is considered as a better solution in time series research community because it allows time series to match even if they are out of phase in the time axis [12,19]. It has been proved that the 1-NN classifier with DTW is the best approach for small datasets. However, as the number of time series increases, the classification accuracy of DTW will converge to ED [5]. As we mentioned in

Sect. 1, these methods have two drawbacks: high time and space complexity for large datasets and long time series and results uninterpretable.

2.3 Local-Based Methods

To address the limitations of 1-NN classifier, a new shapelets-based classification algorithm is proposed by *Ye* in 2009 [21]. Informally, shapelets are time series subsequences which can maximally represent a class in some sense. The algorithm completes the classification task by constructing a decision tree classifier. The important point is that shapelet offers interpretable features to domain experts. The utility of shapelets has been confirmed and extended by many researchers [7]. Nevertheless, since all subsequences of time series could be a shapelet, finding a good shapelet is a time-consuming task. Although there are several speedup strategy, e.g., early abandon pruning, entropy pruning [21], intelligent caching and reuse of intermediate results (called Logical Shapelets) [14], etc., it still takes a long running time, especially in case of large datasets and long time series. To speed up the process of shapelets discovery, *chang et al.* [3] propose to parallelize the distance computations using GPUs. *Rakthanmanon et al.* [17] propose a heuristic strategy, called Fast-Shapelets (FSH), to speed up the searching process by exploiting a random projection method on the SAX representation of time series to find the potential shapelet candidates [13]. It is up to orders of magnitudes faster than Logical Shapelets, which reduces the time complexity from $O(n^2m^3)$ to $O(nm^2)$ (n is the number of time series in the dataset, and m is the length of the longest time series). Except for decision tree classifier, some off-the-shelf classifiers like SVM are also used on shapelets data [8,9], which are transformed by measuring distances between the original time series and discovered shapelets. It can improve prediction accuracy while still maintaining the explanatory power of shapelets.

Another way to find shapelets called Learning Time Series (LTS) is to learn (not search for) shapelets by optimizing an objective function [6]. The LTS algorithm enables learning near-to-optimal shapelets directly without the need to try out lots of candidates and can learn true top-k shapelets. It is an entirely new perspective on time series shapelets. The algorithm can also gain the higher accuracy in some time series dataset than other methods, but it is space and time-consuming.

3   Principal Shape Model

3.1   Principal Shapes Extraction from Time Series

If we take a time series as a vector, the vectors of a training set can form a distribution in m dimensional space. If we model this distribution, we can generate new series which are similar to those in the original training set, and also can examine the test time series to decide whether they are plausible examples. To simplify the problem, we wish to reduce the dimensionality of the training data from m to a reasonable dimension k. An effective approach is to apply Principal Component Analysis (PCA) method. Given a training set of n_i time series with the same class label, the mean time series $\overline{x_i}$ is calculated as: $\overline{x_i} = \frac{1}{n_i} \sum_{j=1}^{n_i} (x_j)$.

The principal shapes of variation in each class, i.e. the ways in which some points of time series tend to move together, can be found by using PCA. In order to remove the meaningless distortions, we first calculate the deviations $dx_j = x_j - \overline{x_i}$ of each class. From these deviations, we can get the matrix M of the i-th class in training set. And then the eigenvectors (also called principal components), $p_1, \ldots, p_{n_i}$, and the corresponding eigenvalues, $\lambda_1, \ldots, \lambda_{n_i}$, can be obtained by singular value decomposition method. The variations of time series points can be described by the principal components and eigenvalues. Each principal component gives a pattern of variation of time series points. The first principal component, which is associated with the largest eigenvalue, λ_1, describes the largest part of the time series variation. The proportion of total variance described by the j-th principal component is equal to the λ_j.

Generally, most of the time series variation can be represented by a small number k of principal components. The parameter k can be assigned a certain number, but a more common practice is to choose the first k principal components from a sufficiently large proportion of the total variance of the training set. The proportion can be decided by the cross validation method for each dataset. The total variance S is defined as: $S = \sum_{j=1}^{n_i} \lambda_j$.

By this way, all time series of a class can be approximated by taking the mean time series and a weighted sum of the first k principal components:

$$x \approx \overline{x} + Pb \tag{1}$$

where $P = (p_1, p_2, \ldots, p_k)$ is the matrix that made up by the first k eigenvectors; $b = (b_1, b_2, \ldots, b_k)^T$ is a weight vector.

By only choosing the first k principal components instead of the whole eigenvectors, we can get the principal shapes contained in each class. Besides, the distortions occasionally occur in one or several time series can be eliminated, because they are usually related to small principal components.

These above equations allow us to generate new instance of time series by varying the parameter b within a suitable limit, which should be related with the training set. Note that the variance of b_j on the training set can be given by λ_j, so a reasonable limit [16] is

$$\|b_j\| \leq 3\sqrt{\lambda_j} \tag{2}$$

The coefficient 3 is selected because most of the variations lie in 3 standard deviation of the mean. By applying limits of $\pm 3\sqrt{\lambda_j}$ to the parameter b_j, we can ensure that the generated time series is similar to those in the training set.

3.2   An Example of Principal Shapes on Toy Dataset

We use Car dataset as our toy dataset, which is one of time series datasets from UCR archive[1], to exhibit the extracted principal shapes and their variations. It is composed of four classes, which contains 60 training time series and 60 test time series. Each time series contains 577 real observations. We extract the principle shapes of each class from training set and the first four principal shapes of each class are shown in Fig. 1.

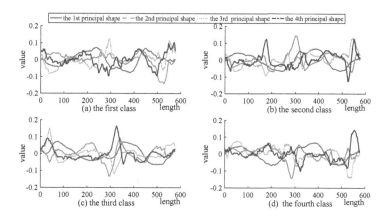

Fig. 1. The principal shapes of 4 classes in toy dataset. The 4 principal shapes of each class are presented in (a), (b), (c) and (d), respectively.

From this figure we can see, the principal shapes of each class are different, so we can build a model based on these principal shapes for time series classification. As mentioned above, the vector b defines a set of parameters of a deformable model. By varying the elements of b, we can vary the principal shapes. All extracted principal shapes of each class are added up after varying by an admissible parameter, and then we get the overall shape variation of each class. Figure 2 shows the range of variation of each class. The shape is a superposition of all admissible principal shapes. The dotted arrow lines indicate that all time series of this class can only vary in these scope.

It is clearly that time series of different classes have different shape variations. The class information of a new time series that we want to predict can be obtained by fitting these shape variations. The one with the minimum distance is the best match with the test time series.

3.3   Classifying New Time Series

The principal shapes of class i and the corresponding parameter b^i form a linear model which we call it principal shape model[2]. After getting the models of all

---

[1] http://www.cs.ucr.edu/~eamonn/time_series_data/.
[2] The superscript i denotes that we are dealing with the i-th class.

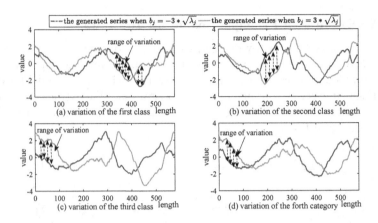

Fig. 2. The overall shape variation of each class in toy dataset.

classes, the training time series can be expressed by these models. When doing the classification task, for a new time series, our purpose is to fit these models to this time series. More specifically, if a time series belongs to a specific class i, the corresponding model (P^i, b^i) must fit it well, otherwise, the model has a large distance to this time series. Assume y is the time series to be classified, an objective function to measure the fitting degree of class i is defined by

$$f(b^i) = \frac{1}{2}\|y - (\overline{x_i} + P^i b^i)\|^2 \qquad s.t. \quad \|b_j^i\| \le 3\sqrt{\lambda_j^i} \tag{3}$$

where $i \in C$ and $\overline{x_i}$ is the mean of i-th class time series, P^i is a matrix of the first k eigenvectors generated by the i-th class, b^i is the weight vector.

Formula (4) is a least squares problem with an inequality constraint. It has been proved that f (without constraint) is a strictly downward convex function and has the only extremum [2]. The best b^{i*} is:

$$b^{i*} = (P^{iT}P^i)^{-1}P^{iT}(y - \bar{x}_i) \tag{4}$$

As we can find that if we want to get the minimum value of objective function, an inverse operation is necessary. Due to the complexity of matrix inverse operation and the inequality constraint, we cannot easily get the best parameter vector b^i. An alternative method to solve this problem is the Gradient Descent (GD). With the help of GD, we can gain the minimum value between a new time series y and the current model $(\bar{x}_i, P^i, b^i)$. Algorithm 1 describes this process.

For gradient descent algorithm, the initial value of b^i is usually set to a random number and the algorithm should run many times to obtain the best value. But in this algorithm, due to the characteristic of f: $\frac{\partial^2 f(b_i)}{\partial b^i{}^2} = P^{iT}P^i$, there's no need to repeat the experiment many times to find the minimum value by setting different initial values of b^i. In Algorithm 1, the lines 1 gives a default value of b^i and the distance between y and the model by using the default b^i is obtained in line 3. The constants A and B in line 7 are used to calculate the

---

Algorithm 1. Fitting the model to a new time series

Input: y: the time series to be classified; P^i: the matrix of eigenvectors of class i
   $\overline{x}_i$: the mean of time series of class i; λ^i: eigenvalues of class i
Output: dis: the minimum distance between y and the current model
1: initialize $b^i = 0$;
2: $r = 3\sqrt{\lambda^i}$;
3: $dis = \frac{1}{2}\|y - (\overline{x}_i + P^i b^i)\|^2$;
4: **if** $dis < \varepsilon$ **then**
5:     return dis;
6: **end if**
7: $A = P^{iT}(y - \overline{x}_i); B = P^{iT}P^i$;
8: **while** true **do**
9:     $\Delta = A + Bb^i$;
10:     $b^i = b^i - \alpha\Delta$;
11:     **if** $\|b^i_j\| > r^i_j$ **then**
12:         $b^i_j = r^i_j$;
13:     **end if**
14:     $dis2 = \frac{1}{2}\|y - (\overline{x}_i + P^i b^i)\|^2$;
15:     **if** $dis2 < \varepsilon$ or $\|dis - dis2\| < \varepsilon$ **then**
16:         return $dis2$;
17:     **end if**
18: **end while**

---

gradient which is completed in line 9 for current b^i. The iterative process for finding the best b^{i*} is from line 8 to line 18. Lines 11–13 limit the range of b^i, i.e. $\|b^i_j\| \leq 3\sqrt{\lambda^i_j}$. Parameter α in line 10 is the step size. If its value is very small, the algorithm will converge slowly. Our objective function is a strictly downward convex function, so we do not need a small step size. In our experiments, α is set to 0.5 and only after a few iterations, we can get the minimum distance.

For each class, we can get a distance from the objective function. If y is an instance of this class, the function value must be smaller than other class. After we find the best parameter vector b^{i*} to f for each class and the label of y is obtained by:

$$label(y) = \arg\min_i f(b^{i*}) \qquad i \in C \qquad (5)$$

3.4   Effectiveness and Efficiency

In PSM, the model is generated by the principal shapes extracted from the training set. So the ability of these principal shapes is very important for our model. If the training data is skewed, the principal shapes gained by this training set can not express the variation of the test time series. In this case, the accuracy of PSM decreases like other methods.

For the first step of our method, we get the eigenvalues and eigenvectors of each class by using SVD. It takes $O(min\{mn_i{}^2, m^2 n_i\})$ time, where m is the length of the time series, n_i is the number of time series in i-th class of training set. In Algorithm 1, the parameter matrix A requires $O(km)$ computation time where k is the number of the selected principal shapes and $k \leq min(m, n_i)$. And B needs a matrix multiplication, which requires $O(k^2m)$ computation time. In each iterative process of fitting the model, the most time-consuming step is the distance calculation for a new b. It take $O(km)$ time. So the Algorithm 1 needs $O(max\{k^2m, kmt\})$ time in total where t is the the number of iterations. Usually, we only use a small number of principal shapes to build our model, so k is very small. As we mentioned above, t is also very small because the objective function f is strictly convex. Thus, the total time complexity is $O(min\{mn_i{}^2, m^2n_i\} + max\{k^2m, kmt\})$.

4   Experiments and Results

4.1   Datasets and Baselines

In order to test the classification performance of our PSM method, we first perform the experiments on the commonly used UCR (See footnote 1) and UEA[3] datasets (called *common datasets* in this paper) as other literatures [6,8,9,17]. They provide diverse characteristics with different lengths and number of the classes and different numbers of time series instances and Table 1 gives the details. We use the default train and test data splits, which is the same as the baselines.

Table 1. *Common datasets* used in the experiments

Name	#Train/test	Length	#Classes	Name	#Train/test	Length	#Classes
Adiac	390/391	176	37	Lighting7	70/73	319	7
Beef	30/30	470	5	MedicalImages	381/760	99	10
Chlorine.	467/3840	166	3	MoteStrain	20/1252	84	2
Coffee	28/28	286	2	MP_Little	400/645	250	3
Diatom.	16/306	345	4	MP_Middle	400/645	250	3
DP_Little	400/645	250	3	Herring	64/64	512	2
DP_Middle	400/645	250	3	PP_Little	400/645	250	3
DP_Thumb	400/645	250	3	PP_Middle	400/645	250	3
ECGFive.	23/861	136	2	PP_Thumb	400/645	250	3
FaceFour	24/88	350	4	SonyAIBO.	20/601	70	2
Gun_Point	50/150	150	2	Symbols	25/995	398	6
ItalyPower.	67/1029	24	2	Synthetic.	300/300	60	6

To verify the performance of our method on large and long time series datasets, we use extra 8 datasets from UCR archives (see footnote 1) (called *large datasets*) to do our experiments. The details are shown in Table 2.

---

[3] https://www.uea.ac.uk/computing/machine-learning/shapelets/shapelet-data.

Table 2. *Large datasets* for experiments

Name	#train/test	Length	#classes	Name	#train/test	Length	#classes
CinC_ECG_torso	40/1380	1639	4	MALLAT	55/2345	1024	8
ECG5000	500/4500	140	5	NonInvas.1	1800/1965	750	42
FordB	810/3636	500	2	NonInvas.2	1800/1965	750	42
HandOutlines	370/1000	2709	2	wafer	1000/6174	152	2

We compare our method with other 8 baselines. One is 1-NN classifier with DTW which has been proved it is very difficult to beat [1]. The other methods are: standard shapelet-based classifier with information gain (IG) [21], Kruskal-Wallis statistic (KW) and F-statistic (FS) [8]; The Fast-shapelets algorithm with the help of SAX representation (FSH) [17]; Learning Time Series which is the state-of-the-art algorithm for accuracy (LTS) [6]; Shapelet-transform algorithm which transforms the shapelets into feature vectors for the first time (IGSVM) [8] and FLAG which is known as the fastest shapelet-based algorithm [9]. All experiments are performed on a computer with intel i7 CPU and 16GB memory.

4.2   Accuracy and Running Time on *Common Datasets*

We first compare our method against the selected baselines in terms of classification accuracy and running time on *common datasets* of Table 1. The results are shown in Tables 3 and 4 respectively and the best method of each dataset is highlighted in **bold**. The symbol "-" denotes that we cannot get the method's result in a reasonable time (over 24 h).

To tell the significant difference in accuracy of different methods of Table 3, a non-parametric Friedman test based on ranks [4] described as a critical difference diagram is employed. Figure 3 shows the results on *common datasets*. The horizontal line, called *cliques*, denotes that the related methods has no significant difference [4]. As can be seen, PSM has a better classification accuracy than most baselines. Although PSM has a slight low accuracy than LTS and FLAG, there is no significant difference in rank with these two methods based on the *cliques* of critical difference diagram recommend.

Time-consuming is another evaluation criterion for different methods. Based on the running time shown in Table 4, PSM is fastest on most datasets and it is 3–4 orders of magnitude faster than most baselines. Recall that LTS has the best accuracy, but it is much slower than PSM in all datasets.

In general, our method PSM is fastest and outperforms almost all of baselines according to the running time while it has comparable accuracy. More precisely, PSM has a clear higher accuracy and less time consuming than NNDTW, IG, KW, FS and FSH methods in terms of the Friedman test and running time. When compared with IGSVM, PSM also has a slight accuracy advantage while it is 3–4 orders of magnitude faster on most datasets. LTS and FLAG is better than PSM on classification accuracy, but they are slow, especially that LTS takes much more running time for slightly larger datasets.

Table 3. Classification accuracy (%) on *common datasets*. The best method for each dataset is highlighted in **bold**.

Name	NNDTW	IG	KW	FS	FSH	IGSVM	LTS	FLAG	PSM
Adiac	58.8	29.9	26.2	15.6	57.5	23.5	51.9	**75.2**	72.4
Beef	56.7	50	33.3	56.7	50	**90**	76.7	83.3	83.3
Chlorine.	62.7	58.8	52	53.5	58.8	57.1	73	76	**86.1**
Coffee	96.4	96.4	85.7	**100**	92.9	**100**	**100**	**100**	**100**
Diatom.	95.8	76.5	62.1	76.5	87.3	93.1	94.2	**96.4**	93.5
DP_Little	47.9	-	-	-	60.6	66.6	**73.4**	68.3	61.1
DP_Middle	62	-	-	-	58.8	69.5	**74.1**	71.3	71
DP_Thumb	57.5	-	-	-	63.4	69.6	**75.2**	70.5	74.1
ECGFive.	78.9	77.5	87.2	99	99.8	99	**100**	92	95.7
FaceFour	84.1	84	44.3	75	92	**97.7**	94.3	90.9	83
Gun_Point	92	89.3	94	95.3	94	**100**	99.6	96.7	90.7
ItalyPower.	94.7	89.2	91	93.1	91	93.7	95.8	94.6	**97.3**
Lighting7	78.1	49.3	48	41.1	65.2	63	**79**	76.7	71.2
MedicalImages	**77.2**	48.8	47.1	50.8	64.7	52.2	71.3	71.4	64.9
MoteStrain	87.3	82.5	84	84	83.8	88.7	**90**	88.8	87.2
MP_Little	64.7	-	-	-	56.9	70.7	**74.3**	69.3	71
MP_Middle	63.1	-	-	-	60.3	76.9	**77.5**	75	74.6
Herring	54.7	**67.2**	60.9	57.8	60.9	64.1	59.4	64.1	65.6
PP_Little	63.1	-	-	-	57.6	**72.1**	71	67.1	71.2
PP_Middle	61.6	-	-	-	61.6	**75.9**	74.9	73.8	74
PP_Thumb	56.7	-	-	-	55.8	**75.5**	70.5	67.4	71.6
SonyAIBO.	71	85.7	72.7	**95.3**	68.6	92.7	91	92.9	87.9
Symbols	94.4	78.4	55.7	90.1	92.4	84.6	**94.5**	87.5	92.9
Synthetic.	98.7	94.3	90	95.7	94.7	87.3	97.3	**99.7**	97.3

4.3   Accuracy and Running Time on *Large Datasets*

We now test the performance of our method PSM on *large datasets*. In this experiments, we discard the comparison with IG, KW, FS and IGSVM methods because these methods are too slow to get the results in a reasonable time on these datasets. The results of PSM and other baselines are shown in Table 5 and Figure 4 gives the critical difference diagram.

Unlike the results on *common datasets*, PSM not only has the least time-consuming but also gets the best classification accuracy than other baselines on these large datasets. The reason is that the model built by PSM greatly depends on the training set. If the training set gives a good variation representation of the whole dataset, PSM can get a higher classification accuracy. For these large

Table 4. Running time (in seconds) on *common datasets*. The best method for each dataset is highlighted in **bold**

Name	NNDTW	IG	KW	FS	FSH	IGSVM	LTS	FLAG	PSM
Adiac	11.5	3287	1349	1513	288	706	80596	2.78	**1.43**
Beef	0.4	471	484	576	154	435	414	1.15	**0.03**
Chlorine.	148	9751	3213	3050	617	1181	556	**6.88**	10.51
Coffee	0.1	22.3	21.8	21.9	16.5	15.9	76	0.13	**0.01**
Diatom.	1.2	9.3	8.99	9.2	13.7	9.3	88	0.41	**0.28**
DP_Little	42.1	-	-	-	1131	9516	803	1.81	**0.6**
DP_Middle	43.2	-	-	-	1286	7041	6082	1.78	**0.45**
DP_Thumb	41.8	-	-	-	1105	11353	1006	3.14	**0.42**
ECGFive	1.1	19	18.4	18.6	4.6	11353	9.1	**0.08**	0.25
FaceFour	0.6	1021	1012	1010	75	410	116	0.26	**0.05**
Gun_Point	0.4	116.4	112	148.9	7.6	74.8	14.1	0.07	**0.04**
ItalyPower.	0.4	0.38	0.22	0.22	0.5	0.22	7.3	**0.03**	0.23
Lighting7	1.2	3442	3438	3584	307	1473	965	0.31	**0.17**
MedicalImages	9.2	4347	2625	2616	164	1547	2199	1.69	**1.01**
MoteStrain	0.6	1.41	1.29	1.29	1.4	0.84	6	**0.04**	0.29
MP_Little	42.1	-	-	-	1297	7394	5233	2.95	**0.38**
MP_Middle	42.9	-	-	-	1186	12102	724	2.23	**0.28**
Herring	2.1	17864	18166	17789	183	8536	264	1.25	**0.03**
PP_Little	42.2	-	-	-	1107	8142	4805	3.91	**0.45**
PP_Middle	43	-	-	-	1135	4753	3406	2.19	**0.34**
PP_Thumb	41.9	-	-	-	1172	8209	6638	4.2	**0.42**
SonyAIBO.	0.2	1.17	1.02	1.02	1.1	0.75	25.4	**0.04**	0.13
Symbols	7.5	3325	3318	3622	59.4	1263	528	0.85	**0.56**
Synthetic.	1.6	291	164	161	39.4	922	293	0.26	**0.25**

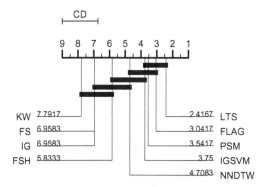

Fig. 3. Critical difference diagram for different methods on *common datasets*.

Table 5. Classification accuracy (%) and running time (s) on *large datasets*. The best method for each dataset is highlighted in **bold**

Name	Accuracy (%)					Time (s)				
	NNDTW	FSH	LTS	FLAG	PSM	NNDTW	FSH	LTS	FLAG	PSM
CinC_ECG.	75.2	56.5	69.9	91.1	**93.4**	483	1995	4911	34.69	**2.41**
ECG5000	92.8	92	93.7	91.9	**94.1**	1581	1681	13571	6.72	**3.22**
FordB	59	77.4	**89.8**	77.6	83.4	2121	8953	4757	95.68	**5.44**
HandOutlines	79.4	86.5	-	83.1	**86.7**	10382	11652	-	56.27	**0.75**
MALLAT	91.4	93.2	-	**96.2**	94.3	351	1042	-	11.99	**3.04**
NonInvasive.1	77.4	73.9	-	**93.6**	90.3	38602	47045	-	83.15	**33.57**
NonInvasive.2	84.8	75.9	-	**94.2**	93.1	3826	41836	-	79.62	**25.52**
wafer	98.6	**99.7**	**99.7**	99.2	99.5	371	214	396	11.31	**3.13**

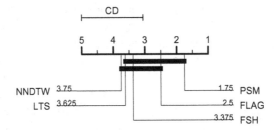

Fig. 4. Critical difference diagram for different methods on *large datasets*.

datasets, the abundant training samples basically contain the main variation of time series of datasets, so we get good results.

4.4 Scalability

In this section, we use the time series dataset *StraLightCurves* in the UCR archives (see footnote 1) to test the scalability of PSM. This dataset contains 9,236 starlight time series of length 1,024 and 3 types of star objects. The training set has 1,000 time series in default partition. Two key factors for testing the scalability are the number of training sample and the length of each sample.

We first fix the length to 1,024 and vary the number of training set from 100 to 1,000. In this case, IG, KW, FS and IGSVM methods are too slow to be run and LTS also cannot be run because of the large memory requirement. So we drop them from the comparison. We have only 4 results for NNDTW because it is also to slow when the number over 400. The running time and classification accuracy of different methods are shown in Fig. 5. As the number of training samples increasing, our method has a distinct advantage while other methods take much more running time, which has the same trends as Tables 4 and 5. One thing we should note that the time consuming of PSM is almost the same when the number of training samples increasing. The reason is that the principal shapes are extracted from the same size of covariance matrix when the training time series have the same length. This suggests that the time consumption of PSM is mainly related to the number of test samples.

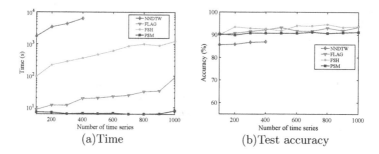

Fig. 5. Time and test accuracy when varying the number of training time series.

Now, we test the effect of the length of time series for different methods. In this experiment, the number is fixed to 1,000 and the length is varied from 100 to 1,024. When length over 300, LTS cannot be run in our computer for the memory constraint and we only get 3 results. For NNDTW, it is too slow when length over 800. Figure 6 gives the results. PSM is still fastest and it is almost linear increase with the length of time series. It's worth noting that PSM can still get a good accuracy when the length is very short while LTS, FSH and FLAG have low accuracy.

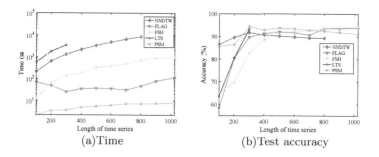

Fig. 6. Time and test accuracy when varying the length of time series.

5   Conclusion and Future Work

In this work, we propose a statistical method, PSM, to find the principal shapes of a time series dataset. With the principal shapes of each class, we can predict the class information of unlabeled time series quickly. Experimental results show that PSM is faster than other time series classification methods while it gains comparable classification accuracy compared with the state-of-the-art method on the commonly used time series datasets. Moreover, our method gets highest accuracy on large time series datasets, which fully demonstrates that the extracted principal shapes represent the intrinsic shape variation effectively when we have sufficient training data. Note that we still employ ED as the metric and do not consider the phase shift in time dimension. We propose to consider this situation in the future.

Acknowledgements. This work was supported by the National 863 Program of China [grant numbers 2015AA015401]; Research Foundation of Ministry of Education and China Mobile [grant number MCM20150507].

References

1. Batista, G.E., Wang, X., Keogh, E.J.: A complexity-invariant distance measure for time series. In: SDM, vol. 11, pp. 699–710. SIAM (2011)
2. Björck, Å.: Numerical Methods for Least Squares Problems. SIAM, Philadelphia (1996)
3. Chang, K.W., Deka, B., Hwu, W.M.W., Roth, D.: Efficient pattern-based time series classification on GPU. In: 2012 IEEE 12th International Conference on Data Mining (ICDM), pp. 131–140. IEEE (2012)
4. Demšar, J.: Statistical comparisons of classifiers over multiple data sets. J. Mach. Learn. Res. **7**(Jan), 1–30 (2006)
5. Ding, H., Trajcevski, G., Scheuermann, P., Wang, X., Keogh, E.: Querying and mining of time series data: experimental comparison of representations and distance measures. Proc. VLDB Endowment **1**(2), 1542–1552 (2008)
6. Grabocka, J., Schilling, N., Wistuba, M., Schmidt-Thieme, L.: Learning time-series shapelets. In: Proceedings of the 20th ACM SIGKDD International Conference on Knowledge Discovery and Data Mining, pp. 392–401. ACM (2014)
7. He, Q., Dong, Z., Zhuang, F., Shang, T., Shi, Z.: Fast time series classification based on infrequent shapelets. In: 2012 11th International Conference on Machine Learning and Applications (ICMLA), vol. 1, pp. 215–219. IEEE (2012)
8. Hills, J., Lines, J., Baranauskas, E., Mapp, J., Bagnall, A.: Classification of time series by shapelet transformation. Data Min. Knowl. Disc. **28**(4), 851–881 (2014)
9. Hou, L., Kwok, J.T., Zurada, J.M.: Efficient learning of timeseries shapelets. In: Thirtieth AAAI Conference on Artificial Intelligence, AAAI, pp. 1209–1215 (2016)
10. Jeong, Y.S., Jeong, M.K., Omitaomu, O.A.: Weighted dynamic time warping for time series classification. Pattern Recogn. **44**(9), 2231–2240 (2011)
11. Keogh, E., Kasetty, S.: On the need for time series data mining benchmarks: a survey and empirical demonstration. Data Min. Knowl. Disc. **7**(4), 349–371 (2003)
12. Keogh, E., Ratanamahatana, C.A.: Exact indexing of dynamic time warping. Knowl. Inf. Syst. **7**(3), 358–386 (2005)
13. Lin, J., Keogh, E., Wei, L., Lonardi, S.: Experiencing Sax: a novel symbolic representation of time series. Data Min. Knowl. Disc. **15**(2), 107–144 (2007)
14. Mueen, A., Keogh, E., Young, N.: Logical-shapelets: an expressive primitive for time series classification. In: Proceedings of the 17th ACM SIGKDD International Conference on Knowledge Discovery and Data Mining, pp. 1154–1162. ACM (2011)

15. Petitjean, F., Forestier, G., Webb, G.I., Nicholson, A.E., Chen, Y., Keogh, E.: Dynamic time warping averaging of time series allows faster and more accurate classification. In: 2014 IEEE International Conference on Data Mining (ICDM), pp. 470–479. IEEE (2014)
16. Baldock, R., Tim, C.: Model-Based Methods in Analysis of Biomedical Images. Oxford University Press, Oxford (1999)
17. Rakthanmanon, T., Keogh, E.: Fast shapelets: a scalable algorithm for discovering time series shapelets. In: Proceedings of the Thirteenth SIAM Conference on Data Mining (SDM), pp. 668–676. SIAM (2013)
18. Ratanamahatana, C.A., Keogh, E.: Making time-series classification more accurate using learned constraints. In: Proceedings of the 2004 SIAM International Conference on Data Mining, pp. 11–22. SIAM (2004)
19. Ratanamahatana, C.A., Keogh, E.: Three myths about dynamic time warping data mining. In: Proceedings of SIAM International Conference on Data Mining (SDM05), pp. 506–510. SIAM (2005)
20. Ueno, K., Xi, X., Keogh, E., Lee, D.J.: Anytime classification using the nearest neighbor algorithm with applications to stream mining. In: Sixth International Conference on Data Mining, ICDM 2006, pp. 623–632. IEEE (2006)
21. Ye, L., Keogh, E.: Time series shapelets: a new primitive for data mining. In: Proceedings of the 15th ACM SIGKDD International Conference on Knowledge Discovery and Data Mining, pp. 947–956. ACM (2009)

Effective Caching of Shortest Travel-Time Paths for Web Mapping Mashup Systems

Detian Zhang[1], An Liu[2(✉)], Gangyong Jia[3], Fei Chen[1], Qing Li[4], and Jian Li[2]

[1] School of Digital Media, Jiangnan University, Wuxi, China
detian.cs@gmail.com, chenf@jiangnan.edu.cn
[2] School of Computer Science, Soochow University, Suzhou, China
anliu@suda.edu.cn, 20164227045@stu.suda.edu.cn
[3] Department of Computer Science, Hangzhou Dianzi University, Hangzhou, China
gangyong@hdu.edu.cn
[4] Department of Computer Science, City University of Hong Kong, Kowloon Tong, Hong Kong
itqli@cityu.edu.hk

Abstract. For location-based services (LBS), the path with the shortest travel time is much more meaningful than the one with the shortest network distance, as it considers the live traffic situation. However, not every LBS provider has enough resources to compute the shortest travel-time paths by themselves. A cost-effective way for LBS providers is retrieving the shortest travel-time paths from Web mapping services (e.g., Google Maps) through external requests. Due to the high cost of processing such external requests and the usage limits, we design an effective cache of shortest travel-time paths for LBS providers in this paper, to reduce the number of external requests to Web mapping services and the query response time to users. Experimental results on real Web mapping service and datasets confirm the effectiveness of the proposed techniques.

Keywords: Shortest paths · Caching · Mapping mashup · Travel time · Road networks

1 Introduction

Shortest path queries are widely used in road networks with the development of location-based services (LBS). Usually, there are two kinds of the shortest paths in a road network, i.e., the path with the shortest network distance and the path with the shortest travel time. In reality, the path with the shortest travel time is much more meaningful than the one with the shortest network distance [13,19], as it considers the live traffic situation. To calculate the shortest travel-time path, an LBS provider not only needs the topology data of the underlying road network, but also should have the real-time traffic data, which can be collected through vehicles, roadside cameras and sensors, etc. However, not every LBS provider has enough resources to do such expensive deployment.

© Springer International Publishing AG 2017
A. Bouguettaya et al. (Eds.): WISE 2017, Part I, LNCS 10569, pp. 422–437, 2017.
DOI: 10.1007/978-3-319-68783-4_29

The Web mapping (or spatial/GIS) mashup provides a cost-effective way for an LBS provider to provide the shortest travel-time path service for its users [7, 19–23]. When the LBS provider receives a shortest travel-time path query from a user, instead of computing the path by itself, it retrieves the path from Web mapping services, like Google Maps, Bing Maps, MapQuest Maps, Baidu Maps and so on, by issuing an external request. These mapping services are owned by large and specialized companies, e.g., Google, which have enough resources to compute the shortest travel-time paths based on live traffics. Besides, they provide friendly and easy-to-use APIs, such as the Google Maps Directions API, the Bing Maps Routes API and the Baidu Maps Direction API. The LBS provider can easily retrieve the shortest travel-time path between two locations through these APIs.

However, retrieving the shortest travel-time path from Web mapping services suffers from the following two critical limitations [19]: (1) It is costly to access the shortest travel-time path information from a Web mapping service through external requests, e.g., retrieving travel time from the Bing Maps Routes API takes 502 ms while the time needed to read a cold and hot 8 KB buffer page from disk is 27 ms and 0.0047 ms, respectively [6]. (2) There is a charge on the number of external requests to a Web mapping service, e.g., the Google Maps Directions API allows only 2,500 requests per day for evaluation users and 100,000 requests per day for business license users [3]. An LBS provider needs to pay for higher usage limits. Therefore, when an LBS provider endures high workload, e.g., a large number of concurrent path queries, it needs to issue a large number of external requests to Web mapping services, which not only results in high business operation cost, but also leads long response time to its querying users.

In this paper, we design a *dynamic* path cache for LBS providers to reduce the number of external requests to Web mapping services and the query response time to users. When an LBS provider receives a path query from a user, it performs a cache lookup first by examining the cache structure. If there is a cache hit, i.e., the cache contains the shortest-travel time path of the query, it returns the query result to the user immediately; otherwise, i.e., a cache miss occurs, the LBS provider retrieves the shortest travel-time path from the Web mapping service through an external request, and tries to *update* (with *delete* and *insert* operations) the cache to keep it up-to-date.

2   Related Work

Shortest path query processing in road networks has been extensively studied both based on the network distance [11, 14, 26] and travel time [2, 13]. Different from these works that focus on path computation in local server, our work employs an LBS provider to access the shortest travel-time paths through external requests to Web mapping services, e.g., Google Maps. There are also a lot of researches about service computing [8–10], social computing [15–17], and caching [4, 5]. The most relevant studies are as follows.

The authors proposed k-NN [7, 19–21] and shortest path [18, 22] query processing algorithms, where the distance metric (i.e., travel time) and path information are retrieved from external Web mapping services. To reduce the number of external requests and query response time to users, pruning [19, 20], grouping [19, 20], direction sharing [18, 20, 22], parallel requesting [21, 22], route log [7] and waypoint [18, 22] techniques are fully exploited. However, none of these papers utilizes path cache for optimizing their system performance.

The authors in [12] proposed a path cache SPC to reduce the path computation cost for a server. However, SPC is a static cache, i.e., it does not support real-time cache update because of its high time complexity. Therefore, a dynamic path cache PPC is design in [24], where the authors employ road types to measure the weight of a path for cache replacement, e.g., the path over major roads has a higher chance to remain in cache than the path over branch roads. However, such detailed road information may be not easy to access for a cache system. More importantly, both SPC and PPC focus on path cache in static networks, which are not suitable for caching shortest travel-time paths.

In this paper, we design a dynamic path cache for an LBS provider, with a new cache replacement policy based on the impact and valid-time of each path, to reduce the number of external requests to Web mapping services and query response time to users.

3   Preliminaries

In this paper, we only consider undirected paths. Our techniques can be easily applied to directed paths. The shortest path and shortest path query are defined as follows:

Definition 1 (Shortest path). *In time dependent road networks, a shortest path $Path(v_o \rightarrow v_d)$ is a path with the minimum travel time from its origin node v_o to its destination node v_d, which is constituted by a sequence of nodes.*

To identify a path, each shortest path is associated with an unique ID p, which can be assigned by the system, i.e.,

$$Path(p) = Path(p.v_o \rightarrow p.v_d) = \{p.v_o, \cdots, v_i, \cdots, p.v_d\},$$

where $p.v_o$ and $p.v_d$ are path origin and destination of p, respectively.

Definition 2 (Shortest path query). *A shortest path query $q = (v_o, v_d)$, which is sent from a user, consists of an origin node v_o and a destination node v_d.*

Given a path query q, the LBS provider answers the query by returning the shortest travel time path from $q.v_o$ to $q.v_d$, i.e., $Path(q.v_o \rightarrow q.v_d)$ (denoted as $Path(q)$ for simplicity).

A cache and its size are defined as follows:

Definition 3 (Cache and cache size). *A cache C contains a collection of shortest paths. The cache size of C (denoted as $|C|$) is measured by the total number of nodes of the cached paths in C (as in [12]), and $|C|$ should not be larger than the system's maximum cache capacity Ψ, i.e.,*

$$|C| \leq \Psi.$$

The shortest paths exhibit the optimal sub-path property [1], i.e., every subpath of a shortest path is also a shortest path (see Lemma 1). In other words, the shortest path(s) in the cache can be taken as an answer for any query as long as both its origin and destination locate in the path(s), i.e., a cache hit happens (see Definition 4).

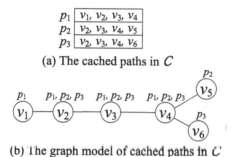

(a) The cached paths in C

(b) The graph model of cached paths in C

Fig. 1. The shortest paths in a cache C.

Lemma 1 (Optimal sub-path property). *Given a shortest path $Path(v_o \rightarrow v_d)$, if $v_i \in Path(v_o \rightarrow v_d)$ and $v_j \in Path(v_o \rightarrow v_d)$, then $Path(v_i \rightarrow v_j)$ is also a shortest path.*

Proof. See the proof of Lemma 24.1 in [1].   □

Definition 4 (Cache hit and cache miss). *Given a cache C and a path query $q = (v_o, v_d)$, a cache hit means there is at least one path in the cache that contains the query origin $q.v_o$ and destination $q.v_d$ at the same, i.e.,*

$$\exists \, p_i \in C, \; q.v_o \in Path(p_i) \wedge q.v_d \in Path(p_i);$$

otherwise, a cache miss happens.

For example, there are three paths a cache C as shown in Fig. 1a, i.e., $Path(p_1) = \{v_1, v_2, v_3, v_4\}$, $Path(p_2) = \{v_2, v_3, v_4, v_5\}$ and $Path(p_3) = \{v_2, v_3, v_4, v_6\}$. Figure 1b is a graph model of the cached paths in C. For this cache, given a user query $q = (v_3, v_5)$, since both its query origin v_3 and destination v_5 are in $Path(p_2)$, the cache can answer q immediately by returning $Path(q) = Path(v_3 \rightarrow v_5) = \{v_3, v_4, v_5\}$.

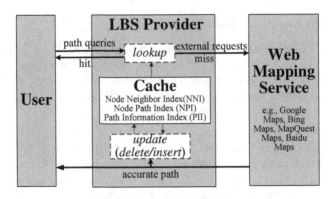

Fig. 2. System overview.

4 System Overview

Figure 2 gives an overview of the proposed system in the paper. It consists of three entities:

User. Users send their path queries to the LBS provider in the form of $q = (v_o, v_d)$, to retrieve the shortest path information from the origin v_o to destination v_d based on the current travel time.

The LBS Provider. After receiving a path query $q = (v_o, v_d)$ from a user, the LBS provider will try to answer it directly through its path cache via a cache *lookup* operation. If there is a hit in the cache, i.e., the cache contains the shortest travel-time path information from v_o to v_d, it returns the query result to the user immediately; otherwise, i.e., a miss occurs, the LBS provider retrieves the shortest travel-time path from the Web mapping service through an external request, and tries to *update* (with *delete* and *insert* operations) the cache.

The Web Mapping Service. Typical Web mapping services are Google Maps, Bing Maps, MapQuest Maps, and Baidu Maps. These mapping services are owned by large and specialized companies, such as Google, Microsoft and Baidu, which have enough resources to compute the shortest travel-time paths based on live traffics.

Since there are two critical limitations to access the path information from the Web mapping service, i.e., high cost and usage limits, our objectives are to maximize the cache hit ratio to reduce the number of external requests issued by the LBS provider and the query response time to users. The shortest travel-time path between two nodes may vary significantly during different time of a day due to dynamic traffics on road networks, which may cause the cached paths out of date. Hence, our another objective is to maximize the accuracy of the retrieved paths from the cache.

5   Cache Management

5.1   Cache Structure

To support efficient cache operations (e.g., lookup and update) and make the cache more compact, our cache structure is based on the following three indexes:

Node Path Index (NPI). NPI, also known as inverted lists, is used to store the path lists of nodes, i.e., each node has a list of path IDs whose paths contain the node. Take Fig. 1 for example, since only path p_1 contains v_1, we have $NPI(v_1) = \{p_1\}$. Similarly, we can easily compute the NPIs of other nodes as shown in Fig. 3a. With the help of NPI, the cache can support efficient lookup (see Sect. 5.3.1).

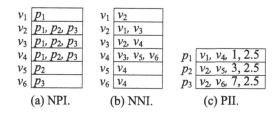

(a) NPI.   (b) NNI.   (c) PII.

Fig. 3. The cache structure of $\mathcal{C} = \{p_1, p_2, p_3\}$.

Node Neighbor Index (NNI). Instead of using arrays to store cached paths (i.e., Fig. 1a), we employ the Node Neighbor Index (NNI) (as the subgraph model used in [12]) to store the adjacency lists of nodes, i.e., each node has an adjacency list to store its neighbor nodes that appear in the cache. In the running example, since node v_1 only has one neighbor (i.e., v_2), while v_4 has three neighbors (i.e., v_3, v_5 and v_6), we have $NNI(v_1) = \{v_2\}$ and $NNI(v_4) = \{v_3, v_5, v_6\}$, as shown in Fig. 3b. Therefore, each neighbor node pair is stored at most once in the cache by NNI, which is more compact than the path array. The saved space can be used for accommodating additional paths into the cache, in turn improving the cache hit ratio.

Path Information Index (PII). PII is used to store essential information of the cached paths during cache lookup and update, including path origins, path destinations, arrival time, and path impacts. The arrival time of each path is recorded to evict outdated paths. The impact of a path is a brand-new indicator to reflect the popularity and importance of the path, which can be calculated based on NPI. These two indicators (i.e., arrival time and impact) determinate the weight of a path that is used for updating the cache (see Sect. 5.2). Figure 3c is an example of PII based on our running example, where $\langle v_1, v_4, 1, 2.5 \rangle$ is the entry of path p_1, which means the origin, destination, arrival time, and path impact of p_1 are v_1, v_4, 1, and 2.5, respectively.

5.2   Path Weight

The weight of a path indicates the importance of a path in the cache. When the cache is full, the path with the lower weight has a higher chance to be replaced by the new one. In [24], the authors employ road types to measure the weight of a path, e.g., the path over major roads has a higher weight than the path over branch roads. By this way, the cache not only needs the topology of the underlying road network, but also needs the road type information of each road segment. However, such detailed road information may be not easy to access for a cache system. Besides, a large extra space in the cache is needed to store these information.

Therefore, we propose another totally new indicator to measure the popularity and importance of a path in this paper, namely *path impact*. Path impact is calculated only based on the node path index of the cache, i.e., NPI. It does not need additional road information and extra cache space, which is much simpler than [24].

Before introducing the path impact, we first bring in the definition of the *node impact*:

Definition 5 (Node impact). *The impact of a node v_i is the number of paths in the cache that contains the node, i.e., $|NPI(v_i)|$.*

The impact of a path is measured on its average node impact:

Definition 6 (Path impact). *The impact of a path (denoted as χ) is the average node impact of the path, i.e.,*

$$p.\chi = \frac{\sum_{v_i=p.v_o}^{p.v_d} |NPI(v_i)|}{|Path(p)|}. \tag{1}$$

Based on Definition 6, we can see the larger impact of a path the more other travel paths that connect with it, i.e., the path has a higher popularity and importance (like major road and good traffic situation); therefore, it has a higher probability to be re-used in the future and should be placed into the cache.

Besides the path impact, another important factor of a path is its *valid-time* (see Definition 7). The path with the smaller valid-time has a higher probability to be evicted from the cache, as it will be outdated soon.

Definition 7 (Path valid-time). *The valid-time of a path (denoted as t_v) is the time period during which a path is valid and accurate in a time dependent road network.*

In this paper, the valid-time of a path p is calculated as follows:

$$p.t_v = t_{max} - (t_{current} - p.t_a), \tag{2}$$

where t_{max} is the maximum valid-time for all the cached paths which is set by the system, $t_{current}$ stands for the system's current time, and $p.t_a$ is the arrival time of the path to the cache recorded in PII.

To this end, the weight of a path is calculated upon its normalized impact and valid-time:

$$p.w = \theta \times \frac{p.\chi}{\chi_{max}} + (1 - \theta) \times \frac{p.t_v}{t_{max}}, \tag{3}$$

where χ_{max} is the maximum impact of all the cached paths that is used to normalize the impact of each path, t_{max} is the maximum valid time that is used to normalize the valid-time of each path, and θ ($0 \le \theta \le 1$) is a parameter to adjust the importance of the path impact (or valid-time).

In our running example, suppose $t_{max} = 10$ and $t_{current} = 9$, it is very easy to calculate the impact, valid-time and weight (when $\theta = 0.5$) of each cached path in $\mathcal{C}$ as shown in Fig. 4.

path ID	path impact χ ($\chi_{max} = 2.5$)	path valid-time t_v ($t_{max} = 10, t_{current} = 9$)	path weight w ($\theta = 0.5$)
p_1	(1+3+3+3)/4 = 2.5	10-(9-1)=2	0.6
p_2	(3+3+3+1)/4 = 2.5	10-(9-3)=4	0.7
p_3	(3+3+3+1)/4 = 2.5	10-(9-7)=8	0.9

Fig. 4. The weight of each cached path in $\mathcal{C}$ when $t_{max} = 10$, $t_{current} = 9$, and $\theta = 0.5$.

5.3 Cache Operations

A dynamic cache system not only needs to support efficient cache lookup, but also needs to continuously update the cache by deleting and inserting paths to make it fresh and effective. Therefore, there are four basic cache operations in the system, i.e., lookup, update, delete and insert.

5.3.1 Lookup

After having received a path query $q = (v_o, v_d)$ from a user, the cache performs a lookup operation immediately by examining NPIs of $q.v_o$ and $q.v_d$. If these two NPIs have an empty intersection, i.e., $NPI(q.v_o) \cap NPI(q.v_d) = \emptyset$, which means there is no path that contains $q.v_o$ and $q.v_d$ at the same time, a cache miss occurs. Otherwise, the cache can answer the query directly. If there are more than one path from $q.v_o$ to $q.v_d$ in the cache, we select the latest one (say p_s) based on the arrival time of these paths by PII. To retrieve the result path $Path(q.v_o \rightarrow q.v_d)$ for q from the cache, we start from the query origin $q.v_o$, and visit a neighbor node v_i whenever the NPI of v_i contains p_s until reach the query destination $q.v_d$.

For example, given a path query $q_1 = (v_2, v_4)$, because $NPI(v_2) \cap NPI(v_4) = \{p_1, p_2, p_3\}$ (Fig. 3a), i.e., there are up to three paths from the query origin to the query destination in the cache, the latest one p_3 is selected. To access the result path $Path(v_2 \rightarrow v_4)$, we start from the query origin v_2 and check all of its neighbors (i.e., v_1 and v_3, Fig. 3b), to find the next node of the result path. Since

$NPI(v_3)$ contains p_3 while $NPI(v_1)$ does not, v_3 is the next node in the path. We use the same way to continuously examine the unvisited neighbor of v_3, i.e., v_4. Since v_4 is the query destination, the whole result path has been found, i.e., $Path(q_1) = Path(v_2 \rightarrow v_4) = \{v_2, v_3, v_4\}$.

5.3.2 Update

When a new path $Path(p_{new})$ is returned from the Web mapping service, since it is up to date and not exists in the cache, we should use it to update the cache to make the cache fresh and effective continuously.

As shown in Algorithm 1, we first collect all the sub-paths of $Path(p_{new})$ from the cache to see whether there is enough space for the new path or not (Lines 6 to 6). If it is, we delete all of these sub-paths from the cache (see Sect. 5.3.3), then insert the new path into it (see Sect. 5.3.4), and finally return the updated cache (Lines 7 to 11). If it is not, we keep collecting the paths whose weight is less than that of p_{new} (Lines 13 to 20). By this time, if the space is still not enough, we discard $Path(p_{new})$ and keep the cache unchanged; otherwise, we delete these collected paths from the cache, and then insert the new one into it (Lines 21 to 24). Finally, we return the cache.

For example, given another query $q_2 = (v_7, v_4)$, since there is no entry for v_7 in NPI (Fig. 3a), i.e., $NPI(v_7) \cap NPI(v_4) = \emptyset$, no shortest path information from v_7 to v_4 can be found in the cache. The LBS provider retrieves the path information from the Web mapping service through an external request. Assume the retrieved path is $\{v_7, v_3, v_4\}$, its assigned ID is p_4 and the current time is 9, i.e., $Path(p_4) = Path(v_7 \rightarrow v_4) = \{v_7, v_3, v_4\}$ and $p_4.t_a = t_{current} = 9$, besides being returned as the query answer of q_2, p_4 is also employed to update the cache.

Suppose the maximum cache capacity is 40 (i.e., $\Psi = 40$), because no cached path is the sub-path of p_4, and the size of the cache after adding p_4 is larger than Ψ (i.e., $|\mathcal{C} + Path(p_4)| = 43 > \Psi$), we need to collect and remove the cached path(s) whose weight is smaller than that of p_4 to get enough space. To do this, we first calculate the path impact and valid-time of p_4:

$$p_4.\chi = \frac{|NPI(v_7)| + |NPI(v_3)| + |NPI(v_4)|}{|Path(p_4)|}$$
$$= \frac{0 + 3 + 3}{3} = 2 \text{ (Fig. 3a)},$$

$$p_4.t_v = t_{max} - (t_{current} - p_4.t_a) = 10 - (9 - 9) = 10;$$

then, we use Eq. 3 calculate its weight ($\theta = 0.5$):

$$p_4.w = 0.5 \times \frac{p_4.\chi}{\chi_{max}} + 0.5 \times \frac{p_4.t_v}{t_{max}} = 0.5 \times \frac{2}{2.5} + 0.5 \times \frac{10}{10} = 0.9.$$

Currently, because the path with the minimum weight is p_1 and $p_1.w < p_4.w$ (Fig. 4), and we calculate out the space is enough now (i.e., $|\mathcal{C} - Path(p_1) + Path(p_4)| = 33 < \Psi$), we stop collecting the cached paths here. Then, we delete

Algorithm 1. Cache update.

Input: a cache $\mathcal{C}$, a path $Path(p_{new})$, the cache capacity Ψ, θ
Output: a cache $\mathcal{C}$

1: $\mathcal{S} \leftarrow \emptyset; \mathcal{D} \leftarrow \emptyset;$
2: **for** each path p_i in the cache **do**
3:     **if** both $p_i.v_o$ and $p_i.v_d$ in $Path(p_{new})$ **then**
4:         insert p_i into $\mathcal{S};$
5:     **end if**
6: **end for**
7: **if** $|\mathcal{C} - \mathcal{S} + Path(p_{new})| \leq \Psi$ **then**
8:     delete $\mathcal{S}$ from $\mathcal{C};$
9:     insert $Path(p_{new})$ into $\mathcal{C};$
10:     **return** $\mathcal{C};$
11: **end if**
12: $p_{new}.w \leftarrow$ the path weight of $p_{new};$
13: **while** $|\mathcal{C} - \mathcal{S} - \mathcal{D} + Path(p_{new})| > \Psi$ **do**
14:     $p_{min} \leftarrow$ the path with the minimum weight in $\mathcal{C} - \mathcal{S} - \mathcal{D};$
15:     **if** $p_{min}.w \leq p_{new}.w$ **then**
16:         insert p_{min} into $\mathcal{D};$
17:     **else**
18:         **break**;
19:     **end if**
20: **end while**
21: **if** $|\mathcal{C} - \mathcal{S} - \mathcal{D} + Path(p_{new})| \leq \Psi$ **then**
22:     delete $\mathcal{S}$ and $\mathcal{D}$ from $\mathcal{C};$
23:     insert $Path(p_{new})$ into $\mathcal{C};$
24: **end if**
25: **return** $\mathcal{C};$

$Path(p_1)$ from $\mathcal{C}$ (as shown in Fig. 5) and add the new path $Path(p_4)$ into it (as shown in Fig. 6). Since NPI has changed, we also need to update the impact of the cached paths as shown in Fig. 6c. In the next two sections, we will discuss how to delete a path from the cache and insert a new path into the cache.

5.3.3 Delete

Given a path ID p_{del}, to remove its corresponding path from the cache, we first find the origin $p_{del}.v_o$ and the destination $p_{del}.v_d$ of the path by PII. Then, we retrieve its path $Path(p_{del})$ through a cache lookup (Sect. 5.3.1). Next, we delete $Path(p_{del})$ from the cache by clearing the three indexes:

Clear NNI. For each node v_i in $Path(p_{del})$, we check its every neighbor node v_j in the path. If $Path(p_{del})$ is the only path that contains them both (makes them becoming neighbors), i.e., $NPI(v_i) \cap NPI(v_j) = \{p_{del}\}$, v_j can be safely removed from the neighbor list of v_i, i.e., $NNI(v_i) = NNI(v_i) - \{v_j\}$. Otherwise, $Path(p_{del})$ is not the only path that makes them becoming neighbors, so we can not delete v_j from $NNI(v_i)$.

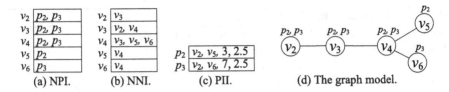

(a) NPI.   (b) NNI.   (c) PII.   (d) The graph model.

Fig. 5. The cache structure and graph model of $\mathcal{C}$ after deleting $Path(p_1) = \{v_1, v_2, v_3, v_4\}$.

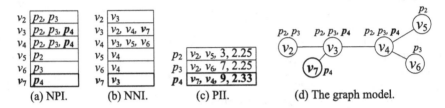

(a) NPI.   (b) NNI.   (c) PII.   (d) The graph model.

Fig. 6. The cache structure and graph model of $\mathcal{C}$ after inserting $Path(p_4) = \{v_7, v_3, v_4\}$ (with updated path impact).

Clear NPI. For each node v_i in $Path(p_{del})$, p_{del} is deleted from the NPI of v_i, i.e., $NPI(v_i) = NPI(v_i) - \{p_{del}\}$.

Clear PII. The whole entry of p_{del} is removed from PII.

Figure 5 shows the cache structure and its graph model after having deleted $Path(p_1)$ from the cache of Fig. 3.

5.3.4   Insert

To insert a new path into a cache, we only need to insert corresponding data into the three indexes which constitute the cache, i.e., NNI, NPI and PII.

Figure 6 shows the updated cache structure and graph model after having inserted a new path $Path(p_4) = \{v_7, v_3, v_4\}$ into the cache of Fig. 5.

6   Experimental Evaluation

6.1   Experimental Setting

We take SPC [12] as the baseline approach in our experiments, and evaluate the performance of SPC and our proposed shortest travel-time path cache (denoted as TPC) in terms of three metrics: (1) cache hit ratio, which reflects the number of external requests to the Web mapping service, (2) the average query response time per user query, and (3) the accuracy of cached paths.

As in [12], we utilize the Beijing road network and the Geo-Life trajectory dataset [25]. We generate nearly 12k path queries from the Geo-Life trajectory dataset by extracting the start and end locations of each trajectory as the query origin and destination, and compress them into four hours with original order.

The paths of the first 5k queries are employed for constructing caches, the rest 7k queries are utilized to evaluate the performance of SPC and TPC. Bing Maps is taken as the Web mapping service in our experiments, and the maximum valid-time for all the cached paths is 30 mins (i.e., $t_{max} = 30$). Unless mentioned otherwise, the number of user path queries is 7k, the cache size is 5k nodes, and the value of θ is 0.5.

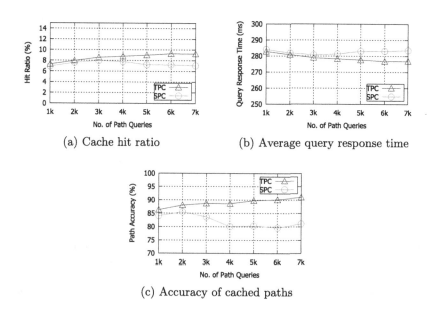

(a) Cache hit ratio

(b) Average query response time

(c) Accuracy of cached paths

Fig. 7. Effect of the number of path queries.

6.2 Experimental Results

6.2.1 Effect of the Number of Path Queries

In this section, we evaluate the performance of SPC and TPC with respect to various numbers of queries (from 1,000 to 7,000) as depicted in Fig. 7. We can see that TPC outperforms SPC in all performance metrics at any number of queries. This is because that SPC is designed as a static path cache, which is not suitable for caching dynamic travel-time paths. A cached path in SPC will never be evicted, even though it may be not popular or the shortest any more, while TPC continuously updates the cache based on path impact and valid-time.

6.2.2 Effect of the Cache Size

Figure 8 shows the performance of the two methods with respect to the cache size from 1,000 to 10,000 nodes. When the cache size gets larger, i.e., more paths

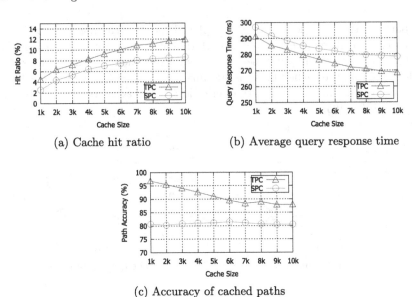

(a) Cache hit ratio

(b) Average query response time

(c) Accuracy of cached paths

Fig. 8. Effect of the cache size.

can be cached, it is expected that the cache hit ratio increases for both SPC and TPC (Fig. 8a), which makes the average query response time getting shorter consequently (Fig. 8b). The path accuracy of TPC gets smaller along with the increase of the cache size, as the updating frequency is slow when the cache is large that makes more cached paths invalid. Because there is no updating in SPC, the path accuracy of SPC remains almost the same as shown in Fig. 8c. Similarly, TPC performs much better than SPC, i.e., the cache hit ratio increased by 36%, the average query response time reduced by 3%, and the path accuracy increased by 13%.

6.2.3   Effect of the Value of θ

As presented in Sect. 5.2, the value of θ indicates the importance of the path impact (or valid-time) during cache update. When the value of θ gets larger, i.e., the path impact gets more important than the valid-time, the cache hit ratio of TPC increases (Fig. 9a), which proves that path impact can accurately reflect the popularity and importance of a path. Consequently, the valid-time gets less important when θ gets larger, so the accuracy of cached paths of TPC gets lower as shown in Fig. 9c. The performance metrics of SPC are unchanged with respect to the value of θ, as it has no connection with θ.

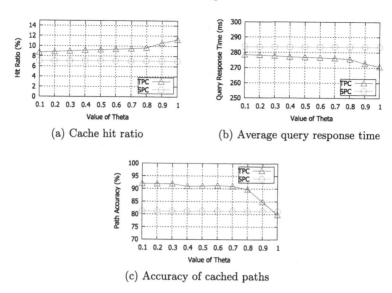

(a) Cache hit ratio

(b) Average query response time

(c) Accuracy of cached paths

Fig. 9. Effect of the value of θ.

7 Conclusion

In this paper, we propose an effective cache of the shortest travel-time paths for LBS providers, to reduce the number of external requests to Web mapping services and the query response time to users. We first design an effective and compact cache structure, including node path index (NPI), node neighbor index (NNI) and path information index (PII), and then devise four cache operations, i.e., lookup, update, delete and insert. In particular, we develop an effective cache update mechanism by considering the impact and valid-time of each path. Comprehensive experiments on real Web mapping service and datasets confirm the effectiveness of our proposed techniques.

Acknowledgments. This work was supported in part by the National Natural Science Foundation of China under Project 61702227, Project 61472337, Project 61572336, Project 61602214, and Project 61602137, in part by the Natural Science Foundation of Jiangsu Province under Project BK20160191.

References

1. Cormen, T.H., Leiserson, C.E., Rivest, R.L., Stein, C.: Introduction to Algorithms. MIT Press, Cambridge (2009)
2. Demiryurek, U., Banaei-Kashani, F., Shahabi, C., Ranganathan, A.: Online computation of fastest path in time-dependent spatial networks. In: Pfoser, D., Tao, Y., Mouratidis, K., Nascimento, M.A., Mokbel, M., Shekhar, S., Huang, Y. (eds.) SSTD 2011. LNCS, vol. 6849, pp. 92–111. Springer, Heidelberg (2011). doi:10.1007/978-3-642-22922-0_7

3. Google Maps APIs Terms of Service. https://developers.google.com/maps/terms?hl=en

4. Jia, G., Han, G., Jiang, J., Liu, L.: Dynamic adaptive replacement policy in shared last-level cache of DRAM/PCM hybrid memory for big data storage. IEEE Trans. Ind. Inform. **13**, 1951–1960 (2016)

5. Jia, G., Han, G., Wang, H., Wang, F.: Cost aware cache replacement policy in shared last-level cache for hybrid memory based fog computing. Enterp. Inf. Syst. 1–17 (2017)

6. Levandoski, J.J., Mokbel, M.F., Khalefa, M.E.: Preference query evaluation over expensive attributes. In: CIKM, pp. 319–328 (2010)

7. Li, Y., Yiu, M.L.: Route-saver: leveraging route apis for accurate and efficient query processing at location-based services. IEEE TKDE **27**(1), 235–249 (2015)

8. Liu, A., Li, Q., Huang, L., Xiao, M.: Facts: a framework for fault-tolerant composition of transactional web services. IEEE Trans. Serv. Comput. **3**(1), 46–59 (2010)

9. Liu, A., Li, Q., Huang, L., Ying, S., Xiao, M.: Coalitional game for community-based autonomous web services cooperation. IEEE Trans. Serv. Comput. **6**(3), 387–399 (2013)

10. Liu, A., Liu, H., Li, Q., Huang, L.S., Xiao, M.J.: Constraints-aware scheduling for transactional services composition. J. Comput. Sci. Technol. **24**(4), 638–651 (2009)

11. Sommer, C.: Shortest-path queries in static networks. ACM Comput. Surv. **46**(4), 45:1–45:31 (2014)

12. Thomsen, J.R., Yiu, M.L., Jensen, C.S.: Effective caching of shortest paths for location-based services. In: SIGMOD (2012)

13. U, L.H., Zhao, H.J., Yiu, M.L., Li, Y., Gong, Z.: Towards online shortest path computation. IEEE TKDE **26**(4), 1012–1025 (2014)

14. Wu, L., Xiao, X., Deng, D., Cong, G., Zhu, A.D., Zhou, S.: Shortest path and distance queries on road networks: an experimental evaluation. VLDB **5**, 406–417 (2012)

15. Xie, H.R., Li, Q., Cai, Y.: Community-aware resource profiling for personalized search in folksonomy. J. Comput. Sci. Technol. **27**(3), 599–610 (2012)

16. Xie, H., Li, Q., Mao, X., Li, X., Cai, Y., Rao, Y.: Community-aware user profile enrichment in folksonomy. Neural Netw. **58**, 111–121 (2014)

17. Xie, H., Li, X., Wang, T., Chen, L., Li, K., Wang, F.L., Cai, Y., Li, Q., Min, H.: Personalized search for social media via dominating verbal context. Neurocomputing **172**, 27–37 (2016)

18. Zhang, D., Chow, C.-Y., Li, Q., Liu, A.: Efficient evaluation of shortest travel-time path queries in road networks by optimizing waypoints in route requests through spatial mashups. In: Li, F., Shim, K., Zheng, K., Liu, G. (eds.) APWeb 2016. LNCS, vol. 9931, pp. 104–115. Springer, Cham (2016). doi:10.1007/978-3-319-45814-4_9

19. Zhang, D., Chow, C.-Y., Li, Q., Zhang, X., Xu, Y.: Efficient evaluation of k-NN queries using spatial mashups. In: Pfoser, D., Tao, Y., Mouratidis, K., Nascimento, M.A., Mokbel, M., Shekhar, S., Huang, Y. (eds.) SSTD 2011. LNCS, vol. 6849, pp. 348–366. Springer, Heidelberg (2011). doi:10.1007/978-3-642-22922-0_21

20. Zhang, D., Chow, C.Y., Li, Q., Zhang, X., Xu, Y.: SMashQ: spatial mashup framework for k-NN queries in time-dependent road networks. Distrib. Parallel Databases **31**(2), 259–287 (2013)

21. Zhang, D., Chow, C.Y., Li, Q., Zhang, X., Xu, Y.: A spatial mashup service for efficient evaluation of concurrent k-NN queries. IEEE Trans. Comput. **65**(8), 2428–2442 (2016)

22. Zhang, D., Chow, C.Y., Liu, A., Zhang, X., Ding, Q., Li, Q.: Efficient evaluation of shortest travel-time path queries through spatial mashups. GeoInformatica 1–26 (2017)
23. Zhang, D., Liu, Y., Liu, A., Mao, X., Li, Q.: Efficient path query processing through cloud-based mapping services. IEEE Access **5**, 12963–12973 (2017)
24. Zhang, Y., Hsueh, Y.L., Lee, W.C., Jhang, Y.H.: Efficient cache-supported path planning on roads. IEEE TKDE **28**(4), 951–964 (2016)
25. Zheng, Y., Zhang, L., Xie, X., Ma, W.Y.: Mining interesting locations and travel sequences from GPS trajectories. In: WWW (2009)
26. Zhu, A.D., Ma, H., Xiao, X., Luo, S., Tang, Y., Zhou, S.: Shortest path and distance queries on road networks: towards bridging theory and practice. In: SIGMOD (2013)

Graph Theory

Discovering Hierarchical Subgraphs
of K-Core-Truss

Zhen-jun Li[1], Wei-Peng Zhang[1], Rong-Hua Li[1(✉)], Jun Guo[1], Xin Huang[2],
and Rui Mao[1]

[1] College of Computer Science and Software Engineering, Shenzhen University,
Shenzhen, China
{15323940,wpzhang1991,332730268}@qq.com,{rhli,mao}@szu.edu.cn
[2] Hong Kong Baptist University, Hong Kong, China
370336203@qq.com

Abstract. Discovering dense subgraphs in a graph is a fundamental graph mining task, which has a wide range of applications in social networks, biology and graph visualization to name a few. Even the problems of computing most dense subgraphs (e.g., clique, quasi-clique, k-densest subgraph) are NP-hard, there exists polynomial time algorithms for computing k-core and k-truss. In this paper, we propose a novel dense subgraph, k-core-truss, that leverages on a new type of important edges based on the concepts of k-core and k-truss. Compared with k-core and k-truss, k-core-truss can significantly discover the interesting and important structural information outside the scope of the k-core and k-truss. We study two useful problems of k-core-truss decomposition and k-core-truss search. In particular, we develop a k-core-truss decomposition algorithm to find all k-core-truss in a graph G by iteratively removing edges with the smallest degree-support. In addition, we propose a k-core-truss search algorithm to identify a particular k-core-truss containing a given query node such that the core-number k is the largest. Extensive experiments on several web-scale real-world datasets show the effectiveness and efficiency of the k-core-truss model and proposed algorithms.

1 Introduction

Graph model is widely used to represent connection relationships between entities in a wide variety of domains such as social and web networks, biology, communication networks, and so on [20]. In the analysis of massive graphs, it is important to discover various dense subgraphs for efficient and effective analysis of a network, due to the large size of the network [24]. Identifying cohesive subgraphs is a fundamental graph-theoretic problem, which lies in the heart of many graph-mining applications, ranging from community mining in social networks [4,7,9,10,14,19,26], to real-time story identification in streaming news [2,12], detecting regulatory motifs in DNA [8], keyword search [15], graph visualization [1], and distance oracle indexing [6,11].

© Springer International Publishing AG 2017
A. Bouguettaya et al. (Eds.): WISE 2017, Part I, LNCS 10569, pp. 441–456, 2017.
DOI: 10.1007/978-3-319-68783-4_30

Recently, the definitions of k-core [5,25] and k-truss [9,24] have been widely used for a good balance of cohesive structure and efficient computations. A k-core of a graph G is the largest subgraph of G such that every vertex has at least k neighbors in this subgraph. On the one hand, k-core and k-truss both are hierarchical subgraphs that represent the cores of a network at different levels of granularity, with regard to number k. In this sense, k-core and k-truss are similar. On the other hand, basic elements of constructing k-core and k-truss are different. k-core is defined on the important vertices having degree at least k; whereas k-truss is defined on the important edges that are involved in several stable and strong triangle relationships. Thus, let us reconsider the importance of a relationship between two endpoints. Intuitively, if two vertices have more common neighbors, their relationship is stronger, which is overlooked in k-core model; Meanwhile, vertices tend to be more important if they have a higher degree in graphs. Thus, if two vertices with high degrees, their relationship is also regarded as a strong connection. Such important relationships are neglected in the k-truss model.

In this paper, we study a novel dense subgraph, k-core-truss, which are based on a new concept of important edges. Specifically, given a parameter $\alpha > 0$, the importance of an edge $e = (u, v)$ in a graph G is defined as the maximum one between the value equaling α times minimum degree of v and u, and the number of triangles containing e plus 2. Then, the k-core-truss of a graph G is the largest subgraph of G that every edge has the importance value at least k in this subgraph. For instance, consider the graph H_1 in Fig. 1, which is a 3-core-truss for $\alpha = 1$. It is because that for every edge e, e is contained in at least one triangle, or its two endpoints have degree at least 3. In addition, the whole graph of 3-core-truss contains two overlapping subgraphs of 3-core and 3-truss. By the definition of 3-core, it is obvious that the vertex v_5 with the degree of 2 does not belong to 3-core. Meanwhile, the edge $e = (v_{12}, v_{13})$ is not contained in any triangle, indicating that it does not belong to 3-truss. In light of the above, mining and querying k-core-truss in graphs is a pressing need, which is not simply dominated by the k-core and k-truss.

To summarize, we make the following contributions:

- We give a novel dense subgraph, k-core-truss, and motivate two problems of k-core-truss decomposition and k-core-truss search.
- We devise a k-core-truss decomposition algorithm to find all k-core-truss in a graph G by iteratively remove edges with the smallest degree-support. For the application of community search, we also design a k-core-truss search algorithm for identifying a particular k-core-truss containing a given query node with the highest k.
- We conduct extensive experiments on five large scale real-world datasets. The results show that our k-core-truss algorithms can efficiently and effectively find cohesive substructures over real-world networks, which can significantly discover the interesting and important relationships outside the scope of k-core and k-truss.

The rest of this paper is organized as follows. We present the definition of the k-core-truss model, and formulate our problems in Sect. 2. The core-truss decomposition and k-core-truss search algorithms are developed in Sect. 3. Extensive performance studies are reported in Sect. 4. We review the related work in Sect. 5, and conclude this paper in Sect. 6.

2   Problem Statement

Consider an undirected graph $G = (V, E)$ where V and E denote the node set and edge set respectively. We denote the number of nodes by $n = |V|$ and the number of edges in G by $m = |E|$. The set of neighbors of a vertex v is denoted by $N(v)$, i.e., $N(v) = \{u \in V : (v, u) \in E\}$. We use d_{max} to represent the maximum vertex degree in graph G.

Given a graph $H = (V_H, E_H)$, H is a subgraph of G iff $V_H \subseteq V$ and $E_H = \{(u, v)|u, v \in V_H, (u, v) \in E\}$. For a vertex $v \in V(H)$, the set of neighbors of vertex v is denoted by $N_H(v) = \{u \in V_H : (v, u) \in E_H\}$. Thus, the degree of v in H is defined as $\deg_H(v) = |N_H(v)|$. A triangle is a cycle of length 3 in graph. Let v, u, w be the three vertices on the cycle, then we use $\triangle_{uvw}$ to represent this triangle. For an edge $e(u, v) \in E(H)$, the support of an edge e, is defined as the number of triangles containing e, denoted by $\sup_H(e) = |\{\triangle_{uvw} : (u, w), (v, w) \in E_H\}|$. In this paper, w.l.o.g, we assume that the graph G we consider is connected, which implies $m \geq n - 1$. In the following, we define the degree for an edge based on the definition of vertex degree.

Definition 1 (Degree of an Edge). *Given a subgraph $H \subseteq G$, the* degree *of an edge $e(u, v) \in E_H$ is denoted by $\deg_H(e) = \min\{\deg_H(v), \deg_H(u)\}$.*

Example 1. Consider the graph G in Fig. 1. The vertex v_5 has 3 neighbors as $N(v_5) = \{v_3, v_8, v_9\}$, and the degree of vertex v_5 in graph G is $\deg_G(v_5) = 3$. The graph H_1 in Fig. 1 is a subgraph of G. For a vertex $v_5 \in V_{H_1}$, the degree of vertex v_5 in H_1 is 2, i.e., $\deg_{H_1}(v_5) = 2$. For an edge $e = (v_5, v_8)$ in H_1, the degree of an edge e is $\deg_H(e) = \min\{\deg_H(v_5), \deg_H(v_8)\} = 2$ by the definition 1, as $\deg_{H_1}(v_8) = 6$ holds.

Based on the definitions of degree and support for an edge, we give a new definition of degree-support as follow.

Definition 2 (Degree-Support). *For a subgraph $H \subseteq G$ and a given number $\alpha \geq 0$, the* degree-support *of an edge $e(u, v) \in E_H$ is denoted by $\mathsf{degsup}_H(e) = \max \{\sup_H(e) + 2,\ \alpha \cdot \deg_H(e)\}$.*

The degree-support of an edge $e(u, v)$, $\mathsf{degsup}_H(e)$, represents the strength of the connection between vertices v and u in graph topology. The underlying principles of $\mathsf{degsup}_H(e)$ contains two folds. On the one hand, a triangle indicates two vertices have a common neighbor, which shows a strong and stable connection among three vertices. Intuitively, if two vertices have more common neighbors with a larger $\sup_H(e)$, their relationship is stronger. On the other hand, if one

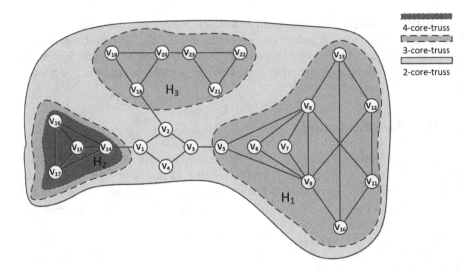

Fig. 1. A running example of graph G. The parameter $\alpha = 1$. (Color figure online)

vertex have a higher degree with more connections, the vertex tends to be more important in this graph. Thus, two endpoints of an edge both have high degree, their relationship is also regarded as a strong connection. Due to different measurements of support and degree, we invoke a parameter α to adjust the relative weight of degree, w.r.t. the support. The larger α is, the more important the degree is. Unless otherwise specified, we assume $\alpha = 1$ throughout this whole paper.

Example 2. In Fig. 1, there are several triangles such as $\triangle_{v_5,v_8,v_9}$, $\triangle_{v_6,v_8,v_9}$, $\triangle_{v_7,v_8,v_9}$, and so on. For each one of these triangles, two vertices have a common neighbor. For an edge $e = (v_5, v_8)$ in H_1, the edge support of e is $sup_{H_1}(e) = 1$. For $\alpha = 1$ and $\deg_{H_1}(e) = 2$, the **degree-support** of edge e is $\mathsf{degsup}_{H_1}(e) = \max \{sup_{H_1}(e) + 2, \alpha \cdot \deg_{H_1}(e)\} = 3$. Then, if we adjust the parameter α to a higher value as $\alpha = 2$, $\mathsf{degsup}_{H_1}(e) = \max \{sup_{H_1}(e) + 2, 2 \cdot \deg_{H_1}(e)\} = 4$.

On the basis of the definition of **degree-support**, we define the k-core-truss in a graph G as follows.

Definition 3 (K-Core-Truss). *Given a subgraph $H \subseteq G$, a parameter $\alpha \geq 0$ and an integer $k \geq 2$, a* k-core-truss *H is the maximal subgraph of G, in which each edge e satisfies* $\mathsf{degsup}_H(e) \geq k$. *Let CT_k represents the k-core-truss of G for a specific k.*

Based on the definition of k-core-truss, we can make a definition of core-truss number as follows.

Definition 4 (Core-Truss Number). *For an edge e in graph $G = (V, E)$, the core-truss number of e, denote by $ct(e) = max\{k : e \in E_{CT_k}\}$.*

For an given edge e with $ct(e) = k$, we have $e \in E_{CT_k}$, but $e \notin E_{CT_{k+1}}$. We use k_{max} to represent the maximum core-truss number of any edge in G, i.e., $k_{max} = \max\{ct(e) : e \in E\}$. We use the following example to illustrate the concept of k-core-truss and core-truss number.

Example 3. Consider an undirected graph G shown in Fig. 1. The graph G has 23 nodes named from v_1 to v_{23}. Assume that the parameter α is set to 1. We can see that every vertex has degree at least 2. Thus, for each edge $e \in E$, the degree-support of e as $\mathsf{degsup}_G(e) \geq 2$ holds. By the definition 3, the entire graph G is 2-core-truss, i.e. $G = CT_2$. In addition, the edge (v_3, v_4) is not contained in any triangle, and the degree of (v_3, v_4) has $\deg(v_3, v_4) = 2$. Thus, the edge (v_3, v_4) does not belong to 3-core-truss, indicating the core-truss number $ct(v_3, v_4) = 2$. The 3-core-truss of G is depicted in light green color in Fig. 1, which is consisted of three connected components H_1, H_2, and H_3. Moreover, we can see that the subgraph H_2 of G is 4-core-truss as CT_4, because H_2 is a 4-clique and every edge e of H_2 is contained in 2 triangles with $\mathsf{degsup}_{H_2}(e) \geq \sup_{H_2}(e) = 2 + 2 = 4$.

Parameter α. Consider the variants of our k-core-truss with the parameter α. The parameter α can make the graph size of k-core-truss flexible by changing its value. If more nodes need to be contained in the subgraph, we can increase the value of α. On the contrary, we can reduce it for finding the subgraphs whose vertices are more closely related for only the edges with higher $\deg_H(e)$ will be contained in the CT_k with the same k. Moreover, k-core-truss with suitable settings of parameter α can be equivalent to the definition of k-core or k-truss. Assume that $\alpha = 0$, our k-core-truss is equivalent to k-truss. On the other hand, if $\alpha = \frac{k}{k-1}$, our k-core-truss is equivalent to $(k-1)$-core.

In this paper, we study two different but related problems of k-core-truss decomposition and k-core-truss search in a graph.

The first problem is k-core-truss decomposition. The problem is to find all k-core-truss CT_k for $2 \leq k \leq k_{max}$ in graph G. As a cohesive subgraph of k-core-truss, k-core-truss decomposition identifies various cohesive subgraphs for efficient and effective analysis of a complex network. The problem is formulated as follows.

Problem 1. Given a graph $G = (V, E)$, a parameter α, the problem of k-core-truss decomposition is to find all k-core-truss CT_k in G for $2 \leq k \leq k_{max}$.

The second problem to study is k-core-truss search. For a given query node, the problem is to find a particular k-core-truss containing this query node. k-core-truss search can benefit the recent attractive and important task of community search, that is to find k-core-truss-based communities. The problem formulation is shown below.

Problem 2. Given a graph $G = (V, E)$, a parameter α and a query node q, the problem of k-core-truss search is to find a connected maximal k-core-truss with the highest k such that it contains node q.

3   K-Core-Truss Algorithms

In this section, we focus on developing efficient algorithms for Problems 1 and 2. Specifically, we first propose a k-core-truss decomposition method for solving Problem 1, which intuitively follows the definition of k-core-truss. In addition, to solve Problem 2, we design a query search algorithm to find a k-core-truss with the largest k such that this k-core-truss contains the input query node. Moreover, we analyze the complexity of two proposed algorithms. Finally, we use running examples to introduce how these two algorithms worked on graph in Fig. 1.

3.1   K-Core-Truss Decomposition Algorithms

Here, we introduce a basic algorithm for k-core-truss decomposition that is to find k-core-truss for all possible k. Similar with the core decomposition [21] and truss decomposition [9], the core idea of our algorithm is to start from $k = 2$ and then iteratively find k-core-truss with the increasing k by one each time. To find a k-core-truss, the algorithm iteratively remove edges violating the constraint of k-core-truss.

The outline of our basic algorithm is represented in Algorithm 1. The algorithm starts with an initialization by computing the degree and the support of every edge in graph G (line 1 to 6). Let H represent the graph G in the following decomposition process. After initialization, for each k starting from $k = 2$, the algorithm iteratively deletes every edge $e = (u, v)$ with the degree-support no greater than k, because e cannot be in the (k+1)-core-truss by definition. Let the core-truss number of e as k, i.e., $ct(e) = k$ (line 15). Obviously, the deletion of $e = (u, v)$ will respectively decrease the degree of u and v by one (line 8 to 9). Moreover, the deletion of e may also lead to the invalidation of all triangles consisting of e, i.e., $\forall \triangle_{uvw}$ where $w \in W = N_H(u) \cap N_H(v)$, the triangle $\triangle_{uvw}$ is no longer valid any more after the deletion of $e = (u, v)$ (line 16). The $sup_H(e)$ and $deg_H(e)$ are needed to recompute(line 10 to 14). This process is repeated iteratively until all the remaining edges in G have degree-support at least $k + 1$, which is the $(k + 1)$-core-truss. If there still exists some edges not yet deleted in G, we increase the k by one and continue repeating the above process, i.e., Steps 7–17(line 18 to 19). The algorithm returns the core-truss numbers of all edges in G as shown in Fig. 2 (line 20).

In the following, we use an example to simulate the process of k-core-truss decomposition.

Example 4. Consider an graph $G = (V, E)$ shown in Fig. 1 and $\alpha=1$. We apply Algorithm 1 on G for k-core-truss decomposition. From line 2 to 6, we get all $deg_H(e)$ and $sup_H(e)$ of 35 edges in G. Now, we start from $k = 2$ to find all edges with the core-truss of k.

Case $k = 2$: since (v_1, v_2), (v_1, v_4), (v_1, v_{14}), (v_2, v_3), (v_2, v_{18}), (v_3, v_4), (v_3, v_5) are satisfied $deg_H(e) \leq k/\alpha \wedge sup_H(e) \leq k - 2$ directly or indirectly,e.g.(v_1, v_{14}) is not satisfied the conditions for $deg_H(v_1, v_{14}) = 3$ at first,but when (v_1, v_4) is deleted,which will update $deg_H(v_1, v_{14})$ to 2. All these seven edges will be deleted

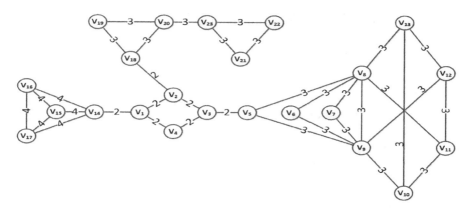

Fig. 2. The $ct(e)$ in graph G where $\alpha = 1$

and assigned with the core-truss number of 2. In addition, the algorithm updates the $deg_H(e)$ or $sup_H(e)$ for the remaining 28 edges, i.e., $deg_H(v_5, v_8) = 2$. The remaining graph of CT_3 consists of three component: H_1, H_2, and H_3.

Case $k = 3$: all edges except (v_{14}, v_{15}), (v_{14}, v_{16}), (v_{14}, v_{17}), (v_{15}, v_{16}), (v_{15}, v_{17}), (v_{16}, v_{17}) are satisfied $deg_H(e) \leq k/\alpha \land sup_H(e) \leq k - 2$, which are deleted from graph. We assign the core-truss number of 3 to each deleted edge.

Case $k = 4$: all remaining edges will be deleted in this loop. We assign the core-truss number of 4 to each deleted edge, and terminated the algorithm. Finally, the core-truss numbers of all edges in G are shown in Fig. 2.

We analyze the time and space complexity of Algorithm 1 in the following theorem.

Theorem 1. *Algorithm 1 takes $O(m^{1.5})$ time using $O(n+m)$ space, where $n = |V|$ and $m = |E|$.*

Proof. In the Algorithm 1, the most time-consuming step is to compute $\sup(e)$ for every $e \in E$. This step takes $O(m^{1.5})$ time complexity [24]. Similarly, updating the support of all edges (in lines 11–12) also consume $O(m^{1.5})$ time. The removal of all edges and the computation and updating the degree of all edges take $O(m)$ time in total. As a consequence, the total time cost of Algorithm 1 is $O(m^{1.5})$.

In addition, we analyze the space cost of Algorithm 1. Clearly, Algorithm 1 needs to store the graph G using $O(n + m)$ space. For each edge $e \in E$, it also use $O(m)$ space to store the edge degree $deg_H(e)$, support $sup_H(e)$, and core-truss number $ct(e)$. Thus, the space complexity of Algorithm 1 is $O(m + n)$ in total.

3.2   Querying k-core-truss

In this section, we investigate Problem 2, that is, given a query node q, to find a connected maximal k-core-truss containing q with the largest k. We develop a k-core-truss search algorithm in Algorithm 2 to solve Problem 2 as follows.

Algorithm 1. Basic k-core-truss Decomposition

Input: Graph $G = (V, E)$, Number $\alpha > 0$
Output: $ct(e)$ for each $e \in E$
1: $k \leftarrow 2$; $H \leftarrow G = (V, E)$;
2: $\forall u \in V_H$, compute the degree of u as $\deg_H(u)$;
3: **for** $(u, v) \in E_H$ **do**
4:    $\deg_H((u, v)) \leftarrow min(\deg_H(u), \deg_H(v))$
5: **for each** $e = (u, v) \in E_H$ **do**
6:    $sup_H(e) \leftarrow |N_H(u) \cap N_H(v)|$;
7: **while** $(\exists e = (u, v)$ such that $\deg_H(e) \leq k/\alpha \wedge sup_H(e) \leq k - 2)$ **do**
8:    $\deg_H(u) \leftarrow \deg_H(u) - 1$;
9:    $\deg_H(v) \leftarrow \deg_H(u) - 1$;
10:    **for all** $w \in N_H(u) \cap N_H(v)$ **do**
11:       $sup_H((u, w)) \leftarrow sup_H((u, w)) - 1$;
12:       $sup_H((v, w)) \leftarrow sup_H((v, w)) - 1$;
13:       $\deg_H((u, w)) \leftarrow min\{\deg_H(u), \deg_H(w)\}$;
14:       $\deg_H((v, w)) \leftarrow min\{\deg_H(v), \deg_H(w)\}$;
15:    $ct(e) \leftarrow k$;
16:    $H \leftarrow$ remove e from H
17: **if** $E_H \neq \emptyset$ **then**
18:    $k \leftarrow k + 1$;
19:    **goto** 7;
      return $\{ct(e)|e \in E\}$

We first outline the framework of Algorithm 2, which consists of the following two main stages. The first stage is to identify the largest k such that there exists a k-core-truss containing node q. We apply the k-core-truss decomposition method in Algorithm 1 to compute the maximum core-truss number of an edge containing q, denoted by the largest k as $K_{max} = \max_{u \in N(q)} ct(q, u)$. In the second stage, we start from the query node q and expand the answer graph in BFS (breadth-first search) manner, which collect all adjacent edges having the core-truss number at least K_{max} into the answer. Specifically, for each vertex u in the neighborhood of q as $u \in N(q)$, if edge (q, u) is unvisited and $ct(q, u) = K_{max}$, we add the edge $e = (q, u)$ into an empty search queue Q and mark it as visited (line 5 to 8). Then, we process the BFS search from the nonempty queue Q. We iteratively pick an edge (u, v) from Q, and then add all adjacent edges of nodes u and v with the core-truss number no less than K_{max} into Q (line 9 to 16).

In the following, we show a running example for Algorithm 2.

Example 5. Consider the graph shown in Fig. 1 and the parameter $\alpha = 1$. We test query node $q = v_8$, and run Algorithm 2 for finding the k-core-truss with the largest k such that it contains v_8.

Identifying K_{max}: We apply Algorithm 1 to compute the core-truss numbers of all edges in G. The results are shown in Fig. 2. We calculate the maximum core-truss number $K_{max} = \max_{u \in N(v_8)} ct(v_8, u) = 3$, since $ct(v_8, v_5) = ct(v_8, v_6) = ct(v_8, v_7) = ct(v_8, v_9) = ct(v_8, v_{11}) = ct(v_8, v_{13}) = 3$.

Algorithm 2. Query algorithm

Input: $G = (V, E), \alpha > 0, q$
Output: a connected maximal k-core-truss containing q
1: $k \leftarrow 2, K_{max} \leftarrow 0, visited \leftarrow \emptyset, l \leftarrow 0, H = (V', E') \leftarrow G = (V, E)$;
2: The same as line 2 to 19 in Algorithm 1;
3: **for all** $u \in N_G(q)$ **do**
4:     $K_{max} \leftarrow max\{ct((u, q)), K_{max}\}$;
5: **for** $u \in N_G(q)$ **do**
6:     **if** $(q, u) \notin visited$ and $ct(q, u) = K_{max}$ **then**
7:         $l \leftarrow l + 1, C_l \leftarrow \emptyset, Q \leftarrow \emptyset$;
8:         $visited \leftarrow visited \cup \{(q, u)\}, Q.push((q, u))$;
9:         **while** $Q \neq \emptyset$ **do**
10:             $(x, y) \leftarrow Q.pop(); C_l \leftarrow C_l \cup \{(x, y)\}$;
11:             **for** $w \in N(x)$ **do**
12:                 **if** $(x, w) \notin visited$ and $ct(x, w) \geq K_{max}$ **then**
13:                     $Q.push((x, w)), visited \leftarrow visited \cup \{(x, w)\}$;
14:             **for** $w \in N(y)$ **do**
15:                 **if** $(y, w) \notin visited$ and $ct(y, w) \geq K_{max}$ **then**
16:                     $Q.push((y, w)), visited \leftarrow visited \cup \{(y, w)\}$;
17: **return** $\{C_1, \ldots, C_l\}$;

Graph Expansion in BFS Manner: Then it starts from v_8 by expanding the graph in the BFS manner. The answer graph includes all edges that are connected to v_8 with $ct(e) \geq 3$. The subgraph H_1 is the 3-core-truss. For, Since the edge (v_3, v_5) with $ct(v_3, v_5) = 2$ is not in 3-core-truss, all edges of H_2 with $ct(e) \geq 3$ are disconnected to q. The final result of connected 3-core-truss containing v_8 is H_1 in Fig. 1.

We analyze the time and space complexity of Algorithm 2 as follows.

Theorem 2. *The time complexity of Algorithm 2 is $O(m^{1.5})$. The space complexity of Algorithm 2 is $O(n + m)$.*

Proof. In the Algorithm 2, the most time-consuming step is the same as Algorithm 1(line 2), whose time complexity is $O(m^{1.5})$. Then, the algorithm takes $O(m + n)$ time to compute K_{max} (line 2 to 4) and perform BFS process (line 5 to 16) in worst case. Also, we can easily derive that the space complexity of Algorithm 2 is $O(m + n)$ as Algorithm 1.

4   Performance Studies

In this section, we conduct extensive experiments to evaluate the efficiency and quality of proposed algorithms. Our experiments include 3 parts. The first experiment tests the runtime of three different graph decompositions: k-core decomposition, k-truss decomposition and k-core-truss decomposition. The second experiment shows the query processing time for three different models: k-core, k-truss, and k-core-truss. The running time results are averaged in 1000 tested queries. In the third experiment, we use case studies on real DBLP networks to evaluate the effective of k-core-truss model.

All algorithms are implemented in C++. All experiments are conducted on a computer with 3.20 GHz Intel Core(TM) i5-6500 CPU and 8 GB memory running Windows 7 professional (64-bit). In all experiments, both graph storage and query processing are conducted in the main memory.

4.1 Datasets

We use five web-scale real-world graphs in our experiments. All of the datasets except DBLP are downloaded from (http://snap.stanford.edu). Among the five graphs, Gowalla is a location-based online social network. NotreDame is a web graph. wiki-Talk is a communication network. LiveJournal1 is a social network. DBLP is a co-author network from the computer science bibliography website (http://dblp.uni-trier.de/).The statistical details of all dataset are listed in Table 1, in terms of vertex size, edge size, average clustering coefficient, and the total number of triangles.

Table 1. Datasets

Dataset	Nodes	Edges	Average clustering coefficient	Number of triangles
Gowalla	196,591	950,327	0.2367	2,273,138
DBLP	234,879	541,814	0.3742	285,730
NotreDame	325,729	1,497,134	0.2346	8,910,005
wiki-Talk	2,394,385	5,021,410	0.0526	9,203,519
LiveJournal	4,847,571	68,993,773	0.2742	285,730,264

In the following two experiments, we evaluate the efficiency of proposed graph decomposition and query processing algorithms. All detailed value of running times are reported in Table 2.

Table 2. Running time of graph decomposition and query processing in three different models: k-core, k-truss and k-core-truss (in seconds)

Dataset	Decomposition			Query Search		
	k-core	k-truss	k-core-truss	k-core	k-truss	k-core-truss
Gowalla	11.71	141.71	269.09	35.76	8.90	46.12
NotreDame	24.83	191.18	388.42	67.81	12.88	63.46
wiki-Talk	242.63	5664.33	7589.88	1058.02	436.32	1448.68
LiveJournal	3845.62	38694.40	92614.00	6731.81	3971.25	6176.80

4.2 Performance Evaluation of k-core, k-truss and k-core-truss Decompositions

We first compare the time consumption of graph decompositions on five datasets: Gowalla, NotreDame, wiki-Talk, and LiveJournal1. Figure 3(a) reported all results. It shows that k-core decomposition is the most effective among all three methods. In addition, the time consumptions of k-core-truss decomposition take much more than the k-core decomposition, but achieves the same order of the running time consumed by k-truss decomposition in all datasets. The reasons of differences are that k-core will take the linear-time [3], k-truss will the $O(m^{1.5})$ time [24] and k-core-truss will take $O(m^{1.5}) + O(m)$ time by Theorem 1 for their decompositions.

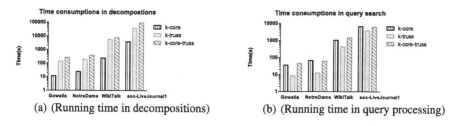

(a) (Running time in decompositions)   (b) (Running time in query processing)

Fig. 3. Performance evaluation of k-core, k-truss and k-core-truss

4.3 Efficiency Evaluation of Querying Processing for Finding k-core, k-truss and k-core-truss

In this experiment, we compare the performance of querying processing for finding k-core, k-truss and k-core-truss on all datasets. For a given query node, three methods respectively find the k-core, k-truss, and k-core-truss with the largest value k containing this query node. For each dataset, we generate 1000 sample of queries by randomly select one vertex in graph. The average query time are reported in Fig. 3(b). As we can see that, the query processing of finding k-truss is the most efficient among all three methods, due to the smallest size of k-truss. In addition, the query processing of finding k-core-truss achieve nearly the same time as finding k-core, which shows a good efficiency performance of our query processing algorithm.

4.4 Case Study on DBLP Network

In this section, we use a real-world DBLP network to test the effectiveness of our new model k-core-truss. In this DBLP network, each node represents an author, and an edge is added between two authors if they have co-authored at least three times. The parameter α is set as 1. For a given author in DBLP and number k, we apply Algorithm 2 to find the connected k-core-truss containing this query author. For comparsion, we also report the connected subgraphs of k-core and k-truss, which both contain this query author with the same input k.

First, we use the query Q = "Homare Murakami" and number $k = 9$ to test our k-core-truss for finding cohesive groups. In this example, we can see the superiority of k-core-truss against k-core and k-truss models. Figure 4 (a) shows the results of k-core model, which has 21 nodes, 105 edges and the average-degree of 10. Every vertex has 9 neighbors in 9-core. In addition, k-truss containing Q is represented in Fig. 4 (b). It has 10 nodes, 45 edges and the average-degree of 9. Every edge is contained in $7(= 9 - 2)$ triangles, indicating every vertex also has degree at least 8. We show the result of k-core-truss in Fig. 4 (c), which contains 24 nodes and 126 edges. The average-degree of k-core-truss is 10.5, which is higher than the k-core and k-truss results. This is because k-core-truss discovers more important edges than k-core and k-truss. To clearly show the difference of our k-core-truss with the results of k-core and k-truss, we scale up Fig. 4 (c) into Fig. 7, where the nodes and edges in red color present in k-core-truss but not in k-core and k-truss. As we can see, k-core-truss contains 3 red nodes which are authors "Jongsik Lim", "Mohammadali Khosravifard", and "I Gusti Bagus". Clearly, these three authors should be considered as members in the community of "Homare Murakami", because all of them densely connect with the coauthors of "Homare Murakami". These three authors, however, cannot be found by using the k-core and k-truss model.

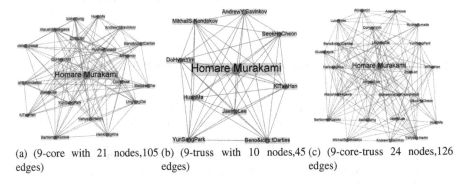

(a) (9-core with 21 nodes,105 edges)   (b) (9-truss with 10 nodes,45 edges)   (c) (9-core-truss 24 nodes,126 edges)

Fig. 4. Query node "Homare Murakami" in different models

In summary, the case studies on DBLP network indicate that our k-core-truss model indeed includes more nodes and edges and discovers a denser substructure than both k-core and k-truss. This additional structural information could assist a relatively comprehensive understanding of community structures in complex networks (Figs. 5 and 6).

5   Related Work

In this paper, we firstly propose a novel model of dense subgraph called k-core-truss. Our work is closely related with k-core and k-truss, which are extensively

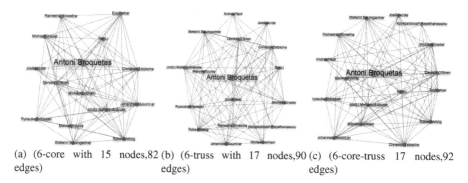

(a) (6-core with 15 nodes,82 (b) (6-truss with 17 nodes,90 (c) (6-core-truss 17 nodes,92 edges)   edges)   edges)

Fig. 5. Query node "Antoni Broquetas" in different models

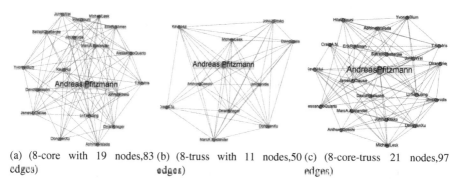

(a) (8-core with 19 nodes,83 (b) (8-truss with 11 nodes,50 (c) (8-core-truss 21 nodes,97 cdgcs)   edges)   edges)

Fig. 6. Query node "Andreas Pfitzmann" in different models

studied in the literature. In [21], Seidman first introduced the concept of k-core to measure the group cohesion in a network. The cohesiveness of k-core increases as k increases. Recently, the k-core decomposition in graphs has been used in many applications. From an algorithmic perspective, Batagelj and Zaversnik proposed an $O(n + m)$ algorithm for k-core decomposition in general graphs [3]. Their algorithm recursively deletes the node with the lowest degree and uses the bin-sort algorithm to maintain the order of the nodes. However, this algorithm has to randomly access the graph, thus it could be inefficient for the disk-resident graphs. To overcome this issue, Cheng et al. [5] proposed an efficient k-core decomposition algorithm for disk-resident graphs. Their algorithm works in a top-down manner to calculate k-core. To make the k-core decomposition more scalable, Montresor et al. [18] proposed a distributed algorithm for k-core decomposition by exploiting the locality property of k-core. All the mentioned algorithms focus on k-core decomposition in static graphs except for [17], in which the authors proposed an efficient k-core maintenance algorithm.

In order to mine transient stories and their correlations implicit in social streams, Lee et al. [12] proposed a new model of cohesive subgraph named (k, d)-core, in which every node has at least k neighbors and two end nodes of every

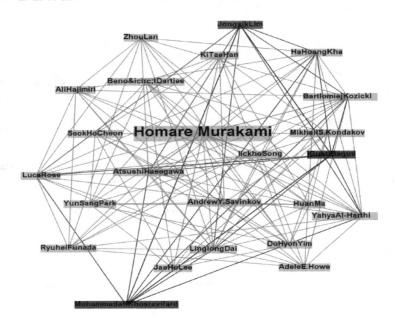

Fig. 7. The case study on 9-core-truss of "Homare Murakami" (Color figure online)

edge have at least d common neighbors. In other words, each edge in the subgraph is included by a k-core and a $(d + 2)$-truss at the same time. The constraints of (k, d)-core is much stricter than k-core-truss. To capture the influence, Li et al. [14, 16] introduced an influential community model based on k-core, and they also proposed several efficient algorithms to compute the influential communities. To provide structural relations among cliques, Sariyuce et al. [20] defines the nucleus decomposition of a graph, which represented the graph as a forest of nuclei. Each nucleus is a subgraph where smaller cliques are present in many larger cliques. With the right parameters, the nucleus decomposition generalizes the classic notions of k-cores and k-truss decompositions. Both (k, d)-core and nucleus are different from our k-core-truss model in terms of structural constraints.

6   Conclusions

In this paper, we propose a new dense subgraph of k-core-truss that combines the nice structural properties of k-core and the k-truss. We study two useful problems of k-core-truss decomposition and k-core-truss search. Extensive experiments on five large scale real-world datasets show that our k-core-truss algorithms can efficiently and effectively find cohesive substructures over real-world networks. Moreover, the proposed k-core-truss model can significantly discover the interesting and important relationships outside the scope of k-core and k-truss.

Our work takes an important first step toward enriching dense subgraph models in the network analysis. It opens up several interesting directions for further research. One of intuitive open problem is to study k-core-truss decomposition and search in the environment of stream or dynamic graphs, that is, the nodes/edges are frequently inserted/deleted. In the future, we plan to study the value of our model for other types of network, e.g., spatial networks [22,23] and location-based social networks [13].

Acknowledgement. We thank anonymous reviewers for their insightful comments. The work was supported in part by NSFC Grants (61402292, U1301252, 61033009), NSF-Shenzhen Grants (JCYJ20150324140036826, JCYJ20140418095735561), and Startup Grant of Shenzhen Kongque Program (827/000065).

References

1. Alvarez-Hamelin, J.I., Dall'Asta, L., Barrat, A., Vespignani, A.: Large scale networks fingerprinting and visualization using the k-core decomposition. In: Advances in neural information processing systems (2005)
2. Angel, A., Sarkas, N., Koudas, N., Srivastava, D.: Dense subgraph maintenance under streaming edge weight updates for real-time story identification. Proc. VLDB Endow. **5**(6), 574–585 (2012)
3. Batagelj, V., Zaversnik, M.: An O(m) algorithm for cores decomposition of networks. CoRR cs.DS/0310049 (2003)
4. Buehrer, G., Chellapilla, K.: A scalable pattern mining approach to web graph compression with communities. In: WSDM (2008)
5. Cheng, J., Ke, Y., Chu, S., Özsu, M.T.: Efficient core decomposition in massive networks. In: ICDE (2011)
6. Cohen, E., Halperin, E., Kaplan, H., Zwick, U.: Reachability and distance queries via 2-hop labels. SIAM J. Comput. **32**(5), 1338–1355 (2003)
7. Dourisboure, Y., Geraci, F., Pellegrini, M.: Extraction and classification of dense communities in the web. In: WWW (2007)
8. Fratkin, E., Naughton, B.T., Brutlag, D.L., Batzoglou, S.: MotifCut: regulatory motifs finding with maximum density subgraphs. Bioinformatics **22**(14), e150–e157 (2006)
9. Huang, X., Cheng, H., Qin, L., Tian, W., Yu, J.X.: Querying k-truss community in large and dynamic graphs. SIGMOD (2014)
10. Huang, X., Lakshmanan, L.V., Yu, J.X., Cheng, H.: Approximate closest community search in networks. Proc. VLDB Endow. **9**(4), 276–287 (2015)
11. Jin, R., Xiang, Y., Ruan, N., Fuhry, D.: 3-HOP: a high-compression indexing scheme for reachability query. In: SIGMOD (2009)
12. Lee, P., Lakshmanan, L.V.S., Milios, E.: CAST: a context-aware story-teller for streaming socail content. In: CIKM (2014)
13. Li, R., Liu, J., Yu, J.X., Chen, H., Kitagawa, H.: Co-occurrence prediction in a large location-based social network. Front. Comput. Sci. **7**(2), 185–194 (2013)
14. Li, R., Qin, L., Yu, J.X., Mao, R.: Influential community search in large networks. Proc. VLDB Endow. **8**(5), 509–520 (2015)
15. Li, R., Qin, L., Yu, J.X., Mao, R.: Efficient and progressive group steiner tree search. In: SIGMOD (2016)
16. Li, R., Qin, L., Yu, J.X., Mao, R.: Finding influential communities in massive networks. VLDB J. (2017)

17. Li, R., Yu, J.X., Mao, R.: Efficient core maintenance in large dynamic graphs. IEEE Trans. Knowl. Data Eng. **26**(10), 2453–2465 (2014)
18. Montresor, A., Pellegrini, F.D., Miorandi, D.: Distributed k-core decomposition. IEEE Trans. Parallel Distrib. Syst. **24**(2), 288–300 (2013)
19. Qin, L., Li, R., Chang, L., Zhang, C.: Locally densest subgraph discovery. In: KDD, pp. 965–974 (2015)
20. Sariyuce, A.E., Seshadhri, C., Pinar, A., Catalyurek, U.V.: Finding the hierarchy of dense subgraphs using nucleus decompositions. In: WWW (2015)
21. Seidman, S.B.: Network structure and minimum degree. Soc. Netw. **5**(3), 269–287 (1983)
22. Shang, S., Chen, L., Wei, Z., Jensen, C.S., Wen, J.R., Kalnis, P.: Collective travel planning in spatial networks. IEEE Trans. Knowl. Data Eng. **28**(5), 1132–1146 (2016)
23. Shang, S., Ding, R., Zheng, K., Jensen, C.S., Kalnis, P., Zhou, X.: Personalized trajectory matching in spatial networks. VLDB J. l Int. J. Very Large Data Bases **23**(3), 449–468 (2014)
24. Wang, J., Cheng, J.: Truss decomposition in massive networks. Proc. VLDB Endow. **5**(9), 812–823 (2012)
25. Wen, D., Qin, L., Zhang, Y., Lin, X., Yu, J.X.: I/O efficient core graph decomposition at web scale. In: ICDE (2016)
26. Zheng, D., Liu, J., Li, R., Aslay, Ç., Chen, Y., Huang, X.: Querying intimate-core groups in weighted graphs. In: 11th IEEE International Conference on Semantic Computing, ICSC 2017, San Diego, CA, USA, 30 January–1 February 2017 (2017)

Efficient Subgraph Matching on Non-volatile Memory

Yishu Shen and Zhaonian Zou[✉]

School of Computer Science and Technology, Harbin Institute of Technology,
Harbin 150001, Heilongjiang, China
znzou@hit.edu.cn

Abstract. The emerging non-volatile memory (NVM) technologies have attracted much attention due to its advantages over the existing DRAM technology such as non-volatility, byte-addressability and high storage density. These promising features make NVM a promising replacement of DRAM. Although the reading cost of NVM is close to that of DRAM, the writing cost is significantly higher than that of DRAM. Existing algorithms designed on DRAM treat read and write equally and thus are not applicable to NVM. In this paper, we investigate efficient algorithms for subgraph matching, a fundamental problem in graph databases, on NVM. We first give a detailed evaluation on several existing subgraph matching algorithms by experiments and theoretical analysis. Then, we propose our write-limited subgraph matching algorithm based on the analysis. We also extend our algorithm to answer subgraph matching on dynamic graphs. Experiments on an NVM simulator demonstrate a significant improvement in efficiency against the existing algorithms.

Keywords: Non-volatile memory · Subgraph matching · Graph database

1 Introduction

Emerging non-volatile memory (NVM) technologies are considered as replacement of the current DRAM technology because of the byte-addressability, non-volatility and high storage density. Existing NVM technologies include PCM [1], memristor [2], STT-MRAM [3] and so on. Table 1 compares these different NVM technologies with the current DRAM technology in terms of volatility, addressability, storage capacity, read latency and write latency.

While NVM has many promising features, its read and write latency is quite different from that of DRAM. Take PCM for example, its read latency is about 50 ns, which is close to DRAM's read latency. However, the write latency of PCM is about 150 ns, which is around 3 times of DRAM's write latency. Therefore, unlike DRAM, the read and write latency of PCM is asymmetric.

Most of the algorithms are designed based on the RAM memory model, which assumes that read and write of the main memory have similar cost. As a result,

© Springer International Publishing AG 2017
A. Bouguettaya et al. (Eds.): WISE 2017, Part I, LNCS 10569, pp. 457–471, 2017.
DOI: 10.1007/978-3-319-68783-4_31

Table 1. Differences between NVM and DRAM technologies.

Characteristic	DRAM	PCM	Memristor	STT-MRAM
Volatility	Yes	No	No	No
Addressability	Byte	Byte	Byte	Byte
Storage capacity	GB	TB	TB	TB
Read latency	60 ns	50 ns	120 ns	20 ns
Write latency	60 ns	150 ns	100 ns	20 ns

these algorithms may not perform well on NVM, where writes are much more expensive than reads. In this paper, we study subgraph matching on NVM. Our goal is to design a subgraph matching algorithm that runs faster on NVM than the existing ones. To the best of our knowledge, no such work has been done on NVM in the literature. The main contributions of this paper are as follows.

- We conduct experiments on an NVM simulator to test the performance of the existing subgraph matching algorithms on NVM. The experimental results verify our assumption that the existing subgraph matching algorithms designed on DRAM are not suitable for NVM.
- We conduct detailed analysis on the experimented subgraph matching algorithms. We abstract the generic backtracking framework of the algorithms and identify the write-intensive procedures in the framework. After that, we investigate the optimization techniques these algorithms adopt and estimate their performance on NVM.
- Based on the analysis above, we propose an efficient subgraph matching algorithm on NVM and extend the algorithm to handle dynamic graphs.
- Extensive experiments on two real datasets and various queries demonstrate that our algorithm is much faster than most existing subgraph matching algorithms on NVM.

The rest of the paper is organized as follows. In Sect. 2, we define our problem and briefly introduce related work on NVM technologies and subgraph matching research. In Sect. 3, we evaluate the existing subgraph matching algorithms by experiments. We give our detailed analysis based on the experimental results in Sect. 4. Section 5 proposes our subgraph matching algorithm on NVM. Section 6 extends our algorithm to dynamic graphs. Experiments on our algorithms are reported in Sect. 7. We give our conclusions in Sect. 8.

2   Preliminaries and Related Work

In this section, we introduce some concepts used throughout the paper and briefly summarize the related work.

In this paper, we consider single-labeled undirected graphs, which can be represented by a tetrad (V, E, L, Σ), where V is the set of vertices, E is the set of edges, Σ is the set of labels, and $L : V \rightarrow \Sigma$ is the function that maps each vertex to a label. Given a vertex v, its degree, denoted by $Deg(v)$, is the number of vertices that are adjacent to v.

Definition 1. *Given two graphs* $G = (V_G, E_G, L_G, \Sigma_G)$ *and* $H = (V_H, E_H, L_H, \Sigma_H)$, G *is isomorphic to* H *if there exists a bijection* $f : V_G \rightarrow V_H$ *such that* $L_G(v) = L_H(f(v))$ *for all* $v \in V_g$ *and* $(f(u), f(v)) \in E_H$ *for all* $(u, v) \in E_G$.

Given a data graph G and a query graph Q, the subgraph matching problem finds all subgraphs of G that are isomorphic to Q. Any subgraph of G that is isomorphic to Q is called an embedding of Q in G.

Most recent subgraph matching algorithms follow the backtracking framework proposed by Ullman [4]. The algorithm maps the vertices of the query graph to the vertices in the data graph one by one. During the mapping process, the vertices already matched form a partial mapping function, which is also called a *partial embedding*. The algorithm terminates when the partial embedding grows into a *full embedding*, or backtracks when the partial embedding is checked unqualified to make a full embedding. To optimize the performance, many recent algorithms uses different strategies such as matching order optimization [5,6] and auxiliary-index-based pruning [7,10]. Recently, TurboIso [8] and BoostIso [9] take advantages of the similar vertices in data and query graphs to compress the graphs and achieve good results.

Algorithms on NVM have also been studied recently. An asymmetric memory model is proposed in [13], where the memory system consists of a DRAM with capacity M and an NVM with unbounded size. Algorithms such as breadth-first search (BFS), depth-first search (DFS) and minimum spanning trees (MST) are investigated under this memory model. Reference [14] proposes write-efficient algorithm for sorts on NVM. [15,16] study the problem of minimizing writes on NVM. To the best of our knowledge, the problem of subgraph matching on NVM has not been studied in the literature.

3   Evaluation of Existing Algorithms on NVM

In this section, we evaluate the performance of seven representative subgraph matching algorithms: Ullman [4], VF2 [5], QuickSI [6], GraphQL [7], TurboIso [8], BoostIso [9] and CFL-Match [10] on NVM. All of the algorithms were implemented in C++. The source codes of the first six algorithms were obtained from the authors of [9], and the source code of CFL-Match was directly obtained from its authors. We first evaluate the performance of these algorithms on both NVM and DRAM through carefully designed experiments. Then, we give our analysis on these algorithms based on the experimental results.

All the experiments were conducted on a 64-bit Ubuntu operating system with 8 GB of DRAM. Especially, we simulate NVM with an architectural level main memory simulator NVMain [11], integrated with the CPU simulator gem5 [12]. We use the configuration script provided by NVMain to simulate 20 nm 1.8 V 8 GB of PCM with 50 ns read latency and 150 ns write latency.

We use two real datasets, namely Human and Yeast, for experiments on DRAM. Both Human and Yeast are protein interaction networks and have been used in many applications. Due to the limited performance of the NVM simulator, we use two relatively small datasets, Cage6 and Hamming6-4, for experiments on NVM. Both graphs are obtained from [19]. The vertex labels of these two graphs are randomly generated. Table 2 shows some properties of the graphs, where d_{avg} indicates the average degree of the graph.

Table 2. Summary of datasets.

Data graph	$|V|$	$|E|$	d_{avg}	$|\Sigma|$
Yeast	3112	12519	8.1	71
Human	4674	86282	36.9	44
Cage6	93	346	7.4	6
Hamming6-4	64	704	22	10

We generate multiple query graphs with different sizes (the number of vertices). All query graphs are generated by randomly selecting connected subgraphs of a specific size from the data graph. For each data graph, we generate query sets of sizes from 4 to 10 with 20 query graphs for each size.

We ran the seven algorithms listed above on both DRAM and NVM to process all the generated queries and examined the average running time for each query set. For time-saving purpose, we limit the number of embeddings found for each query graph to 1000, that is, once 1000 embeddings are found, the searching process terminates for the current query graph. Figures 1(a) and (b) show the results of the algorithms on DRAM and NVM, respectively. Note that TB in Fig. 1 represents the TurboIsoBoosted algorithm, which is TurboIso boosted by the compressing technique of BoostIso.

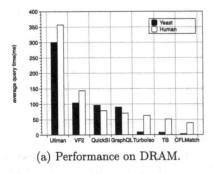

(a) Performance on DRAM.

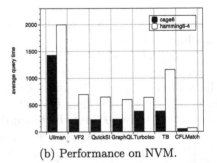

(b) Performance on NVM.

Fig. 1. Performance of existing subgraph matching algorithms on DRAM and NVM.

Algorithm 1. GenericBacktrackingFramework

Input: data graph G, query graph Q;
Output: all embeddings of Q in G;
1: $M := \emptyset$
2: **for each** $u \in V(G)$ **do**
3:     $C(u) := FilterCandidates(Q, G, u, ...)$
4:     **if** $C(u) = \emptyset$ **then**
5:         **return**
6:     **end if**
7: **end for**
8: $SubgraphSearch(Q, G, M, ...)$

According to the experimental results, the performance of these algorithms on NVM are significantly different from that on DRAM. TurboIso and TurboIso-Boosted are quite fast on DRAM; however, they perform poorly on NVM. VF2 is pretty slow on DRAM; while it is considerably fast on NVM. These observations verify our assumption that the algorithms designed for DRAM are not suitable to be directly applied to NVM. Since writes are more expensive than reads on NVM, the key to efficient subgraph matching on NVM is reducing write operations. In next section, we further explore the details of these algorithms to better understand their performance on NVM.

4 Detailed Analysis of Existing Algorithms

In this section, we give detailed analysis of the existing subgraph matching algorithms. We first evaluate the general framework of the backtracking-based algorithms. Then, we investigate different optimization techniques they use and estimate their performance on NVM.

4.1 General Backtracking Framework

As mentioned in Sect. 2, most of the recent subgraph matching algorithms are based on Ullman's backtracking framework [4]. It tries to find all embeddings by incrementally completing a partial embedding or discarding it if it is unpromising to be extended to a full embedding. Reference [17] gives the procedure of the backtracking framework, which is shown in Algorithm 1. Variable M at line 1 stores a partial embedding and is set to be empty at the beginning. The algorithm first finds a candidate set $C(u)$ for each vertex u in the query graph, where $C(u)$ is a subset of the data graph vertex set. Generally, the data vertices with the same label as u are extracted from the data graph and then the vertices with degree less than $deg_Q(u)$ are filtered to make the final candidate set for u. Some algorithms, such as GraphQL [7], uses more complicated strategies to further reduce the sizes of candidate sets.

After all candidate sets are generated, the general backtracking framework invokes recursive procedure $SubraphSearch$ to find all embeddings. The details

---

Algorithm 2. SubgraphSearch

Input: data graph G, query graph Q, partial embedding M;
Output: all embeddings of Q in G;

1: **if** $|M| = |V(Q)|$ **then**
2:     **report** M
3: **else**
4:     $u := NextQueryVertex(...)$
5:     $C_R := RefineCandidates(M, u, C(u), ...)$
6:     **for** each $v \in C_R$ such that v has not been matched **do**
7:         **if** $isJoinable(Q, G, M, u, v, ...)$ **then**
8:             $UpdateState(M, u, v, ...)$
9:             $SubgraphSearch(Q, G, M, ...)$
10:            $RestoreState(M, u, v, ...)$
11:         **end if**
12:     **end for**
13: **end if**

---

of *SubraphSearch* are presented in Algorithm 1. The procedure tries to match the query vertices one by one based on the given matching order. When vertex u is ready to be matched, it calls subroutine *RefineCandidates* to obtain a refined set C_R from $C(u)$ with specific pruning methods. Then, for each vertex v in C_R, it calls *isJoinable* to further checks v's qualification with the connected information between v and the vertices that are already matched. If v is qualified, *UpdateState* is called to add vertex pair (u, v) into M, and the procedure recursively calls itself until all query vertices are matched. After that, *RestoreState* is called to remove all the changes *UpdateState* made.

We illustrate the generic backtracking process by the example given in Fig. 2. Figure 2(a) shows the query graph, and Fig. 2(b) shows the data graph. We assume that the matching order is u_1, u_2. The candidate set of u_1 is $\{v_3, v_4, v_5\}$ and the candidate set of u_2 is $\{v_1, v_2\}$. First, u_1 is matched to v_3. Then, we match next query vertex u_2. We search the candidate set of u_2 and try to match u_2 to v_1 and report an embedding $\{(u_1, v_3), (u_2, v_1)\}$. Next, we try to match u_2 to v_2 and find that v_2 is not qualified. The reason is that u_2 is adjacent to u_1, which is matched to v_3 but v_2 is not adjacent to v_3. As a result, we backtrack to the previous query vertex u_1 and match u_1 to v_4. Next, we try to match u_2 to v_1 and v_2, respectively, and find none of them are qualified for a similar reason. Thus, we backtrack again and match u_1 to v_5 and find no embeddings either. Finally, the process terminates and only one embedding is reported.

The generic backtracking framework contains three main procedures: *Filter-Candidates, RefineCandidates* and *isJoinable*. *FilterCandidates* builds candidate sets for each query vertex and thus involves a large number of write operations. *RefineCandidates* refines candidate sets with different pruning rules and contains massive reads on auxiliary data structures. *FilterCandidates* further checks the qualification of the current data vertex using the connection information of the current matched vertices. Clearly, only *FilterCandidates* involves

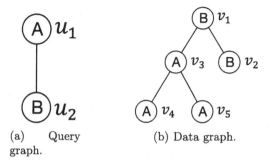

(a)   Query
graph.

(b) Data graph.

Fig. 2. Example of general backtracking process.

constructing new data structures, which means writing new data to the main memory, while the others just perform scans over existing data.

The generic backtracking framework is originally designed for DRAM, where read and write operations are regarded to have the same latency. As we know, writes are more expensive than reads on NVM. In order to apply the framework to NVM, we have to reduce write operations it incurs. According to our analysis above, *FilterCandidates* procedure incurs the most write operations in the framework. To make the generic backtracking framework efficient on NVM, it is crucial to reduce write operations in this procedure.

4.2   Optimization Techniques

Based on the generic backtracking framework, many algorithms adopt their unique optimization techniques to further improve efficiency. Generally, these techniques can be categorized into three classes. Although all of them are reported to be effective on DRAM, their performance on NVM are inconclusive. Next, we evaluate these three techniques.

Matching Order Optimization. It is reported that the matching order of query vertices could have a great impact on the efficiency of a subgraph matching algorithm. A good matching order could prune more unpromising intermediate results at an early stage. While the generation of a matching order may involve many reads on the data graph and computations, writing operations are relatively limited (with respect to the size of the query graph, intuitively). According to the experimental results, the algorithms using this technique (QuickSI and CFL-Match) are faster on NVM than those do not.

Compression Technique. This technique aims to combine similar vertices in query or data graphs together to reduce intermediate results of the matching process. Extra data structures need to be built to record these combining information, which may lead to many write operations. Moreover, this technique may not be effective on every data graph since the compressing ratio varies for different data graphs. In our experiments, the TurboIsoBoosted algorithm, which integrates two algorithms using this technique, performs poorly on NVM.

Auxiliary Data Structure. This technique is used in almost every algorithm except the original Ullman's algorithm [4]. Each algorithm builds their unique auxiliary index to prune intermediate results and reduce reads on data and query graphs. Most of these data structures are large because they need to record the information of both the query graph and the data graph. Obviously, it may cause large number of writes (with respect to the size of the data graph). Intuitively, this technique may not work that well on NVM, compared to matching order optimization. Our experiments in Sect. 7 will show that.

5   Our Algorithm

In this section, we propose our efficient subgraph matching algorithm on NVM based on the analysis in Sect. 3. We first give our revised backtracking framework and then state our choices of optimization strategies.

5.1   Revised Backtracking Framework

According to our analysis in Sect. 3.1, in order to improve the efficiency of the generic backtracking framework on NVM, the construction of candidate sets in procedure *FilterCandidates* needs to be revised. In order to get the candidates sets efficiently, we build an auxiliary data structure called *Inverse Label-Degree List* (ILD-List) as illustrated in Fig. 3. All the data vertices are stored in the ILD-List, and the vertices with the same label are stored consecutively in the non-increasing order of their degrees. The starting position of each label's corresponding vertex sequence is recorded. Note that the structure of an ILD-List is query-independent, which means we only have to build it once, and it can be used to deal with any query graph.

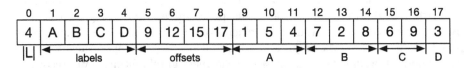

Fig. 3. Example of ILD-List. The vertices with the same label are stored consecutively in the non-increasing order of their degrees.

Now, we use the example in Fig. 4 to explain how to get the candidate set of a specific query vertex using an ILD-List. Suppose that we want to get the candidate set of query vertex u_1 in Fig. 4(c). The label of u_1 is A, so we scan the IDL-List and find that the offset of the first vertex with label A is 5. We check the vertex at position 5, which is v_3, and find that its degree is larger than u_1's degree. Therefore, v_3 is a candidate matching vertex of u_1. Then, we keep checking next vertex v_4 and find that its degree is less than u_1's degree. Thus, v_4 cannot be a matching vertex of u_1. The process terminates, and we get the candidate set of u_1, that is, $\{v_3\}$. Note that we do not have to record these candidate vertices in the real searching process because we can directly use the sublist as the candidate set.

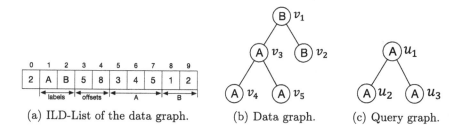

(a) ILD-List of the data graph.   (b) Data graph.   (c) Query graph.

Fig. 4. Example of using IDL-List to retrieve candidate sets.

Algorithm 3 shows our revised backtracking framework. We skip the process of building candidate sets for query vertices. Instead, when trying to match a specific query vertex u, we can easily find the sequence that stores all data vertices with the same label as u's. Then, we scan from the beginning of the sequence and try to match u to each data vertex we run into until the degree of the vertex is less than that of u. In this way, all candidate vertices of u can be retrieved without any write operations. We also skip the *RefineCandidates* procedure because we do not build any auxiliary structure for pruning. Obviously, the overall process includes only a few write operations.

5.2   Choices of Optimization Strategies

Based on our analysis in Sect. 3.2, we only choose matching the order optimization to enhance our revised backtracking framework. Intuitively, we adopt the matching order generation strategy of Algorithm CFL-Match [10] since it is the fastest on NVM according to our experiments. We modify its Core-Forest-Leaf strategy to further reduce write operations. Especially, we do not adopt the technique used in its leaf nodes matching because it involves computing candidate sets and generating NEC classes, both of which involve many write operations.

6   Subgraph Matching on Dynamic Graphs

In this section, we extend our algorithm to handle dynamic data graphs with vertex/edge updates. Given a data graph G, a query graph Q and all embeddings of Q in G, we need to recompute the embeddings of Q in G after an updating request

---

Algorithm 3. SubraphSearchNVM

Input: data graph G, query graph Q, partial embedding M, ILD-List of G;
Output: all embeddings of Q in G;
1: **if** $|M| = |V(Q)|$ **then**
2:     **report** M
3: **else**
4:     $u := NextQueryVertex(...)$
5:     $C_R :=$ the sublist of ILD-List that contains all vertices with label $L_Q(u)$
6:     **for** each $v \in C_R$ such that v has not been matched **do**
7:         **if** $Deg_G(v) < Deg_Q(u)$ **then**
8:             **break**
9:         **end if**
10:        **if** $isJoinable(Q, G, M, u, v, ...)$ **then**
11:            $UpdateState(M, u, v, ...)$
12:            $SubgraphSearch(Q, G, M, ...)$
13:            $RestoreState(M, u, v, ...)$
14:        **end if**
15:    **end for**
16: **end if**

---

comes up. A typical update may be an edge/vertex insertion/deletion or a vertex label changing. In this section, we assume that the query graph is connected, since a query over a non-connected graph could be solved by performing independent queries over its connected components and combining non-overlapping embeddings.

First, we introduce the updating strategy for our auxiliary data structure. With the update of the data graph, the ILD-List also needs to be modified. Some vertices in the list may be moved to other positions because their degrees have changed. For the IDL-List illustrated in Sect. 5, it will incur a lot of write operations. In order to facilitate the updating process, we should avoid too much write operations. Interestingly, what we have to do is just implementing our ILD-List as a linked list with head pointers of each sublist recorded. Since there is no need to perform random accesses to the ILD-List during the entire query process, our subgraph matching algorithm does not have to make any changes. Figure 5 shows an example of the linked IDL-List. There are two labels in the data graph in Fig. 5(b). Vertices v_3, v_4 and v_5 are labeled by A and are stored in the upper linked list in Fig. 5(a). Vertices v_1 and v_2 are labeled by B and are stored in the lower linked list. The head pointers of both linked lists are stored in the left array. Suppose an edge between v_2 and v_5 is inserted into the data graph in Fig. 5(b). The degree of v_2 becomes 2 and is not larger than v_1's degree, so v_2 does not have to be moved. The degree of v_5 becomes 2 and is now larger than v_4's degree. Since v_3's degree is 3, we put v_5 right after v_3 and before v_4. Obviously, with any vertex/edge inserted or deleted, the ILD-List can be updated within $O(1)$ writes. Dealing with label updates is quite similar. The only difference is that vertices can be moved to another label's sublist. Next, we introduce our updating strategy for deletions, insertions and label updates respectively.

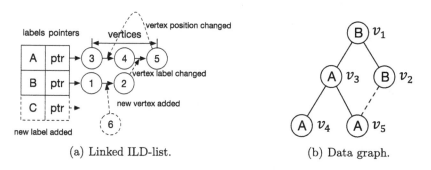

(a) Linked ILD-list.   (b) Data graph.

Fig. 5. Example of linked IDL-List.

Dealing with Deletions. The problem of handling vertex/edge deletions is trivial. We only have to search the original embeddings of Q and discard any embedding that contains the vertex/edge deleted. The remaining embeddings are the results.

Dealing with Insertions. It becomes more complex to deal with insertions. We need to find all new embeddings of Q in G that are not qualified before the insertion. Performing the query on the whole updated graph from scratch will be inefficient. Our main idea is to bound the region where we can find new embeddings. Here, we give the definition of a connected graph's diameter first.

Definition 2. *Given a connected non-weighted graph G, the diameter of G is the maximal distance between any two vertices u, v in G, where the distance between u and v is the number of edges in the shortest path between u and v.*

When the query graph Q is connected, we can easily get the diameter of Q by Floyd-Warshall algorithm [18]. Note that all qualified new embeddings must contain the vertex/edge inserted. Based on this observation, we give the following theorem.

Theorem 1. *Assume that v is the vertex inserted or one of the two vertices of the edge inserted. The distance between any vertex in the new embeddings and v is no more than d, where d is the diameter of the query graph.*

Theorem 1 gives a bound of the region where we can search for new embeddings. Based on this, we give our updating algorithm, which is shown in Algorithm 4. The algorithm first computes the diameter d of the query graph. If a vertex is inserted, the algorithm chooses it as the root vertex. If an edge is inserted, one of its vertices is selected as the root vertex. Then, it performs a d-level breadth-first search (BFS) on the data graph from the root vertex and get

---

Algorithm 4. Update

Input: data graph G, query graph Q, all embeddings of Q in G, the vertex or edge inserted v/e;
Output: all new embeddings of Q in G;
1: $d := Floyd\text{-}Marshall(Q)$
2: **if** vertex inserted **then**
3:     $root := v$
4: **else**
5:     $root :=$ one vertex of e
6: **end if**
7: $V_r :=$ all vertices visited through d-level BFS search over G from $root$
8: **if** $|V_r| < |V_Q|$ **then**
9:     **return**
10: **end if**
11: $G_r :=$ the subgraph induced by V_r
12: **return** $SubgraphSearchNVM(Q, G_r)$

---

the bounded search region. Finally, *SubgraphSearch* is performed on the search region to find all new embeddings.

Figure 6 shows an example of our updating algorithm. The diameter of the query graph in Fig. 6(a) is 2. Once an edge between v_2 and v_5 is inserted, we perform a 2-level BFS starting from v_2. All the data vertices visited during BFS make up the search region, which is surrounded by the dotted circle. Any new embeddings will not contain v_4, because it is not in the search region. Instead of processing the query on the whole data graph, we only need to answer the query on the search region and get the new embedding $\{(u_1, v_5), (u_2, v_3), (u_3, v_2)\}$.

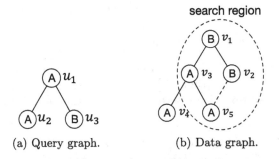

(a) Query graph.   (b) Data graph.

Fig. 6. Example of dealing with edge insertions.

Dealing with Label Updates. When the label of a data vertex v changes, all embeddings that contain v should be removed immediately. There may be new embeddings coming up. It is easy to prove that all new embeddings must contain v, and the problem becomes similar to that in Sect. 6.2. With the algorithm proposed in Sect. 6.2, we can easily get all new embeddings of the query graph, and the problem is solved.

7   Experiments

In this section, we evaluate our proposed algorithms against the existing subgraph matching algorithms on NVM. We use the same datasets and query sets described in Sect. 3. We implemented three variants of our algorithms.

- **PCME:** the algorithm using the revised backtracking framework without the optimization techniques.
- **PCME_OO:** PCME with matching the order optimization.
- **PCME_UP:** the updating algorithm in Sect. 6.

7.1   Efficiency and Scalability Evaluation

In this section, we compare PCME and PCME_OO with the existing algorithms. We only compare them with CFL-Match, since it is the fastest existing algorithm on NVM according to our preliminary studies in Sect. 3. We vary the size of the query graph from 4 to 10 and perform 20 queries for each size. We record the average processing time of a query with respect to the query size. Figure 7 shows the results of the experiments.

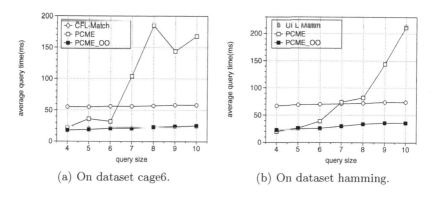

(a) On dataset cage6.   (b) On dataset hamming.

Fig. 7. Compare with CFL-Match (vary query size).

As is shown in Fig. 7, both PCME and PCME_OO are much faster than CFL-Match when the query size is small (with number of vertices less than 7). When the query size increases, the processing time of PCME increases much faster than the other two algorithms and exceeds the processing time of CFL-Match when query size increases to 7. This is because when the number of query vertices increases, the matching order of query vertices becomes more crucial to reducing intermediate results and total query time. The query time of PCME_OO, which adopts the matching order optimization, grows slowly when the query size increases and is always faster than CFL-Match.

7.2   Evaluation of Update Algorithm

In this section, we present our experiments on the update algorithm proposed in Sect. 6. For each query graph, we randomly generate 10 updating requests and compute the average updating time. According to Sect. 6, we do not consider edge/vertex deletions. We also avoid generating the insertion of a single isolated vertex. We compare the average update time with the query time, and the results are shown in Fig. 8.

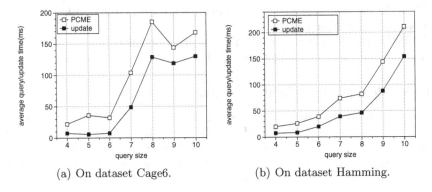

(a) On dataset Cage6.   (b) On dataset Hamming.

Fig. 8. Update time compared with query time (vary query size).

As is shown in Fig. 8, the average update time is much less than the average query time when the query size is small. However, when the query size increases, the update time grows fast and becomes closer to the query time. This is because when the query size increases, the search region in Algorithm 4 becomes larger. As a result, more write operations are done to store the information of the search region, which are expensive. Actually, when the query size is larger than 6, it is even more efficient to perform a query on the whole data graph with PCME_OO algorithm.

8   Conclusions

In this paper, we study the problem of subgraph matching on NVM, where reads and writes have asymmetric costs. We discover that most of the existing subgraph matching algorithms improve performance by increasing write operations (e.g. constructing auxiliary structures) as a trade-off for less reads and thus do not fit NVM. We propose a new algorithm that speeds up subgraph matching on NVM by trading more reads for less writes. We also study the subgraph matching problem on dynamic graphs. Extensive experiments on real datasets verify the efficiency of our algorithm. As future work, we can extend our study to approximate subgraph matching on NVM.

Acknowledgments. This work was partially supported by the National Natural Science Foundation of China (Nos. 61532015 and 61672189).

References

1. Raoux, S., Burr, G.W., Breitwisch, M.J., Rettner, C.T., Chen, Y.-C., Shelby, R.M., Salinga, M., et al.: Phase-change random access memory: a scalable technology. IBM J. Res. Dev. **52**(4.5), 465–479 (2008)
2. Strukov, D.B., Snider, G.S., Stewart, D.R., Williams, R.S.: The missing memristor found. Nature **453**(7191), 80–83 (2008)
3. Driskill-Smith, A.: Latest advances and future prospects of STT-RAM. In: Non-volatile Memories Workshop, pp. 11–13 (2010)
4. Ullmann, J.R.: An algorithm for subgraph isomorphism. J. ACM **23**(1), 31–42 (1976)
5. Cordella, L.P., Foggia, P., Sansone, C., et al.: A (sub)graph isomorphism algorithm for matching large graphs. IEEE Trans. Pattern Anal. Mach. Intell. **26**(10), 1367–1372 (2004)
6. Shang, H., Zhang, Y., Lin, X., et al.: Taming verification hardness: an efficient algorithm for testing subgraph isomorphism. Proc. VLDB Endow. **1**(1), 364–375 (2008)
7. He, H., Singh, A.K.: Query language and access methods for graph databases. In: ACM SIGMOD International Conference on Management of Data, SIGMOD 2008, Vancouver, BC, Canada, pp. 405–418. DBLP, June 2010
8. Han, W.S., Lee, J., Lee, J.H.: Turbo ISO: towards ultrafast and robust subgraph isomorphism search in large graph databases. In: ACM SIGMOD International Conference on Management of Data, pp. 337–348. ACM (2013)
9. Ren, X., Wang, J.: Exploiting vertex relationships in speeding up subgraph isomorphism over large graphs. Proc. VLDB Endow. **8**(5), 617–628 (2015)
10. Bi, F., Chang, L., Lin, X., et al.: Efficient subgraph matching by postponing cartesian products. In: International Conference, pp. 1199–1214 (2016)
11. Poremba, M., Zhang, T., Xie, Y.: NVMain 2.0: architectural simulator to model (non-)volatile memory systems. IEEE Comput. Archit. Lett. **14**(2), 140–143 (2015)
12. Binkert, N., Beckmann, B., Black, G., et al.: The gem5 simulator. ACM SIGARCH Comput. Archit. News **39**(2), 1–7 (2011)
13. Blelloch, G.E., Fineman, J.T., Gibbons, P.B., et al.: Efficient algorithms with asymmetric read and write costs. arXiv preprint arXiv:1511.01038 (2015)
14. Blelloch, G.E., Fineman, J.T., Gibbons, P.B., et al.: Sorting with asymmetric read and write costs. In: ACM on Symposium on Parallelism in Algorithms and Architectures, pp. 1–12. ACM (2016)
15. Viglas, S.D.: Write-limited sorts and joins for persistent memory. Proc. VLDB Endow. **7**(5), 413–424 (2014)
16. Carson, E., Demmel, J., Grigori, L., et al.: Write-avoiding algorithms. In: IEEE International Parallel and Distributed Processing Symposium, pp. 648–658. IEEE (2016)
17. Lee, J., Han, W.S., Kasperovics, R., et al.: An in-depth comparison of subgraph isomorphism algorithms in graph databases. Proc. VLDB Endow. **6**(2), 133–144 (2012)
18. Floyd, R.W.: Algorithm 97: shortest path. Commun. ACM **5**(6), 345 (1962)
19. Rossi, R., Ahmed, N.: The network data repository with interactive graph analytics and visualization. In: AAAI (2015)

Influenced Nodes Discovery in Temporal Contact Network

Jinjing Huang[1,2], Tianqiao Lin[1], An Liu[1], Zhixu Li[1], Hongzhi Yin[3],
and Lei Zhao[1(✉)]

[1] School of Computer Science and Technology, Soochow University, Suzhou, China
huangjj@siit.edu.cn,lintim111@126.com,{anliu,zhixuli,zhaol}@suda.edu.cn
[2] Suzhou Vocational Institute of Industrial Technology, Suzhou, China
[3] School of ITEE, University of Queensland, Brisbane, Australia
h.yin1@uq.edu.au

Abstract. Information diffusion has been studied for many years to understand how information diffuse in social network or real world. However, which nodes and when they will get influenced are unpredictable because of the uncertainty of information diffusion even we know the initial influenced nodes and diffusion network. Verification is the only way to make sure if a node is influenced or not. The target of discovering influenced nodes is to find more influenced nodes under the limited amount of verifications. In this paper, the temporal contact network is modeled. Then influenced nodes discovery problem in temporal contact network are studied based on the Independent Cascade (IC) model. A path length limited approach is proposed to calculate the infection probability approximately. Experimental results on real and synthetic data sets show our approach has better performance than BFS and Random Walk algorithm.

Keywords: Temporal contact network · Information diffusion · Influenced nodes discovery

1 Introduction

Discovering influenced nodes is important in applications. Advertising agencies want to know who will be influenced by their advertisement to evaluate advertisement and make decisions. Potentially infected individuals should be found out to prevent the spread of virus for disease controlling in epidemiology. However, who and when will they become influenced is unpredictable because of the uncertainty of information diffusion. So it is necessary to track and verify the result of information diffusion.

Information is diffused via contacts. However, contacts are varying over time and it takes time to propagate a message. So a diffusion network is modeled as a temporal contact network, which records when and between whom a contact happens without knowing what are diffused, to study information diffusion, i.e.

© Springer International Publishing AG 2017
A. Bouguettaya et al. (Eds.): WISE 2017, Part I, LNCS 10569, pp. 472–487, 2017.
DOI: 10.1007/978-3-319-68783-4_32

how to discover nodes influenced by nodes in initial influenced nodes set $\mathbb{L}$ before check time t in a given temporal contact network $\mathbb{G}$.

With the development of network technology, social networks have become the main medium of information diffusion. However, it is difficult to discover influenced nodes in social networks which usually contains billions of users because influenced nodes only account for a small percentage in social networks. So it is inefficient to find out influenced nodes with traditional approaches like BFS (Breadth First Search). Najar et al. predicts the final state of information diffusion from initial state on static network [14]. However, this approach only predict the final state of diffusion. This approach cannot solve the problem of influenced nodes discovery at any check time.

It is intuitive that only reachable nodes can be influenced nodes under Closed World assumption [6]. So we construct the candidate node set by reachable nodes that have temporal diffusion path from influenced nodes before check time t in a given temporal contact network $\mathbb{G}$. Considering the uncertainty of information diffusion, it is necessary to verify the states of selected nodes. In order to find more influenced nodes under the limited times of verifications, infection probabilities of candidate nodes are calculated and the node with a maximal infection probability is verified.

The challenges of the problem are: (1) It is #p-hard to find out all diffusion paths and compute infection probability exactly. (2) Positive verification increases the influence probabilities of subsequent nodes while negative result decreases them. Obviously, it is necessary to update candidate set after verification. However, it is costly to recompute infection probability after each verification.

The contributions of this paper are as follows: (1) We model the influenced nodes discovery problem in temporal contact network. (2) We present effective and efficient approximate algorithms to solve this #p-hard problem. (3) We measure the performance of our algorithms against BFS and Random Walk algorithm by precision and recall.

The rest of this paper is organized as follows. Section 2 gives some literature study. We formalize the influenced nodes discovery problem and infection probability in temporal contact network based on IC model in Sect. 3. Section 4 proposes the approach to calculate the infection probability. Then we propose our HDS (Heuristic Diffusion Simulation) algorithm in Sect. 5. The extension of our Heuristic Diffusion Simulation algorithm is proposed in Sect. 6. In Sect. 7, we compare the performance of HDS algorithm with BFS and Random Walk. We conclude the paper in Sect. 8.

2   Related Work

Information diffusion has attracted researchers from different fields. Underlying diffusion network inference [1,4], diffusion probability inferring [5,10,15], information diffusion modeling [8] and phenomenon analysis [19] have been studied to understand information diffusion in social networks. Independent Cascade

(IC) and Linear Threshold (LT) are two seminal diffusion models in information diffusion predicting [9]. Gomez et al. proposed NETINF, NETRATE and INFOPATH to describe how infections occur over time [4,5]. Study of information diffusion with time factor is booming. Besides, path problems and graph traversals like BFS and DFS in temporal networks have been studied in [7,18]. Our method is proposed based on those studies.

Influence maximization focus on finding a k node set to maximize the expected number of influenced nodes is hot in social network [3,9,16]. Discovering influenced nodes is similar to influence maximization that they both predict the result of information diffusion, but the former tries to find out the influenced nodes while the later tries to maximize the expected number of influenced nodes. To deal with the np-hard of influence estimate, Chen et al. proposed heuristic methods like PMIA [2] to approximate the result. Our method is equal to PMIA when path length is set to 1 in discovering influenced nodes.

Sampling and graph traversal can be used to discover influenced nodes. Leskovec and Faloutsos [11] compare sampling methods in large static and temporal graph, Random Walk and "forest fire" have best performance. Mehdiabadi et al. [12] compared the structure and diffusion based BFS with Random Walk for infected nodes sampling and showed diffusion-based Random Walk is the best. We compare our method with BFS and Random walk algorithm on synthetic and real data, experiment result shows our method has better performance.

3   Problem Formulation

3.1   Notations

Table 1 defines the notations used in this paper.

Table 1. Notations

Notation	Description
$\mathbb{G}(\mathbb{V}, \mathbb{E})$	Temporal contact network with node set $\mathbb{V}$ and edge set $\mathbb{E}$
$\mathbb{L}$	Set of nodes with definite status
$\mathbb{C}$	Candidate set of infected nodes
$\mathbf{Pr}(e)$	Diffusion probability of edge e
$\mathbf{Pr}(v|\mathbb{L}, t)$	Infection probability of node v
$\mathbf{Pr}_u(v|\mathbb{L}, t)$	Infection probability of node v through node u
$\mathbf{St}(v|t)$	State of node v at time t
$\mathbf{It}(v)$	Infection time of node v

3.2   Preliminary

The social influence of an message is enlarged from a certain number of nodes. And message diffusion in social networks relies on the contacts between nodes in viral marketing strategies. As the contacts between nodes are varying over time and it takes time to influence nodes, the information diffusion network is modeled as temporal contact network.

Definition 1. *Contact: A contact is a 5-tuple (v_s, v_e, t, d, p), in which v_s means start node, v_e means end node, t means start time, d means delay of information diffusion, and p represents the diffusion probability.*

Definition 2. *Temporal Contact Network (TCN): A given network $\mathbb{G} = (\mathbb{V}, \mathbb{E})$ with node set $\mathbb{V}$ and edge set $\mathbb{E}$ is a temporal contact network, in which each edge represents a contact $e_{x,y,i} = (v_x, v_y, t_i, d_i, p_i)$ where $i = 1, 2, 3, \ldots$, for more than one contact may appear between any two nodes.*

Diffusion probability inferring has been studied in [5,10,15], so we consider diffusion probability $\mathbf{Pr}(e)$ of edge e as known quantity in our study.

Information diffusion in temporal contact network must hold the time constraint [18] which has been well studied in temporal graph. For example, consider two nodes v_1 and v_2, if v_1 gets a information from other nodes after a contact with node v_2, then v_2 cannot get this information from v_1. This constraint, defined as follows, is called a temporal diffusion path.

Definition 3. *Temporal Diffusion Path (TDP): Given a temporal contact network $\mathbb{G} = (\mathbb{V}, \mathbb{E})$, a temporal diffusion path is a sequence of m edges, denoted $P = <e_{0,1,*}, e_{1,2,*}, \ldots, e_{m-1,m,*}>$ where $\forall e_{i,i+1,*} \in P(1 \leq i < m)$, it follows that $t_i + d_i \leq t_{i+1}$. $|P|$ is called the length of path P.*

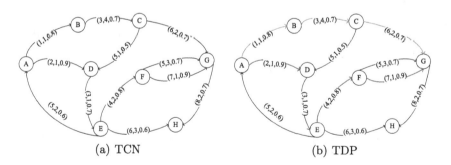

Fig. 1. Illustration of TCN and TDP (Color figure online)

Figure 1 illustrates a TCN and a TDP, respectively. Figure 1(a) shows a TCN is composed of nodes linked by edges with start time of influence, delay of information diffusion, and influence probability. In Fig. 1(b) the blue path is a TDP, but the red path is not, for the start time of the edge between C and G is later than the influenced time of C which does not satisfy the time constraint of TDP.

Definition 4. *Infection Time:* *Given a temporal contact network* $\mathbb{G} = (\mathbb{V}, \mathbb{E})$, *suppose node* v *has* m *incoming edges* $<v_1, v, t_1, d_1, *>, <v_2, v, t_2, d_2, *>, \ldots,$ $<v_m, v, t_m, d_m, *>$, *if* $\exists 1 \leq i \leq k, \forall 1 \leq j \leq k(j \neq i)$, *it follows that* $t_j + d_j \geq t_i + d_i$, *then the infection time of node* v *is* $t_i + d_i$, *denoted* ***It(v)***.

Definition 5. *Infection Probability:* *Given an influenced node set* $\mathbb{L}$, **Pr** $(v|\mathbb{L}, t)$ *is the infection probability of* v *infected by nodes in* $\mathbb{L}$ *at time* t.

Definition 6. *State Function:* *Given an node set* $\mathbb{V}$, *the state function* $\mathbf{St}(v|t)$ *fetches the state of node* $v \in \mathbb{V}$. $\mathbf{St}(v|t) = 1$, *if* v *is influenced by message* m *at time* t, *otherwise* $\mathbf{St}(v|t) = 0$.

3.3   Problem Statement

As information diffusion is a stochastic process, verification is the only way to confirm if a node is influenced or not. However, verifying a whole graph is too costly. Therefore, we take selecting-verifying strategy in our work. We evaluate nodes in a candidate set $\mathbb{C}$ by temporal infection probability $\mathbf{Pr}(v|\mathbb{L}, t)$ and select a node which has the maximum infection probability to verify, i.e. the selected node is

$$v = \{v_i | \forall v_i, v_j \in \mathbb{C}, v_j \neq v_i, \mathbf{Pr}(v_i|\mathbb{L}, t) > \mathbf{Pr}(v_j|\mathbb{L}, t)\}. \tag{1}$$

The problem studied in this paper is defined as follows.

[Influenced Nodes Discovery]. Given a temporal contact network $\mathbb{G} = (\mathbb{V}, \mathbb{E})$ and an initial influenced node set $\mathbb{L} \in \mathbb{V}$, the target of influenced nodes discovery is to find influenced nodes as much as possible out of $\mathbb{V} - \mathbb{L}$ under k times of verifications.

Nodes can be influenced by external events and factors outside the network [13]. In this paper, we assume information diffusion is under closed world assumption [6] in which information can only diffuse via the edges in the given network. So we construct candidate node set $\mathbb{C}$ by reachable nodes of influenced nodes in $\mathbb{L}$, because unreachable nodes could not be infected by influenced nodes. Verified influenced nodes will be added into $\mathbb{L}$ after verifications.

4   Infection Probability

In this section, we will formalize how to compute temporal infection probability $\mathbf{Pr}(v|\mathbb{L}, t)$ based on IC model.

4.1   Accurate Infection Probability

[Single Path]. Given a temporal diffusion path $P = <e_1, e_2, \ldots, e_n>$, the infection probability of node v_n that infected by v_1 through path P can be figured out as follows based on IC model:

$$\mathbf{Pr}(P) = \prod_{i=1}^{n} \mathbf{Pr}(e_i). \tag{2}$$

[Multiple Path Without Overlaps]. If temporal diffusion paths $P_1, P_2, P_3,$ $\ldots, P_m$ from influenced nodes $\mathbb{L}$ to node v before time t do not have overlaps, then the infection probability of node v can be simplified as:

$$\mathbf{Pr}(v|\mathbb{L}, t) = 1 - \prod_{i=1}^{m}(1 - \mathbf{Pr}(P_i)) \tag{3}$$

[Multiple Path with Overlaps]. If temporal diffusion paths from influenced nodes $\mathbb{L}$ to node v before time t have overlaps, and all these paths are consisted of $e_1, e_2, \ldots, e_n$, then the infection probability of node v should be figured out through inclusion-exclusion theorem. The details are as follows:

$$\mathbf{Pr}(v|\mathbb{L}, t) = \sum_{i=1}^{n}(-1)^{i-1} \sum_{1 \leq j_1 \leq j_2 \leq \ldots \leq j_m \leq n} \mathbf{Pr}(e_{j_1})\mathbf{Pr}(e_{j_2})\ldots\mathbf{Pr}(e_{j_m}) \tag{4}$$

If $\mathbf{Pr}(v|\mathbb{L}, t_i)$ is already known, then $\mathbf{Pr}(v|\mathbb{L}, t_{i+1})$ can be computed by Eq. 5 proposed by Zhang et al. in [20] which is also based on the inclusion-exclusion theorem.

$$\mathbf{Pr}(v|\mathbb{L}, t_{i+1}) = \mathbf{Pr}(v|\mathbb{L}, t_i) + (1 - \mathbf{Pr}(v|\mathbb{L}, t_i))\mathbf{Pr}(v|\mathbb{L}, (t_i, t_{i+1})) \tag{5}$$

4.2 Infection Probability Approximation

All contacts, i.e. all temporal diffusion paths, should be included when calculating temporal infection probability. However, it is #p-hard to find out all temporal diffusion paths from influenced nodes in temporal contact network [17]. Therefore, presenting an approximative algorithm to compute the infection probability approximatively is necessary.

Suppose there are two temporal diffusion paths P_1 and P_2 from u to v and each edge has the same diffusion probability of $\mathbf{Pr}(e), 0 \leq \mathbf{Pr}(e) < 1$, then the infection probability of P_1 and P_2 can be calculated as $\mathbf{Pr}(P_1) = \mathbf{Pr}(e)^{|P_1|}$ and $\mathbf{Pr}(P_2) = \mathbf{Pr}(e)^{|P_2|}$ according to Eq. 2 in IC model. It is obvious that $\mathbf{Pr}(P_1) < \mathbf{Pr}(P_2)$ if $|P_1| > |P_2|$. Besides, node that has more paths for each path length $|P|$ will have higher infection probability when diffusion probability is set uniformly. So we can get following observations.

Observation 1. *Shorter temporal diffusion path that constructed by less edges affect more than longer paths in calculating temporal infection probability when diffusion probability is set uniformly.*

Observation 2. *Node that has more temporal diffusion paths at each path length has higher temporal infection probability if the diffusion probability is set uniformly.*

Inspired by the above observations, we propose a hop limited approximative method, named Limited Hops Approximation (LHA), to compute the temporal diffusion probability effectively and efficiently. Obviously, hop value is a trade-off of efficiency and effectiveness. The infection probability will be computed effectively but loss precision with a small hop value.

[**Limited Hops Approximation (LHA)**]. For a given node v and an integer $h > 0$, the temporal diffusion probability $\mathbf{Pr}(v)$ is computed approximatively through TDPs whose lengths are all less than h.

5   HDS Algorithm

In this section, we will show the implementation of our Heuristic Diffusion Simulation (HDS) algorithm which consists of infection probability calculation (U2V) and candidate set updating (CSU) to construct candidate set $\mathbb{C}$. HDS algorithm is based on LHA method to compute the temporal infection probability. Node has maximum infection probability will be selected to verify after computation.

5.1   Point-to-Point Infection Probability Calculation

According to the LHA method, we propose the point-to-point infection probability calculate (U2V) algorithm to find out h path length limited temporal diffusion paths from each node in $\mathbb{L}$ to form candidate set and compute temporal infection probability approximatively.

There are some conditions should be considered while finding temporal diffusion paths. (1) The length of each path should be smaller than h according to the LHA method. (2) Influenced nodes should only appear once in each diffusion paths as the start node; otherwise paths from influenced nodes will be considered repeatedly in infection probability computation. (3) One node can only appear once in each diffusion path, because each node can be influenced only once in information diffusion. As DFS can record already-visited nodes in stack while traversing, U2V is implemented under DFS method to find temporal diffusion paths that satisfies these conditions.

As we only care about the rank of nodes in candidate set according to the temporal infection probability, and the probability computed by Eq. 3 is approximate to the result computed by Eq. 5. So we simplify the calculation of temporal infection probability as follows.

Suppose the infection probability of node v is $\mathbf{Pr}(v|\mathbb{L}, t)$ and there is another temporal diffusion path from u to v and $<u, v, *, *, *> \in \mathbb{E}$, then the infection probability of node v can be updated as:

$$\mathbf{Pr}(v|\mathbb{L}, t) = 1 - (1 - \mathbf{Pr}(v|\mathbb{L}, t)) \times (1 - \mathbf{Pr}(u) \times \mathbf{Pr}(e)). \tag{6}$$

If u is influenced node, then the infection probability of v is $\mathbf{Pr}(v|\mathbb{L}, t) = 1 - (1 - \mathbf{Pr}(v|\mathbb{L}, t)) \times (1 - \mathbf{Pr}(e))$.

Algorithm 1 is the implementation of U2V algorithm. If a node is first added into candidate set, the probability is computed by Eq. 2 at line 9; otherwise, probability will be updated by Eq. 6 at line 11 for nodes in $\mathbb{C}$.

Algorithm 1. Point-to-point Infection Probability: U2V($\mathbb{G}, u, t, \mathbb{C}, h$)

Require: $\mathbb{G}$: temporal contact network, u: source node, t: start time, $\mathbb{C}$: candidate set, h: path length parameter

Ensure: No duplicate value in set $\mathbb{C}$

```
 1: function U2V(𝔾, u, t, ℂ, h)
 2:     Stack 𝕊 ← {< u, t >}
 3:     Set 𝕄 ← φ
 4:     while (𝕊 ≠ φ and 𝕊.size() < h) do
 5:         < u', t >← 𝕊.pop()
 6:         𝔼' ← {e|e = (u', *, tₛ, *, *) and tₛ ≥ t}
 7:         if 𝔼' − (𝔼' ∩ 𝕄) ≠ φ then
 8:             e = (u', v, t, d, *) ← earliest(𝔼' − (𝔼' ∩ 𝕄))
 9:             p ← Pr(u) × Pr(e)
10:             if (v ∈ ℂ) then
11:                 p ← 1 − (1 − Pr(v)) × (1 − p)
12:                 Pr(v) ← p
13:             else
14:                 ℂ ← ℂ ∪ {v}
15:             end if
16:             𝕊.push(< v, t + d >)
17:             𝕄 ← 𝕄 ∪ {e}
18:         else
19:             𝕊.pop()
20:         end if
21:     end while
22: end function
```

5.2 Candidate Set Updating

A naive way to update candidate set is recomputing candidate set by calling U2V algorithm for each influenced node in $\mathbb{L}$ after verification. However, it is costly to get diffusion paths from influenced nodes. So we propose candidate set update (CSU) algorithm to update candidate set. The main idea of our CSU algorithm is getting nodes affect by the verification and updating the probability.

Suppose v is the verified node and v' is node affected by v, diffusion paths from v to v' which begin after $\mathbf{It}(v)$ are $\{P_1, P_2, \ldots, P_m\}$, and there are n paths from $\mathbb{L}$ to v with infected probabilities of $\{\mathbf{Pr}(v)_1, \mathbf{Pr}(v)_2, \ldots, \mathbf{Pr}(v)_n\}$, then the infection probability of v' influenced through v can be computed as:

$$\mathbf{Pr}_v(v'|\mathbb{L}, t) = 1 - \prod_{j=1}^{n} \prod_{i}^{m} (1 - \mathbf{Pr}(v)_j \times \mathbf{Pr}(P_i)) \tag{7}$$

If v is verified as uninfluenced node, the infection probability of affected node v' is updated as:

$$\mathbf{Pr}(v'|\mathbb{L}, t) = 1 - \frac{1 - \mathbf{Pr}(v'|\mathbb{L}, t)}{\mathbf{Pr}_v(v'|\mathbb{L}, t)}. \tag{8}$$

If v is verified as influenced node, then the infection probability of affected node v' is updated as follows if $v' \in \mathbb{C}$. Otherwise, v' will be added into candidate set with infection probability $\mathbf{Pr}_v(v'|\mathbb{L}, t)$.

$$\mathbf{Pr}(v'|\mathbb{L}, t) = 1 - \frac{1 - \mathbf{Pr}(v'|\mathbb{L}, t)}{\mathbf{Pr}_v(v'|\mathbb{L}, t)} \prod_{i=1}^{m}(1 - \mathbf{Pr}(P_i)). \tag{9}$$

The candidate set update (CSU) algorithm is outlined in Algorithm 2. U2V algorithm is called to find out vertices affected by the selected node. If the selected node is verified as influenced node, candidate set is updated by Eqs. 8 and 9 to except diffusion paths before verified node is influenced.

Algorithm 2. Candidate Set Updating: CSU()

1: **function** CSU($\mathbb{G}, u, t, \mathbb{C}, h$)
2:     $\mathbb{C}_1 \leftarrow$ U2V($\mathbb{G}, u, t, \mathbb{C}, h$)
3:     Update the candidate set $\mathbb{C}$ with $\mathbb{C}_1$ by Eq. 8
4:     **if** $\mathbf{St}(u, t)$ **then**
5:         $\mathbb{C}_2 \leftarrow \{v | v \in \mathbb{C}_1$ and $\forall v \neq u$, it follows that $\mathbf{It}(v) \geq \mathbf{It}(u)\}$
6:         Update the candidate set $\mathbb{C}$ with $\mathbb{C}_2$ by Eq. 9
7:     **end if**
8:     $\mathbb{C} \leftarrow \mathbb{C} - \{u\}$
9: **end function**

Heuristic Diffusion Simulation (HDS) algorithm is outlined in Algorithm 3. HDS is constructed by initialization phase (line 4–6) and verification phase (line 7–15). Node with maximum infection probability is selected to verify after computation at line 7. Verified nodes L is returned in HDS algorithm.

5.3   Algorithm Analysis

HDS algorithm includes initialization and updating phases. Candidate set is computed by U2V algorithm from initial influenced nodes in initialization phase. CSU algorithm is called to update candidate set after each verification. As path length is limited to h in U2V algorithm under LHA method, the complexity of U2V algorithm is $O((d_{out} + |\mathbb{C}|)^h)$, d_{out} is the average outdegree of the nodes in the TCN. As CSU algorithm calls U2V algorithm to update candidate set, the complexity of CSU algorithm is also $O((d_{out} + |\mathbb{C}|)^h)$. So the complexity of HDS algorithm is $O((|\mathbb{L}| + k) \times (d_{out} + |\mathbb{C}|)^h)$, $\mathbb{L}$ is the initial influenced node set and k is the number of verification.

6   AHDS Algorithm

Heuristic Diffusion Simulation (HDS) algorithm selects node in all h reachable nodes that has maximum infection probability. However, with verification goes

Algorithm 3. Heuristic Diffusion Simulation: HDS()

```
 1: function HDS(𝔾, 𝕃, t, k, h)
 2:     l ← |𝕃|
 3:     ℂ ← φ
 4:     for < u_i, t_i >∈ 𝕃 do
 5:         ℂ ← U2V(𝔾, u_i, t_i, ℂ, h)
 6:     end for
 7:     while |𝕃| − l < k and ℂ ≠ φ do
 8:         v ← v ∈ ℂ and ∀v' ∈ ℂ Pr(v) ≥ Pr(v')(v ≠ v')
 9:         if St(v|t) then
10:             𝕃 = 𝕃 ∪ < v, t >
11:         else
12:             𝕃 = 𝕃 ∪ < v, −1 >
13:         end if
14:         CSU(𝔾, v, t, ℂ, h)
15:     end while
16:     return 𝕃
17: end function
```

on, the infection probability of candidate nodes are getting close. So the precision will decrease fast while recall increase slowly after verifications. HDS algorithm does not work well when the number of verification is large.

As information can only be diffused to adjacent nodes of influenced nodes under Close World assumption, we develop the AHDS algorithm to overcome the disadvantages. The main idea of AHDS algorithm is selecting adjacent nodes of influenced nodes to construct candidate set. The computation of infection probability and update is similar with HDS algorithm in AHDS.

The complexity of AHDS algorithm is also $O((|𝕃| + k) \times (d_{out} + |ℂ|)^h)$. As only adjacent nodes of influenced nodes will be selected to construct candidate set, AHDS algorithm will be more efficient than HDS algorithm.

7   Experiments

We compare the precision and recall of our algorithms with several existing algorithms from experiments on real-world and synthetic datasets to validate the performance of our algorithms. We will show how precision and recall change when verification increase.

[**Compared Algorithms**]. The algorithms we compare in this paper are as follows:

1. Breadth First Search (BFS). BFS is a basic algorithm in graph and can be used to verify influenced nodes. As the network is temporal network, we use temporal BFS algorithm to select nodes [7].
2. Random Walk (RW). RW is well-known in graph sampling that select edge uniformly at random based on the prior probability. In this paper, diffusion probability is the prior probability in discovering influenced nodes.

3. Maximum Influence Path (MIP). Chen et al. proposed PMIA heuristic method that using Maximum Influence Path (MIP) to estimate the influence from one node to another [2].
4. Heuristic Diffusion Simulation (HDS). HDS is proposed in Sect. 5. As HDS algorithm use path length limited algorithm to compute infection probability, we use x-HDS to denote the x-length limited HDS algorithm.
5. Adjacent Heuristic Diffusion Simulation (AHDS). AHDS is described in Sect. 6. Similar to HDS algorithm, we use x-AHDS to denote the x-length limited AHDS algorithm.

[**Information Diffusion Simulation**]. It is difficult to track information diffusion without loss in real diffusion, so we simulate information diffusion in given graph from initial influenced nodes. We select initial influenced nodes randomly to have a comprehensive evaluation. Exponential model is widely used in information diffusion, we assign diffusion probability based on the exponential model. The diffusion probability is computed as $\beta \cdot e^{-d \cdot (t-\tau)}$ based on [4], β is the topic performance, τ is the start time of information diffusion, t is the time stamp of current edge and d is the diffusion delay.

[**Influenced Nodes Verification**]. Verification is necessary to check a node is influenced or not. We assign state label to indicate a node is influenced or not in this diffusion. So verification is to check the state label of selected node.

[**Evaluation Methodology**]. Precision and recall are used to evaluate the performance of the algorithms. If the verification number is set as k, then we will have $|\mathbb{C}| = k$ for verified node set $\mathbb{C}$. The precision and recall are computed as

$$Precision = \frac{1}{|\mathbb{C}|} \sum_{u_i \in \mathbb{C}} \mathbf{St}(u_i|t) \text{ and } Recall = \sum_{u_i \in \mathbb{C}} \mathbf{St}(u_i|t) / \sum_{u_i \in \mathbb{V}} \mathbf{St}(u_i|t).$$

7.1   Experiments on Synthetic Datasets

In this section, we will show the recall and precision of algorithms with different number of verification and path length in synthetic datasets.

To show how recall and precision change with the increase of verification, we set the number of verification from ten percentage of the number of all influenced nodes with the growth of ten percentage of the number of all influenced nodes until finding all influenced nodes or the number of verification is large than 3 times of influenced nodes. The limited path length approximate algorithms are set to 1-HDS, 2-HDS, 2-AHDS, 3-AHDS to illustrate how path lengths effect the performance of our algorithms.

[**Synthetic Datasets**]. We use Kronecker Graph model to generate synthetic datasets based on [3]. Kronecker graphs are generated as follows: random ([0:5; 0:5; 0:5; 0:5]), hierarchical ([0:9; 0:1; 0:1; 0:9]) and coreperiphery ([0:9; 0:5; 0:5; 0:3]). We generate those synthetic graphs based on the algorithms in SNAP (See http://snap.stanford.edu/). 1K nodes and 10K temporal edges are generated in each synthetic graph. We have generated lots of synthetic temporal graphs to avoid accidental factors.

[**Experiment Results**]. Figure 2 illustrates the recall performance under different number of verification. It is nearly a liner function between recall and the ratio of verified nodes to all influenced nodes at the beginning, but slows down when verification increases. AHDS, BFS and Random Walk performance better than HDS algorithm when verification goes on. Because HDS selects node from all h reachable nodes, while other algorithms only consider adjacent nodes.

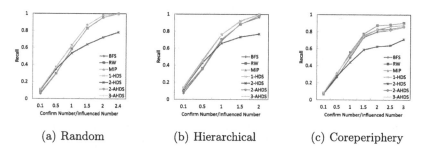

(a) Random   (b) Hierarchical   (c) Coreperiphery

Fig. 2. Recall on synthetic graphs.

Figure 3 is the precision result. HDS algorithm has best precision performance when the number of verification is small. The precision of HDS algorithm and AHDS algorithm is 20% higher than BFS and RW when verification is 10% of influenced nodes. Precision decreases when verification increases. AHDS algorithm has similar performance with BFS and RW algorithm when verification is large.

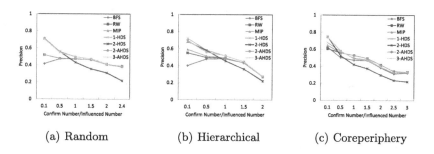

(a) Random   (b) Hierarchical   (c) Coreperiphery

Fig. 3. Precision on synthetic graphs.

Precision-recall is showed in Fig. 4. It is obvious that our algorithms are better than traditional algorithms. AHDS algorithm has best performance in those algorithms that have close performance with HDS algorithm when recall is small and is stable as BFS algorithm when recall is large. We can find larger h has better performance that 2-AHDS and 2-HDS are better than 1-HDS, but h does not effect the performance very much.

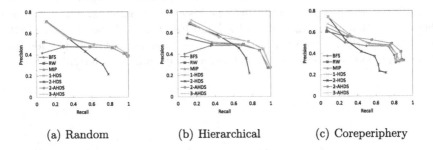

(a) Random   (b) Hierarchical   (c) Coreperiphery

Fig. 4. Precision-recall on synthetic graphs.

From the above experiment results, we can find our method is better than BFS, Random Walk, and MIP. AHDS, BFS and Random Walk algorithm are more stable than HDS.

7.2 Experiments on Real-World Datasets

In this section, we will show how our methods perform on real datasets under the same experimental setup in synthetic dataset experiment.

Table 2. Real-world datasets

Dataset	Digg	Slashdot	Infectious	Facebook
Nodes	30K	51K	410	63K
Edges	87K	140K	17K	817K
Average degree	5.8	5.5	84.4	25.6
Maximal degree	310	3,357	294	1,098
Edge type	Reply	Reply	Contact	Friendship

[**Real-World Datasets**]. We use four real-world datasets Digg, Slashdot, Infectious and Facebook in this experiment. Table 2 is the detail of those datasets. Edges in those four real temporal graph only have time stamps but do not have delay information, so we assign a random value as delay to compute diffusion probability. Then information diffusion is simulated on those datesets with different initial influenced nodes.

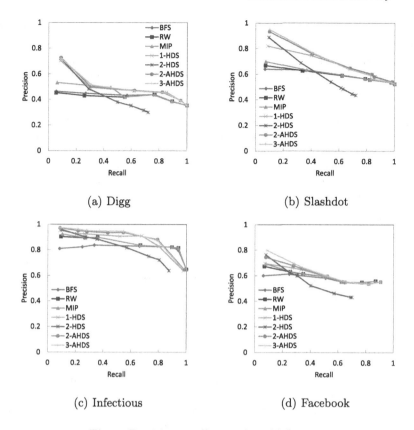

(a) Digg

(b) Slashdot

(c) Infectious

(d) Facebook

Fig. 5. Precision-recall on real-world datasets

[**Experiment Results**]. Precision and recall result on real datasets is similar with result on synthetic datasets and precision-recall figure can illustrate the performance of our algorithms, so we only show the precision-recall result in this experiment. Figure 5 is the experiment result on real-world datasets. The precision of HDS and AHDS is obviously higher than BFS and Random Walk at beginning on real-world datasets. And larger h has better precision performance, especially in Slashdot and Facebook as Fig. 5(b) and (d).

Experiment results on real-world and synthetic datasets show our methods are better than traditional methods BFS and Random Walk in precision and recall. The characteristic of our algorithms is having high precision at the beginning but reduce to the BFS and Random Walk algorithm when more and more influenced nodes were discovered.

8   Conclusion

In this paper, the diffusion network is modeled as TCN and formulate the computation of temporal infection probability based on the inclusion-exclusion theorem and IC model. The goal of influenced nodes discovery in TCN is to discover influenced nodes as much as possible before check time t under k verification. To tackle this #p-hard problem, path length limited approximate method is proposed to compute probability efficiently. Experiments show that algorithms presented in this paper have better performance for influenced nodes discovery.

Acknowledgement. This work was supported by the National Natural Science Foundation of China under Grant Nos. 61572335 and 61572336, the Natural Science Foundation of Jiangsu Province of China under Grant No. BK20151223, the Natural Science Foundation of Jiangsu Provincial Department of Education of China under Grant No. 12KJB520017, and Collaborative Innovation Center of Novel Software Technology and Industrialization, Jiangsu, China.

References

1. Abrahao, B., Chierichetti, F., Kleinberg, R., Panconesi, A.: Trace complexity of network inference. In: Proceedings of the 19th ACM SIGKDD International Conference on Knowledge Discovery and Data Mining, pp. 491–499. ACM (2013)
2. Chen, W., Wang, C., Wang, Y.: Scalable influence maximization for prevalent viral marketing in large-scale social networks. In: Proceedings of the 16th ACM SIGKDD International Conference on Knowledge Discovery and Data Mining, pp. 1029–1038. ACM (2010)
3. Du, N., Liang, Y., Balcan, M.-F., Song, L.: Influence function learning in information diffusion networks. In: Proceedings of the 31st International Conference on Machine Learning, vol. 2014, p. 2016. NIH Public Access (2014)
4. Gomez Rodriguez, M., Balduzzi, D., Schölkopf, B., Scheffer, G.T. et al.: Uncovering the temporal dynamics of diffusion networks. In: 28th International Conference on Machine Learning (ICML 2011), pp. 561–568. International Machine Learning Society (2011)
5. Gomez Rodriguez, M., Leskovec, J., Krause, A.: Inferring networks of diffusion and influence. In: Proceedings of the 16th ACM SIGKDD International Conference on Knowledge Discovery and Data Mining, pp. 1019–1028. ACM (2010)
6. Guille, A., Hacid, H., Favre, C., Zighed, D.A.: Information diffusion in online social networks: a survey. ACM SIGMOD Rec. **42**(2), 17–28 (2013)
7. Huang, S., Cheng, J., Wu, H.: Temporal graph traversals: definitions, algorithms, and applications. arXiv preprint arXiv:1401.1919 (2014)
8. Iwata, T., Shah, A., Ghahramani, Z.: Discovering latent influence in online social activities via shared cascade poisson processes. In: Proceedings of the 19th ACM SIGKDD International Conference on Knowledge Discovery and Data Mining, pp. 266–274 (2013)
9. Kempe, D., Kleinberg, J., Tardos, É.: Maximizing the spread of influence through a social network. In: Proceedings of the Ninth ACM SIGKDD International Conference on Knowledge Discovery and Data Mining, pp. 137–146. ACM (2003)
10. Kutzkov, K., Bifet, A., Bonchi, F., Gionis, A.: STRIP: stream learning of influence probabilities. In: Proceedings of the 19th ACM SIGKDD International Conference on Knowledge Discovery and Data Mining, pp. 275–283. ACM (2013)
11. Leskovec, J., Faloutsos, C.: Sampling from large graphs. In: Proceedings of the 12th ACM SIGKDD International Conference on Knowledge Discovery and Data Mining, pp. 631–636 (2006)

12. Mehdiabadi, M.E., Rabiee, H.R., Salehi, M.: Sampling from diffusion networks. In: Proceedings of the 2012 International Conference on Social Informatics, pp. 106–112. IEEE Computer Society (2012)

13. Myers, S.A., Zhu, C., Leskovec, J.: Information diffusion and external influence in networks. In: Proceedings of the 18th ACM SIGKDD International Conference on Knowledge Discovery and Data Mining, pp. 33–41. ACM (2012)

14. Najar, A., Denoyer, L., Gallinari, P.: Predicting information diffusion on social networks with partial knowledge. In: Proceedings of the 21st International Conference Companion on World Wide Web, pp. 1197–1204. ACM (2012)

15. Tang, J., Sun, J., Wang, C., Yang, Z.: Social influence analysis in large-scale networks. In: Proceedings of the 15th ACM SIGKDD International Conference on Knowledge Discovery and Data Mining, pp. 807–816. ACM (2009)

16. Tang, Y., Shi, Y., Xiao, X.: Influence maximization in near-linear time: a martingale approach. In: Proceedings of the 2015 ACM SIGMOD International Conference on Management of Data, pp. 1539–1554. ACM (2015)

17. Valiant, L.G.: The complexity of enumeration and reliability problems. SIAM J. Comput. **8**, 410–421 (1979)

18. Wu, H., Cheng, J., Huang, S., Ke, Y., Lu, Y., Xu, Y.: Path problems in temporal graphs. Proc. VLDB Endow. **7**(9), 721–732 (2014)

19. Zhang, J., Wang, C., Wang, J., Yu, J.X.: Inferring continuous dynamic social influence and personal preference for temporal behavior prediction. Proc. VLDB Endow. **8**(3), 269–280 (2014)

20. Zhang, M., Dai, C., Ding, C., Chen, E.: Probabilistic solutions of influence propagation on social networks. In: Proceedings of the 22nd ACM International Conference on Information and Knowledge Management, pp. 429–438. ACM (2013)

Tracking Clustering Coefficient on Dynamic Graph via Incremental Random Walk

Qun Liao, Lei Sun, Yunpeng Yuan, and Yulu Yang[(✉)]

College of Computer and Control Engineering, Nankai University,
Tianjin, China
{liaoqun, sunleier, yuanyp}@mail.nankai.edu.cn,
yangyl@nankai.edu.cn

Abstract. Clustering coefficient is an important measure in complex graph analysis. Tracking clustering coefficient on dynamic graphs, such as Web, social networks and mobile networks, can help in spam detection, community mining and many other applications. However, it is expensive to compute clustering coefficient for real-world graphs, especially for large and evolving graphs. Aiming to track the clustering coefficient on dynamic graph efficiently, we propose an incremental algorithm. It estimates the average and global clustering coefficient via random walk and stores the random walk path. As the graph evolves, the proposed algorithm reconstructs the stored random walk path and updates the estimates incrementally. Theoretical analysis indicates that the proposed algorithm is practical and efficient. Extensive experiments on real-world graphs also demonstrate that the proposed algorithm performs as well as a state-of-art random walk based algorithm in accuracy and reduces the running time of tracking the clustering coefficient on evolving graphs significantly.

Keywords: Clustering coefficient · Graph mining · Incremental algorithm · Random walk

1 Introduction

In the last decades, clustering coefficient [1] has emerged as an important measure of the homophily and transitivity of a network. Tracking clustering coefficient for networks which are large and evolving rapidly, such as World Wide Web and online social networks, is an important issue for many real-world applications, such as detecting spammer for search engines [2] or online video networks [3] and detecting events in phone networks [4].

However, there are two challenges in tracking clustering coefficient. First, computing clustering coefficient for evolving graphs with millions or billions of edges is time-consuming. Especially, the efficient algorithms for static graphs [5, 6] are not sufficient for evolving graphs. Second, for most real-world networks, such as online social networks, the network topology and the size of network are evolving and can not be known beforehand. Moreover, it is always limited restrictedly to access the underlying network for the sake of performance or safety. Thus, the basic assumption of "independent random sampling" in most random sampling based algorithms [7, 8] is not realistic.

© Springer International Publishing AG 2017
A. Bouguettaya et al. (Eds.): WISE 2017, Part I, LNCS 10569, pp. 488–496, 2017.
DOI: 10.1007/978-3-319-68783-4_33

Aiming to track clustering coefficient on large and evolving networks, we propose an incremental algorithm. The proposed algorithm follows a random walk based sampling method to estimate clustering coefficient [9], only relying on the public interface and requiring no prior knowledge. The proposed algorithm reuses previous random walk and gets result updated via reconstructing partial random walk as the graph evolves, instead of recomputing from scratch. Both theoretical analysis and experimental evaluations on real-world graphs demonstrate that the proposed algorithm reduces the running time significantly comparing with a state-of-art random walk sampling based algorithm without sacrificing accuracy.

2 Tracking Clustering Coefficient

2.1 Clustering Coefficient

Let $G = (V, E)$ stand for a simple graph with n vertices and m undirected edges. The ith vertex is donated by v_i, $1 \leq i \leq n$. The edge between v_i and v_j is donated by e_{ij}, or e_{ji}. The degree of vertex v_i is donated by d_i. The adjacency matrix of G, donated by A, is a $n \times n$ symmetric matrix, $A_{i,j} = A_{j,i} = 1$ if and only if there is an edge between v_i and v_j, and $A_{i,j} = A_{j,i} = 0$ otherwise. We assume that there are no edges from any arbitrary vertex v_i pointing to itself, thus $A_{i,i} = 0$, $1 \leq i \leq n$.

We define a wedge as a triplet (v_j, v_i, v_k) for any i, j and k $\epsilon[1, n]$, if and only if $A_{i,j} = A_{i,k} = 1$, and $j < k$. For any wedge (v_j, v_i, v_k) with $A_{j,k} = 1$, we define it a triangle. Let l_i stand for the number of triangles with v_i lying in the middle. It is obvious that l_i is equal to the number of connected edges between v_i's neighbors.

The local clustering coefficient for any node v_i, donated by c_i, is the ratio of the number of edges between v_i's neighbors to the maximal possible number of such edges. For node v_i where $d_i \geq 2$, we define c_i as $2l_i/d_i(d_i - 1)$. For node v_i with less than 2 neighbors, c_i is defined as 0.

In this paper, we focus on two popular versions of clustering coefficient: the average clustering coefficient [10] and the global clustering coefficient [10]. The average clustering coefficient of a graph, donated by a, is the average of the local clustering coefficient over the set of nodes. It is defined as $\sum_{i=1}^{n} c_i/n$. The global clustering coefficient, also called transitivity in some previous works, is a metric measuring the probability that two neighbors of a node are connected to one another in global. In this paper, we donate it by g and define it as $2\sum_{i=1}^{n} l_i/ \sum_{i=1}^{n} d_i(d_i - 1)$.

2.2 Problem Definition

Let $G(t)$ stand for a snapshot of an evolving graph at time t, $0 \leq t$. $G(0)$ is the initial graph. For $t > 0$, $G(t)$ is a graph with an arbitrary edge inserted into (or removed from) $G(t - 1)$. The average and global clustering coefficient of $G(t)$ are donated by $a(t)$ and $g(t)$ respectively. Their estimates are donated by $\widehat{a}(t)$ and $\widehat{g}(t)$ respectively. Our goal is to compute $\widehat{a}(t)$ and $\widehat{g}(t)$ efficiently, $0 < t$.

It is supposed that $\widehat{a}(0)$ and $\widehat{g}(0)$ are computed via the algorithm in [9] (hereinafter referred to as baseline algorithm). It generates a random walk with r steps, donated by $R(0) = \{x_1(0), x_2(0),\ldots, x_r(0)\}$. A variable $\varphi_k(t)$ is defined as $A_{x_{k-1}(t),x_{k+1}(t)}$, $2 \leq k \leq r - 1, 0 \leq t$. There are another four variables defined as follows. $\Phi_a(0)$ is defined as $(r - 2)^{-1}\sum_{k=2}^{r-1} \phi_k(0)\left(d_{x_k(0)} - 1\right)^{-1}$, $\Psi_a(0)$ is defined as $r^{-1} \sum_{k=1}^{r} \left(d_{x_k(0)}\right)^{-1}$, $\Phi_g(0)$ is defined as $(r - 2)^{-1}\sum_{k=2}^{r-1} \phi_k(0)d_{x_k(0)}$ and $\Psi_g(0)$ is defined as $r^{-1} \sum_{k=1}^{r} \left(d_{x_k(0)} - 1\right)$. The approximate average and global clustering coefficient for $G(0)$, are defined as $\Phi_a(0)/\Psi_a(0)$ and $\Phi_g(0)/\Psi_g(0)$ respectively. It is assumed that for any t $(0 < t)$, $R(t-1)$, $\Phi_a(t-1)$, $\Psi_a(t-1)$, $\Phi_g(t-1)$ and $\Psi_g(t-1)$ are stored.

2.3   Tracking Clustering Coefficient Incrementally

Without loss of generality, we present how the proposed algorithm works at time t $(0 < t)$. It is supposed that e_{uv} is added at time t. The proposed algorithm checks whether v_u and v_v accessed by the stored path $R(t - 1)$. If neither v_u nor v_v is accessed, it returns the previous results $\widehat{a}(t - 1)$ and $\widehat{g}(t - 1)$ without any updating. Otherwise, the proposed algorithm finds set of steps in $R(t - 1)$ where either v_u or v_v accessed, donated by $X(t - 1)$. $X(t - 1) = \{x_k(t - 1) \mid x_k(t - 1) = v_u \| x_k(t - 1) = v_v\}$. For each entry $x_k(t - 1)$ in $X(t - 1)$, it takes a chance to reroute the random walk and update the estimates. The probability of rerouting the random walk is $1/d_{x_k(t-1)}$.

As rerouting the random walk $R(t-1)$ from the kth step, there are two phases, as shown in Fig. 1. Firstly, the proposed algorithm undoes the random walk from $x_k(t- 1)$, removing $r - k$ steps in $R(t - 1)$ from $x_k(t - 1)$. Secondly, the proposed algorithm regenerates random walk with $(r - k)$ steps starting form the $(k + 1)$th step and gets the estimates updated. As v_u (or v_v) is accessed at the kth step, the proposed algorithm picks v_v (or v_u) at the regenerated $(k + 1)$th step, then moves to one of its neighbors uniformly at random and repeats such random movement for $(r - k - 1)$ times.

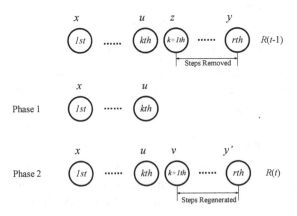

Fig. 1. Two phases in rerouting the random walk

We donate the set of removed steps by $R^- = \{x_h(t-1)\}$ and the added steps by $R^+ = \{x_h(t)\}$, where $k < h \leq r$. $R(t) = R(t-1) - R^- + R^+$. And the variables are updated as follows.

$$\Phi_a(t) = \Phi_a(t-1) - (r-2)^{-1}\sum\nolimits_{x_k \in R^-} \phi_k(t-1)(d_{x_k}-1)^{-1} + (r-2)^{-1}\sum\nolimits_{x_k \in R^+} \phi_k(t)(d_{x_k}-1)^{-1} \quad (1)$$

$$\Psi_a(t) = \Psi_a(t-1) - r^{-1}\sum\nolimits_{x_k \in R^-} (d_{x_k})^{-1} + r^{-1}\sum\nolimits_{x_k \in R^+} (d_{x_k})^{-1} \quad (2)$$

$$\Phi_g(t) = \Phi_g(t-1) - (r-2)^{-1}\sum\nolimits_{x_k \in R^-} \phi_k(t-1)d_{x_k} + (r-2)^{-1}\sum\nolimits_{x_k \in R^+} \phi_k(t)d_{x_k} \quad (3)$$

$$\Psi_g(t) = \Psi_g(t-1) - r^{-1}\sum\nolimits_{x_k \in R^-} (d_{x_k}-1) + r^{-1}\sum\nolimits_{x_k \in R^+} (d_{x_k}-1) \quad (4)$$

To make how the proposed algorithm works clear, we provide an example. As depicted in Fig. 2, $G(t) = G(t-1) + \{e_{25}\}$ and $R(t-1) = \{v_3, v_6, v_2, v_1\}$. The proposed algorithm reroutes $R(t-1)$ at the third step (where v_2 accessed) with $1/3$ probability. Supposing we reroutes $R(t-1)$, we remove $R^-=\{v_1\}$ and pick $R^+=\{v_5\}$ to take the place of the removed steps. Finally we get the random walk updated $R(t) = \{v_3, v_6, v_2, v_5\}$ and the estimates are updated according to the forementioned equations respectively.

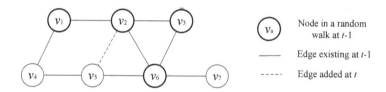

At t the random walk may be updated as $R(t) = (v_3, v_6, v_2, v_5)$.

Fig. 2. An example of a random walk on a graph with an edge added

There's a little difference in the case of edge removal. It is supposed that e_{uv} is removed at time t. The proposed algorithm updates the stored random walk if and only if e_{uv} is passed by the previous random walk. Approach of updating the random walk and estimates is analogous with the case of edge addition.

3   Correctness and Complexity

3.1   Correctness

Here we prove the proposed algorithm is correct via comparing the proposed algorithm with the baseline algorithm on a same graph. We demonstrate that the set of random walks generated by the proposed algorithm on graph $G(t)$ is equal to the set of random walks generated by the baseline algorithm.

Proof. First, we prove that for any random walk $R(t)$ generated by the baseline algorithm on $G(t)$, there is an equal random walk $R(t)$' generated by the proposed algorithm on the same graph. There are two cases. In the first case, we suppose that $R(t)$ doesn't access either v_u or v_v. The proposed algorithm inherits the random walk $R(t-1)$ generated by baseline algorithm on $G(t-1)$. And we could always find a random walk path $R(t)$' $= R(t-1)$ which is equal to $R(t)$. In the second case, we consider situations that $R(t)$ accesses either v_u or v_v. Let's suppose that $R(t)$ accesses v_u at its kth step. Similar as the first case, we could always find a path $R(t-1)$ who is equal to $R(t)$ at the first k steps. At the $(k+1)$th step, it takes $1/d_u$ probability to access each neighbor of v_u (including v_v) in $R(t)$. At the $(k+1)$th step of $R(t)$', it takes $1/d_u$ probability to access v_v (rerouting $R(t-1)$ from the $(k+1)$th step) and $(d_u-1)/d_u$ probability to access one of v_u's neighbors except v_v (inheriting from $R(t-1)$ without rerouting). Thus it is easy to find a random walk $R(t)$' which is equal to $R(t)$ from the $(k+1)$th step to the end of the random walk path. In summary, for any random walk generated by the baseline algorithm there's always an equal random walk generated by the proposed algorithm on the same graph.

In a similar way, it is easy to prove that for any random walk generated by the proposed algorithm, there's an equal random walk generated by the baseline algorithm on the same graph. Thus, the sets of random walk path generated by the proposed algorithm and the baseline algorithm respectively are equal. So the proposed algorithm is correct, due to it is demonstrated that the baseline algorithm is correct in [9, 11].

3.2   Computing Complexity

Here we analysis the complexity of updating estimate of clustering coefficient as graph evolves all the time. Assuming e_{uv} is the edge added at time t and M_t is the number of random walks rerouted. Then we have $E[M_t] \leq \alpha n/m$, where $\alpha n = r$, the length of a random walk path.

Proof. It is intuitive that most edges in the graph are not accessed by the random walk. A random walk needs to be updated only if it accesses v_u (or v_v) at time $t-1$ and chooses v_v (or v_u) as the next step. For an arbitrary vertex v_u, the expectation of the number of times v_u is accessed by a random walk is $\pi_u r$. The probability that node v_u is accessed at each step of a random walk, donated by π_u, is equal to d_u/D, which is proved in [9].The probability of choosing v_v as the next step of v_u is $1/d_u$. Thus, consider the expectation of M_t, which can be expressed as Eq. (5). As a random walk updated, the amount of work is $O(r)$ at most. In summary, we could get that the upper bound of expected amount of work that the proposed algorithm needs to do is $O((\alpha n)^2/m)$ as an arbitrary edge arrives randomly.

$$E[M_t] = r\sum_{i,j}\left(\pi_i/d_i + \pi_j/d_j\right)\Pr[i=u,j=v] \approx 2r\sum_i (d_i/d_iD)\Pr[i=u] = 2r/D = \alpha n/m \quad (5)$$

4   Evaluations

4.1   Experiment Setup

In experiments, we mainly compare the accuracy and performance of the proposed algorithm with the baseline algorithm [9], a state-of-art algorithm estimating the clustering coefficient via random walk. We implement both of the algorithms by Java. Our experiments all run on a machine with Intel(R) Core (TM) i7-2600 CPU and 8 GB RAM. The graphs used in the experiments are some public datasets downloaded from SNAP [12]. We deal with all of them as undirected graphs by ignoring the direction of directed edges. Table 1 lists the main characteristics of the graphs.

Table 1. Main parameters of data sets

Graph	Background	Nodes	Edges	a	g
amazon0601	Amazon product co-purchasing	403K	3.4M	0.4177	0.1656
as-Skitter	Internet topology	1.6M	11.1M	0.2581	0.0054
cit-Patents	Citation network among US Patents	3.7M	16.5M	0.0757	0.0671
com-DBLP	DBLP collaboration network	317K	1.0M	0.6324	0.3064
com-Youtube	Youtube online social network	1.1M	3.0M	0.0808	0.0062
higgs-twitter	Followers graph on Twitter	457K	14.9M	0.1887	0.0102
soc-Pokec	Pokec online social network	1.6M	30.6M	0.1094	0.0468
web-Google	Web graph from Google	875K	5.1M	0.5143	0.0552
wiki-Talk	Wikipedia talk network	2.4M	5.0M	0.0526	0.0033

For generating the evolving graphs, we randomly remove 100000 edges from each graph and use the rest part as the initial graph in our experiments. The initial estimates are computed by the baseline algorithm. Then we add the removed edges one by one and estimate the average and global clustering coefficient respectively via the proposed algorithm and its competitor. For all experiments, we set $r = 0.02n$ by default, which is enough for getting accurate approximations [9, 11]. Moreover, we run the experiments for 100 times on each graph and use the average results in our evaluations.

4.2   Accuracy

We use RMSE (Root Mean Square Error) to measure the accuracy, which is defined as Eq. (6), where c stands for the exact clustering coefficient and $\hat{c}$ stands for its estimate. We computed the exact clustering coefficient by a node-iteration based method [13]. Table 2 provides a comparison of the average RMSE.

$$RMSE = \sqrt{E\left[(\hat{c}/c - 1)^2\right]} \tag{6}$$

Table 2. Comparison of RMSE

Graph	Average clustering coefficient		Global clustering coefficient	
	Proposed	Baseline	Proposed	Baseline
amazon0601	0.0424	0.0424	0.0959	0.0912
as-Skitter	0.0477	0.0486	0.1928	0.1989
cit-Patents	0.0285	0.0297	0.0328	0.0313
com-DBLP	0.0385	0.0409	0.1373	0.1300
com-Youtube	0.0708	0.0808	0.1622	0.1649
higgs-twitter	0.0782	0.0790	0.1881	0.1920
soc-Pokec	0.0481	0.0512	0.0395	0.0390
web-Google	0.1025	0.1117	0.1125	0.5312
wiki-Talk	0.0399	0.0377	0.3526	0.3483

The results demonstrate that the proposed algorithm performs as well as the baseline algorithm in accuracy. The proposed algorithm achieves smaller RMSE in two-thirds of the experiments. RMSE of the proposed algorithm is about 12% smaller than the baseline algorithm in the best case and it is about 6% bigger than the baseline algorithm in the worst one. In summary, the proposed algorithm achieves close (or even smaller) RMSE comparing with the baseline algorithm for all experiments, which means the proposed algorithm performs as well as the baseline algorithm.

4.3   Performance

We measure the running time and the number of random walk rerouting of the proposed algorithm to evaluate the performance. We also define the speedup as the ratio of the running time of the baseline algorithm to the running time of the proposed algorithm. Table 3 provides a detailed comparison of the running time, the number of random walk rerouting and the speedup of the proposed algorithm on each graph.

Table 3. Comparison of the running time and number of times the random walk rerouted

Graph	Average clustering coefficient			Global clustering coefficient		
	Running time (s)	Number of rerouting	Speedup	Running time (s)	Number of rerouting	Speedup
amazon0601	1.97	28661	186.9	1.97	28865	187.9
as-Skitter	22.41	26250	356.2	23.43	26000	353.3
cit-Patents	22.80	37800	217.9	22.83	37456	216.4
com-DBLP	1.62	40080	209.6	1.74	40169	194.6
com-Youtube	12.52	43595	305.1	12.63	42915	180.6
higgs-twitter	6.20	6692	531.1	6.22	6825	524.1
soc-Pokec	6.41	13204	520.6	6.23	13144	549.1
web-Google	4.88	29991	163.9	5.02	30194	181.8
wiki-Talk	125.45	73340	129.5	125.88	73385	115.8

It is obvious that the proposed algorithm reduces the running time significantly for dealing with large and evolving graphs. The speedup of the proposed algorithm comparing with the baseline algorithm for dealing with graph with 100000 evolving edges is in the range of 115 to 549. We also find that estimating the average and global clustering coefficient on a graph via the proposed algorithm cost similar running time.

5 Conclusion

In this paper, we propose an incremental algorithm for tracking approximate clustering coefficient on dynamic graph via random walk method. As an arbitrary edge is added and/or removed, our algorithm replaces partial random walk path around the evolving part and updates the estimate based on previous result. The accuracy and performance improvement of the proposed algorithm are verified through analysis in theory and experiments on real-world graphs. It is demonstrated that the proposed algorithm improves the performance effectively comparing with the state-of-art algorithm based on random walk.

References

1. Watts, D.J., Strogatz, S.H.: Collective dynamics of 'small-world' networks. Nature 393 (6684), 440–442 (1998)
2. Shen, G., Gao, B., Liu, T.Y., Feng, G., Song, S., Li, H.: Detecting link spam using temporal information. In: 6th IEEE International Conference on Data Mining, pp. 1049–1053. IEEE Press, New York (2006)
3. Benevenuto, F., Rodrigues, T., Almeida, V., Almeida, J., Gonçalves, M.: Detecting spammers and content promoters in online video social networks. In: 32nd International ACM SIGIR Conference on Research and Development in Information Retrieval, pp. 620–627. ACM, New York (2009)
4. Akoglu, L., Dalvi, B.: Structure, tie persistence and event detection in large phone and SMS networks. In: 8th Workshop on Mining and Learning with Graphs, pp. 10–17. ACM, New York (2010)
5. Becchetti, L., Boldi, P., Castillo, C., Gionis, A.: Efficient algorithms for large-scale local triangle counting. ACM Trans. Knowl. Discov. Data (TKDD) 4(3), 13 (2010)
6. Park, H.M., Chung, C.W.: An efficient mapreduce algorithm for counting triangles in a very large graph. In 22nd ACM International Conference on Information & Knowledge Management, pp. 539–548. ACM, New York (2013)
7. Tsourakakis, C.E., Kang, U., Miller, G.L., Faloutsos, C.: DOULION: counting triangles in massive graphs with a coin. In: 15th ACM SIGKDD International Conference on Knowledge Discovery and Data mining, pp. 837–846. ACM, New York (2009)
8. Seshadhri, C., Pinar, A., Kolda, T.G.: Wedge sampling for computing clustering coefficients and triangle counts on large graphs. Stat. Anal. Data Min. ASA Data Sci. J. 7(4), 294–307 (2014)
9. Hardiman, S.J., Katzir, L.: Estimating clustering coefficients and size of social networks via random walk. In: 22nd International Conference on World Wide Web, pp. 539–550. ACM, New York (2013)

10. Costa, L.D.F., Rodrigues, F.A., Travieso, G., Villas Boas, P.R.: Characterization of complex networks: a survey of measurements. Adv. Phys. **56**(1), 167–242 (2007)
11. Katzir, L., Hardiman, S.J.: Estimating clustering coefficients and size of social networks via random walk. ACM Trans. Web (TWEB) **9**(4), 19 (2015)
12. Stanford large network dataset collection. http://snap.stanford.edu/data/index.html
13. Schank, T.: Algorithmic aspects of triangle-based network analysis. Ph.D. thesis, Universität Karlsruhe (TH) (2007)

Event Detection

Event Cube – A Conceptual Framework for Event Modeling and Analysis

Qing Li[1,2(✉)], Yun Ma[1], and Zhenguo Yang[2,3]

[1] Department of Computer Science, City University of Hong Kong,
Hong Kong SAR, China
itqli@cityu.edu.hk, yunma3-c@my.cityu.edu.hk
[2] Multimedia Software Engineering Research Centre,
City University of Hong Kong, Hong Kong, China
yzgcityu@gmail.com
[3] School of Computer Science and Technology,
Guangdong University of Technology, Guangzhou, China

Abstract. The publicly available data such as the massive and dynamically updated news and social media data streams (a.k.a. big data) covers the various aspects of social activities, personal views and expressions, which points to the importance of understanding and discovering the knowledge patterns underlying the big data, and the need of developing methodologies and techniques to discover real-world events from such big data, to manage and to analyze the discovered events in an effective and elegant way. In this paper we present an event cube (EC) model which is devised to support various queries and analysis tasks of events; such events include those discovered by techniques of untargeted event detection (UED) and targeted event detection (TED) from multi-sourced data. Specifically, based on the essential event elements of 5W1H (i.e., When, Where, Who, What, Why, and How), the EC model is developed to organize the discovered events from multiple dimensions, to operate on the events at various levels of granularity, so as to facilitate analyzing and mining hidden/inherent relationships among the events effectively. Case studies are provided to illustrate the usages and show the benefits of EC facilities in on-line analytical processing of events and their relationships.

Keywords: Event modeling · Event cube · Event relationship analysis · On-line analytical processing

1 Introduction

Web 2.0 and social media websites result in massive amounts of online data (a.k.a. big data) being generated and consumed by users. Such big data streams capture people's experiences, expressions, activities, etc., associated directly with many real-world events and public/personal views. This points to the importance of understanding and discovering the knowledge patterns underlying the big data, particularly to the discovery of real-world event patterns. Real-world events are the events happened or taking place in the real world, such as celebrations, parades, protests, gatherings,

© Springer International Publishing AG 2017
A. Bouguettaya et al. (Eds.): WISE 2017, Part I, LNCS 10569, pp. 499–515, 2017.
DOI: 10.1007/978-3-319-68783-4_34

disasters, sports events, etc., which have received or may incur much public attention, and are of public concerns.

Event detection has received much research attention, and the seminal work which is known as topic detection and tracking (TDT) can be traced back to the late nineties [1, 2], aiming to discover and track real-world events from online news media. In recent years, with the popularity of social media, there have been increasing works proposed to detect real-world events from Twitter-like social media [3–5] and Flickr-like social media [6–8]. Due to the large amounts of real-world events happening at any time anywhere, it is a great challenge for us to access and understand the massive real-world events discovered from the underlying big data. Furthermore, discovering and organizing events without a conceptual structure is only applicable to mini-scaled information space. Indeed, a mechanism able to organize the massive events with a conceptual structure will be very useful and highly desired in the era of big data.

In this paper, we propose an even cube (EC) model to organize and manage the massive real-world events with a conceptual structure, which can benefit the analysis and discovery of the knowledge patterns underlying the events. The proposed EC model is based on the primary event elements that are generally used for describing events, which is a natural fit to the perception of human users. In particular, the EC model is designed to organize the event-related data, and model the underlying relationships among the events. To this end, a number of operations are defined to facilitate the analysis of the detected events in multiple perspectives and granularity, and various event relation types are defined to enable the discovery of the underlying relationships among the events.

The paper is organized as follows. Section 2 reviews related work. Section 3 introduces the EC modelling framework, including the event cube model and associated operations. Section 4 presents the online analytical processing of event cubes. Section 5 shows a case study for the proposed event cube framework. Section 6 concludes this paper.

2 Related Work

Over the past two decades, there have been many existing works for event modelling and analysis, most of which focus on event detection, event relationship or evolution analysis. To the best of our knowledge, no work has integrated these components into a complete conceptual framework with formal data definition and data manipulation. In this section, we briefly review the existing works related to the various aspects of event modelling and analysis.

2.1 Event Detection

Social event detection (SED) has become a hot topic along with the increasing amounts of user-generated data on social media sites like Twitter and Flickr. Primarily, SED can be distinguished into targeted event detection (TED) and untargeted event detection (UED). TED exploits supervised classification models to identify specific events based on predefined features. For example, Reuter and Cimiano [9] classified social media data in an incremental manner. Takeshi et al. [3] detected earthquake from Twitter social media by exploiting SVM classifier, and devised temporal and spatial models to this end. However, none of these approaches has taken the multi-sourced nature into account. UED, without the labelling information, relies on unsupervised clustering models to discover possible events. Chen and Roy [6] exploited a wavelet transformation to analyze the temporal and locational distributions of the tags, and clustered them into real-world events. Petkos et al. [11] presented a graph-based multimodal clustering approach for UED tasks. Kaneko and Yanai [12] detected Twitter keyword bursts and extracted visual features from the associating images; image clustering was conducted to discover event clusters. However, these approaches either exploit only a few features or require high computational cost to fuse different modalities. We have earlier dealt with UED as a clustering task from the multimodal data on Flickr, which are seamlessly fused by methods from the perspectives of manifold learning and feature coding [13].

2.2   Event Relationship Analysis

Event relationship analysis has been mostly developed based on textual data. Mei et al. [14] modelled the subtopic and spatiotemporal theme patterns from weblogs, and investigated the evolutionary patterns of events by comparative analysis of theme life cycles and theme snapshots. Yang et al. [15] discovered event evolution relationships based the temporal relationship, event similarity, and document distributional proximity among event episodes. Nallapati et al. [16] represented the content dependencies between stories in a news topic in a hierarchical structure, called event threading. Focusing on finer granularity of news stories, Feng and Allan [17, 18] extended the event threading to passage threading and incident threading. Deng et al. [19] extracted atomic events from documents and explored event evolution patterns by identifying their co-reference and measuring their relationship. Cai et al. [20] and Huang et al. [10] explored temporal relationship, content dependence relationship and event reference relationship to analyze the inter-dependencies between component events of a complex target event.

2.3   Online Analytical Processing

Our proposed event cube model is inspired by the concept of data cube, or OLAP cube, in data warehousing and mining field [21]. To recap, a data cube is used to store a measure of interest along multiple dimensions. To facilitate data analysis, a set of operations are defined on the cube, most common ones are: slice and dice, drill down, roll up, and pivot. Here, slice and dice extract a subset of data with constraint on a single or multiple dimensions; roll up and drill down summarize or specialize the data along a dimension; pivot rotates the cube to show various view-points to users. For our proposed EC framework, things become much more complicated. Instead of simple transactional records, each cell in our EC stores an event, the dimensions of which include both structured and unstructured data. Furthermore, the EC model requires more operations to enable relationship analysis among the events, other than the traditional slice and dice, drill down, roll up, etc., as to be presented in the next section.

3 The Modelling Framework

As mentioned earlier, it is important to have a conceptual structure to organize and manage the massive amount of real-world events discovered from the big data. To this end, we present in this section the EC model and associated operations, upon which event relationship analysis can be conducted effectively.

3.1 Event Cube (EC)

From the viewpoint of journalism, a real-world event can be described from multiple dimensions, such as when and where it happens, who are involved, what it is about, how and why it happens, a.k.a. 5W1H. Our EC model is thus proposed to organize real-world events according to the conceptual structure formulated below:

$$E\text{-}Cube = \{E, R, H\} \tag{1}$$

where $E = \{e = (A, D)\}$ denotes the set of events, in which A represents the values of the primary event elements (i.e., A_1 for when, A_2 for where, A_3 for who, and A_4 for what), and the last two dimensions (i.e., how and why) will be discussed in the stage of event relation analysis; D represents the collection of data for the events; $R = \{R_{cd},$ $R_{cr}, R_{cs}, ...\}$ denotes the relationships among the events, e.g., content dependence relationship, content reference relationship, content similarity reference, etc.; $H = \{H_T, H_L, H_o, H_e\}$ is the concept hierarchy which defines a sequence of mappings from low-level concepts to high-level ones according to the dimensions or event knowledge. Specifically, dimension-level and event-level concept hierarchies are defined respectively. For A_1 (When) dimension, the hierarchy H_T follows the basic time units, e.g., a year consists of 12 months, etc. Therefore, H_T is built for the time units based on their conversions. For A_2 (Where) dimension, the hierarchy H_L is a tree, the root of which could be a continent (e.g., Asia, Antarctica, Europe, etc.,) and the child nodes are the countries, provinces, respectively. For A_3 (Who) dimension, the hierarchy H_O could be Minor (infant (0–1 years), toddler (2 years), child (3–13 years), teenager (14–18 years)), and Adult (adults (18–45 years), middle adulthood (45–65 years), old age (65-), etc.). Note that for A4 (What) dimension, there may or may not necessarily exist any form of hierarchy, thus we do not assume A4 to inherently hold a hierarchy in general. Finally, H_e is an event-level hierarchy, where each node is an event, and a child node is a sub-event, which can be obtained by hierarchical clustering. The concept hierarchy is beneficial to the operations to be presented subsequently in Sect. 3.2.

As an illustration, we show a toy example in Fig. 1 to demonstrate how an event cube looks like from the perspective of tables in a database. As shown in box D of Fig. 1, three events, distinguished by the event labels, have been discovered from the data. After extracting each event's primary event elements, we store them in tables (i.e., box A of Fig. 1). For the three events, we can identify the relationships among them, such as content dependence relationship (as shown in box R of Fig. 1). The concept hierarchies, such as the time unit hierarchy (as shown in H of Fig. 1), will be exploited during conducting roll-up and other operations defined in Sect. 3.2. Therefore, the primary event elements, the data documents related to the events, and the relations

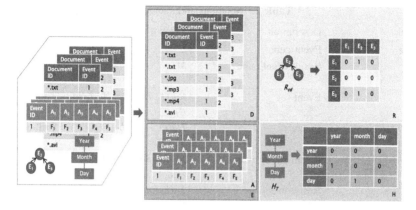

Fig. 1. Visual depiction of the event cube model

among the events, are all specified and captioned by the EC model; operations and relationship analyses are to be presented subsequently.

3.2 EC Operations

To facilitate analyzing events in multiple perspectives and granularity, intra-cube and inter-cube operations are defined, depending on whether the operations are conducted on a single event cube or cross multiple ones. In particular, the inter-cube operations are designed to enable exploring cross analysis based two or more event cubes of different types (e.g., a targeted event cube and an untargeted event cube). For clarity, Table 1 summarizes the notations and symbols used throughout this paper.

Intra-cube Operations: *Roll-up, Drill-down, Slice, Dice, Pivot, and X-validate.*

- **Roll-up** is done by traversing upwards through a concept hierarchy, where the concept hierarchy maps a set of low level concepts to higher level. *Roll-up* is formalized as $Ru(Q_e, \textbf{E-Cube}, H_k, L, M)\colon R_E$, where R_E is the resulting events returned by the operation, Q_e is the set of start events (or query events), *E-Cube* is the manipulated event cube, H_k is an element-level or event-level hierarchy from H, L is the targeted level in absolute or relative mode M. For instance, $Ru(Q_e, \textit{E-Cube}, H_T, 1, \textit{'rel'})\colon R_E$ rolls up the events by one-level on the When dimension (e.g., expanding the time units from month to year, etc.) using H_T.
- **Drill-down** acts in reverse compared with *Roll-up*, allowing to go to more details by traversing down the hierarchy, so as to provide more specific and detailed views of an event cube. Similar with *Roll-up*, *Drill-down* is formalized as $Dd(Q_e, \textbf{E-Cube}, H_k, L, M)\colon R_E$.
- **Slice** selects events based on a specific dimension, formalized as $Sl(\textbf{E-Cube}, A_i\text{–}C_i)\colon R_{E\text{-}Cube}$, where $R_{E\text{-}Cube}$ is the resulting event cube returned by the operation, *E-Cube* is the manipulated event cube, and C_i specifies the condition A_i needs to satisfy. For instance, $Sl(\textit{E-Cube}, A_1 = [\text{May } 1, 2016; \text{June } 1, 2016]\colon R_{E\text{-}Cube}$ obtains a sub-cube of *E-Cube*, where the events happened between *May* 1, 2016 and *June* 1, 2016.

Table 1. Notations and symbols

Symbol	Description	Symbol	Description
E-Cube	Event cube	L	Level
$R_{E\text{-}Cube}$	Result event cube	M	Mode: 'rel' or 'abs'
E	Event set	$Ru(Q_e, E\text{-}Cube, H_k, L, M)$: R_E	Roll-up
e	Event	$Dd(Q_e, E\text{-}Cube, H_k, L, M)$: R_E	Drill-down
R_E	Result set of events	$Sl(E\text{-}Cube, A_i\text{-}C_i)$: $R_{E\text{-}Cube}$	Slice
R_e	Result event	$Di(E\text{-}Cube, \{A_i\text{-}C_i\})$: $R_{E\text{-}Cube}$	Dice
Q_e	Set of start/query events	$Pivot(E\text{-}Cube, A_i, A_j, A_k)$: $R_{E\text{-}Cube}$	Pivot
$A = \{A_1, A_2, A_3, A_4\}$	Event element values of {when, where, who, what}	$Xv(E\text{-}Cube, e, A_i, \{H_k\})$: R_e,	X-validate
$H = \{H_T, H_L, H_O, H_e\}$	Concept hierarchies of {when, where, who, what}	$Un(E\text{-}Cube1, E\text{-}Cube2, [\{A_i\}])$: $R_{E\text{-}Cube}$	Union
A_i	One particular element in A	$In(E\text{-}Cube1, E\text{-}Cube2, [\{A_i\}])$: $R_{E\text{-}Cube}$	Intersect
H_k	One particular element in H	$Sb(E\text{-}Cube1, E\text{-}Cube2, [\{A_i\}])$: $R_{E\text{-}Cube}$	Subtract
C_i	Condition on A_i	$Sc(e, E\text{-}Cube1, E\text{-}Cube2, \{A_i\text{-}C_i\})$: $R_{E\text{-}Cube}$	Scoping

- **Dice** selects events based on multiple dimensions, which is formalized as **Di (E-Cube, {A_i-C_i}): $R_{E\text{-}Cube}$**, which is similar with *Slice* but specifies conditions on multiple dimensions.
- **Pivot** permutes the dimensions and reshapes the event cube to show the corresponding view of interest, which is formulized as **Pivot(E-Cube, A_i, A_j, A_k): $R_{E\text{-}Cube}$**, where A_i, A_j, A_k are three main dimensions of interest for visualization.
- **X-validate** supports an event's validation within a cube through cross checking the multiple dimensions of related events, e.g., checking the consistency of the event elements and fills the missing values along the concept hierarchy. *X-validate* is formularized as **Xv(E-Cube, e, A_i, {H_k}): R_e**, where the A_i dimension of the target event e is to be validated based on the referred concept hierarchies {H_k}, resulting the validated event R_e. For instance, $Xv(E\text{-}Cube, e, A_1, \{H_e\})$: R_e, checks A_1 dimension (i.e., When) of event e based on H_e (i.e., event-level hierarchy). Specifically, the operation first checks whether e is consistent with the of its parent and child events along H_e on A_1 dimension. If the values of A_1 is missing, an estimation is made based on the A_1 dimensions of the parent and child events.

Inter-cube Operations: *Union, Intersect, Subtract,* and *Scoping.*

- *Union* reorganizes the events underlying either of two event cubes, which is formalized as $Un(E\text{-}Cube1, E\text{-}Cube2, [\{A_i\}])$: $R_{E\text{-}Cube}$, where $R_{E\text{-}Cube}$ is the resulting event cube by incorporating the events underlying *E-Cube1* or *E-Cube2*, and $[\{A_i\}]$ are optional to specify the dimensions considered for the events. The default of $[\{A_i\}]$ is all the dimensions of the events.
- *Intersect* reorganizes the events underlying both of two event cubes, formalized as $In(E\text{-}Cube1, E\text{-}Cube2, [\{A_i\}])$: $R_{E\text{-}Cube}$, which is syntactically similar to *Union*.
- *Subtract* keeps the difference set between two cubes, formalized as $Sb(E\text{-}Cube1, E\text{-}Cube2, [A_i])$: $R_{E\text{-}Cube}$, which is syntactically similar to *Union* and *Intersect*.
- *Scoping* retrieves the events from an event cube in the restricted scope defined for a query event in another event cube, which is formalized as $Sc(e, E\text{-}Cube1, E\text{-}Cube2, \{A_i\text{-}C_i\})$: $R_{E\text{-}Cube}$, where e is the query event in *E-Cube1*, and *E-Cube2* is the target event cube for retrieval. $\{A_i\text{-}C_i\}$ are the possible constraints on dimensions, and $R_{E\text{-}Cube}$ is the resulting event cube which is a sub-cube of *E-Cube2*. For instance, *Sc* ("2015 Tianjin explosion", E-Cube1, E-Cube2, {Time = August 2015, Location = "Tianjin"}): $R_{E\text{-}Cube}$ retrieves from *E-Cube2* the events related to "2015 Tianjin explosion" of *E-Cube1*, i.e. events in *E-Cube2* happened at the same time in Tianjin.

4   Online Analytical Processing of ECs

In this section, we define the online analytical processing (OLAP) operations on the EC model to support event relationship analysis (ERA), explaining the Why and How properties for events. We categorize these operations into intra-cube and inter-cube event relationship analysis operations.

4.1   Intra-cube Event Relation Analysis

Since each event is described by a set of dimensions, we first define atomic operations for ERA on single dimensions, based on which we derive more complex relationships on multiple dimensions subsequently.

Single dimension event relationship analysis is defined as the similarity of two events on a specific dimension with respect to a given similarity measure, as follows:

Definition 1. Given two events e_i and e_j, $i \neq j$, the degree of relationship between e_i and e_j on dimension k is defined as $R_k(e_i, e_j) = Sim_k\left(e_i^k, e_j^k\right)$, where $Sim_k(\cdot, \cdot)$ is a similarity measure on dimension k.

Therefore, for each dimension, a similarity measure is required to compute the relationship between two events. Considering more than one kind of relationship may exist, multiple similarity measures are allowed. In the following, we propose the desired similarity measures for the defined dimensions, i.e., when, where, who and what.

When: Following [15], the time similarity between two events, e_i and e_j, is defined using power function:

$$Sim_T\left(e_i^T, e_j^T\right) = \begin{cases} e^{-\mu\frac{\left(e_j^T - e_i^T\right)}{Z}}, & e_i^T \le e_j^T \\ -e^{\mu\frac{\left(e_i^T - e_j^T\right)}{Z}}, & e_i^T > e_j^T \end{cases} \tag{2}$$

where Z is for normalization, μ controls the sensitivity of the measurement and the temporal order is kept by the sign of the result.

Where: The spatial similarity between two events, e_i and e_j, is defined using the Haversine formula [22]:

$$Sim_L\left(e_i^L, e_j^L\right) = \left(2r \arcsin\left(\sqrt{sin^2\left(\frac{e_j^{L\_lat} - e_i^{L\_lat}}{2}\right) + \cos\left(e_i^{L\_lat}\right)\cos\left(e_j^{L\_lat}\right)\sqrt{sin^2\left(\frac{e_j^{L\_lon} - e_i^{L\_lon}}{2}\right)}}\right)\right)^{-1} \tag{3}$$

where $e_i^{L\_lat}$ and $e_i^{L\_lon}$ are e_i's latitude and longitude, and r is the radius of the earth.

Who: The similarity between two events, e_i and e_j, on the who dimension (which is a set of entities) is measured by the Jaccard's coefficient [23]:

$$Sim_O\left(e_i^O, e_j^O\right) = \frac{\left|e_i^O \cap e_j^O\right|}{\left|e_i^O \cup e_j^O\right|} \tag{4}$$

What: Since the *What* describes specific content for an event, we consider three subunits for this information, i.e., textual information (TI), visual information (VI), and event category (C). Similarity measures are defined on these three subunits.

(i) ***Textual information***: *TI* further contains three levels of text describing the event, namely, keywords (*TI_k*), summary (*TI_s*), and source documents (*TI_d*). To extract different kinds of relationships, we define several similarity measures on *TI*.

KeywordsOverlap (KO): Keywords are the core features of an event, which are representative for the event contents and distinguishable from other events. KeywordsOverlap can be evaluated by the Jaccard's coefficient.

$$Sim_{TI\_KO}\left(e_i^{TI\_k}, e_j^{TI\_k}\right) = \frac{\left|e_i^{TI\_k} \cap e_j^{TI\_k}\right|}{\left|e_i^{TI\_k} \cup e_j^{TI\_k}\right|} \tag{5}$$

KeywordsCooccurrence (KC): Two events may have dependent but not identical keywords, inferring a content dependence relationship. Following [20], *KC* is defined as the averaged mutual information between each pair of words in the keywords set:

$$Sim_{TI\_KC}\left(e_i^{TI-k}, e_j^{TI-k}\right) = \frac{\sum_{f_i \in e_i^{TI-k}} \sum_{f_j \in e_j^{TI-k}} I(f_i, f_j)}{\left|e_i^{TI-k}\right| \left|e_j^{TI-k}\right|} \tag{6}$$

where $I(f_i, f_j)$ is the mutual information for a word pair (f_i, f_j).

KeywordReference (KR): Since some related events may have no similar or dependent keywords, for example, the events with causality relationship, following [20], we define the *KR* from event e_i to e_j based on e_i's keywords and e_j's source documents:

$$Sim_{TI\_KR}\left(e_i^{TI-k}, e_j^{TI-d}\right) = \frac{\sum_{m=0}^{\left|e_j^{TI-d}\right|} N_{jm}^i}{\left|e_i^{TI-k}\right|} \times \frac{1}{\left|e_j^{TI-d}\right|} \tag{7}$$

where N_{jm}^i is the number of keywords of e_i existing in the m-th document in e_j.

SummaryProximity (SP): A summary is a longer description for events than keywords. *SP* is defined as the cosine similarity between their fixed-length representations:

$$Sim_{TI\_SP}\left(e_i^{TI-s}, e_j^{TI-s}\right) = \frac{v\left(e_i^{TI-'}\right) \cdot v\left(e_j^{TI-'}\right)}{\left\|v\left(e_i^{TI-s}\right)\right\| \left\|v\left(e_j^{TI-s}\right)\right\|} \tag{8}$$

where the fixed-length vector $v(e_i^{TI-s})$ can be a TF-IDF feature vector or representations modeled by Doc2Vec [24], LSTM [25] or other more complex models.

(ii) **Visual information (VI):** the visual information contains a set of images describing the event. *VI* is defined as the averaged cosine similarities between each pair of images of the events:

$$Sim_{VI}\left(e_i^{VI}, e_j^{VI}\right) = \frac{1}{\left|e_i^{VI}\right|\left|e_j^{VI}\right|} \sum_{m=1}^{\left|e_i^{VI}\right|} \sum_{n=1}^{\left|e_j^{VI}\right|} \frac{v\left(e_{im}^{VI}\right) \cdot v\left(e_{jn}^{VI}\right)}{\left\|v\left(e_{im}^{VI}\right)\right\| \left\|v\left(e_{jn}^{VI}\right)\right\|} \tag{9}$$

where the representation for each image can be extracted by the widely used convolutional networks (such as GoogleNet [26] and VGG [27]).

(iii) **Category:** the category information labels an event in a more abstract level (which can be obtained by event ontology models), and its domain is in the form of a tree structure. The category proximity is defined as:

$$Sim_C\left(e_i^C, e_j^C\right) = \frac{1}{d(e_i^C) + d\left(e_j^C\right) - 2d\left(LCA\left(e_i^C, e_j^C\right)\right) + 1} \qquad (10)$$

where $d(\cdot)$ denotes the distance of current node from the root, and $LCA(\cdot, \cdot)$ denotes the lowest common ancestor of two nodes in the tree.

Multi-dimension event relationship analysis evaluates the relationships by considering a set of relevant dimensions, which can be defined on the basis of the single dimension relationship analysis operations. Formally,

Definition 2. For a multi-dimensional event relationship type r, given two events e_i and e_j, the degree of r from e_i to e_j, denoted as $C_r(e_i, e_j)$, is defined as:

$$C_r(e_i, e_j) = \begin{cases} \sum_{m \in M_r} \alpha_m Sim_m\left(e_i^m, e_j^m\right), & \text{if for } \forall u \in U_r, Sim_u\left(e_i^u, e_j^u\right) > \lambda_u \\ 0, & otherwise \end{cases} \qquad (11)$$

where U_r is the set of dimensions used as conditions for the relationship type r, λ_u is the similarity threshold for dimension u, M_r is the set of dimensions participating in the final relation degree computation, and α_m is the weight for dimension $m(\in M_r)$. For two events not satisfying the conditions, a zero value will be assigned.

Based on this generalized formula, various event relationships can be defined. Since the component events, or sub-events, in a complex event are rather important, we define the following relationships (we omit assigning values for λ and α for simplicity):

- *Content similarity relationship*: sub-events in a complex event may have similar contents. In this case, U_r and M_r in the above formula are defined as:

$$U_r = \{When, Where, TI\_SummaryProximity\}$$
$$M_r = \{TI\_KeywordOverlap, TI\_SummaryProximity, Category\}$$

- *Content dependence relationship*: sub-events in a complex event may have dependent contents. In this case, U_r and M_r are defined as:

$$U_r = \{When, Where, TI\_SummaryProximity\}$$
$$M_r = \{TI\_KeywordCooccurrence, TI\_SummaryProximity, Category\}$$

- *Content reference relationship*: in a complex event, a sub-event e_i may reference another sub-event e_j. This kind of relationship is likely to explain the causality, or the Why element, for an event. In this case, U_r and M_r are defined as:

$$U_r = \{When, Where, TI\_SummaryProximity\}$$
$$M_r = \{TI\_KeywordReference\}$$

Note that the above definition of multi-dimension event relationship is not fixed but adaptive to the user requirement. A user may derive a new relationship as long as the required parameters, i.e., U_r, M_r, λ and α, are indicated.

4.2 Inter-cube Event Relation Analysis

For events in different cubes, the intra-cube relationship analyzing operations (ref. Sect. 4.1) can still be applied for cubes sharing the required dimensions. In this section, we mainly focus on discovering relationships from two heterogeneous cubes, which share few features. Borrowing the term and technique from data mining, we coin these relations as association relationships and extract them using association rule learning [28]; the following workflow is proposed for inter-cube event relation analysis.

- Step 1. 'Transaction' dataset construction. Each record is a combination of selected dimensions s_a from event e_i in cube a, and selected dimensions s_b of each event e_j in cube b retrieved by a Scoping operation defined on e_i with constraint on dimensions c_a, denoted by $\{e_i^u | \forall u \in s_a\} \cup \{e_j^u | \forall u \in s_b, \forall e_j \in Sc(e_i, a, b, c_a)\}$. For dimensions whose values are a set, we may further break it.
- Step 2. Conduct association analysis with thresholds for support and confidence to extract rules existing in the 'Transaction' dataset. Only rules with dimensions in the two sides coming from different event cubes are considered.
- Step 3. Given any event e from event cube a, retrieve the rules with all items in either side existing in event e. Using the items in the other side to retrieve related events in event cube b of e from the Scoping result $Sc(e, a, b, c_a)$.

Exploiting the association rule learning method enables us to mine the relationships dependent on different dimensions for events in different cubes. In particular, for a targeted event cube and an untargeted event cube, which may have limited common dimensions and much distinction in contents, such analytical processing can help extract some underlying important but not obvious associations. For example, through learning the rules for stock price movement events and their contextual social events, we may find the triggering event for a specific stock market fell/rise.

5 Case Study

To better understand our event cube framework, we show several usage cases in social events for some of the defined operations. In particular, we illustrate the *Roll-up/ Drill-down* operations and event relationship analysis for intra-cube operation, and the *Scoping* and event association analysis for inter-cube operation. In this paper, we focus on conceptualizing the proposed EC model and leave the proto-type system presentation and experimental evaluation for future work.

5.1 Case Study for Intra-cube Operations

Roll-up. As defined in Sect. 3.2, *Roll-up* summarizes events along a dimension, a dimension-level concept hierarchy, or an event-level concept hierarchy. Since *Roll-up*

Table 2. A set of events with corresponding initial and after-rollup category sub-dimension.

Id	Event	Category	Category after roll-up
1	Sichuan Earthquake in 2008	Earthquake	Disasters
2	2016 China floods	Flood	Disasters
3	United States presidential election, 2016	Presidential election	Politics
4	The United Kingdom European Union membership referendum	Referendum	Politics
5	2016 Summer Olympics in Rio, Brazil	Olympics	Sports

along a dimension has no difference than a set union, we omit this scenario here. The *Roll-up* along a dimension-level concept hierarchy is demonstrated in Table 2, which shows five events with their initial category sub-dimension and the category sub-dimension after *Roll-up* (the category is a subunit for the What dimension). Both the "earthquake" and "flood" reach the "disasters" category, while both "presidential election" and "referendum" reach the "politics" category, and "Olympics" reaches the "sports" category. Moreover, a user may define either a relative or an absolute level for *Roll-up*. *Roll-up* along an event-level concept hierarchy abstracts events to a larger granularity, i.e., from a child event to a parent event. Figure 2 (Left) shows a hierarchical evolution graph of component events in three levels for "Sichuan Earthquake in 2008". The event hierarchy, is available in H_e, which is constructed during event detection. The 1st-level event is the complex event "Sichuan Earthquake in 2008", while the 2nd-level events, such as "Earthquake struck" and "Casualties" focus on several aspects of the earthquake, and the 3rd-level events, such as "Rescuers reach epicentre", are events in finer granularities. *Roll-up* each component event can easily reach to the higher-level events in its path to the root. For example, the 3rd-level event "Girl pulled from rubble after 50 h" can reach the 2nd-level event "Earthquake relief" after *Roll-up* by one level, and reach the root event "Sichuan Earthquake in 2008" after *Roll-up* by two levels.

Drill Down. The *Drill-down* operates on events in the opposite way to *Roll-up* with only one exception, i.e. *Drill-down* along the event-level concept hierarchy. Since a complex event can include a set of component events, *Drill-down* to a finer granularity requires more structures than returning just a set of components. Specifically, a hierarchical evolution graph should be presented based on event relationship analysis. For example, drill down the 1st-level event "Sichuan Earthquake in 2008" by two levels can arrive at the whole hierarchical evolutional picture with events in three levels together with their relationships as shown in Fig. 2 (Left).

Event Relationship Analysis. For event relationship analysis, we focus on the relationships among a complex event's component events w.r.t. time. As shown in Fig. 2 (Left), we arrive at an evolutional graph for the component events of the complex event "Sichuan Earthquake in 2008" by considering the defined relationship analysis for content similarity, content dependence and content reference. The analysis is conducted on events with the same parent event. For the 2nd-level events, since "Casualties",

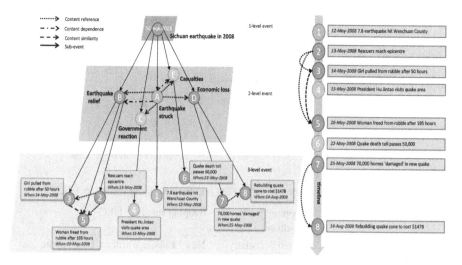

Fig. 2. Left: hierarchical evolution graph of the component events of "Sichuan Earthquake in 2008". Right: temporal evolution graph of the 3rd-level component events of "Sichuan Earthquake in 2008"

"Earthquake relief", "Government reaction", and "Economic loss" can all be regarded as consequences of "Earthquake struck", a content reference relationship exists between the "Earthquake struck" and each of these four events. Meanwhile, both "Casualties" and "Earthquake relief" have content dependence on "Earthquake struck" due to the frequent keyword co-occurrence. For the 3rd-level events, take the event 2, event 3, and event 5, with the same parent event "Earthquake relief" as an example. Event 3 "Girl pulled from rubble after 50 h" and event 5 "Woman freed from rubble after 195 h", both talking about specific cases in rescue, have high degree of content similarity relationship; in addition, both these two events have a content reference relationship with event 2 "The rescuers reach epicenter". Note that although in many cases, content-similarity, content-dependence, and content reference may be bi-directional, we choose to draw the edge along the timeline to organize these sub-events and their relationships from the evolutional perspective. Moreover, we can provide a temporal evolutional graph for a specific level of sub-events as an illustration of the How property for a complex event. Figure 2 (Right) shows the temporal evolutional graph for the 3rd-level events for "Sichuan Earthquake in 2008". The events are arranged in timeline with their relationships annotated, and events in the same color means that they have the same parent event. As a result, users can have a clear conception how a complex event evolves and how the sub-events relate with each other.

5.2 Case Study for Inter-cube Operations

Scoping. Given a query event in an event cube, *Scoping* defines a restricted area parameterized by a set of dimensions and retrieves the events in another event cube within this area. Suppose we have a TEC for stock market events and an UEC with

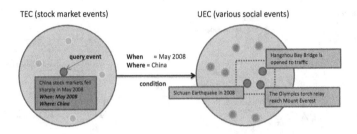

Fig. 3. An illustrative example for the inter-cube operation Scoping

Table 3. A set of query events in TEC and their Scoping results in UEC

Query event in TEC	Scoping condition	Scoping result in UEC
China stock markets fell sharply in January 2008 (**Label:** stock fell sharply)	When = January 2008 Where = China	New "Labour Contract Law" come into effect (**Category: law**)
		2008 Chinese winter storms (**Category: natural disasters**)
		"water cube" is delivered (**Category: social engineering**)
China stock markets fell sharply in May 2008 (**Label:** stock fell sharply)	When = May 2008 Where = China	Hangzhou Bay Bridge is opened to traffic (**Category: social engineering**)
		The Olympics torch relay reach Mount Everest (**Category: Olympics**)
		Sichuan Earthquake in 2008 (**Category: natural disasters**)
Japanese stock markets fell sharply in March 2011 (**Label:** stock fell sharply)	When = March 2011 Where = Japan	Seiji Maehara resigns as Foreign Minister due to a political donation scandal (**Category: political scandal**)
		2011 Tōhoku earthquake and tsunami (**Category: natural disasters**)

various social events. As shown in Fig. 3, for a query event "China stock markets fell sharply in May 2008", *Scoping* defines the restricted area with (user-specified) dimensions {When = May 2008, Where = China}. As a result, a set of events are retrieved, i.e. "The Olympics torch relay reach Mount Everest", "Sichuan Earthquake in 2008", and "Hangzhou Bay Bridge is opened to traffic".

Event Association Analysis. As defined in Sect. 4.2, event association analysis targets at the relationships difficult to be discovered by the similarity measures as for the intra-cube case. A typical example is the relationship mining between an event in a TEC and an event in an UEC. Similar with the case in *Scoping*, suppose we have a TEC for stock market events and an UEC for various social events. Table 3 shows a set of query events from the TEC, and the *Scoping* results in UEC with specified condition.

Table 4. Transactions dataset constructed for the query events in Table 3

Transactions
{stock fell sharply, law, natural disasters, social engineering}
{stock fell sharply, social engineering, Olympics, natural disasters}
{stock fell sharply, political scandal, natural disasters}

To extract the association relationship between events, we construct the "transaction" dataset following the steps in Sect. 4.2. As shown in Table 4, based on the labels (state of stock price movements) in TEC and the Category dimension in UEC, three records are generated. The association rules can then be applied on this dataset to mine the latent relationships between dimensions in TEC and UEC. As of result, we may get the rule "natural disasters -> stock fell sharply", which infers the association relationship between events with label "stock fell sharply" and natural disasters events.

6   Conclusion

In this paper, we have presented an even cube (EC) model which is devised to organize and manage the massive real-world events with a conceptual structure, with an aim to benefit the analysis and discovery of the knowledge patterns (viz., intra- and inter-event relationships) underlying the events. The proposed EC model is based on the primary event elements generally used for describing events, thus it is a natural fit to the perception of human users. As a part of this model, a number of operations are defined to facilitate the analysis of the detected events in multiple perspectives and granularity, and various event relation types are defined to facilitate analyzing and mining hidden/inherent relationships among the events effectively. Case studies have been provided to illustrate the usages and show the benefits of EC facilities in on-line analytical processing of events and their relationships. In our further research, we plan to incorporate the EC model into a domain specific application platform, with semi-supervised and unsupervised event discovery facilities and an ontology-based event classification module, so as to make the platform fully-fledged and functionally complete.

Acknowledgement. The authors are thankful to the useful comments and suggestions made by our partners Prof. Lei Chen (HKUST), Prof. Ho-fung Leung (CUHK), Dr. Hong-Va Leong (PolyU), and members of our research group. This work has been supported by a Strategic Research Grant from City University of Hong Kong (project no. 7004420) and a General Research Fund by the Hong Kong Research Grant Council (project no. CityU 11211417).

References

1. Allan, J., Papka, R., Lavrenko, V.: On-line new event detection and tracking. In: Proceedings of the 21st Annual International ACM SIGIR Conference on Research and Development in Information Retrieval, pp. 37–45. ACM (1998)

2. Yang, Y., Pierce, T., Carbonell, J.: A study of retrospective and on-line event detection. In: Proceedings of the 21st Annual International ACM SIGIR Conference on Research and Development in Information Retrieval, pp. 28–36. ACM (1998)

3. Sakaki, T., Okazaki, M., Matsuo, Y.: Earthquake shakes Twitter users: real-time event detection by social sensors. In: Proceedings of the 19th International Conference on World Wide Web, pp. 851–860. ACM (2010)

4. Zhou, X., Chen, L.: Event detection over twitter social media streams. VLDB J. **23**(3), 381–400 (2014)

5. Xie, W., Zhu, F., Jiang, J., Lim, E.P., Wang, K.: TopicSketch: real-time bursty topic detection from Twitter. IEEE Trans. Knowl. Data Eng. **28**(8), 2216–2229 (2016)

6. Chen, L., Roy, A.: Event detection from Flickr data through wavelet-based spatial analysis. In: Proceedings of the 18th ACM Conference on Information and Knowledge Management, pp. 523–532. ACM, November 2009

7. Reuter, T., Papadopoulos, S., Petkos, G., Mezaris, V., Kompatsiaris, Y., Cimiano, P., Geva, S.: Social event detection at MediaEval 2013: challenges, datasets, and evaluation. In: Proceedings of the MediaEval 2013 Multimedia Benchmark Workshop, Barcelona, Spain, 18–19 October 2013 (2013)

8. Yang, Z., Li, Q., Lu, Z., Ma, Y., Gong, Z., Liu, W.: Dual structure constrained multimodal feature coding for social event detection from Flickr data. ACM Trans. Internet Technol. (TOIT) **17**(2), 19 (2017)

9. Reuter, T., Cimiano, P.: Event-based classification of social media streams. In: Proceedings of ICMR. Article No. 22 (2012)

10. Huang, D., Hu S., Cai Y., Min, H.: Discovering event evolution graphs based on news articles relationships. In: Proceedings of ICEBE, pp. 246–251 (2014)

11. Petkos, G., Papadopoulos, S., Schinas, E., Kompatsiaris, Y.: Graph-based multimodal clustering for social event detection in large collections of images. In: Gurrin, C., Hopfgartner, F., Hurst, W., Johansen, H., Lee, H., O'Connor, N. (eds.) MMM 2014. LNCS, vol. 8325, pp. 146–158. Springer, Cham (2014). doi:10.1007/978-3-319-04114-8_13

12. Kaneko, T., Yanai, K.: Event photo mining from twitter using keyword bursts and image clustering. Neurocomputing **172**, 143–158 (2016)

13. Yang, Z., Li, Q., Liu, W., Ma, Y.: Learning manifold representation from multimodal data for event detection in Flickr-like social media. In: Gao, H., Kim, J., Sakurai, Y. (eds.) DASFAA 2016. LNCS, vol. 9645, pp. 160–167. Springer, Cham (2016). doi:10.1007/978-3-319-32055-7_14

14. Mei, Q., Liu, C., Su, H., Zhai, C.: A probabilistic approach to spatiotemporal theme pattern mining on weblogs. In: Proceedings of the 15th International Conference on World Wide Web, pp. 533–542. ACM (2006)

15. Yang, C.C., Shi, X., Wei, C.P.: Discovering event evolution graphs from news corpora. IEEE Trans. Syst. Man Cybernetics-Part A: Syst. Hum. **39**(4), 850–863 (2009)

16. Nallapati, R., Feng, A., Peng, F., Allan, J.: Event threading within news topics. In: Proceedings of the Thirteenth ACM International Conference on Information and Knowledge Management, pp. 446–453. ACM (2004)

17. Feng, A., Allan, J.: Finding and linking incidents in news. In: Proceedings of the Sixteenth ACM Conference on Information and Knowledge Management, pp. 821–830. ACM (2007)

18. Feng, A., Allan, J.: Incident threading for news passages. In: Proceedings of the 18th ACM Conference on Information and Knowledge Management, pp. 1307–1316. ACM (2009)

19. Deng, L., Ding, Z., Xu, B., Zhou, B., Jia, Y., Zou, P.: Exploring event evolution patterns at the atomic level. In: 2011 International Conference on Cyber-Enabled Distributed Computing and Knowledge Discovery (CyberC), pp. 40–47. IEEE (2011)

20. Cai, Y., Li, Q., Xie, H., Wang, T., Min, H.: Event relationship analysis for temporal event search. In: Meng, W., Feng, L., Bressan, S., Winiwarter, W., Song, W. (eds.) DASFAA 2013. LNCS, vol. 7826, pp. 179–193. Springer, Heidelberg (2013). doi:10.1007/978-3-642-37450-0_13
21. Gray, J., Chaudhuri, S., Bosworth, A., Layman, A., Reichart, D., Venkatrao, M., Pirahesh, H.: Data cube: a relational aggregation operator generalizing group-by, cross-tab, and sub-totals. Data Min. Knowl. Disc. 1(1), 29–53 (1997)
22. Sinnott, R.W.: Virtues of the Haversine (1984)
23. Guha, S., Rastogi, R., Shim, K.: ROCK: a robust clustering algorithm for categorical attributes. Inf. Syst. 25(5), 345–366 (2000)
24. Le, Q.V., Mikolov, T.: Distributed representations of sentences and documents. In: ICML, vol. 14, pp. 1188–1196 (2014)
25. Hochreiter, S., Schmidhuber, J.: Long short-term memory. Neural Comput. 9(8), 1735–1780 (1997)
26. Szegedy, C., Liu, W., Jia, Y., Sermanet, P., Reed, S., Anguelov, D., Erhan, D., Vanhoucke, V., Rabinovich, A.: Going deeper with convolutions. In: Proceedings of the IEEE Conference on Computer Vision and Pattern Recognition, pp. 1–9 (2015)
27. Simonyan, K., Zisserman, A.: Very deep convolutional networks for large-scale image recognition. arXiv preprint arXiv:1409.1556 (2014)
28. Agrawal, R., Imieliński, T., Swami, A.: Mining association rules between sets of items in large databases. ACM SIGMOD Rec. 22(2), 207–216 (1993). ACM

Cross-Domain and Cross-Modality Transfer Learning for Multi-domain and Multi-modality Event Detection

Zhenguo Yang[1,3], Min Cheng[2], Qing Li[2,3(✉)], Yukun Li[1], Zehang Lin[1],
and Wenyin Liu[1(✉)]

[1] School of Computer Science and Technology,
Guangdong University of Technology, Guangzhou, China
yzgcityu@gmail.com, gdutkelvin@outlook.com, gdutlin@outlook.com,
liuwenyin@gmail.com
[2] Department of Computer Science, City University of Hong Kong,
Hong Kong, China
designer357@hotmail.com, itqli@cityu.edu.hk
[3] Multimedia-Software Engineering Research Center,
City University of Hong Kong, Hong Kong, China

Abstract. Online news media and social media are popular domains
for people to acquire real-world event knowledge. In this work, the problem of multi-domain and multi-modality event detection (MMED) is
elaborated. We wish to organize the multi-modality data from multiple domains based on real-world events. To this end, a cross-domain and
cross-modality transfer learning (CDM) model is proposed. The CDM
model aligns the data by exploiting a dictionary-based alignment strategy, and identifies the event labels of the data samples based on the
class-specific reconstruction residual. Extensive experiments conducted
on real-world data demonstrate the effectiveness of the proposed models.
In particular, a benchmark dataset, denoted as MMED100, is released,
which can hopefully be used to promote the research on this topic and
advance related applications.

Keywords: Social media analytics · Multimedia analysis · Transfer
learning · Event detection

1 Introduction

Internet platforms provide new ways and platforms for people to publish information, leading to amounts of application scenarios [1, 4, 9, 11–13]. Internet platforms (e.g., online news media, and social media, etc.) are challenging the traditional ways (e.g., TV, radio, and print, etc.) in which people consume news, which
allow people to read news quickly and conveniently. Real-world event detection
from Internet platforms has become an important problem and received much
research attention in recent years.

© Springer International Publishing AG 2017
A. Bouguettaya et al. (Eds.): WISE 2017, Part I, LNCS 10569, pp. 516–523, 2017.
DOI: 10.1007/978-3-319-68783-4_35

Different data domains (e.g., online news media and social media) provide various viewpoints about events. For instance, news articles from online news media domain are published by professional journalists, while social posts on social media domain are shared by non-professional users. To some extent, the cross-domain data is complementary, which may hold different viewpoints for the same events. Multi-domain and multi-modality event detection (MMED) aims to discover events (e.g., protests, parades, gatherings, natural disasters, incidents, etc.) underlying both domains, showing comprehensive perspectives about the events.

In this paper, we address the problem of MMED from the perspective of transfer learning to organize the multi-domain and multi-modality data based on real-world events they depict. Given the data collected from a domain (e.g., textual articles from news media), MMED identifies the event labels of the data from a different domain even in different data modalities (e.g., images from social media). To this end, modality-specific dictionaries is constructed [15], which will be exploited to align the cross-domain data. Furthermore, a cross-domain and cross-modality transfer learning (CDM) model is proposed, based on which the reconstruction residuals can be used to identify the event labels of the data samples. The main contributions are as follows:

- We elaborate the problem of multi-domain and multi-modality event detection (MMED) to organize the cross-domain data based on real-world events.
- We propose a cross-domain and cross-modality transfer learning (CDM) model to discover event patterns underlying the multi-domain data.
- We release a real-world dataset consisting of textual articles and visual images from online news media and social media for MMED.

The rest of the paper is organized as follows. In Sect. 2, related work is reviewed. In Sect. 3, the proposed CDM model is presented. In Sect. 4, extensive experiments are conducted and analyzed. Finally, Sect. 5 offers some concluding remarks.

2   Related Work

In this section, we investigate the research work on event detection and transfer learning, respectively.

2.1   Event Detection from Internet Platforms

The work on event detection can be classified into two categories according to the data domains, i.e., online news media, and social media. Event detection from online news media is derived from the topic detection and tracking (TDT) task [1], which aims to monitor a stream of news stories. Krstajic et al. [4] proposed an incremental time-series visualization technique to deal with time-based event data. In addition, quite a few researchers discover events from social media platforms. Sakaki et al. [6] detected earthquake events from Twitter streams. Timo et al. [9] published a list of event detection challenges on Flickr data. However, these approaches have not taken the advantage of multiple domains.

2.2   Transfer Learning

Transfer learning [5] aims to adapt different feature spaces, and exploit the labels in the source domain for classification tasks in the target domain where the labeled samples are inadequate. Xiao et al. [10] proposed a feature space independent semi-supervised kernel matching method for domain adaptation. Zhang et al. [19] proposed to learn the latent projection and sparse reconstruction coefficient for subspace transfer. However, these models address the in-modality data from different data sources, which cannot address the heterogeneity of the data modalities. Yan et al. [14] proposed to align the textual and visual data based on 346 semantic objects. However, the objects do not cover the basic event elements, such as When, Where, Who, How, Why, etc. It remains unknown whether 346 concepts are enough for identifying the massive amount of real-world event concepts.

3   Cross-Domain and Cross-Modality Transfer Learning

In this section, we introduce the dictionary-based data alignment firstly, followed by the introduction on the cross-domain and cross-modality transfer learning (CDM) model.

3.1   Dictionary-Based Alignment

Unlike the previous transfer learning models that have not considered the event-related factors, we propose to construct a dictionary consisting of event elements to align the multi-domain and multi-modality data. In particular, the bases in the dictionaries [16] can be initialized as the labelled data samples. Furthermore, we extract event elements and align the data based on similarity measurements.

(1) Event element extraction. For event detection tasks, it is natural to exploit the event element related factors, such as When, Where, and the semantic elements (e.g., Who, What, How, Why, etc.). In the context of MMED, we take the textual news articles from online news media and visual images from Flickr-like social media as two domains for illustration.

In terms of the news articles, the metadata includes reporting time, and title, which correspond to the event elements well. Particularly, we obtain location information of the documents by extracting the city names and their GPS information from the textual content. We exploit multiple feature including time, location, title, and textual content. In terms of the Flickr images, the metadata includes time-taken, GPS-tags, user, tags, title, textual description, and the visual content.

(2) Similarity measurement based alignment. Given the extracted event elements (i.e., feature modalities) for each new article and image, we adopt a similar manner [15] to deal with the heterogeneity by defining their respective similarity measurement. The measurement on time, location, user, tags, title is the same as in [15], based on Gaussian function, Haversine formula, binary indicator, Jaccard index, WordNet, respectively. For the news articles, we exploit the tf-idf model to measure their similarities. To measure the similarities between the images, we exploit cosine similarity on the fully connected layers in vector forms of the VGG net [8] and Places205-AlexNet [20], and the feature modalities are denoted as $VisualO$ (O) and $VisualP$ (P), respectively. For clarity, we exploit X^m to denote a unified feature modality for the textual news articles, and use Y^m to denote the one for the Flickr images.

3.2 Cross-Domain and Cross-Modality Transfer Learning (CDM) Model

In the context of MMED, there are several transfer scenarios given domains X and Y, and we devise their objective functions respectively.

(1) **Y→ Y** (in-domain and in-modality, cross social media users).

$$\min_{Z_{YY}^m} \frac{1}{2} \sum_{m=1}^{M_Y} \|Y^m - D_Y^m Z_{YY}^m\|_F^2 + \alpha \sum_{m=1}^{M_Y} \|P_Y^m Z_{YY}^m\|_F^2 \tag{1}$$

where the first regularization term aims to achieve the smallest reconstruction error, and the second one exploits the geometry structure underlying the dictionary. The symbols Y^m and D_Y^m are consistent with the previous. M_Y is the number of feature modalities in domain Y. P_Y^m is the distance matrix among the bases of dictionary D_Y^m on feature modality m. Z_{YY}^m is the coefficient of the data samples corresponding to the dictionary D_Y^m.

(2) **X→ X** (in-domain and in-modality, cross news media sites). The objective function can be obtained by replace "Y" with "X" of Eq. (1).

(3) **X→ Y** (cross domains and modalities).

$$\min_{Z_{XY}^m} \frac{1}{2} \sum_{m=1}^{M_C} \|Y^m - D_X^m Z_{XY}^m\|_F^2 + \alpha \sum_{m=1}^{M_C} \|P_X^m Z_{XY}^m\|_F^2 \tag{2}$$

where the symbols Y^m, D_X^m are consistent with the previous. M_C is the common feature modalities between the two domains, and Z_{XY}^m denotes the specific coefficient in the scenario of $X \rightarrow Y$. Note that there is not any label knowledge in the target domain Y being exploited in this scenario.

(4) **Y→ X** (cross domains and modalities). The objective function can be obtained by replace "Y" with "X" and "X" with "Y" of Eq. (2).

The objective functions of the CDM models are convex corresponding to Z_{YY}^m, Z_{XX}^m, Z_{XY}^m, Z_{YX}^m, respectively, which can be solved by using gradient decent.

3.3 CDM for Event Discovery

The event labels can be identified based on the smallest residual [16] corresponding to the event-class-specific dictionaries. For illustration, we take the scenario of $Y \rightarrow Y$ as an example. More specifically, we separate the dictionary bases according to their event labels, and obtain the event-class-specific dictionaries, $(D_Y^m)_{E_1} \in \mathbb{R}^{d \times n_E}$, $(D_Y^m)_{E_2} \in \mathbb{R}^{d \times n_E}$,..., $(D_Y^m)_{E_K} \in \mathbb{R}^{d \times n_E}$, where K is number of events, and n_E is the number of data samples being exploited for each event. For a given test sample $Y_{(t)}$ whose representation on feature modality m is denoted by $Y_{(t)}^m$, the event-class-specific coefficient corresponding to an event-class-specific dictionary can be denoted by $(Z_{YY}^m)_{E_i}$, which can be achieved as follows:

$$\min_{(Z_{YY}^m)_{(E_i)}} \frac{1}{2} \sum_{m=1}^{M_Y} \|Y^m - (D_Y^m)_{(E_i)}(Z_{YY}^m)_{(E_i)}\|_F^2 + \alpha \sum_{m=1}^{M_Y} \|(P_Y^m)_{(E_i)}(Z_{YY}^m)_{(E_i)}\|_F^2$$

(3)

Furthermore, we calculate the reconstruction residual $R_{(E_i)}(Y_{(t)})$ for $Y_{(t)}$ as follows:

$$R_{(E_i)}(Y_{(t)}) = \sum_{m=1}^{M_Y} \|Y_{(t)}^m - (\widetilde{Y_{(t)}^m})_{(E_i)}\|_F^2, \ where \ (\widetilde{Y_{(t)}^m})_{(E_i)} = (D_Y^m)_{(E_i)}(Z_{YY}^m)_{(E_i)}$$

(4)

where $(\widetilde{Y_{(t)}^m})_{(E_i)}$ is the event-class-specific reconstruction of the test sample based on the event-class-specific dictionary and coefficient.

Finally, we can identify the event labels according to the smallest reconstruction residual as follows:

$$E_{(t)} = \min_{E_i} R_{(E_i)}(Y_{(t)}), e \in \{1, ..., K\}$$

(5)

4 Experiments

In this section, we introduce the dataset collected for the MMED tasks, specify the experiments settings and performance metrics, and evaluate the performance of the approaches.

4.1 Dataset

The dataset consists of data samples from two domains, i.e., online news media and social media. The preliminary version consists of 100 real-world events, denoted as MMED100, which will be released subsequently (it is available upon request now). MMED100 includes 9,722 news articles collected from hundreds of data sources, such as Yahoo News, Google News, Huffington Post, CNN News, New York Times, NBC News, Fox News, Washington Post, The Guardian, etc., and 23,500 images shared by 269 Flickr social media users. In particular, we divide MMED100 into two subsets, i.e., a seed set and a test set. The seed set consists of 10 data samples from each event and each domain, resulting in 1,000 news articles and 1,000 Flickr images. The rest of the data samples are included in the test set.

4.2   Evaluation Metrics and Baselines

We exploit NMI and F1 metrics for evaluations, which range from 0 to 1, and a large value is preferred. The baselines include a number of recently proposed approaches on data representation learning, transfer learning, and event detection, such as MMF [18], SDR [2], LLRR [17], LSDT [19], DGNMF [15], etc. The approaches can be applied seamlessly on the feature vectors achieved after aligned by using the dictionaries. For the different transfer scenarios, the corresponding dictionaries will be exploited.

4.3   The Best Performance Achieved by the Approaches

The best performance of the proposed approach and the baselines are summarized in Table 1. From the table we can observe that the proposed CDM model outperform the state-of-the-art baselines on the MMED task in terms of both NMI and F1 metrics in the four scenarios. The experimental results demonstrate the effectiveness of the proposed CDM model for the MMED tasks.

Table 1. The best performance of the approaches

Scenario	$Y \rightarrow Y$		$X \rightarrow X$		$X \rightarrow Y$		$Y \rightarrow X$	
Metric	NMI	F1	NMI	F1	NMI	F1	NMI	F1
MMF	0.5139	0.2288	n.a	n.a	n.a	n.a	n.a	n.a.
SDR	0.5403	0.3811	0.3594	0.2918	0.3919	0.1892	0.2278	0.0824
LSDT	0.6757	0.4295	0.4420	0.2065	0.6960	0.4608	0.3964	0.1490
LLRR	0.7265	0.4521	0.3951	0.1165	0.7313	0.4658	0.4496	0.1631
DGNMF	0.7534	0.5617	0.7034	0.4600	0.6630	0.5305	0.4156	0.1986
CDM	**0.8394**	**0.6815**	**0.8176**	**0.6615**	**0.7772**	**0.5508**	**0.5214**	**0.2432**

4.4   Evaluations on the Transfer Scenarios of Cross-Domain and Cross-Modality

To simply the discussion, we only evaluate the performance on NMI in this and the following subsections. In the transfer scenario of cross-domain and cross-modality, domain X (textual articles from online news media) is used to recognize the events in domain Y (visual images from social media), and vice versa. The challenging problems are two-fold. Firstly, the data modalities (i.e., textual data and images) are heterogeneous. Secondly, there are not any label knowledge in the target domain being exploited. We evaluate the impact of n_E on CMD in the two scenarios in Fig. 1, from which we can observe that *Time* feature is the most effective one in the two transfer scenarios. The reason relies that *Time* feature can be aligned more accurately. *Location* is not discriminative due to its incompleteness, because more than 86% of them in domain Y are missing.

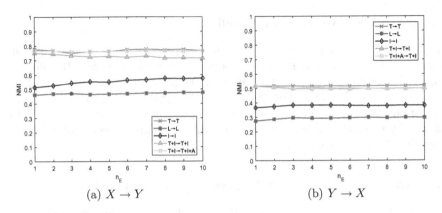

(a) $X \rightarrow Y$ (b) $Y \rightarrow X$

Fig. 1. Evaluating CDM in $X \rightarrow Y$ and $Y \rightarrow X$.

5   Conclusion

In this work, we elaborate the task of MMED, aiming to organize the multi-domain and multi-modality data according to real-world events. To this end, a CDM model is proposed from the perspective of transfer learning to make full use of the heterogeneous feature modalities possessed by the multi-modality data that are contributed by professional journalists and amateur social media users. Entensive experiments demonstrate the effectiveness of the proposed MMED models in different transfer scenarios. In particular, we release a real-world dataset, denoted as MMED100, which can hopefully be used to promote the research on the topic of event detection, especially from multiple data sources, and to inspire more real-world applications.

Acknowledgments. We would like to thank Mr. Shentao Yang and Ms. Yue Wu for the help on data collection, and our group members for the discussions. This work is partially supported by the National Natural Science Foundation of China (No.61703109), the Guangdong Innovative Research Team Program (No.2014ZT05G157), a General Research Fund by the Hong Kong Research Grant Council (project no. CityU 11211417), and the National Natural Science Foundation of China (No.61472337).

References

1. Allan, J., Papka, R., Lavrenko, V.: On-line new event detection and tracking. In: ACM SIGIR Conference on Research and Development in Information Accessed, pp. 37–45 (1998)
2. Jiang, X., Lai, J.: Sparse and dense hybrid representation via dictionary decomposition for face recognition. IEEE Trans. Pattern Anal. Mach. Intell. **37**(5), 1067–1079 (2015)
3. Koutra, D., Bennett, P.N., Horvitz, E.: Events and controversies: influences of a shocking news event on information seeking. In: 24th International Conference on World Wide Web, pp. 614–624 (2015)

4. Krstajic, M., Bertini, E., Keim, D.: Cloudlines: compact display of event episodes in multiple time-series. IEEE Trans. Vis. Comput. Graph. **17**(12), 2432–2439 (2011)
5. Pan, S.J., Yang, Q.: A survey on transfer learning. IEEE Trans. Knowl. Data Eng. **22**(10), 1345–1359 (2009)
6. Sakaki, T., Okazaki, M., Matsuo, Y.: Tweet analysis for real-time event detection and earthquake reporting system development. IEEE Trans. Knowl. Data Eng. **25**(4), 919–931 (2013)
7. Shao, M., Kit, D., Fu, Y.: Generalized transfer subspace learning through low-rank constraint. Int. J. Comput. Vis. **109**(1–2), 74–93 (2014)
8. Simonyan, K., Zisserman, A.: Very deep convolutional networks for large-scale image recognition. arXiv preprint (2014). arXiv:1409.1556
9. Reuter, T., Papadopoulos, S., Petkos, G., Mezaris, V., Kompatsiaris, Y., Cimiano, P., de Vries, C., Geva, S.: Social event detection at mediaeval 2013: challenges, datasets, and evaluation. In: MediaEval 2013 Multimedia Benchmark Workshop (2013)
10. Xiao, M., Guo, Y.: Feature space independent semi-supervised domain adaptation via kernel matching. IEEE Trans. Pattern Anal. Mach. Intell. **37**(1), 54–66 (2015)
11. Xie, H., Li, Q., Mao, X., Li, X., Cai, Y., Rao, Y.: Community-aware user profile enrichment in folksonomy. Neural Networks **58**, 111–121 (2014)
12. Xie, H., Li, X., Wang, T., Chen, L., Li, K., Wang, F.L., Min, H., Cai, Y., Li, Q.: Personalized search for social media via dominating verbal context. Neurocomputing **172**, 27–37 (2016)
13. Xie, H., Zou, D., Wang, F.L., Wong, T.L., Rao, Y., Wang, S.H.: Discover learning path for group users: a profile-based approach. Neurocomputing (2017)
14. Yan, Y., Yang, Y., Meng, D., Liu, G., Tong, W., Hauptmann, A.G., Sebe, N.: Event oriented dictionary learning for complex event detection. IEEE Trans. Image Process. **24**(6), 1867–1878 (2015)
15. Yang, Z., Li, Q., Liu, W., Ma, Y., Cheng, M.: Dual graph regularized NMF model for social event detection from Flickr data. World Wide Web **20**(5), 995–1015 (2017)
16. Yang, Z., Li, Q., Lu, Z., Ma, Y., Gong, Z., Liu, W.: Dual structure constrained multimodal feature coding for social event detection from Flickr data. ACM Trans. Internet Technol. **17**(2), 19 (2017)
17. Yin, M., Gao, J., Lin, Z.: Laplacian regularized low-rank representation and its applications. IEEE Trans. Pattern Anal. Mach. Intell. **38**(3), 504–517 (2016)
18. Zhang, Z., Zhao, K.: Low-rank matrix approximation with manifold regularization. IEEE Trans. Pattern Anal. Mach. Intell. **35**(7), 1717–1729 (2013)
19. Zhang, L., Zuo, W., Zhang, D.: LSDT: latent sparse domain transfer learning for visual adaptation. IEEE Trans. Image Process. **25**(3), 1177–1191 (2016)
20. Zhou, B., Lapedriza, A., Xiao, J., Torralba, A., Oliva, A.: Learning deep features for scene recognition using places database. In: Advances in Neural Information Processing Systems, pp. 487–495 (2014)

Determining Repairing Sequence
of Inconsistencies in Content-Related Data

Yuefeng Du[1,2], Derong Shen[1(✉)], Tiezheng Nie[1], Yue Kou[1],
and Ge Yu[1]

[1] School of Computer Science and Engineering, Northeastern University,
Shenyang 110004, Liaoning, China
dr.duyuefeng@gmail.com,
{shenderong,nietiezheng,kouyue,yuge}@ise.neu.edu.cn
[2] PLA Troops 65154, Lingyuan 122513, China

Abstract. Data consistency is one of the central issues of data quality management. Content-related conditional functional dependencies (CCFDs) are practical techniques for data consistency. CCFDs catch inconsistencies by putting content-related data together. Specially, repairing sequence plays a key role in consistency repairing. Some repairing sequences may bring unexpected results (e.g., incorrect repairs and results with extra repairing-cost). Hence, reasonable repairing sequences are advocated and readily supported by commercial system for better performance. To meet this need, this paper present a method of determining repairing sequence of inconsistencies in content-related data. (1) We present repairing sequence graph about CCFDs to select the inconsistencies which should be repaired preferentially. (2) We analyze the repairing mutex and discuss the interaction between repairing sequence and repairing mutex. (3) We proof that the problem of determining repairing sequence with minimum repairing-cost is NP-complete so that our method heuristically finds the appropriate repairing sequence. Our solution performs to be effective by empirical evaluation on three datasets.

Keywords: Data quality management · Content-related data · Repairing sequence · Consistency repairing

1 Introduction

Data consistency is one of the central issues about data quality management [1, 2]. Dirty data inflicts a daunting cost, *e.g.*, it costs US businesses over \$600 billion annually [3]. Errors in the data are typically detected as violations of constraints such as conditional functional dependencies (CFDs) [4, 5] $(X \to A, tp)$ which catch violations under specified conditions by using enforced patterns of semantically related constants. Further, Real-life data are content-related that can be utilized for data consistency analysis. Du *et al.* [6] presented content-related conditional functional dependencies (CCFDs) for inconsistencies detection by putting content-related data together.

Among techniques for solving inconsistencies, optimal repairing computation is well studied. It aims to repair an inconsistent instance by minimally modifying it *w.r.t.*

© Springer International Publishing AG 2017
A. Bouguettaya et al. (Eds.): WISE 2017, Part I, LNCS 10569, pp. 524–539, 2017.
DOI: 10.1007/978-3-319-68783-4_36

some cost measure, so as to get a new instance satisfying constraints. Existing algorithms [7] for optimal repairing computation mainly concern about how to select target-values for repairs. On the other hand, repairing sequences are regarded as another factor impact on the repairing quality. For example, an inappropriate repairing sequence may bring incorrect repairing results, extra repairing cost or repairing mutex. Hence, we contend that it is necessary to develop algorithms for repairing content-related data with reasonable repairing sequences.

Example 1. Figure 1 shows the relation schema R with records from the 1994 US Adult Census database that contains name (NM), Year, marital status (MS), gender (Gen), family relationship (FR), spouse's name (SN), work class (WC), and salary level (SL). The highlighted cells are error values violating the actual world and the values in brackets are truth values. The following CCFDs are defined over R.

t_{id}	NM	Year	MS	Gen	FR	SN	WC	SL
t_1	Joe	1971	married (single)	male	unmarried	N/A	lawyer	50K~70K
t_2	Joe	1972	married	male	husband	Ada	lawyer	50K~70K
t_3	Joe	1973	married	male	husband	Ada	lawyer	50K~70K
t_4	Joe	1974	divorced	male	unmarired	N/A	lawyer	50K~70K
t_5	Ada	1971	single	female	unmarried	N/A	employee	20K~30K
t_6	Ada	1971	single	female	unmarried	N/A	employee	30K~50K
t_7	Ada	1971	married (single)	female	wife (unmarried)	N/A	manager	50K~70K
t_8	Ada	1971	single	female	wife (unmarried)	N/A	manager	50K~70K
t_9	Ada	1972	married	female	wife	Joe	manager	50K~70K

repairing sequence rs_1 : ψ_1 $(t_{1,7}[MS]\rightarrow"single")$ $\rightarrow$ ψ_2 $(t_{7,8}[FR]\rightarrow"unmarried")$

rs_2 : ψ_3 $(t_8[MS]\rightarrow"married")$ $\rightarrow$ ψ_1 $(t_{5,6}[MS]\rightarrow"married")$ $\rightarrow$ ψ_2 $(t_1[FR]\rightarrow"husband",t_{5,6}[FR]\rightarrow"wife")$

rs_3 : ψ_2 $(t_1[FR]\rightarrow"husband")$ $\rightarrow$ ψ_1 $(t_{1,7}[MS]\rightarrow"single")$ $\rightarrow$ ψ_2 $(t_{7,8}[FR]\rightarrow"unmarried")$

Fig. 1. The records from 1994 US Adult Census database

ψ_1:(NM | Year $\rightarrow$ MS, Sc) $Sc\{Sc_0\{$"Joe", "Ada"$\}$

ψ_2:(MS | Gen $\rightarrow$ FR, Sc) $Sc\{Sc_0\{$"married"$\}, Sc_1\{$Single$\}\}$

ψ_3:(FR| $\rightarrow$ MS, Sc) $Sc\{Sc_0\{$"husband", "wife"$\}\}$

ψ_1 shows that, for the couple Joe and Ada, their "Year" determines their marital status. For example, their marital status in 1971 must be with the same value. Due to the spousal relationship of Joe and Ada, ψ_1 puts their records together for consistency analysis. So do ψ_2 and ψ_3. Hence, CCFDs contribute to consistency analysis in content-related data.

ψ_1–ψ_3 are employed to detect inconsistencies. Tuples t_1, t_5–t_8 violate ψ_1: $t_1[NM]$, $t_{5-8}[NM] \in Sc_0$ of ψ_1, $t_{1,5-8}[Year] = $ "1971", but $t_{1,7}[MS] = $ "married" $\neq t_{5,6,8}[MS] = $ "single". With the same way, ψ_2 detects two group of inconsistencies as (1) t_1–t_3 and (2) t_5, t_6, t_8. ψ_3 detects the inconsistencies about t_2, t_3, t_7–t_9.

Next, we employ optimal repairing computation for solving these inconsistencies, *e.g.*, for inconsistencies t_1, t_5–t_8 detected by ψ_1, we correct t_1[MS] with "single" rather than correct t_{5-8}[MS] with "married". Because it makes fewer modifications for repairing t_1 than t_5–t_8.

On the other hand, repairing sequences have influence on repairing results. Different repairing sequences may lead to disparate repairing results. For illustration, we assume three repairing sequences rs_1–rs_3 shown in Fig. 1. rs_1: $\psi_1 \rightarrow \psi_2$ firstly modifies $t_{1,7}$[MS] with "single". It also corrects the inconsistencies violating ψ_3. After that, inconsistencies t_5–t_8 are detected by ψ_2 and modified with "unmarried". rs_1 corrects all the inconsistencies to truth values and makes 2 cell-modifications. rs_2: $\psi_3 \rightarrow \psi_1 \rightarrow \psi_2$ makes a repairing result as $t_{5,6,8}$[MS] = "married", t_1[FR] = "husband" and $t_{5,6}$[FR] = "wife". Although the result is consistent, it is incorrect because Joe and Ada are single respectively in 1971. Compared to rs_1, rs_3: $\psi_2 \rightarrow \psi_1 \rightarrow \psi_2$ leads to a correct result, but needs 2 more cell-modifications that rs_3 modifies t_1[FR] with "husband" temporarily and recovers it to "unmarried". Hence, rs_3 makes extra repairing cost.

Besides, constraints themselves may be mutexes in repairing. ψ_1, ψ_3 share the common inconsistencies t_7, t_8. ψ_1 modifies $t_{7,8}$[MS] with "single". While ψ_3 modifies them with "married". The repairing strategies of ψ_1 and ψ_3 are mutexes. And either strategy won't singly satisfies both ψ_3 and ψ_3 at the same time.

As above, inappropriate repairing sequences will reduce the accuracy and efficiency of data repairing. Hence, reasonable repairing sequences are necessary to solve these problems.

Although some previous works have been proposed to manage minimum-cost repairing in [7], they ignored the content-relationship of data and repairing sequences for inconsistencies. In contrast, by analyzing the interaction of CCFDs, we propose a method of determining reasonable repairing sequences for inconsistencies repairs in content-related data.

In this paper, we research in determining repairing sequence with the following challenges.

(1) *Determining Repairing Sequence.* Which inconsistencies should be repaired preferentially is a core issue since inappropriate repairing sequences may bring problems (*e.g.*, incorrect repairing, extra-cost repairing and repairing mutex). And it is intractable to determine reasonable repairing sequences based on numerous of CCFDs.

(2) *Solving Repairing Mutex.* For the common error tuples over CCFDs, repairing mutex will emerge if the repairing target values are conflict. However, it is difficult to select appropriate target values satisfying all the CCFDs.

Furthermore, these two challenges are not independent. They can be mixed to become a more complicated problem.

Contribution. We focus on both accuracy and efficiency of determining repairing sequence for inconsistencies repairing.

– We investigate fundamental problems associated with determining repairing sequence in minimum-cost repairing, *e.g.*, we proof that the problem of minimum-cost repairing is NP-complete.

- We present definition of repairing sequence graph by analyzing the relationship of CCFDs. It helps to compute the inconsistencies which should be repaired preferentially.
- We analyze the problem of repairing mutex. Further we discuss the problem mixed with repairing sequence determination and repairing mutex.
- Using 2 real-life datasets and 1 synthetic dataset for large scale data, we show the effectiveness and efficiency of our solution.

Organization. We introduce the related works in Sect. 2. Next, we introduce some basic notions and analyze the problem of minimum-cost repairing with CCFDs in Sect. 3. In Sect. 4, we propose a workflow for determining repairing sequence. Then we give the detail algorithms of determining repairing sequence (including repairing sequence graph and solving repairing mutex) and computing target-values for repairs respectively in Sects. 5, 6. The experiments results are shown in Sect. 7. Finally We draw a conclusion in Sect. 8.

2   Related Work

Our work finds similarities to three lines of the works: (1) integrity constraints, (2) content-related data, (3) repairing strategy.

First, integrity constraints [1] are critical techniques of data quality management. Functional dependencies (FDs) [8], conditional functional dependencies (CFDs) [4, 5] have been already proposed and proved effective in consistency cleaning. Interlandi and Tang [9] presented a method how to prove which data was positive or negative with Sherlock rules and reference table. eCFDs [10] demonstrates the CFDs can be combined. Moreover, to solve inconsistencies in content-related, Du *et al.* [6] presented content-related conditional functional dependencies (CCFDs) which solved consistency by putting content-related data together.

Second, the content-relationship of data can be adopted to catch potential errors. Volkovs *et al.* [11] researched in continuous data cleaning that permitted both the data and its semantics to evolve and suggested repairs based on accumulated evidence to date. Prokoshyna *et al.* [12] proposed a cleaning method in quantitative and logical data with metric FDs. Besides, the content-relationship can be applied in distributed data [6, 13], even big data [14].

Third, cleaning strategy explains the detail of implementation in repairing and has direct influence on repairing performance. Cong *et al.* [15] resolved violations by changing values for attributes in both the premise and conclusion of constraints. Wang and Tang [16] presented repairing tags to control repairs which modified errors only once. Geerts *et al.* [17] proposed a method of computing repairing sequences according to cells-distribution.

Additionally, consistency technique are widely employed in data cleaning systems such as [18, 19].

So far, works about determining repairing sequence are still from sufficiency, even in content-related data over CCFDs. To meet this need, we do researches associated with this problem.

3 Sequential Repairing

In this section, we introduce some basic notions including (1) a review of CCFDs, a class of consistency constraints for content-related data, (2) the raise of repairing cost, a measurement method for content-related data, and (3) the definition of repairing sequence, a strategy for valid repairing. Finally, we propose the problem statement of determining repairing sequence for minimum-cost repairing over CCFDs.

3.1 Content-Related Conditional Functional Dependencies (CCFDs)

Given a relation schema R, all of the attributes sets over R are $attr(R)$ and the domain of attribute A is $dom(A)$.

CCFDs. A content-related conditional functional dependency over R is defined by $\psi: (C|Y \rightarrow A, Sc)$, where (1) C is the conditional attributes, Y is the variable attributes, C and Y are separated by "|", $C, Y \subseteq attr(R)$ and $C \cap Y = \varnothing$. $C \cup Y$ is denoted as LHS (ψ), and single attribute A is denoted as RHS (ψ); (2) $Y \rightarrow A$ is a standard FD; (3) For tuple sample $t_0 - t_s$, we denote content-related value conditional set by $Sc_i = \bigcup_{i=0\,to\,s} \{t_i[C]\}$. And Sc is the set of Sc_i, $Sc = \cup \{Sc_i\}$. The number of tuples which support ψ is denoted by sup (ψ).

The instances of CCFDs are shown in Example 1. CCFDs show that, for content-related data, they may express consistent semantics although they have different conditional values.

Semantics. To explain how a relation D over schema R satisfies CCFDs, we formalize the semantics of CCFDs.

A relation D satisfies a CCFD $\psi: (C|Y \rightarrow A, Sc)$ if and only if for all pairs of tuples $t_i, t_j \in D$, $t_i[C]$, $t_j[C] \in Sc_i$, $t_i[Y] = t_j[Y]$, and $t_i[A] = t_j[A]$. Then we denote D satisfying ψ as $D \vDash \psi$, otherwise ψ is a violated CCFD. Σ is a CCFDs set over R. For $\forall \psi \in \Sigma$, if $D \vDash \psi$, we say D satisfies Σ and denote it as $D \vDash \Sigma$.

Remark. In this paper, we only consider the CCFD ψ with disjoint Sc_i of Sc where, for $\forall Sc_i, Sc_j \in Sc$, $Sc_i \cap Sc_j = \varnothing$. Hence, the conditional values will not be combined duplicately.

3.2 Repairing Cost for Content-Related Data

Repairing cost is one of central factors of evaluating data repairing. For one thing, it increases by considering content-related data which enhances the confidence of the data. For another, for the data dominated by different CCFDs, their contentrelated data may result in varieties of repairing costs. In this paper, we present a repairing-cost model to measure the cell-modifications for content-related data over CCFDs.

Given a relation D and a CCFD $\psi: (C|Y \rightarrow A, Sc)$, t is an inconsistent tuple violating $Sc_i \in Sc$. v is a target tuple with only distinguished value on A from t. We will explain how to select v in Sect. 6. Next, we will discuss the repairing cost of repairing t with v.

Repairing Weight. For an inconsistent tuple t violating Sci, we define the weight ω of repairing t as

$$\omega(t, \psi) = \frac{|\sigma_{C=t[C], Y=t[Y]}(D)|}{\sum\limits_{t_i[C] \in Sc_i} |\sigma_{C=t_i[C], Y=t[Y]}(D)|}. \tag{1}$$

Here, $|\sigma_{C=t[C], Y=t[Y]}(D)|$ is the frequency of the inconsistencies with $C = t[C]$ in D. $\sum\limits_{t_i[C] \in Sc_i} |\sigma_{C=t_i[C], Y=t[Y]}(D)|$ is the number of all the content-related data about t. Repairing weight describes the impact of content-related data on t. Higher $\omega(t, \psi)$ is, more confident t is and more cost it takes for repairing.

Repairing Cost. In practice, an inconsistent tuple may violate several CCFDs at once. Σ_t is the violated CCFDs set about t. Repairing cost for t over Σ_t is defined as

$$cost(t, v, \Sigma_t) = \sum\limits_{\psi_i \in \Sigma_t} \omega(t, \psi_i) \times Isrepair(t, v) \tag{2}$$

where $Isrepair(t,v)$ is a discriminant function:

$$Isrepair(t, v) = \begin{cases} 1 & \text{if } t[A] \neq v[A], \\ 0 & \text{otherwise.} \end{cases} \tag{3}$$

$Isrepair(t, v)$ shows that if $t[A] \neq v[A]$, $t[A]$ will be corrected to $v[A]$.

For relation D and CCFDs set Σ, vio (D, Σ) returns all the inconsistent tuples violating Σ in D and D' is the repaired relation for D. The repairing cost for D is

$$cost(D, D', \Sigma) = \sum\limits_{t \in vio(D, \Sigma)} cost(t, \text{findc}(\Sigma, t), \text{findt}(D', t)). \tag{4}$$

Here, *findc* (Σ, t) is used to find violated CCFDs set Σ_t about t from Σ. *findt*$(D'; t)$ finds the target value t' for t from D'.

Example 2. In a review of Example 1, inconsistent tuple t_7 violates $\Sigma_{t_7} = \{\psi_1, \psi_3\}$ and $\omega(t_7, \psi_1) = 2/5 = 0.4$, $\omega(t_7, \psi_3) = 0.6$. If we consider v_7 with "single" as the target value, then $cost$ $(t_7, \Sigma_{t_7}, v_7) = 0.6$.

3.3 Problem Statement

In actual world, it is difficult to correct all the error data completely, especially with unknown ground truth, so that optimal repairing computation becomes a valid solution instead. Based on the repairing-cost model in preceding, in this paper, we formally state our problem in terms of determining repairing sequence about CCFDs with minimum-cost repairing.

Definition 1. *For repairing relation D, a repairing sequence (rs) about CCFDs Σ is defined by rs: $\psi_i \to \ldots \to \psi_m$ where $\psi_i, \psi_m \in \Sigma$.*

Repairing sequence *rs* starts with repairing inconsistencies violating ψ_i till ψ_m. $|rs|$ is denoted as the number of CCFDs in *rs*.

Without lose of generality, we first introduce repairing sequence determination with minimum-cost repairing. Then we state our problem of determining repairing sequence with fixed target values.

Given a relation *D* and CCFDs set Σ, repairing sequence determination with minimum-cost repairing is to find repairing sequence *rs* which repairs *D* to *D'*, for which $D' \vDash \Sigma$ and *cost* (D, D', Σ) is minimum. And this problem is intractable as described in Lemma 1.

Lemma 1. *For a constant C, the problem of repairing, if there exists a repairing sequence rs which makes D', whose cost (D, D', Σ) is at most C, is Σ_2^p-complete (NP^{NP}).*

Fan *et al.* [20] proved that the problem of minimum-cost repairing with CFDs is Σ_2^p-complete. This problem converts to our problem within PTIME. The computation complexity of our problem is $O((n^n \times |\Sigma|)^{|rs|})$ where *n* is the number of tuples in *D*.

Moreover, if the target value could be automatically fixed by repairing strategy, the repairing sequence determination problem can be simplified.

Problem Statement. To find a repairing sequences *rs* which repairs *D* to *D'* with the fixed target values, for which $D' \vDash \Sigma$ and *cost* (D, D', Σ) is minimum.

Theorem 1. *The problem of determining repairing sequence in minimum-cost repairing with fixed target values is NP-complete and the computation complexity is $O(|\Sigma|^{|rs|})$.*

Proof. Minimum-cost repairing with CFDs has been proved as NP-complete in [7], which our problem can be converted into within Θ $(|Sc|)$ since CCFDs extends from CFDs, where $|Sc|$ is the number of conditional values in *Sc*. The proof is provided by reduction from [7].

For illustration, we propose an algorithm of computing target values automatically in Sect. 6. It allows us to concern about only how to determine repairing sequence, besides target value assignment.

4   Inconsistencies Repairing

Our work is to determine a reasonable repairing sequence. In this section, we first describe the workflow for repairing sequence determination. Then we briefly explain how to detect violated CCFDs.

4.1   Overview

Given relation *D* and CCFDs set Σ, as shown in Fig. 2, our workflow contains three components: (1) inconsistencies detection which catches violated CCFDs from Σ,

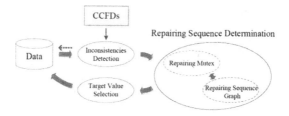

Fig. 2. The workflow for data repairing

(2) repairing sequence determination which computes the CCFDs which should be repaired preferentially using repairing sequence graph and analyzes repairing mutex, (3) target value selection which implements repairing strategy.

Algorithm 1 shows the process of our workflow. Line 2 uses $inDet\,(D', \Sigma)$ to detect violated CCFDs which is introduced in Sect. 4.2. In Line 5–6, $RSDet\,(\Sigma')$ and $inRep$ (D', Π_{RS}) determine which CCFDs need to be repaired and repair the inconsistencies respectively, which will be described in Sects. 5 and 6.

Algorithm 1: *The workflow of Inconsistencies Repairing* inClean

---

Input: Relation D, CCFDs set Σ
Output: Repairing sequence rs

1   $D' = D$, $rs = \Pi_{RS} = null$, $\Sigma' = \emptyset$;
2   **while** *True* **do**
3 | $\Sigma' = inDet(\,D', \Sigma\,)$,
4 | **if** $\Sigma' \neq \emptyset$ **then**
5 | | $\Pi_{RS} = RSDet(\,\Sigma'\,)$,
6 | | $D' = inRep(\,D', \Pi_{RS}\,)$;
7 | | $rs.add(\,\Pi_{RS}\,)$;
8 | **else**
9 | | **return** rs;

Remark. For our workflow, we make descriptions from the following two factors:

(1) *Detection Results.* Our detection returns violated CCFDs Σ', but not inconsistent tuples. We only care the preferentially repaired CCFDs Π_{RS} which contributes to selecting the inconsistencies to be repaired first. So that it is waste to repair inconsistencies violating $\Sigma' - \Pi_{RS}$. To compute the CCFDs which should be repaired preferentially, it is unnecessary to return all inconsistent tuples. And our workflow will terminate if the repaired relation is consistent.

(2) *Iterative Implementation.* Our workflow detects and repairs inconsistencies iteratively. Each iteration returns the CCFDs which should be repaired currently as partial repairing sequence. And we obtain the entire repairing sequence by summing all the iterations.

4.2   Inconsistencies Detection

This component catches the violated CCFDs Σ' from Σ. To accelerate this process, we detect inconsistencies with Lemma 2.

Lemma 2. *Given a CCFD $\psi : (C|Y \to A, Sc)$, a relation $D \vDash \psi$ if and only if, for $\forall Sc_i \in Sc$, $\left| \bigcup\limits_{t[C] \in Sc_i} \pi_Y \sigma_{C=t[C]}(D) \right| = \left| \bigcup\limits_{t[C] \in Sc_i} \pi_{Y \cup A} \sigma_{C=t[C]}(D) \right|$ where $\pi_Y \sigma_{C=t[C]}(D))$ is the projection on Y over D in the condition that $C = t[C]$. $\left| \bigcup\limits_{t[C] \in Sc_i} \pi_Y \sigma_{C=t[C]}(D) \right|$ is the number of distinct tuples in $\pi_Y \sigma_{C=t[C]}(D)$.*

Compared with the semantics in Sect. 3.1, Lemma 2 only analysis the number of tuples on Y and $Y \cup A$, but not all pairs of tuples.

5   Determining Repairing Sequence

To determine a reasonable repairing sequence, we present repairing sequence graph. Then we analysis repairing mutex problem and discuss the interaction between repairing sequence and repairing mutex.

5.1   Repairing Sequence Graph

The detected violated CCFDs are related. One CCFD may dominated by the others so that its violations should be repaired after others. Thus, we present repairing sequence graph to solve this problem.

Definition 2. *Given CCFDs set Σ, the repairing sequence graph on Σ is defined by $G_\Sigma = (V, E)$ where $V = \Sigma$. For $\forall \psi_i, \psi_j \in \Sigma$, if RHS $(\psi_i) \subset$ LHS (ψ_j), there exists an edge $e_{ij} = (\psi_i, \psi_j)$ from ψ_i pointing to ψ_j and $E = \bigcup\limits_{i,j \in |\Sigma|} \{e_{ij}\}$.*

Example 3. The repairing sequence graph G_Σ on Σ is shown in Fig. 3. ψ_2 is dominated by ψ_1 and ψ_3.

In practice, the CCFDs with no indegree are not dominated by any other CCFDs so that they can be repaired preferentially. This contributes to avoiding incorrect repairs and extra-cost repairs. If there exists no CCFDs with no indegree, we heuristically select the CCFDs with current minimal cost from the violated CCFDs by Theorem 1.

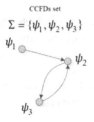

Fig. 3. The repairing sequence graph

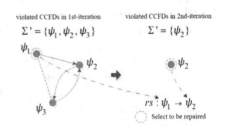

Fig. 4. The repairing sequence determination

Theorem 2. *Given a violated CCFD $\psi \in \Sigma'$, ψ must be a CCFD with no indegree of $G_{\Sigma'}$ if there is no $\psi' \in \Sigma'$ with RHS $(\psi') \subset LHS (\psi)$.*

Proof. $RHS(\psi') \subset LHS (\psi)$ describes that ψ' dominates ψ. There is no ψ' dominating ψ in Σ'. It equals that ψ are not pointed by any edges so that ψ is a CCFD with no indegree.

Theorem 2 discovers CCFDs with no indegree no matter whether we know the detail of $G_{\Sigma'}$ or not. It can facilitate the process for determining repairing sequence. Additionally, repairing sequence graph may change after each iteration completes so that it need to be recalculated.

5.2   Repairing Mutex

For the CCFDs violated by the same inconsistencies, their repairing strategies may be conflict, which results in repairing mutex.

Definition 3. *For $\forall \psi, \psi' \in \Sigma$, RHS $(\psi) = RHS(\psi')$ and inconsistent tuple t, if v; v' are repairing target values about ψ, ψ' on t. If $v[A] \neq v'[A]$, then Σ are repairing mutex.*

As described in Example 1, $\Sigma' = \{\psi_1, \psi_3\}$ is repairing mutex because of the different target values ($v_{\psi_1}[A]$ = "single" $\neq v_{\psi_3}[A]$ = "married").

Repairing mutexes are caused by conflicted repairing strategies, but not error data. In condition of unknown target values for the same inconsistence, the violated CCFDs with same *RHS* will be mutexes potentially. Hence, we need to put these CCFDs together and select a common target value for them.

Remark. Repairing sequence graph and repairing mutex are not independent. The relationship of CCFDs may affect repairing mutex. As shown in Fig. 4, although ψ_1 and ψ_3 connected with dotted line share the common inconsistencies t_7, t_8. While ψ_3 is indirectly dominated by ψ_1 through RHS $(\psi_1) \subset LHS(\psi_2)$ and RHS $(\psi_2) \subset LHS(\psi_3)$. With inconsistencies violating ψ_1 being repaired, the repairing mutex about ψ_1 and ψ_3 will disappear automatically. Hence, it is a fake repairing mutex indeed. In practice, we only need to concern about the relationship of mutex about CCFDs with no indegree since they won't be dominated by others. We select ψ_1 to be repaired in Iteration 1 and ψ_2 in Iteration 2. Consequently the repairing sequence is *rs:* $\psi_1 \rightarrow \psi_2$.

Algorithm 2: *The Repairing Sequence Determination* RSDet

Input: Violated CCFDs set Σ'

Output: Repairing mutex set Π_{RS}

1   $\Pi_{RS} = \Pi_{RM} = null$, $\Sigma_{NI} = \emptyset$;

2   $\Sigma_{NI} = $findNoInd($\Sigma'$);

3   **if** $\Sigma_{NI} \neq \emptyset$ **then**

4   $\quad \Pi_{RS} = \Pi_{RM} = $findRepMut($\Sigma_{NI}$);

5   $\quad$ **for** $\psi \in (\Sigma_{NI} / \Pi_{RM})$ **do**

6   $\quad\quad \Pi_{RS} = \Pi_{RS} \cup \{\{\psi\}\}$;

7   **else**

8   $\quad \Pi_{RS} = $findMaxSup($\Sigma'$);

9   **return** Π_{RS};

Algorithm 2 shows how to select the preferentially repaired CCFDs in one iteration.

According to Lemma 2, Line 2 uses *findNoInd* (Σ') finds all CCFDs with on indegree. In Line 4, *findRepMut* (Σ_{NI}) finds all repairing mutexes in Σ_{NI}. Minimum-cost repairing is NP-complete so that we employ a heuristic method *find-MaxSup* (Σ') to select the CCFD with the maximum support to be repaired in Line 8 and omit the detail of *findMaxSup* (Σ').

Computing Complexity. findNoInd (Σ'); *findRepMut* (Σ_{NI}); *findMaxSup* (Σ') cost $\Theta(|\Sigma'|^2)$, $\Theta(|\Sigma'|^2)$ and $\Theta(n \times |\Sigma'|)$ respectively for computation. The upper bound of Σ' is Σ so that the computing complexity of *RSDet* is $O(n \times |\Sigma|)$.

6 Repairing Target Value

As description in Sect. 3.3, minimum-cost repairing problem is Σ_2^p-complete. To simplify this problem to NP-complete, we propose a method to fix repairing target values. Moreover, it is still intractable to obtain a exact solution so that our method heuristically compute the target values in Algorithm 3.

Algorithm-3:-The Target Value Selection TVSel

Input: Relation D, Repairing mutex set Π_{RS}

Output: Repaired relation D'

1 $D' = D, \psi S = V = \varnothing, \psi v = null, \psi I \psi = findInc(D' \cdot \Pi_{RS})$;

2 **for** $t \in I$ **do**

3   **for** $\Sigma_{RS} \in \Pi_{RS}$ **do**

4          $S = findCD(t, D', \Sigma_{RS})$,

5          $V = \pi_{RHS(\Sigma_{RS})}(S)$,

6          $v = minCTV(S, V)$,

7          $D' = repair(D', S, v)$;

8 **return** D';

---

findInc (D', Π_{RS}) in Line 1 computes all inconsistent tuples violating CCFDs of Π_{RS} in D'. In Line 4–7, $S = findCD(t, D', \Pi_{RS})$ finds all content-related tuples about t over Σ from D'. $\pi_{RHS(\Sigma_{RS})}(S)$ is a projection operation about S over attribute RHS (Σ_{RS}). *minCTV* $(S; V)$ select target value v for S from V. *repair*(D', S, v) repairs S with v and updates D' with S.

Computing Complexity. findInc(D', Π_{RS}), $S = findCD(t, D', \Sigma_{RS})$ and *minCTV* (S, V) cost $\Theta(n^2 \times |\Pi_{RS}|)$, $\Theta(|S| \times |\Sigma_{RS}|)$, and $\Theta(|V|)$ respectively for computation. And the computing complexity of *TVSel* is $O(n^2 \times |\Sigma|^2)$.

Proposition 1. *Based on the fixed target values assigned by the heuristic method, the minimum-cost repairing problem is terminated.*

For the repairing sequence graph with circle (*e.g.*, $\psi_i \rightarrow \psi_j \rightarrow \ldots \rightarrow \psi_i$), the inconsistencies about ψ_i may be modified repeatedly. While the modifications is no more than n and terminate finally.

7 Experimental Results

In this section, we experimentally evaluate the performance of our solution on three datasets.

7.1 Experimental Setting

Dataset. We use two real-life datasets and a synthetic dataset for our experiments including (1) *Adults*[1] which contains 1994 US Census information with 48842 record on 15 attributes, (2) *HOSP*[2] which is taken from US Department of Health & Human Services with more than 200K records on 17 attributes, (3) *hAdults* which is a synthetic dataset over the schema of *Adults* and is composed with 120K records using average domain of attribute $avgdom(A) = 5$.

Rule. We design two rule-sets using CCFDs discovery method in [21] which searches for all CFDs by 2-level lattice and combines CFDs of the same $C|Y \rightarrow A$. The rule-sets *Adults* (*hAdults*) and *HOSP* contain 41 and 57 CCFDs separately. Based on these two rule-sets, the domain experts produced fixing rules artificially according to their understanding of the violations.

Algorithms. In our experiments, we compare three algorithms *CR, CN, CF* and *Fix*. To illustrate the content-related relationship of data, *CR* and *CF* repair the data with repairing sequence determination respectively over CCFDs and CFDs. *CN* is a naive algorithm which repairs all detected inconsistencies by CCFDs in each iteration without using repairing sequence graph. *Fix* employs fixing rules iteratively without using repairing sequences. A cell is a value of a tuple on one attribute. And we adopt *recall* (R), *precision* (P), and *F-measure* for accuracy measurement where $recall = \dfrac{\text{correctly repaired cells}}{\text{actual error cells}}$, $precision = \dfrac{\text{correctly repaired cells}}{\text{total cells}}$ and $F\text{-}measure = \dfrac{2RP}{R+P}$.

Our experiments ran by using Intel Core i7-2600 (3.4 GHz) with 8 GB of memory in Java program. Each experiment were repeated 5 times, and the average is reported here.

7.2 Experimental Performance

We investigate the performance in three factors: accuracy, running time and iteration.

We evaluate our solution from the following two aspects: (1) the overall performance and (2) the scalability.

Exp-1: Overall Performance. Table 1 shows the repairing performance of *CR, CN, CF* and *Fix* for comparison. Totally, all the algorithms make high recall over 0.7. Compared with *CN*, *CR* increases the recall by 10.8% since a reasonable repairing

Table 1. Accuracy of Algorithms on Datasets

	Recall			Precision			F-measure		
	Adults	HOSP	hAdults	Adults	HOSP	hAdults	Adults	HOSP	hAdults
CR	0.876	0.933	0.857	0.058	0.030	0.100	0.109	0.060	0.180
CN	0.768	0.867	0.743	0.051	0.028	0.087	0.096	0.054	0.156
CF	0.739	0.851	0.703	0.043	0.026	0.082	0.081	0.050	0.147
Fix	0.901	0.945	0.889	0.067	0.041	0.112	0.125	0.082	0.199

sequence attributes to avoiding incorrectly repairs. Content-related data also assists in inconsistencies repairing by comparing CR with CF. Note that CR makes a low precision (0.03) and a high recall (0.933) on $HOSP$. It reflects a high-quality dataset itself is great benefit to repairing. Fix makes an excellent performance due to their negative patterns.

We observe the running time of different components. As shown in Fig. 5(a), determining repairing sequence (rsDet) takes nearly 5% of running time. Without repairing sequence, CN makes a big deal of computation for target value selection (TVSel). Some of these computation may be reduplicated, which leads to extra repairing cost and running time. Hence, repairing sequence can facilitate the efficiency. Additionally, inconsistencies detection (IncDet) with CCFDs takes more time than CFDs because of considering content-related data. Fix takes the fewest time repairing inconsistencies with the optional target values in patterns. However, generation of fixing rules consumes large volume of manual work.

Figure 5(b) and (c) show the iteration information. In Fig. 5(b), although CN repairs all detected inconsistencies, these repairs may also bring new inconsistencies attribute to incorrect modifications so that it takes 4 iterations. For $HOSP$, it makes the most iterations due to its most CCFDs which generates the repairing sequence graph with the most vectors. And violations are repaired only once in Fix. Figure 5(c) shows that the recalls of CR and CF trends to be stable respectively in Iterations 7 and 8.

Exp-2 Scalability. As shown in Fig. 5(d)–(i), we investigate the scalability of algorithms with 4 parameters: (1) the number of rules, (2) the number of tuples, (3) noise ratio and (4) average domain of attribute.

Figure 5(d), (e) and 5(f) shows the recall, the F-measure and the running time of repairing sequence determination by varying the number of rules. With the increasing number of rules, triple of recall, F-measure and running time rise smoothly. CR almost keeps a high recall over 0.85 and a high F-measure over 0.03. It shows high accuracy of CR in inconsistencies repairing.

We observe the running time of repairing sequence determination by varying the number of tuples in Fig. 5(g). In our experiment, we varied the number of $HOSP$, $hAdults$ quadratically and uniformly respectively. And the running time ran smoothly, which indicates the stability of our solution.

In Fig. 5(h), we add noise data by modifying the tuples to be incorrect. With tuples on LHS of rules fixed, their attributes on RHS are modified to be conflicted with others. With increasing noise ratio, CR is slightly influenced. Hence, the methods using repairing sequences for content-related data are effectiveness.

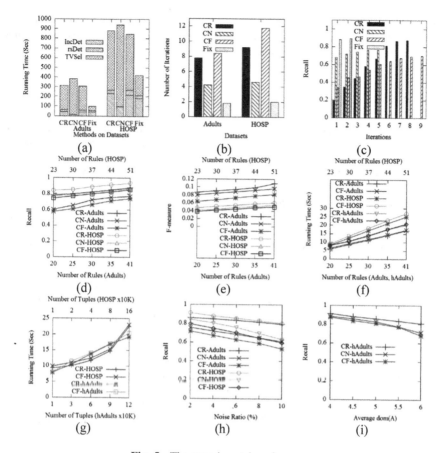

Fig. 5. The experimental performance

For synthetic dataset, we control the distribution of data by varying average domain of attribute. It is useful to analyze big data consistency by using a small data sample. As shown in Fig. 5(i), *CN* and *CF* decrease rapidly when *avgdom(A)* > 5.5. And our solution are fit to be extended to big data.

8 Conclusions

We have studied the problem of determining repairing sequence for inconsistencies repairing in content-related data. This paper discusses fundamental problems of determining repairing sequences with minimum repairing-cost. To compute a reasonable repairing sequence, we present repairing sequence graph and solve the repairing mutex problem. Our future work contains (1) extending our method to big data (*e.g.*, Hadoop), (2) investigation in constraints repairing which allows to correct inappropriate rules rather than records.

Acknowledgement. Our research was supported by, the National Natural Science Foundation of China under Grant Nos. 61672142 and 61472070, and the Fundamental Research Fundation for the Central Universities of China under Grant Nos. N150408001-3 and N150404013.

References

1. Fan, W., Geerts, F.: Foundations of Data Quality Management. M&C, San Rafael (2012)
2. Fan, W.: Data quality: from theory to practice. In: Proceedings of the 36th ACM SIGMOD International Conference, pp. 7–18. ACM (2015)
3. Eckerson, W.W.: Data quality and the bottom line. J. Radioanal. Nucl. Chem. **160**(4), 355–362 (1992)
4. Bohannon, P., Fan, W., Geerts, F., Jia, X., Kementsietsidis, A.: Conditional functional dependencies for data cleaning. In: Proceedings of the 23rd International Conference of Data Engineering, pp. 746–755. IEEE (2007)
5. Fan, W., Geerts, F., Jia, X., Kementsietsidis, A.: Conditional functional dependencies for capturing data inconsistencies. Trans. Database Syst. **33**(2), 6–47 (2008)
6. Du, Y.F., Shen, D.R., Nie, T.Z., Kou, Y., Yu, G.: Content-related repairing of inconsistencies in distributed data. J. Comput. Sci. Technol. **31**(4), 741–758 (2016)
7. Bohannon, P., Fan, W., Flaster, M., Rastogi, R.: A cost-based model and effective heuristic for repairing constraints by value modification. In: Proceedings of the 26th ACM SIGMOD International Conference, pp. 143–154. ACM (2005)
8. Papenbrock, T., Ehrlich, J., Marten, J., Neubert, T., Rudolph, J.P., Schönberg, M., Zwiener, J., Naumann, F.: Functional dependency discovery: an experimental evaluation of seven algorithms. Int. J. Very Large Data Bases **8**(10), 1082–1093 (2015)
9. Interlandi, M., Tang, N.: Proof positive and negative in data cleaning. In: Proceedings of the 31st International Conference of Data Engineering, pp. 18–29. IEEE (2015)
10. Bravo, L., Fan, W., Ma, S.: Extending dependencies with conditions. In: Proceedings of the 33rd International Conference on Very Large Data Bases, pp. 243–254 (2007)
11. Volkovs, M., Fei, C., Szlichta, J., Miller, R.J.: Continuous data cleaning. In: Proceedings of the 30th International Conference of Data Engineering, pp. 244–255. IEEE (2014)
12. Prokoshyna, N., Szlichta, J., Chiang, F., Miller, R.J., Srivastava, D.: Combining quantitative and logical data cleaning. Int. J. Very Large Data Bases **9**(4), 300–311 (2015)
13. Chen, Q., Tan, Z., He, C., Sha, C., Wang, W.: Repairing functional dependency violations in distributed data. In: Renz, M., Shahabi, C., Zhou, X., Cheema, M.A. (eds.) DASFAA 2015. LNCS, vol. 9049, pp. 441–457. Springer, Cham (2015). doi:10.1007/978-3-319-18120-2_26
14. Khayyat, Z., Ilyas, I.F., Jindal, A., Madden, S., Ouzzani, M., Papotti, P., QuianRuiz, J.A., Tang, N., Yin, S.: Bigdansing: A system for big data cleansing. In: Proceedings of the 36th ACM SIGMOD International Conference, pp. 1215–1230. ACM (2015)
15. Cong, G., Fan, W., Geerts, F., Jia, X., Ma, S.: Improving data quality: consistency and accuracy. In: Proceedings of the 33rd International Conference on Very Large Data Bases, pp. 315–326. VLDB (2007)
16. Wang, J., Tang, N.: Towards dependable data repairing with fixing rules. In: Proceedings of the 35th ACM SIGMOD International Conference, pp. 457–468. ACM (2014)
17. Geerts, F., Mecca, G., Papotti, P., Santoro, D.: The llunatic data-cleaning framework. Int. J. Very Large Data Bases **6**(9), 625–636 (2013)
18. Chalamalla, A., Ilyas, I.F., Ouzzani, M., Papotti, P.: Descriptive and prescriptive data cleaning. In: Proceedings of the 35th ACM SIGMOD International Conference, pp. 445–456. ACM (2014)

19. Dallachiesa, M., Ebaid, A., Eldawy, A., Elmagarmid, A., Ilyas, I.F., Ouzzani, M., Tang, N.: Nadeef: a commodity data cleaning system. In: Proceedings of the 34th ACM SIGMOD International Conference, pp. 541–552. ACM (2013)
20. Fan, W., Geerts, F., Tang, N., Yu, W.: Inferring data currency and consistency for conflict resolution. In: Proceedings of the 29th International Conference of Data Engineering, pp. 470–481. IEEE (2013)
21. Du, Y., Shen, D., Nie, T., Kou, Y., Yu, G.: Discovering condition-combined functional dependency rules. In: Chen, L., Jia, Y., Sellis, T., Liu, G. (eds.) APWeb 2014. LNCS, vol. 8709, pp. 247–257. Springer, Cham (2014). doi:10.1007/978-3-319-11116-2_22

Author Index

Akiyama, Mitsuaki   II-278
Alexander, Rukshan   II-75
Alkalbani, Asma Musabah   I-290
Al-Khalil, Firas   II-57
Anagnostopoulos, Marios   II-517
Anwar, Md Musfique   I-59
Anwar, Tarique   I-59
Assy, Nour   I-275
Azé, Jérôme   II-346

Babar, M. Ali   I-315
Badsha, Shahriar   II-502
Bao, Xuguang   I-199
Bennacer, Nacéra   I-49, II-109
Bertino, Elisa   II-502
Bhavsar, Maitry   I-33
Breslin, John G.   II-420
Bringay, Sandra   II-346
Bugiotti, Francesca   I-49, II-109

Cai, Hui   II-37
Cai, Peng   II-205
Cai, Yi   II-117
Cao, Jian   I-259, II-450
Cao, Jinli   II-479
Cao, Yanhua   I-259
Cardinale, Yudith   II-57
Chakraborty, Roshni   I-33
Chandra, Anita   II-90
Chandra, Joydeep   I-33
Charalambous, Theodoros   II-247
Chbeir, Richard   II-57
Chen, Chi   I-165
Chen, Fei   I-422
Chen, Hongmei   I-199
Chen, Lisi   I-299
Chen, Long   II-295
Chen, Yang   I-75
Chen, Yazhong   I-123
Chen, Yifan   II-357
Cheng, Min   I-516
Cheng, Reynold   I-330
Costa, Gianni   I-215

Cristea, Alexandra I.   I-18
Cui, Zhiming   I-91

Dai, Gaokun   I-135
Dai, Qiangqiang   I-123
Dandapat, Sourav   I-33
de Heij, Daan   II-338
Dikenelli, Oguz   II-221
Ding, Xiaofeng   II-295
Ding, Yue   II-329
Dongo, Irvin   II-57
Drakatos, Panagiotis   II-517
Du, Yuefeng   I-524

Fang, Yuan   I-183
Feng, Jianhua   II-19
Feng, Ling   II-313
Frasincar, Flavius   II-338

Gaaloul, Walid   I-275
Galicia, Jorge   II-109
Gao, Xing   I-376
Garg, Himanshu   II-90
Ghamry, Ahmed Mohammed   I-290
Goto, Shigeki   II-278
Grobler, Marthie   II-528
Guo, Jinwei   II-205
Guo, Jun   I-441
Guo, Kaiyang   I-391

Han, Fengling   II-490
Hanaoka, Hiroki   II-159
Hariu, Takeo   II-278
Hatami, Siamak   II-184
He, Yueying   I-359
Hewasinghage, Moditha   I-49
Hoang, My Ly   I-290
Hong, Xiaoguang   II-372
Hu, Fei   II-98
Hu, Yupeng   II-372
Huang, Feiran   I-359
Huang, Jinjing   I-472

Huang, Jiuming   I-3
Huang, Joshua Zhexue   II-148
Huang, Xin   I-441
Hung, Nguyen   I-347
Hussain, Farookh Khadeer   I-290

Inan, Emrah   II-221
Isaj, Suela   I-49

Jasberg, Kevin   I-106
Jia, Gangyong   I-422
Jia, Weijia   II-3
Jia, Yan   I-3
Jin, Hai   II-295
Jin, Li   II-313

Kambourakis, Georgios   II-517
Kapitsaki, Georgia M.   II-247
Karavolos, Michail   II-517
Karmakar, Kallol   II-550
Kashima, Hisashi   II-46
Khalil, Ibrahim   II-502
Kotsilitis, Sarantis   II-517
Kou, Yue   I-524
Kuang, Hongbo   II-540
Kurabayashi, Shuichi   II-159

Labba, Chahrazed   I-275
Lai, Yongxuan   I-376
Lei, Xue   II-117
Leung, Ho-fung   II-117
Li, Bing   II-467
Li, Chaozhuo   I-359
Li, Fan   II-98
Li, Fang   I-299
Li, Guoliang   II-19
Li, Jian   I-422
Li, Jianxin   I-59
Li, Jiyi   II-46
Li, Juan   I-259
Li, Kuan-Ching   I-376
Li, Li   II-98
Li, Minglu   I-259
Li, Qi   II-313
Li, Qing   I-422, I-499, I-516, II-117
Li, Rong-Hua   I-123, I-391, I-441
Li, Wenzhuo   I-243
Li, Yukun   I-516
Li, Zhenjun   I-123, I-391

Li, Zhen-jun   I-441
Li, Zhi   I-3
Li, Zhixu   I-472, II-263
Li, Zhoujun   I-359
Liao, Minghong   I-376
Liao, Qun   I-488
Lie, Hendi   I-150
Lin, Chuang   I-243
Lin, Tianqiao   I-472
Lin, Zehang   I-516
Liu, An   I-422, I-472, II-263
Liu, Chengfei   I-59
Liu, Dongxi   II-502
Liu, Guanfeng   I-91, II-263
Liu, Jiamou   I-75
Liu, Wenyin   I-516
Long, Yan   I-91
Lu, Hongyu   II-450
Lu, Li   I-259
Lu, Minhua   I-391
Lyu, Zheng   I-376

Ma, Wanlun   II-528
Ma, Xiao   I-243
Ma, Yun   I-499
Maiti, Abyayananda   II-90
Mao, Rui   I-123, I-441
Mao, Xuehui   II-435
Moulahi, Bilel   II-346
Murray, David   II-75

Nayak, Richi   I-150
Nepal, Surya   II-467, II-490, II-502
Nie, Tiezheng   I-524
Niu, Lei   II-132
Niu, Zhendong   I-135

Ortale, Riccardo   I-215

Paris, Cecile   II-467
Patricio, Mariana   II-109
Peng, Min   II-562
Peng, Zhaohui   II-372
Phung, Dinh   I-347
Piao, Guangyuan   II-420

Qian, Shiyou   I-259
Qian, Weining   II-205
Qiao, Shaojie   I-123, I-391
Qin, Jianbin   I-231

Quan, Yong II-562
Quercini, Gianluca I-49, II-109

Rafique, Wajid II-479
Rasool, Raihan Ur II-479
Ren, Fenghui II-132
Rezvani, Mohsen II-175
Rezvani, Mojtaba II-175
Ryan, Caspar II-490

Saoud, Narjès Bellamine Ben I-275
Scharff, Milena Zychlinsky II-338
Schouten, Kim II-338
Shang, Shuo I-299
Shen, Derong I-524
Shen, Yishu I-457
Sheng, Quan Z. I-315
Sheng, Victor S. I-91
Shi, Jinyu II-3
Shi, Lei I-18
Sizov, Sergej I-106
Song, Andy II-490
Sun, Lei I-488
Sun, Yue II-263

Takata, Yuta II-278
Tan, Yudong II-450
Tang, Jie I-165
Tang, Lihong II-528
Tang, Yan II-192
Thin, Nguyen I-347
Thompson, Nik II-75
Tian, HongLiang I-165
Tran, Nguyen Khoi I-315
Tran, Vu I-290
Trinh, Thanh II-148
Troyanovsky, Artiom II-338
Tsai, Yi-Chan I-290
Tupakula, Uday II-550

van Schyndel, Ron II-490
Varadharajan, Vijay II-550
Venkatesh, Svetha I-347
Vimalachandran, Pasupathy II-540

Wang, Dong II-329
Wang, Fu Lee II-117
Wang, Hua II-479, II-540
Wang, Lizhen I-183, I-199
Wang, Qian II-229

Wang, Senzhang I-359
Wang, Wangsong II-192
Wang, Wei I-231
Wang, Wenyuan II-303
Wang, Xiaoxuan I-183
Wang, Yaoshu I-231
Wang, Ye II-562
Wang, Yuheng II-357
Wang, Zhong I-259
Wei, Daming II-435
Wei, Lei II-192
Wen, Sheng II-528
Wen, Yanlong I-406
Wu, Dingming II-148
Wu, Jian I-91
Wyeth, Gordon I-150

Xiang, Yang II-528
Xiao, Bing II-205
Xiao, Lin II-387, II-403, II-458
Xiao, Quanwu II-450
Xiao, Weidong II-357
Xie, Haoran II-117
Xing, ChunXiao I-165
Xu, Jiajie I-91
Xu, Jie I-359
Xu, Weimin II-435
Xu, Yang II-372
Xu, Ying I-299
Xu, Yong I-330
Xuan, Pengcheng II-192
Xue, Guangtao I-259
Xue, Yuanyuan II-313

Yagi, Takeshi II-278
Yan, Bo I-75
Yang, Cynthia II-338
Yang, Fan I-376
Yang, Guang II-372
Yang, Jian II-467
Yang, Qing II-229
Yang, Xuechao II-490
Yang, Yulu I-488
Yang, Zhenguo I-499, I-516
Yao, Bin I-299
Yao, Lina I-315
Yau, David K.Y. II-517
Yi, Xun II-490, II-502
Yin, Hongzhi I-472

Yin, Litian   II-329
Yong, Jianming   II-479
Yongchareon, Sira   II-303
Yu, Ge   I-524
Yu, Jiadi   I-259, II-37
Yu, Wenli   II-98
Yuan, Shizhong   II-435
Yuan, Xiaojie   I-406
Yuan, Yunpeng   I-488

Zhang, Bing   I-3
Zhang, Detian   I-422
Zhang, Jingwei   II-229
Zhang, Jinjing   II-98
Zhang, Minjie   II-132
Zhang, Puheng   I-243
Zhang, Tao   I-3
Zhang, Wei-Peng   I-391, I-441
Zhang, Wenjie   II-357
Zhang, Yanchun   II-229, II-540, II-562
Zhang, Ying   I-406
Zhang, Zhenguo   I-406

Zhang, Zizhu   II-467
Zhao, Lei   I-472, II-263
Zhao, Liang   II-313
Zhao, Pengpeng   I-91
Zhao, Weiliang   II-467
Zhao, Xiang   II-357
Zhao, Zhonghua   I-359
Zhaoquan, Gu   II-387, II-403, II-458
Zheng, Kai   I-299, II-263
Zheng, Yudian   I-330
Zhou, Aoying   II-205
Zhou, Bin   I-3, II-562
Zhou, Lihua   I-183, I-199
Zhou, Rui   II-229
Zhou, Wanlei   II-528
Zhou, Yiwei   I-18
Zhu, Shunzhi   I-299
Zhu, Yanmin   I-259, II-37
Zhuang, Yan   II-19
Zhuo, Guangping   II-540
Zou, He   II-329
Zou, Zhaonian   I-457

Printed in the United States
By Bookmasters